Engineering Practice with Oilfield and Drilling Applications

Engineering Practice with Oilfield and Drilling Applications

First Edition

Donald W. Dareing
University of Tennessee

Registered Office
John Wiley & Sons, Inc., 111 River Street, Hoboken, NJ 07030, USA

Editorial Office
111 River Street, Hoboken, NJ 07030, USA

For details of our global editorial offices, customer services, and more information about Wiley products visit us at www.wiley.com.

Wiley also publishes its books in a variety of electronic formats and by print-on-demand. Some content that appears in standard print versions of this book may not be available in other formats.

Library of Congress Cataloging-in-Publication Data has been applied for:

ISBN: 9781119799498

Cover Design: Wiley
Cover Image: © Puneet Vikram Singh, Nature and Concept photographer/Getty

Set in 9.5/12.5pt STIXTwoText by Straive, Pondicherry, India

SKY10032267_123021

To Kristin:

My wonderful companion, whose energy, integrity, support, and reliability go beyond measure.

Contents

Preface

Engineers are trained to understand the fundamental principles of mechanics and mathematics. These tools provide a background of knowledge for making professional decisions. The tools of engineering science apply across most engineering disciplines. The key to their application is visualizing a reasonable mathematical model for the problem at hand. Freebody diagrams are helpful in this regard. Mathematical solutions follow, leading to reasonable engineering results. Typically, there is only one answer, so each problem is closed-ended.

On the other hand, design and problem-solving are open-ended. There are many possible solutions and alternatives must be created. While each engineering design is different, the approach is the same. An objective of this book is to explain the engineering design process and show how to apply basic engineering tools.

The book contains three parts.

Part I	Engineering Design and Problem-Solving
Part II	Power Generation, Transmission, Consumption
Part III	Analytical Tools of Design

Part I gives a systematic process for developing an engineering design. The application of engineering tools is illustrated during the conceptual and preliminary activities of design. Concept evaluation and selection are explained. Visualizing a total device or any system in terms of its subsystems is helpful in creating a design. Key considerations in finalizing a design are implementing feedback from test results or other evaluation sources, finalizing a design and presentation of final manufacturing drawings.

Every machine has (i) a prime mover or power source, (ii) mechanisms to transmit energy and (iii) energy consumed by forming the final product, plus friction. Part II covers Power Generation, Transmission, and Consumption.

Part III contains useful tools of engineering mechanics. Each selected topic goes beyond the traditional tools of design. Mathematical modeling and methods of solution are of historical significance. Each topic is supplemented with key references for additional background information.

Physical responses of engineering systems are predictable through science and mathematics. This one thing makes it possible to design modern structures and machinery to a high degree of reliability. The first scientifically based engineered bridge is the Eads Bridge which spans the Mississippi River at St. Louis. It was designed and constructed by James Eads. Construction began in 1867. It was dedicated in 1874 and is still in use today.

My goal in writing this book was to document the essence of engineering practice. The manuscript is a condensation of lecture notes developed over years of teaching across the mechanical engineering curriculum and industrial practice in the petroleum industry. It is written for undergraduate and graduate students and as a reference for practicing engineers.

Donald W. Dareing
Professor Emeritus, Mechanical Engineering
University of Tennessee, Knoxville
Life Fellow Member, ASME
Knoxville, TN, USA
April 2021

Nomenclature

a	acceleration
BF	buoyancy factor
c	distance to outside beam surface, damping coefficient
c_{cr}	critical damping coefficient
E	modulus of elasticity
E_m	energy per pound
F	applied force, axial internal force at drill pipe-collar interface above hydrostatic
FS	safety factor
f	friction force, vibration frequency
F_B	axial force in pipe (lower end)
F_{cr}	critical buckling force
f_n	natural frequency, cps
G	modulus of rigidity, angular momentum
h	lubrication film thickness, enthalpy
h_f	friction head
H	linear momentum, elevation
I	area moment of inertia
Im	impulse
J	angular moment of inertia of a cross section, angular mass moment of inertia
k, K	local (modal) mechanical spring constant
K_0	stress intensity factor
L	length
m, M	local (modal) mass, bending moment
N	force
N_R	Reynolds number
p	pressure
P	unit force (force per area), power, diametral pitch of gears
Q	moment of area above shear surface, heat, compressive force
$Q_{eff} = Q + (p_iA_i - p_oA_o)$	plus sign means compression
q	roller bearing exponent
r	radial position, frequency ratio (ω/ω_n)

S	section modulus of a cross sectional area, Sommerfeld number, entropy
t, T	time, torque, period of oscillation
$T_{eff} = F_B + wx + (L-x)(A_0\gamma_0 - A_i\gamma_i)$	marine riser (x measured up from bottom)
$T_{eff} = F_B + wx + (L-x)w_m$	drill pipe (x measured up from bottom)
TR	transmissibility
U, V	principal axis of inertia of a cross section, V also indicated shear force
V	velocity, also total potential energy
w, W	distributed load on a beam, weight of a discrete body
x, y, z	reference frame
X, Y, Z	reference frame
$[X]$	modal matrix
$x(t)$	local response
Z	viscosity (cp)

Greek Symbols

δ	displacement, log decrement
μ	viscosity, coefficient of friction
ω	rotational speed, circular frequency
ω_n	natural circular frequency
θ	angular position
σ	normal stress
τ	shear stress
ε	normal strain
γ	shear strain
ζ	damping factor
ν	Poisson's ratio
$\eta(t)$	modal response
σ_a	allowable design stress
σ_{yld}	yield strength
$\zeta = \dfrac{x}{L}$	
$\beta = \dfrac{(F_B + LA_0\gamma_0 - LA_i\gamma_i)L^2}{EI}$	
$\alpha = \dfrac{(w - A_0\gamma_0 + A_i\gamma_i)L^3}{EI}$	
$\Theta = \dfrac{TL}{EI}$	

Part I

Engineering Design and Problem Solving

Engineering design is a logical sequence of activities that solves a problem or achieves a specified objective. Every design project has a beginning and an end. They can be several years long, such as putting a man on the moon and returning safely to earth, or it can be short, such as designing and fabricating a water pump. Successful engineering designs require a clear objective – well thought out and executed. Planning is critical. The design process may also be applied to management or any problem situation.

In conducting design, it is important to understand the difference between "open-end problems" and "closed-ended problems." Engineering tools of design are usually closed ended and based on fundamental laws of engineering science. The answer is unique. Use of engineering tools usually follows certain steps:

1) Develop a mathematical model for the physical element under consideration.
2) Develop a freebody diagram of the element along with forces and moment considering the constraints placed on the element.
3) Solve the mathematical equations leading to a prediction of performance, usually expressed in terms of stress, deflection, vibration, etc.
4) Judge the answer against experience, order of magnitude (believable), and uniformity of dimensions.

On the other hand, open-ended problems have many possible solutions. Each must be generated and evaluated before a design can start. Solving open-ended problems requires imagination and creativity. Part I gives a process for solving open-ended problems, including steps in project work. It also underscores important design principles that may be considered in moving through an engineering project.

1

Design and Problem Solving Guidelines

Engineering design involves management of people, resources, money, and time. Success depends on planning, resource, and time management. Time is usually the driver.

When discussing the importance of teaming with one company, the response was, "teaming isn't important – it is everything." The very success of a company depends upon people skills and the ability to work with others as a team member. Pete Carroll, while head football coach at the University of Southern California, says, "Winning players don't always win. It's the winning plays that win."

Planning is a matter of thinking through the activities and tasks that will be necessary to achieve the stated goal. This is somewhat experience dependent. For large projects, it may be useful to divide tasks into major activities, such as design, fabrication, installation, and commission, which are usually conducted in tandem. In other projects, where several activities are conducted simultaneously, major groupings may be needed. An example would be a military operation involving various branches.

General Dwight Eisenhower, along with his staff, spent months developing a plan for the invasion of Europe. His team of officers generated and evaluated various plans of attack. Eisenhower once said "... the plan itself is not as important as the act of planning." Thinking through the plan is the key.

Plans need to be flexible. As new information is gathered along the way, the plan may need to be modified. A good manager anticipates problems and deals with them early to avoid crises. A crisis is a situation where a critical problem needs to be solved, but there is little time to solve it. A Gantt chart can be useful in this regard.

Design Methodology

Basic steps for developing a product idea (or service) into a profitable venture are given in Figure 1.1. The first few boxes indicate the importance of a preliminary market analysis and input from customers to determine market reaction to a new product. Also, a preliminary market analysis helps define and refine the attributes of the product. Initial feedback from customers is useful in deciding whether to proceed with further development.

Design specifications are based on specific needs and expected performance. Design specifications represent the initial *engineering baseline* for generating design alternatives. In most cases, design specification are legal statements of what is expected. They must be established accurately and in concert with users of the future product.

Engineering Practice with Oilfield and Drilling Applications, First Edition. Donald W. Dareing.
© 2022 John Wiley & Sons, Inc. Published 2022 by John Wiley & Sons, Inc.

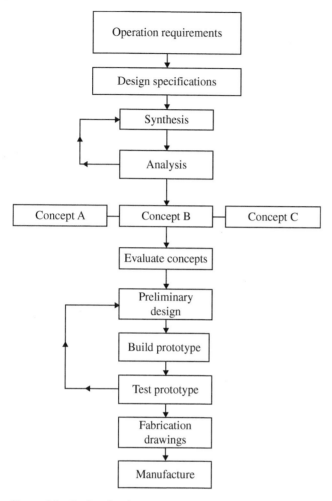

Figure 1.1 Design development process.

Design alternatives are typically generated by a team of professionals with special skills, such as marketing, design, and manufacturing. This activity is sometimes called concurrent engineering where the team considers every aspect of the product from technical feasibility to product life cycle to manufacturing and marketing strategy. Feedback from potential customers is important. The team also evaluates each design alternative and selects the best concept to advance. Depending on the complexity of the concept, technical feasibility studies, requiring advanced computational techniques, may be required during the refinements of design alternatives.

Since design is open ended, there are many possible solutions or design alternatives that satisfy a given set of specifications. Once viable designs have been generated, they need to be ranked so choice can be made. Choosing a preferred concept is based on trade-offs among evaluation metrics identified for a given product; an evaluation method will be described later.

A preliminary design represents an update of the engineering baseline. The preliminary design refines the preferred alternative. It advances the engineering baseline for the final design and fabrication phases.

Developing a final design may require the use of computer-aided-design (CAD), numerical analysis, and other analytical tools to refine dimensions. Prototype testing may also be desirable. Computer simulations may alleviate the high cost of prototype testing.

The product configuration is again evaluated in the marketplace for customer feedback and approval. This is accomplished through market surveys or market focus groups depending on the nature of the product.

The next step is to interface CAD codes with manufacturing (CAM). This requires converting design codes into machine tool codes. Depending on the product and the market, the ability to reconfigure the machining and handling process in a timely manner may be important for "just in time" delivery.

Market Analysis

The purpose of a market analysis is to identify what potential customers want in a new product, establish the size of the market, and determine what price the market is willing to pay for the product. A market analysis will produce a set of product attributes, which more clearly define the main features of the planned product. Using customer input and competitor product features, important features for the new product can be identified and ranked as to their importance. This information identifies customer preferences and competitive differentiation during the conceptual stage of product development.

New products can be either research driven, or market driven. Research-driven products stem from ideas that spawn from basic or fundamental research. In this case, a new technique or device may be the objective or a by-product of the study. The technique or device then becomes a solution looking for a problem, so to speak. The market-driven product is developed in response to a definite market need. In some cases, a market may be developed for a new idea.

Before investing much time and money, it is best to conduct a patent search to make sure the product does not infringe on active patents. This exercise will also give useful information on the state-of-the-art of products as applied to a given market. It may show the patent protection period on a product has expired, offering the opportunity to enter the market with a competitor's product – with improvements.

In recent years, markets have become more demanding on product delivery. Customer needs may change over a short period. Companies that can retool for "just in time manufacturing" in response to this demand have an advantage. One tool company, that makes diamond drill bits for oil and gas well drilling, built its business on making diamond drill bits overnight; each diamond was handset. Each diamond bit was and still is tailored to suit a set of design specifications stipulated by an oil company. The main reason for a quick response capability (or "just in time manufacturing") is moderate demand for high cost of diamond drill bits. It is not good business to stockpile high-cost products for a limited market application. Warehoused products may become outdated. It is costly and risky.

Operational Requirements

Operational requirements or product attributes describe the expected functional performance of a new product. Product description may come out of a business plan for a new product concept, a government need for a new weapon, or an oil company's need to develop an oil field in a given geographic location.

Product Development

Top management may define the operational requirements for a product, based on a market analysis. Company engineers then develop a set of design specifications before proceeding. Product design may be conducted within a company or contracted outside.

Government Procurement Procedure

The federal government has very strict guidelines for procuring products and services [1]. Government procurement is normally through the General Service Administration (GSA). The need for a product may come to the GSA from any government agency, which in turn coordinates the Federal Product Description (FPD). An FPD describes the operational requirements and required functions.

For example, assume that the Joint Chief of Staff decides that the military needs a new type of aircraft. They would make their request to the GSA and describe the aircraft in terms of expected operational requirement, such as:

- Range
- Speed
- Landing capabilities
- Weapon weight
- Weather considerations.

The GSA would expand the requesting agency's description of operational requirements.

> Effective market research and analysis must be conducted to assure that user need is satisfied. During the market research and analysis phase, the preparing activity should advise potential agency users that a FPD is being developed. Ask potential users to provide a statement of their needs in essential functional or performance terms to the maximum practical extent.

In addition to a clear description of operational requirements, FPDs will also develop a comprehensive list of design specifications for a new product. The tender document would be reviewed at various levels with GSA before it is released to contract bidders.

Petroleum Industry Procedure

An economic analysis is conducted on each new oil reservoir to determine its profitability and the best way to develop it. Following this, operational requirements are set before proceeding. Operational requirements may include such factors as

- Ocean water depth
- Size of reservoir
- Oil, gas, water content of reservoir
- Reservoir pressure
- Production rate (barrels per day).

Design specifications would document detailed engineering constraints on the design, such as environmental conditions, ocean floor mud line load-bearing capabilities, material specifications, expected loads, 100-year storm conditions, etc.

Considerable time is spent in gathering this information to establish operational requirements and design specifications. Company engineers build a set of design specifications to form a "tender document" for contractor bid preparation.

There are two contract approaches: turnkey and cost plus. Turnkey simply means that the contractor will deliver a product at a fixed price. The contractor is responsible for every detail, including identifying and satisfying all codes and standards relevant to the design. Since the price is fixed, oil companies would be concerned about delivery dates. Missing the planned delivery date could greatly increase future monetary returns and profit.

A cost plus is based on an agreed hourly rate. The equipment and supplies are additional costs to the buyer. Usually there is a percentage tacked on to these costs. Company representatives are directly involved in day-to-day decisions.

Design Specifications

Design specifications are an itemized set of constraints placed on a design. They identify product performance expectations: what the product is supposed to do and how the product should perform. They are contractual and represent the initial "engineering baseline" from which all concepts are generated. They are an important part of a contract between customer and designer. Usually, the customer signs-off on a set of specifications once they have been documented. Any changes, for any reason, after the development work starts, will cause delays, and increase costs. The cost of making changes is usually written into a contract.

Once operational requirements have been set, design specifications are documented. They may be expanded by outside contractors in conjunction with company engineers. The contract usually puts the burden of completeness on the contractor, such as all relevant Codes and Standards are the responsibility of the design contractor.

Specification Topics

Design specifications are usually subdivided into key topics. Topics normally considered are discussed below.

Performance Requirements

Performance requirements identify specifics, such as loads, motions, flow rates, operating pressures, and temperature limits, to name a few. In addition, the technical specifications may include physical and chemical properties of materials to be used. Material properties may include such items as yield strength and hardness. Weld procedures (including preheating) and welder qualification requirements, special heat treatment and annealing may be specified.

Environmental and climate conditions may affect design. Examples are wind, oceanographic conditions, such as wave height, wind-driven current velocities, and tidal currents.

Performance requirements define the physical constraints in the design. Depending upon the size of the project, the specification document can be as small as a few pages or several volumes.

Sustainability

Sustainability means being good stewards of the resources on planet earth. A 1987 UN report defines sustainable development as: "Meeting the needs of the present without compromising the ability of future generations to meet their own needs." This specification is relevant, ethical, and makes good economic sense. Earlier business strategies were based on planned obsolescence, where products were intentionally designed to wear out after a given period [2].

A few metrics relating to sustainability are:

a) design efficiency in terms of materials, weight, cost, energy consumption
b) quantity and type of waste from fabrication
c) minimal friction, wear, maintenance, reliability
d) use of renewable energy sources, such as wind and solar
e) environmental impact of a possible failure
f) environmental recovery plan in case of an unexpected failure/disaster
g) design for modular replacement instead of product throw-away
h) disposal at end-of-life cycle (computers, TVs)
i) automation verses human control (cause of many disasters)
j) redundancy in monitoring system performance.

Codes and Standards

During the midst of the industrial revolution of the nineteenth century, it became apparent that mechanical components needed to be standardized to assure, for example, bolts made in one place could fit together with nuts made in another place. The American Society of Mechanical Engineers (ASME) took a leading role in standardizing mechanical components, such as pulleys, gears, and key seats. Even shop drawing symbols needed to be standardized.

During the early days of the steam engine, it was common for steam boilers to explode causing personal injury and death. In 1884, the ASME established a uniform test standard for boilers; this was ASME's first standard. This standard was followed by a boiler construction standard, which was published in 1915. Such standards became the foundation of ASME's current Boiler and Pressure Vessel Code. Since the development of this Code, boiler disasters have been essentially eliminated.

What is a Code? A code is a standard that has been mandated by one or more government bodies. A code has the force of law behind it. When a standard is specified in a business contract, it becomes a code, as well. Therefore, the words, codes, and standards, are sometimes used interchangeably.

What is a Standard? Standards are a set of technical definitions and guidelines or a set of instructions for designers and manufacturers. Their use is strictly voluntary, and they do not have the force of law. Standards serve as a vehicle of communications, defining quality, and establishing safety criteria for producers and users.

Many professional organizations develop Standards; however, they must follow procedures accredited by the American National Standards Institute (ANSI). These procedures must reflect openness, transparency, balance of interest, and due process.

Many turnkey contracts make the contractor or design company responsible for applying all relevant codes and standards to a design. While specific codes and standards are not listed in a tender document, they are implied through legal contracts.

Environmental

Designers should consider the environmental impact of a new product throughout the product's life cycle, i.e. from fabrication to product disposal. Even in the early stages of offshore platform design, disposal of a structure at the end of production life (about 20–30 years) is considered.

The most vulnerable or highest risk components in the design should be identified and environmental consequences associated with possible failure considered. The designer needs to consider the "what if" scenarios. If a failure occurs, how would it affect public health, public safety, public image of company, property damage, as well as damage to the environment. Goodwill is an asset on the balance sheet of any company.

Social Considerations

People, who will be affected by the implementation or use of a given product or project, should be consulted and brought into design deliberations as early as possible – the sooner the better. This is not only right but, by doing so, misunderstanding is alleviated, and public resistance will be reduced as the project develops. People simply want to be consulted and given the opportunity to make input on issues that affect their lives. If not given this opportunity, citizens may unite and work against a worthwhile project.

Aside from human reaction, there are legitimate reasons for considering social factors in design, such as safety and preservation of a culture. When oil companies began to develop oil fields in the northern part of the North Sea, oil transportation from the offshore platforms had to be resolved. Crude oil could be off-loaded directly onto tankers or transported to land by a pipeline and then loaded onto tankers in a protected harbor for transport to refineries. The closest land point from the platforms was to the small fishing village of Solom Voe, Shetland Islands. An extensive study was conducted to design a port that would not destroy centuries-old lifestyle of the people living in this area. There were also benefits to local economies.

Reliability

There are two methods of design which relate to the safety and reliability of products: (i) factor of safety and (ii) statistical or reliability.

The factor of safety method is commonly used in engineering design. It is a time-proven design tool and when used properly, safe and reliable designs are developed. Factor of safety (FS) is the ratio of failure stress to an allowable design stress.

Material yield strength, σ_{yld}, is often used as the failure stress. In this case

$$FS = \frac{\sigma_{yld}}{\sigma_{allowable}} = \frac{\sigma_{yld}}{\sigma_a} \tag{1.1}$$

where σ_a is the allowable stress level used throughout a design. When the factor of safety is given, then

$$\sigma_a = \frac{\sigma_{yld}}{FS} \tag{1.2}$$

Designs are configured (size, dimension) according to the allowable stress.

Factors of safety are intended to cover all uncertainties not identified in a set of specifications. In general, the higher the factor of safety, the higher the product weight and cost. So, it is important to keep the FS as low as possible. Computation accuracy affects factor of safety too. Computer software based on numerical techniques, such as finite element methods, provide very accurate predictions of local stress in complex geometries, and thus reduce the uncertainty of stress predictions.

The statistical or reliability approach seeks to establish design parameters so that the product performs to an expected level of reliability [3]. Reliability is a statistical measure of performance. For example, a product reliability of 0.9 means that there is a 90% chance that a given product will perform its proper function without failure. The method requires statistical data on all random variables, such as strength, size, and weight. With this approach, products can be designed to a required level of reliability.

Cost Considerations

Economics should be considered during the early stages of concept generation. Concepts that allow for early return on investments may be critical. Oil companies want to begin oil production as early as possible, maybe prior to the completion of the overall production system. Time is money and if it takes three years to bring a reservoir online that's three years without return on investment if early production is not achieved.

Aesthetics

Product appearance is not usually a concern of engineers. However, aesthetics could be an important marketing feature and should not be overlooked. *Actual* performance capabilities versus *perceived* performance capabilities can be important. The customer may perceive a feature to be weak or strong depending on the history of a product. Perception of certain product features should be considered in some cases.

Aesthetics does not mean that a design must be ornamental or that geometry controls the shape of the design or its components. Manufacturing complex geometries is impractical and costly. However, if a design is balanced (cost effective and functional), it is usually artful.

Product Life Cycle

Every product has a life cycle, which includes periods of:

1) Development
2) Market growth
3) Market maturity
4) Market decline
5) Product disposal

During the first phase, the product is developed and introduced to the market. This creates negative cash flow, so developing the product in a timely and cost-efficient manner is important. During the second phase, the product finds its way into the market and can generate cash flow while establishing itself in the marketplace. Products then reach a level of maturity and are usually able to capture a portion of the market and generate revenue. During this time, the product generates maximum return on investment. At some point in time, the market for a product declines and eventually vanishes. This could be the result of new products entering the market or simply the lack of demand for a product. For these reasons, investors look for new products.

In setting design specifications, remember *cost* and *safety* override everything.

Product Safety and Liability

Currently, there are no federal product liability laws. Each state has developed its own legal approach in this area. Under Tennessee's product liability laws, there are four theories of recovery [4, 5]:

Negligence – In legal terms, negligence is the failure to do what people of ordinary care and prudence would do under the same or similar circumstances. When applied to design and manufacturing, the question becomes: are decisions, that affect product safety, being made professionally, objectively, and fairly? Prior to 1970, many personal injury claims were filed under the negligence theory of law. The plaintiff's attorney had to establish the standard of due care required of engineers and manufacturers.

Strict liability – The emphasis under this theory of recovery is on product defect and not on the person or person's negligence. The focus of the court is solely on the performance of the product. In this case, the plaintiff needs to only prove that the product had an unreasonable dangerous defect when it left manufacturing. The court recognizes three types of defects: manufacturing defects, design defects, and warning defects. Design as used by the courts encompasses the entire process by which a product is created and marketed. This theory is more favorable to the plaintiff because it exposes the entire product development chain to liability, allowing the plaintiff to choose the most advantageous defendant.

Fraud/misrepresentation – Fraud is a false representation of fact, by works, or conduct, which is intended to and does deceive another, who then makes decisions based on the false information and suffers legal injury. The elements required in an action of law are representation, falsity, knowledge of falsity, intent to deceive the plaintiff, justifiable reliance by the plaintiff, and damages. Examples are falsifying test data to deliver a product to the customer to meet a deadline, approving a design that does not meet required codes to eliminate redesign and fabrication costs and misuse of Professional Registration approval.

Breach of warranty – Product does not satisfy the expressed warranty or implied warranty for fitness.

Engineering Ethics

Ethics deals with the principles of human duty, moral principles, and rules of conduct. Engineering ethics deals with the moral conduct of engineers in serving the public, their employers, and their clients. What is at stake can be expressed in terms of public safety, public health, and the environment to mention a few. The challenge is to design and manufacture a product for profit without undue risk to the general welfare of the public and environment. Ethics comes into play when a company and its professional staff knowingly produce a product that has a high risk for personal injury and/or damage to the environment. Unethical decisions are usually made for selfish or monetary reasons.

Engineers make many ethical decisions independent of others. A good check list for these decisions is:

- Is it legal?
- Is it fair?
- Is it morally right; can I live with the outcome?

The ASME has adopted the following Code of Ethics of Engineers for its members[1]

1) Engineers shall hold paramount the safety, health and welfare of the public in the performance of their professional duties.
2) Engineers shall perform services only in the areas of their competence.
3) Engineers shall continue their professional development throughout their careers and shall provide opportunities for the professional and ethical development of those engineers under their supervision.
4) Engineers shall act in professional matters for each employer or client as faithful agents or trustees and shall avoid conflicts of interest or the appearance of conflicts of interest.
5) Engineers shall build their professional reputation on the merit of their services and shall not compete unfairly with others.
6) Engineers shall associate only with reputable persons or organizations.
7) Engineers shall issue public statements only in an objective and truthful manner.
8) Engineers shall consider environmental impact in the performance of their professional duties.

Creating Design Alternatives

Once design specifications have been set, design alternatives can be generated. Since design is open ended, i.e. there are many possible solutions, it is desirable to generate several design concepts, evaluate them as a group before choosing the best direction for the design. The objective at this point is to create design concepts that satisfy the specifications. Realistic concepts are ones that are technically feasible and cost effective.

Innovation is a matter of synthesis and analysis of ideas. It requires time, focus, and effort of thought. The quality of each concept depends on the ability to think conceptually.

Parker [6] lists some traits of the creative personality:

- curiosity
- risk taker
- emotionally stable
- uninhibited
- imaginative, original
- intuitive
- high energy level
- independent
- task committed
- sense of humor.

Innovation is an individual activity. The foundation of science is built on the genius of individuals. Great ideas were not usually generated by committees. Once a fundamental concept or hypothesis is presented, others will add value. Design innovation usually starts with individual thought. Don't be afraid to propose novel ideas. You may be the leader others are waiting to follow.

Tools of Innovation

Ullman [7] discusses useful concept generation tools.

Patents

Some new products are based on a modification or direct extension of established products, in which case, the objective is to provide a higher quality or improved version of what is already on the market. Legally, this is allowed, provided the patent right period has expired. US patent law provides a 20-year life of a patent from the date the patent application was filed, and no shorter than 17 years from issuance. A patent survey will show whether your idea is truly novel or is infringing on an existing patent. A patent search early in the development process is worth the cost, time, and effort. It could eliminate legal problems later.

Reference Books and Trade Journals

Reference books are a good source of information on existing design concepts or designs that are currently in use. They may give analytical discussions of related designs. These discussions sometime suggest alternatives based on direct extensions of current design concepts. Many trade journals feature new products, which may spark new ideas.

Experts in a Related Field

Experts can sometimes provide insight for new concepts. Experts can be research and development oriented in each technical field or they can be a company representative, such as a salesperson or a customer service representative, who have detailed knowledge about a product type.

Brainstorming

Ideas are generated spontaneously. Successive ideas feed off the group discussion; "piggyback" is sometimes used to describe this type of idea generation. All the ideas are reviewed and evaluated for relevance and practicality. Features of the better ideas will merge and come together.

The rules for brainstorming are:

1) Record all the ideas generated.
2) Generate as many ideas as possible.
3) Think wildly. Impractical ideas sometime lead to a useful one.
4) Do not evaluate or criticize ideas while they are being generated.

Existing Products and Concepts

Many hours have gone into developing existing products and most have the benefit of being tested in the marketplace. Existing products can be extrapolated into new product variations. This also eliminates reinventing the wheel so to speak. Types of modifications that may be considered are:

- Geometric modifications
- Energy-flow modifications
  - Change in the path
  - Change in the form
- Materials used in the product.

Remember, there is no need to redesign every component in a new product. Use as many off-the-shelf items as possible. Take advantage of current and established technology, such as gears, bearings, and motors. On the other hand, some components may have to be tailored.

As part of the creative process, new ideas (untried techniques) must be proven technically feasible before they can be accepted as viable alternatives. Feasibility studies may be required to show concepts are fundamentally sound and will work. Some concepts may be evaluated experimentally, analytically, or both. The results may expose risk and cost of untried concepts.

This is a screening activity to move the better ideas forward and remove alternatives that are not technically sound. More detailed analyses may be required during the design stage of the development process. Concept development may include iterations involving synthesis and analysis as shown in Figure 1.1. Also, every aspect of product life cycle needs to be considered by the product development team.

Concurrent Engineering

It is generally accepted that there are three basic activities for developing a profitable product. They are marketing, designing, and manufacturing. Until the 1980s, these activities were conducted sequentially, i.e. market data were passed on to designers who subsequently transmitted their design to manufacturing. These three activities were conducted separately with essentially little or no collaboration among the three disciplines.

World competition brought about competitive pricing and quick response to a dynamic market. In response to this challenge, marketing, designing, and manufacturing issues are now being considered by a team. Marketing seeks and monitors customer input to the new product. Designers seek the latest customer feedback from market analyst. Manufacturing issues are considered throughout the conceptual work to avoid costly redesigns brought about by impractical or inefficient manufacturing requirements. Collaboration of marketing, design, and manufacturing throughout the development process is commonly called Concurrent Engineering. It substantially reduces development cost over the sequential operations approach.

Feasibility of Concept

Each concept must prove to be technically feasible before it is evaluated. This means that some engineering may be required, and even drawings made to advance an idea beyond a hand sketch. A preliminary study may be needed to show an idea is workable, realistic and satisfies all design constraints or specifications, including cost constraints. In engineering practice, feasibility analyses may require extensive computer calculations and/or laboratory testing.

Evaluating Design Alternatives

Innovation will produce several design alternatives, each of which satisfies the given set of design specifications. The problem now is to choose the "best" concept among many alternatives.

Each of us make choices every day. Without realizing it, trade-offs are made leading up to a decision. The basic elements in making decisions are performance, cost, risk, and availability. Consider the purchase of an automobile. The buyer has established mentally a set of performance criteria (city driving, off road, mountain terrain, luxury, etc.), the price compatible with the family budget, maintenance, or track record of the car (is it a new model), and availability. The final choice is a trade-off among these four evaluation elements. The same rationale is used to purchase a suit or buy a house.

Evaluation Metrics

Four metrics (performance, cost, risk, and availability) are basic in choosing a preferred design concept [8]. The following defines these metrics as they relate to mechanical design:

Performance – Capability to achieve needed operational characteristics, plus reliability.
Cost   – Estimated cost of the design, including development and manufacturing costs.
Risk   – Possibility that performance may not be met because of the design approach, absence of testing, or some specific technical consideration.
Availability – Availability of a design depending upon the stage of development.

A procedure for scoring several alternatives is to divide each of these four metrics into key submetrics and give each an appropriate weight. Each design alternative can then be scored under each submetric. The scoring is strictly judgmental, so experience is important. The more experience, the better the judgment of scoring. Once each concept has been scored for each submetric, the numbers are totaled for a composite score for each concept. The total scores provide a means of comparing each concept against the others.

This numerical evaluation scheme has two key objectives. The first is to have a way to quantify one's judgment against a fixed scale so that each design concept can be rated in the same manner, thus, showing their truest level of merit in comparison with each other. The second objective is to give a way to examine the rationale of the final scores by looking at the subelements of each concept to see the strong and weak features of each. Since all design selections are the result of trade-offs, the scoring system aids in selecting a preferred concept.

Scoring Alternative Concepts

Table 1.1 illustrates the evaluation of four design alternatives and a hypothetical score for each. All constituents affected by the outcome of the project work together to set the weight given each metric and scoring. Each of the four basic metrics will have sub-metrics, as well.

This example indicates that concept C is the best because it scored the highest among the four alternatives. Other alternatives may have ranked higher in certain categories, but overall concept C scored the highest. There may a discussion over individual scores in the various categories, but the table provides a good basis for discussing the strengths and weaknesses of each alternative. Results from this scoring method give a rationale for recommending a preferred concept and a justification for the choice.

Table 1.1 Evaluation summary (hypothetical numbers).

	Weight	Concepts			
	(%)	A	B	C	D
Performance	35	20	35	30	25
Cost	20	15	10	18	16
Risk	25	10	20	23	18
Availability	20	20	18	16	12
Total score	100	65	83	87	71

Once each alternative has been evaluated and ranked and the preferred design concept selected, it is always a good idea to revisit the marketplace to see if the customers agree and to get further feedback for the preliminary design. Keep in mind that each stage in the progression along the development process represents an upgrade of the engineering baseline. The description and specifics of the preferred concept are an expansion of the initial set of design specifications and represent the most recent engineering baseline for the remaining steps in the development process. Design specifications have therefore been greatly refined beyond the initial set. The design can now move forward with a detailed description of the product.

Starting the Design

Mechanical devices typically contain a power source and a means of transmitting the power to bring about a desired end effect.

A design may start with the end effect, which establishes the magnitude of loads throughout the device, transmission linkages, and power requirements. This sets the overall size of the device and nominal dimensions of the subparts. Force magnitudes, power requirements (output torque and speed), pump requirements (flow rates and pressure demand), materials, and criteria of failure, fasteners should have been established earlier during the feasibility studies. Forces affect stress magnitude and component dimensions. Types of stresses (bending, torsion, shear, etc.) and how they combine should be visualized and understood. Are stresses the result of static or dynamic loads? These initial calculations can be made by hand. More precise calculations, based on computer models, can be made later.

Design for Simplicity

Keep the design simple. Use off-the-shelf components where possible. Commercial items have already gone through development and testing and have a proven performance record. Vendors for commercial products are happy to discuss performance details and quote prices. Also, keep the number of parts to a minimum. This increases product reliability. One company in East Tennessee rewards individual performance on reduction of design components. In addition, always consider how each part is to be fabricated. Curved shapes look good but are costly to make. Each of these considerations will make the device more reliable and easier to maintain in the end. Underlying principles of good design practice are:

a) Minimize the number of parts. Reliability varies inversely with number of parts.
b) Keep the design simple, Complexity reduces reliability.
c) Use standard parts when possible; there are plenty of statistical performance data.

When configuring the design, it is helpful to consider whether starting the design from the outside-in or inside-out. If the design centers on a specific technical concept, such as a microcantilever sensor, it may be helpful to start there, and work inside-out, i.e. let the cantilever be the center point of the design and build outward from it. The sensor itself becomes the center point of the device and will dictate other features, such as how it is to be held and how information is to be retrieved.

Other projects may be constrained by space or how it links with other devices or subsystems. In this case, it may be useful to start with the geometric constraint and work inward. For example, the design of a house starts with the available space on a lot. Many times, the shape of the lot dictates the shape of the house, including the foundation, exterior walls, and roof. These subsystems, of course, are based on the preferred concept, established earlier. The details of the inside, such as heating and air conditioning, are designed after the size and shape of each room has been determined.

Identify Subsystems

Consider the total design as a combination of subsystems. This is typically done in very large projects, but it is helpful in small projects, too. Breaking the whole design into subsystems helps visualize and organize how the whole design goes together. It also gives a clear division of responsibility for different team members. Subsystems may include

- Frame or fixture
- Power unit, such as motor
- Transmission linkages
- "Use" mechanism, final form of design
- Controls.

This approach is common in industry especially for huge projects involving several subcontractors. It provides a clear division of effort and responsibility. Subsystems may also be broken down into smaller units or sub-subsystems, etc.

The human body is a good example of subsystems. The body contains a frame (skeleton) or bone structure with consideration of joints and flexibility. The skeleton supports other subsystems, such as the lungs, heart, kidneys, brain, and all the plumbing. The gastrointestinal subsystem converts raw fuel into useful energy. Medical doctors specialize in each subsystem.

Innovation continues during this activity and therefore teaming is important. Each team member has something unique to add and the division of responsibility makes efficient use of each team member. Team meetings allow ideas to be integrated and refined.

An objective at this point is to configure the design in terms of its shape and subsystems and how they all fit together. This activity is best made by using classical engineering calculations; simple hand calculations are fine. These initial calculations establish component sizes, which are refined later. It is important to get a feel for the magnitude of loads, stress levels, and deflections as you begin to work through the design.

Development of Oil and Gas Reservoirs

Significant advances in two technologies allowed oil companies to increase production over the past 50 years. They are (i) geophysical mapping of underground rock formations and (ii) directional drilling, which allows navigation through multilayered formations to reach specific locations in deep and complex reservoirs. A third technology, which has almost doubled oil and gas reserves, is horizontal drilling and fracking. The later technology substantially increases oil and gas recovery.

While seismic surveys can map geologic formations and identify possible oil and gas traps, exploratory drilling must be performed to determine the existence of hydrocarbons and chemical composition. Once an oil reservoir has been delineated by directional drilling, an economic evaluation is conducted to determine the best plan for developing the reservoir for maximum recovery.

Oil companies typically identify business activities according to the following categories.

1) Geophysical surveying
2) Exploratory drilling
3) Production drilling
4) Production of oil and gas
5) Transportation of crude oil and gas
6) Refining crude oil and gas
7) Marketing.

Perhaps 80% of a company's budget is directed at the first four (4) categories, which are classified as upstream activities.

Design of Offshore Drilling and Production Systems

The design, fabrication, installation, and operation of offshore production system require an overall plan that may take a few years to complete. The project begins only after an economic analysis has been made, including current cost and future value of the asset.

In many cases, an offshore reservoir may extend across multiple sections licensed by different oil companies. The proportional ownership is determined seismic survey maps. The oil company possessing the largest portion of the reservoir becomes the operator of the field. All companies pay their share of development costs and benefit proportionately.

Planning is usually divided into the major activities mentioned above. One of the first activities is documenting a thorough and complete understanding of design specifications. This may require extensive data gathering on environmental, oceanographic, soil, and other specifics as mentioned earlier. Results of these efforts are compiled in a sizable specifications document often called a tender document. The tender document is used to gather additional information on contractor capabilities as well as for contractor bids on certain portions of the project. The bidding process, involving a few contractors is an opportunity, through discussions, to expand design specifications.

The design, fabrication, installation of offshore oil and gas production systems are multifaceted, high risk, and costly. The total effort involves a team, including operator, design contractors, fabrication contractors, installation contractors, and drilling contractors to mention a few. An Engineering Management contractor is often used to coordinate and interface each contractor's activities. The tow-out date of offshore platforms usually depends on the weather pattern for a specific location. For example, the tow-out and installation window for platforms in the North Sea starts in May and ends in September. Delays in design or fabrication could result in costly losses of early production. Upstream planning for this weather window is of the essence.

Major subsystems in offshore drilling and production platforms include

1) Base structure (steel jacket, concrete, tension leg, compliant structure).
2) Operating facilities, including personnel facilities, oil and gas processing for transportation to land.
3) Drilling and production equipment.
4) Transportation, such as pipelines, of crude to onshore facilities.

Each of these subsystems has its own subparts, etc. Each can be viewed as specific designs, which interface with the total system. For example, the drilling system typically includes

1) Derrick
2) Hydraulic system

3) Blow Out Preventer system
4) Drillstring including bottom hole equipment
5) Power system.

A few considerations in designing oil well drilling programs are

- Location of entry point into the oil and gas reservoir relative to the rig.
- Assessment of directional drilling equipment needed to reach the entry point as well as the well path.
- Formations to be encountered and anticipated drill bit types to penetrate these formations.
- Anticipated formation pressures (normal as well as abnormal pressures).
- Reservoir pressures to be controlled and blow out prevention concerns and strategies.
- Casing programs from surface to total depth.
- Implementing optimum drilling practices.

Connection of Subsystems

An early consideration in any design is the attachment of subsystems. Bolt-type attachments allow removal and replacement of subparts. Bolted connections can also be adjusted. Welded connections can't. It may be desirable in some cases to weld subsystems, such as the frame. Warpage and machining of surfaces should be considered.

Torsion Loading on Multibolt Patterns

Bolts are often used to fasten beams to form a frame or support a given load, such as illustrated in Figure 1.2. Four bolts are shown even though multiple bolts making up various patterns could be used. The design objective is to determine the total shear force in each bolt to establish bolt size and material strength.

The total shear force on each bolt is the vector sum of direct shear force and force caused by the torsion moment at the support. The first force is simply the total shear force divided by the number of bolts (assumes the connection is rigid). The direction of the shear force on each bolt is downward. The second force is caused by a moment on the bolt pattern and is determined as follows.

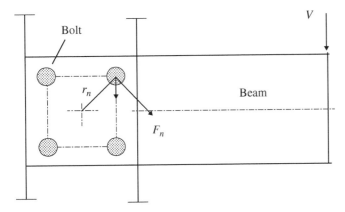

Figure 1.2 Bolted connection.

The total torque, T, applied to the bolted joint relates to shear force in each bolt by

$$T = F_1 r_1 + F_2 r_2 + F_3 r_3 + \cdots \tag{1.3}$$

Assuming the shear force, F, taken by each bolt depends on distance from the centroid of the bolt pattern,

$$\frac{F_1}{r_1} = \frac{F_2}{r_2} = \frac{F_3}{r_3} = \cdots = \frac{F_n}{r_n} \tag{1.4}$$

Combining these two equations gives the force in each bolt.

$$F_n = \frac{T r_n}{r_1^2 + r_2^2 + r_3^2 + \cdots} \tag{1.5}$$

The above formula is very similar to the shaft shear stress formula,

$$\tau = \frac{Tr}{J} \tag{1.6}$$

which states that shear stress is proportional to radius, r. The shear stress formula is modified as follows to match the above bolt analysis.

$$\tau_n = \frac{F_n}{A_n} = \frac{T r_n}{\sum r_i^2 A_i} \tag{1.7}$$

$$F_n = \frac{T r_n A_n}{\sum r_i^2 A_i} \tag{1.8}$$

Only when all bolts are the same size,

$$F_n = \frac{T r_n}{\sum r_i^2} \tag{1.9}$$

which agrees with Eq. (1.5). The direction of shear forces caused by torque on a bolt pattern is perpendicular to the r vectors as shown.

With reference to Figure 1.2, assume

$V = 2000\,\text{lb}$
Length $= 12$ in. (from center of bolt pattern)
Bolt diameter $= 0.25$ in.
Bolt spacing $= 2$ in.

Direct shear force	$F_V = \dfrac{2000}{4} = 500\,\text{lb}$
Torsion shear force	$F_n = \dfrac{T r_n}{4 r_n^2} = \dfrac{T}{4 r_n} = \dfrac{24\,000}{4(1.414)} = 4243\,\text{lb}$
Total shear force	$\overline{F} = \overline{F}_V + \overline{F}_n$ where each force is a vector
	$\overline{F} = -500j + 4243(i \cos 45 - j \sin 45)$
	$\overline{F} = -500j + 3000(i - j) = 3000i - 3500j$
	$F = 4610\,\text{lb}$ (scalar magnitude)

The cross-sectional area of each bolt is $A = \frac{\pi}{4} d^2 = \frac{\pi}{4}(0.25)^2 = 0.0491$ in.^{2} The maximum shear stress in the nth bolt is $\tau_n = \dfrac{4610}{0.0491} = 93\,885$ psi. This shear stress level occurs in both inside bolts.

Make-Up Force on Bolts

A consideration in bolted attachments is the level of pretightening. Often bolted connections are subjected to externally applied forces, which may cause further extension of the bolt leading to possible separation of contacting surfaces.

Consider the two situations illustrated in Figure 1.3. The left drawing shows a bolt compressing a spring of stiffness, k_s. As the bolt is tightened, the shank of the bolt elongates while compressing the spring. The extension of the bolt shank relates to bolt force by

$$\delta_B = \frac{PL}{E_B A_B} = \frac{P}{k_B} \quad \text{where} \quad k_B = \frac{E_B A_B}{L} \tag{1.10}$$

At the same time, the spring is compressed by

$$\delta_S = \frac{P}{k_S} \tag{1.11}$$

The internal force, P, in the bolt and the spring are the same (Figure 1.4a). The force level is established by the make-up torque.

If an external force, F, is applied to the connection, forces in the bolt (P_B) and spring (P_S) are no longer equal (Figure 1.4b). The challenge is to determine the magnitude of these two forces in relation to the externally applied force, F.

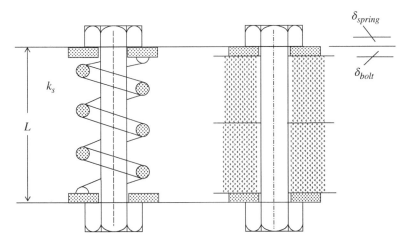

Figure 1.3 Make-up force in bolted connection.

Figure 1.4 Force response to an external load.

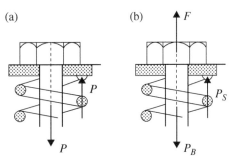

Consider the freebody diagram in Figure 1.4b. For equilibrium

$$F = P_B - P_S \tag{1.12}$$

Since there are two unknown forces in this one equation, it is necessary to consider deflections in the bolt and spring.

$$F = (P + \Delta P_B) - (P - \Delta P_S)$$
$$F = \Delta P_B + \Delta p_S$$

where ΔP_B is increase in bolt tension and ΔP_S is reduction in spring force. F is the externally applied force.

When load, F, is applied, the bolt stretch increases while the spring compression is relaxed by the same amount.

$$\Delta \delta_B = \Delta \delta_S \text{ (deflection equation)} \tag{1.13}$$

In terms of force changes

$$\frac{\Delta P_S}{k_S} = \frac{\Delta P_B}{k_B} \quad \text{where} \quad k_B = \frac{EA}{L} \tag{1.14}$$

By substitution

$$F = \Delta P_S + \frac{k_B}{k_S} \Delta P_S \tag{1.15}$$

giving

$$\Delta P_S = F \left(\frac{k_S}{k_B + k_S} \right) \tag{1.16}$$

Also

$$F = \Delta P_B + \frac{k_S}{k_B} \Delta P_S \tag{1.17}$$

$$\Delta P_B = F \left(\frac{k_B}{k_B + k_S} \right) \tag{1.18}$$

The resulting forces in the spring and bolt are

$$P_S = P - \Delta P_S$$
$$P_B = P + \Delta P_B$$

These relationships are shown in Figure 1.5.

$$P_S = P - F \left(\frac{k_S}{k_B - k_S} \right) \tag{1.19}$$

$$P_B = P + F \left(\frac{k_B}{k_B + k_S} \right) \tag{1.20}$$

The force required to separate the surfaces or created zero force in the spring is determined by setting $P_S = 0$

$$F_{cr} = P \left(1 + \frac{k_B}{k_S} \right) \tag{1.21}$$

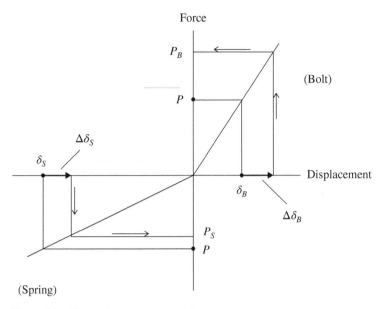

Figure 1.5 Force changes in connection.

In this case, by substitution into Eq. (1.20)

$$P_B = P\left(1 + \frac{k_B}{k_S}\right)$$

which is the same expression as Eq. (1.21) as expected.

When two plates are bolted together as shown in Figure 1.3b, the elastic compression in both plates have been modeled as two springs in series. A common math model for this rather complex state of strain is a truncated cone under uniaxial loading. Various formulations can be found in Ref. [9].

Example The arrangement of a weight (W) is held in place by a bolt tightened against a spring (Figure 1.6). Assume the initial compression of the spring is 0.1 in. The frame experiences base motion defined by

$$u(t) = u_0 \sin \omega t \tag{1.22}$$

This motion may cause the spring to disengage from the frame due to the acceleration of the weight. Ignoring the mass of the bolt and spring, determine the frequency at which the bolt becomes loose from the frame.

Other variables are quantified as

W	–	100 lb
L	–	2 in.
d	–	3/16 in. (bolt diameter)
k_S	–	2000 lb/in.
u_0	–	1/16 in.

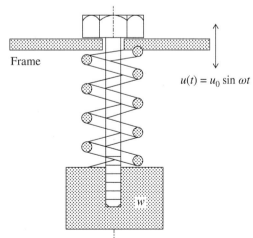

Figure 1.6 Spring-mounted weight.

To solve, we use Eq. (1.21) with the parameters below.

$$P = k_s \delta_s = 2000(0.1) = 200 \, \text{lb (initial spring force)}$$

$$A_B = \frac{\pi}{4}\left(\frac{3}{16}\right)^2 = 0.0276 \, \text{in.}^2$$

$$k_B = \frac{30 \times 10^6 (0.0276)}{2} = 414\,000 \, \text{lb/in.}$$

$$F_{cr} = u_0 \omega^2 M = 0.0625(\omega_{cr})^2 \frac{100}{386} = 0.0162(\omega_{cr})^2$$

Substituting the numbers gives

$$0.0162(\omega_{cr})^2 = 200\left(1 + \frac{414\,000}{2000}\right)$$

$$\omega_{cr} = 1602 \, \text{rad/s}$$

$$f = 255 \, \text{cps}$$

Preload in Drill Pipe Tool Joints

In oil well drilling, joints of drill pipe are connected by "tool joints" welded to the pipe body. At one end is the pin joint, at the other end is a box joint (Figure 1.7). As the total string of pipe is drilled down, new joints of drill pipe are added. The connections are critical. If joints are torqued to lightly, then bending stress may be created within the pin leading to fatigue. Also, if the shoulder in the connection loose, high pressure of the drill fluid could literally cut through the threads causing a washout. Proper make up torque of each joint is therefore critical.

The American Petroleum Institute (API) recommended (or allowable) preload in the shoulders of both box and pin for new pipe/tool joints is 50% of the force required to yield either box or pin. This preload is generated by make-up torque. For used drill pipe tool joints, it is 60%. This level of preload allows for additional pin loading due to direct pull, such as drillstring weight and over-pull.

After the tool joint is made up by the tongs (special torqueing tools), drill pipe is lifted vertically, and the slips removed. At this point, both shoulder and internal pin forces change from their make-up values. The questions then become (i) What is the new shoulder contact force? and (ii) How much direct pull would cause shoulder separation?

The following explains how pin force and box shoulder force change when external pull forces are applied across

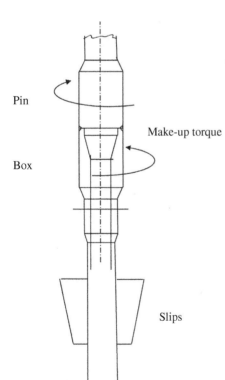

Pin

Box

Make-up torque

Slips

Figure 1.7 Tool joint make-up torque.

tool joints. The problem is statically indeterminate, so deflection equations are required along with the static forces.

For the sake of discussion, let the pin and box be represented by a cylindrical bar and tube as shown in Figure 1.8a. The tube is slightly longer than the bar. If the two heights are brought together and fixed, the preload (tension) in the pin is equal in magnitude to the preload (compression) in the box (Figure 1.8b). The pin is extended by

$$\delta_p = \frac{FL_p}{EA_p} = \frac{F}{k_p} \tag{1.23}$$

The box is compressed by

$$\delta_b = \frac{FL_b}{EA_b} = \frac{F}{k_b} \tag{1.24}$$

where

F	–	preload in pin and box after make up, lb
A_p, A_b	–	cross-sectional area of pin and box at the pitch point, in.^{2}
L_p, L_b	–	active length of pin and box (~0.75 in.)
k_p, k_b	–	effective spring constants in pin and box, lb/in.
E	–	modulus of elasticity, psi.

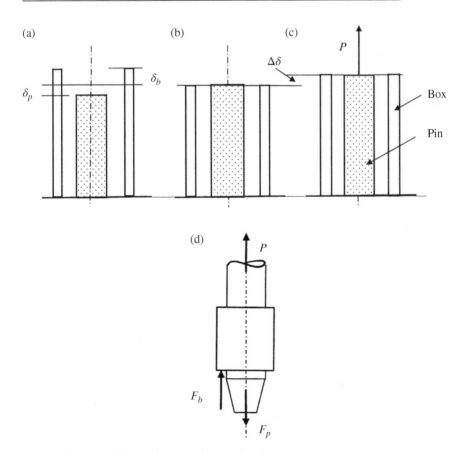

Figure 1.8 Internal forces due to preload and pull.

Now assume that a pull force, P, is applied across the tool joint (Figure 1.8c). The pin is stretched by $\Delta\delta_p$ while the box is relaxed by $\Delta\delta_b$: note that $\Delta\delta_p = \Delta\delta_b = \Delta\delta$ Therefore, the corresponding increase in the pin force is related to the corresponding reduction in the box force by

$$\frac{\Delta F_p}{k_p} = \frac{\Delta F_b}{k_b} \qquad (1.25)$$

where ΔF_p and ΔF_b are changes in pin and box forces.

Using the freebody diagram shown in Figure 1.8d,

$$P = F_p - F_b \qquad (1.26)$$

where F_p and F_b are new forces in pin and box resulting from the force, P, across the tool joint. These forces can be expressed as

$$F_p = F + \Delta F_p \qquad (1.27)$$

$$F_b = F - \Delta F_b \qquad (1.28)$$

Bringing Eqs. (1.25)–(1.28) together gives

$$\Delta F_p = \frac{k_p}{k_p + k_b} P \qquad (1.29)$$

Therefore, the new force in the pin due to both preload and direct pull is

$$F_p = F + \frac{k_p}{k_p + k_b} P \qquad (1.30)$$

where

P	–	pull force across tool joint
F	–	initial preload in box/pin
F_p	–	new force in pin.

Similarly, the new force in the box is

$$F_b = F - \frac{k_b}{k_p + k_b} P \qquad (1.31)$$

Two load conditions are of interest: (i) magnitude of P, which causes shoulder separation and (ii) possible yielding in the pin caused by P.

Shoulder Separation

Shoulder separation occurs when $F_b = 0$. Using Eq. (1.31) with $F_b = 0$ gives

$$P = \frac{k_p + k_b}{k_b} F \qquad (1.32)$$

which is the pull force across the tool joint required for shoulder separation.

Possible Yielding in the Pin

Rewriting Eq. (1.30) in terms of yield force in the pin

$$\left(F_{yld}\right)_{pin} = F + \frac{k_p}{k_p + k_b} P \qquad (1.33)$$

The force across a tool joint causing yielding is

$$P = \frac{k_p + k_b}{k_p}\left(F_{yld} - F\right) \tag{1.34}$$

This equation applies only if there is no shoulder separation.

Combining Eqs. (1.30) and (1.32) and assuming separation when $(\Delta\delta = \delta_b)$ and $F = 0$,

$$F_p = P \tag{1.35}$$

as expected.

Example Consider an oil field drill pipe tool joint as shown in the figure. Assuming the make-up contact force on the two shoulders of the pin and box is 472 000 lb, determine the shoulder force after a hook force of 250 000 lb is applied across the tool joint when the slips are removed. Determine the magnitude of a pull force required to separate the shoulder (zero contact force).

The dimensions for calculating these areas are:

Box OD = 6 3/8 in.
Pin ID (bore) = $3\tfrac{3}{4}$ in.
Pitch diameter = 5.04 in. (first thread on pin, see API RP 7G).

From the given dimensions, $A_p = 8.93$ in.^{2} and $A_b = 11.97$ in.^{2} and the active pin and box lengths $(L_p = L_b \sim \tfrac{3}{4}$ in.$)$ are the same. Then

$$\frac{k_b}{k_p + k_b} = \frac{A_b}{A_p + A_b} = \frac{11.97}{8.93 + 11.97} = 0.573$$

$$\frac{k_p}{k_p + k_b} = \frac{A_p}{A_p + A_b} = \frac{8.93}{8.93 + 11.97} = 0.427$$

Force in the pin is

$$F_p = F + \frac{k_p}{k_p + k_b}P \tag{1.36}$$

$$F_p = 472\,000 + 0.427(250\,000) = 578\,750\,\text{lb}$$

Similarly, the new force in the box is

$$F_b = F - \frac{k_b}{k_p + k_b}P$$

$$F_b = 472\,000 - 0.573(250\,000) = 472\,000 - 143\,250 = 329\,000\,\text{lb} \tag{1.37}$$

This represents the new shoulder force after the 250 000 lb load is applied.

The pull force required to separate the pin/box shoulder is (using Eq. (1.32))

$$P = \frac{1}{0.573}(472\,000) = 823\,700\,\text{lb}$$

This force is felt by the pin at separation. It is less than the 944 000 lb required to yield the tool joint. This number, P, far exceeds the strength capacity of each of the four pipe grades ($4\tfrac{1}{2}$ in.), which means that the drill pipe body would fail before tool joint shoulders separation.

Make-Up Torque

There are various formulas used to predict make-up torque, which account for friction in the threads and shoulder. One that is commonly used to predict make-up torque in drill pipe joints is called the screw jack formula.

$$T = \frac{F}{12}\left[\frac{p}{2\pi} + \frac{R_t f}{\cos\theta} + R_s f\right] \tag{1.38}$$

where

T	–	torque, ft-lb
p	–	lead of threads, inches ($p = \frac{1}{4}$ in. for 4 threads/in.)
R_t	–	average mean radius of threads, in.
f	–	coefficient of friction (~0.08)
R_s	–	mean radius of shoulder, in.
θ	–	½ of included angle of thread ($2\theta = 60°$)
F	–	contact force between mating shoulders, lb

The screw jack formula shows that total make-up torque is distributed among three areas. One component of the torque drives the mating shoulders together creating the contact force. If there were no friction in the connection, all the applied torque would create this force. From energy considerations, the work done by a torque over one revolution is equal to the work to move an axial force, F, over one threat pitch.

$$2\pi T(12) = Fp \tag{1.39}$$

The second component of torque is the torque required to overcome friction in the threads. The third torque component overcomes friction in the shoulder.

If tool joints, connecting drill pipe, are made up too tight, the pin could be overstretched. If tool joints are made up too low, threads can be exposed to washouts or bending fatigue. In practice, tool joints are generally made up too low. According to one drilling engineer, "if tool joints are made up too low, they will not make up downhole. On the other hand, drill collar connections will make up downhole. If drill collar connections are difficult to break out, it usually means they were made up too low initially."

The make-up force or internal contact force between the two mating shoulders directly affects the structural integrity and pressure sealing capacity of the tool joint. Consider, for example, the torque components in a new NC50 ($6\frac{3}{8}$ in. × $3\frac{3}{4}$ in.) tool joint using the following set of numbers:

p	=	0.25 in.
f	=	0.08
θ	=	30°
R_t	=	2.385 in.
R_s	=	2.922 in.

Substituting these numbers into Eq. (1.28) gives

$$T = \frac{F}{12}\left[\frac{0.25}{2\pi} + \frac{2.385(0.08)}{0.866} + (2.922)0.08\right]$$

$$T = \frac{F}{12}(0.0398 + 0.2203 + 0.233)$$

$$T = \frac{F}{12}(0.4931)$$

This calculation shows that only 8% of make-up torque drives the shoulders together. The remaining 92% is used to overcome friction. The coefficient of friction of thread dope is therefore critical. Make-up torque recommendations [10] are based on a coefficient of friction of $f = 0.08$.

If the actual coefficient of friction is less than 0.08, contact force between the mating shoulders will be higher than expected when the recommended API make-up torque is developed. This condition could over stretch the pin or damage the shoulder area. If the actual coefficient of friction is greater than 0.08, contact force between the mating shoulders will be lower than expected when the recommended API make-up torque is developed. This condition allows bending stresses to reach the pen causing fatigue damage and creates an inadequate pressure seal leading to a washout through the threads.

Bolted Brackets

Things to consider in bracket design are indicated in Figure 1.9. The bracket supports a force, F, which is resolved into vertical (F_V) and horizontal (F_H) components. Forces in both bolts (A and B) are the superposition of several types of loads. First, the vertical force component is supported evenly by both bolts. The horizontal component creates shear forces evenly in both bolts as well. In addition, the horizontal component creates a moment (bF_H) about point "a," which adds to bolt forces.

This moment creates a statically indeterminate problem, which requires deflection considerations. Equations to be considered are

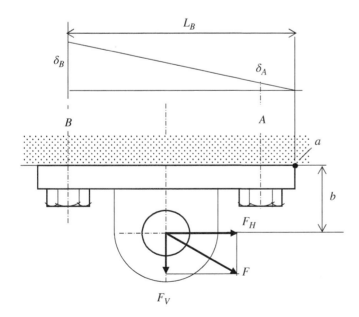

Figure 1.9 Bracket attached by two bolts.

$$bF_H = F_B l_B + F_A l_A \quad \text{(one statics equation, two unknowns)} \tag{1.40}$$

$$\frac{\delta_A}{\delta_B} = \frac{l_A}{l_B} \quad \text{(deflection equation)} \tag{1.41}$$

The stretch in both bolts is

$$\delta_A = \frac{F_A L}{EA} \quad \text{and} \quad \delta_B = \frac{F_B L}{EA} \quad (L \text{ is bolt length}) \tag{1.42}$$

Combining Eqs. (1.41) and (1.42) gives a second equation for determining the forces in each bolt.

$$\frac{F_A}{F_B} = \frac{l_A}{l_B} \tag{1.43}$$

Substituting this result into Eq. (1.40) gives

$$bF_H = F_B l_B \left[1 + \left(\frac{l_A}{l_B} \right)^2 \right] \tag{1.44}$$

Welded Connections

Torsion Loading in Welded Connections

The analysis of welded connections is similar to bolted connections. In this case, total shear stress at any point in a weld is the superposition of direct shear and torsion shear. Direct shear stress is determined by

$$\tau_F = \frac{F}{A_{total}} \tag{1.45}$$

where F is total shear load on the joint and A_{total} is the total throat weld area (tL). Local shear stress due to applied torque, T, is assumed to follow classic shear stress predictions.

$$\tau_T = \frac{Tr}{J_0} \tag{1.46}$$

where J_0 is the polar moment of inertia of the weld pattern.

When the size and shape of each weld pattern is known, these equations can be used to determine both shear stress components. The resulting shear at any point in the weld is the vector sum of both stress components.

Consider for example a steel plate attached to a vertical post by two fillet welds (Figure 1.10). For this weld pattern

$$dJ_0 = r^2 dA = \left(h^2 + x^2 \right) dA \tag{1.47}$$

$$J_0 = \int\limits_{-\ell}^{\ell} h^2 t \, dx + \int\limits_{-\ell}^{\ell} x^2 t \, dx \tag{1.48}$$

$$J_0 = h^2 t(2\ell) + \frac{2}{3} \ell^3 t \tag{1.49}$$

Combining parameters, $l = \frac{L}{2}$ and $A = Lt$.

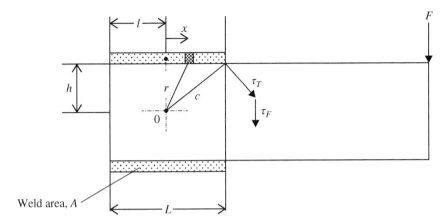

Figure 1.10 Welded connection in torque.

$$J_0 = h^2 A + \frac{2}{3} \left(\frac{L}{2} \right)^3 t \tag{1.50}$$

$$J_0 = h^2 A + \frac{1}{12} L^2 A \quad \text{(one weld)} \tag{1.51}$$

$A = tL$ (t is measured across triangular weld)

Maximum shear stress occurs at the end of the weld. This stress is the vector sum of two components.

$$\tau_{\max} = \tau_T + \tau_F \quad \text{(vector summation)} \tag{1.52}$$

$$\tau_{\max} = \frac{Tc}{2J_0} + \frac{F}{2A} \tag{1.53}$$

Example Consider the specific numbers shown below.

$L = 8$ in.
$t = 0.25$ in.
$h = 3$ in.
$F = 250$ lb
$T_0 = 3000$ ft-lb (assumed)
$A = t L = 0.25(8) = 2$ in.^{2} ($A_{total} = 2A = 4$ in.^{2})
$J_0 = 2 \left[3^2(2) + \frac{1}{12} 8^2(2) \right] = 57.3$ in.^{4} (both welds)
The stress components are

$$\tau_T = \frac{3000(12)5}{57.3} = 3141 \text{ psi (torsion load)}$$

$$\tau_F = \frac{250}{4} = 62.5 \text{ psi (direct load)}$$

The shear stress produced by the direct loading is minor compared to the shear stress produced by the torque.

Attachments of Offshore Cranes

Cranes are essential in the operation of offshore structures. They load and unload necessary equipment and supplies from supply boats. Off-loading cranes are often attached to a base structure, such as a platform or floating vessel by weldments. A weldment with gusset plates is illustrated in Figure 1.11.

The stresses in the weldment are due to a bending moment produced by a hook load in the crane. The crane is rotated about a vertical axis after the equipment has been elevated to deck level. The classic calculation is based on

$$\sigma = \frac{Mc}{I} \tag{1.54}$$

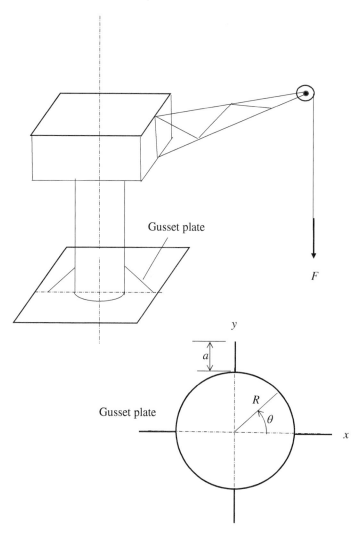

Figure 1.11 Offshore crane weldment.

By inspection we note that moment of inertia, $I_x = I_y$ and each are principal moments of inertia. They are both principal moments of inertia and are represented by a point in the Mohr inertia circle. Furthermore, the moment of inertia about every diametrical axes is the same. Bending stress magnitude varies by distance, c. The c value is maximum when bending occurs about the x or y axis. One of the engineering challenges is to determine the moment of inertia. We choose to use the x axis as the reference line.

The moment of inertia of the ring is determined as follows.

$$dI_x = (R \sin \theta)^2 t \, ds \tag{1.55}$$

$$I_x = 4R^3 t \int_0^{\frac{\pi}{2}} \sin^2 \theta \, d\theta \tag{1.56}$$

$$I_x = 4R^3 t \left[\frac{\theta}{2} - \frac{\sin 2\theta}{4} \right]_0^{\frac{\pi}{2}} \tag{1.57}$$

$$I_x = \pi R^3 t \tag{1.58}$$

The moment of inertia of two gusset plates with respect to the x axis (using the transfer formula) is

$$I_x = 2 \left[\frac{ta^3}{12} + \left(R + \frac{a}{2} \right)^2 at \right] \tag{1.59}$$

Total moment of inertia with respect to the x axis is

$$I_x = \pi R^3 t + 2 \left[\frac{ta^3}{12} + \left(R + \frac{a}{2} \right)^2 at \right] \tag{1.60}$$

If gusset plates are placed evenly around a cylindrical stand, bending moments of inertia about any diameter are the same. This means the Mohr circle of inertia is a point.

Quality Assurance

Performance evaluation is necessary for nearly every product or service. Evaluation can be based on quality assurance at the assembly line, laboratory testing, product performance in the marketplace, and customer feedback, to mention a few. The metrics for any evaluation should be established early. The metrics provide a baseline from which to evaluate performance. It is useful to include the stake holders and end user in setting the metrics. Performance should be given a numerical score and the rationale behind the score.

- Functionality
- Root cause analysis
- Maintenance and reliability
- Market response
- Assessment/feedback

Engineering Education

Engineering education is a good example of program assessment and feedback. A few years ago, engineering programs were based on curricula containing certain required components, such as mathematics, humanities, engineering science, and engineering design. In 2000, the Accreditation Board for Engineering and Technology (ABET) initiated a different set of criteria based on objectives, curricula, and a continuous improvement process.[2] Each academic unit can design or modify existing programs to satisfy these criteria. The flow diagram (Figure 1.12) illustrates such a feedback process.

Mission Statement

The mission statement provides focus. It sets direction. It is much like the operational requirements of Figure 1.1. For example

> To provide a broad-based integration of courses and experience
> That prepares its graduates to practice their profession successfully, to apply their skill to solve current engineering problems collaboratively, and to help advance the knowledge and engineering practice in their field.

This statement is reviewed from time to time with the constituents of the program.

Academic Design Specifications

Design specifications are set by the ABET. Each engineering program must satisfy each of these components. These criteria offer the opportunity to redesign or modify existing undergraduate programs. There is a lot of flexibility in how programs are designed. It is very open ended. Typically, well-established programs are modified to satisfy these criteria and continually improved.

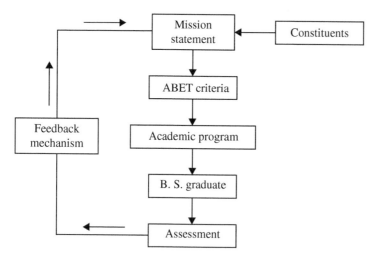

Figure 1.12 Feedback process of continuous improvement.

Design of the Academic Program

An academic program includes the integration of every component (students, curricula, faculty, facilities, and resources) as spelled out by the eight ABET criteria. Curricula are at the center of the academic program.

Perhaps the biggest improvement in engineering education has been the requirement that all seniors must have a major design experience. While the basic engineering science courses are fundamentally the same, the use of computers, modern laboratory equipment, and classroom equipment have brought about big changes in how courses are taught. In my opinion, the greatest improvement to engineering curricula has been the Professional Component (Criteria 4), which requires a major capstone design experience by each student prior to graduation. These projects bring seniors face to face with the realities of applying engineering principles to solve practical engineering problems. These experiences help engineering graduates step directly into an industrial setting and immediately take on projects with confidence.

Graduate engineers are the product of this service. Just as industry evaluates the quality of its service products, academia should do the same. The question becomes one of how prepared are graduates to enter the work force. Have the graduates gained proficiency in the expected outcomes as specified in Criteria 3 and 5?

Outcomes Assessment

What are the tools for measuring the results of the educational program? Evaluation and assessment must be quantified in some way to measure improvements in the educational service from year to year. The outcomes assessment process should involve all constituents involved or those who have a stake in the outcome.

Once the outcome of the educational service has been evaluated, how are the conclusions fed back into the educational process?

Saturn – Apollo Project

A few years ago, I had the pleasure of visiting with Albert C. Martin, Director of Launch Operations for Stage II, Saturn/Apollo Project (Albert C. Martin, personal communication). We discussed the steps leading up to each launch. Testing was a huge part of launch preparation.

During the early 1960s, the digital computer was just emerging as an engineering tool. Software, as we have today, was not yet available for making numerical calculations of critical aspects of the rocket. Thus, reliability of design, along with improvements, depended heavily on testing.

Test results were fed back to the design team for improvements. This cycle was made until each team could say "... that everything that can possibly be done to assure success has been done." Final testing was conducted at the Kennedy Space Center, Cape Canaveral, prior to rocket assembly and launch. Over 400 engineers and technicians conducted multiple tests on critical component of the Saturn Stage II under the direction of Lyle C. Bjorn, Manager of Testing. There were no launch failures during the Saturn/Apollo Project.

Important aspects of design in the Saturn/Apollo project were

Produce ability
Reliability
Safety.

A "Design Review and Change Board" reviewed results after each test. Information gained from testing was evaluated for possible improvement to the overall second stage design. Before any design changes were made, the Change Board scrutinized the recommendations. Representatives from various contributors attended these meetings, including

Manufacturing

Design and engineering

    Structures

    Aerodynamics

    Testing

    Combustion

Quality control

Financial

Contracts

    These types of meetings were held regularly since second-stage testing was conducted around the clock and hundreds of data had to be evaluated. While every component was essential, the turbo pump was central to the overall operation of each engine. To produce the necessary thrust, a high rate of fuel burning was vital. That meant the turbo pumps operated at near 80 000 rpm. Bearings that held the turbine shaft were pushed to the limit at this speed. Also, turbine blades were susceptible to vibration and fatigue, and had to be monitored.

    Since then, computer capability in both size and speed, along with computer software, make it possible to accurately predict expected performance of critical components. Computer simulation of complex systems (airplanes, for example) gives accurate predictions of expected performance, essentially eliminating the need for many tests.

Notes

1 http://www.asme.org/Education/PreCollege/TeacherResources/Code_Ethics_Engineers.cfm.
2 Established by the Accreditation Board for Engineering and Technology (ABET)..

References

1 Federal Standardization Manual. (2000). U.S. General Service Administration, Federal supply Service, http://www.fss.gov/pub.
2 Packard, V. (1960). *The Waste Makers*. David McKay Co.
3 Shigley, J.E. and Mischke, C.R. (1991). *Mechanical Engineering Design*, 5e. Marcel Dekker, Inc.
4 Peters, G.A. (1971). *Product Liability and Safety*. Coiner Publications, Ltd.
5 Heidelang, H. (1991). *Safe Product Design in Law, Management and Engineering*. Marcel Dekker, Inc.
6 Parker, J.P. (1989). *Instructional Strategies for Teaching the Gifted*. Allyn and Bacon, Inc.
7 Ullman, D.G. (1997). *The Mechanical Design Process*, 2e. McGraw-Hill Book Co.
8 Dareing, D.W. (2010). *Engineering Design and Problem Solving*. Tennessee Valley Publishing.
9 Shigley, J.E. and Mitchell, L.D. (1983). *Mechanical Engineering Design*. McGraw-Hill Book Co.
10 API (RP 7G) (1987). *Recommended Practice for Drill Stem Design and Operating Limits*. American Petroleum Institute.

2

Configuring the Design

This chapter reviews classical analytical tools useful for determining shapes and sizes of product components. Configuring a design refers to dimensions and interface between every part.

A good place to start the force analysis is at the utility end of the design. Magnitudes and type of loading will come from the expected result or "use" point. Reactions are transferred through the "transmission" linkages to the frame. Transmission loads will define equipment specifications and size and shape to the transmission linkages (gears) the frame must accommodate. Defining these forces, moments, or torque will also establish the input power required to drive the device. Classical equations are good enough at this point.

Material yield strength, σ_{yld}, is usually taken as the failure stress. The allowable design stress, σ_a, depends on the factor of safety.

$$\sigma_a = \frac{\sigma_{yld}}{FS} \tag{2.1}$$

The allowable force, F_a, is related to allowable design stress by

$$\sigma_a = \frac{F_a}{A} \tag{2.2}$$

Three types of design situations may occur.

1) $F_a = \sigma_a A$ (area given, determine allowable load, F_a),
2) $A = \frac{F}{\sigma_a}$ (force given, determine structural dimensions, A),
3) $\sigma_a = \frac{F}{A}$ (force and dimensions given, determine material strength).

The same scenario applies for bending and torsion loads. Using elementary stress formulas and classic mechanics $\left(\frac{Mc}{I}, \frac{Tc}{J}\right)$ is fine. Computer software can be used later to refine the design.

As components are being configured, always consider: how the part is to be made? A machinist is a good source of information.

Force and Stress Analysis

An important aspect of design is determining external and internal forces and visualizing how forces flow throughout the design. This will help to determine shapes and member sizes consistent with material strengths and factors of safety. Freebody diagrams of subsystems are helpful.

Engineering Practice with Oilfield and Drilling Applications, First Edition. Donald W. Dareing.
© 2022 John Wiley & Sons, Inc. Published 2022 by John Wiley & Sons, Inc.

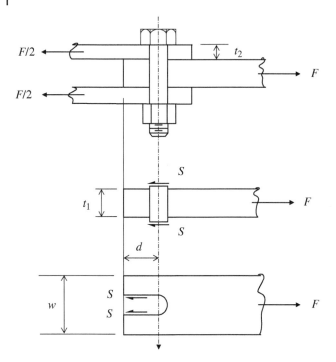

Figure 2.1 Simple stresses.

Example Consider the double strap connection with the following load conditions (Figure 2.1). The problem is to determine

1) Thickness, t_2, and width of outside straps.
2) Width of center strap.
3) Set-back location, d, of bolt.

Design specifications for this example are:

Connection load	$F = 5000\,\text{lb}$
Bolt size	$D = 3/8\,\text{in. (area} = 0.1104\,\text{in.}^2)$
$\sigma_{yld} = 75\,000\,\text{psi}$	$(\tau_{yld} = 43\,275\,\text{psi})$ (straps)
$\sigma_{yld} = 120\,000\,\text{psi}$	$(\tau_{yld} = 69\,240\,\text{psi})$ (bolt)
Factor of safety	$FS = 2$
Thickness of middle strap	$t_1 = ¼\,\text{in.}$

Step 1 – Check shear stress in bolt

$$\tau = \frac{S}{A} = \frac{2500}{(0.1104)} = 22\,645\,\text{psi}$$

$$FS = \frac{\tau_{yld}}{\tau} = \frac{69\,240}{22\,645} = 3.058$$

Step 2 – Check normal stresses in middle strap and determine width, w.

$$FS = \frac{\sigma_{yld}}{\sigma_a}$$

$$\sigma_a = \frac{75\,000}{2} = 37\,500 \text{ psi}$$

$$\sigma_a = \frac{F}{(w-D)t_1}$$

$$37\,500 = \frac{5000}{(w-0.375)(0.25)}$$

$w = 0.9083$ in. (compare with standard strap iron width)

Step 3 – Check tear out shear stresses in middle strap

$$FS = \frac{\tau_{yld}}{\tau_a}$$

$$\tau_a = \frac{43\,275}{2} = 21\,637 \text{ psi} \quad \text{(allowable shear stress)}$$

Equating to

$$\tau_a = \frac{S}{dt_1} = \frac{2500}{d(0.25)} = \frac{10\,000}{d}$$

gives

$$21\,637 = \frac{10\,000}{d}$$

$d = 0.462$ in. (set-back distance, 0.5 in. OK)

Step 4 – Check normal stress in two outside straps and determine their thickness. Since both outside straps carry $\frac{1}{2}F$, each cross-sectional area must be $\frac{1}{2}$ area of middle strap. Therefore

$w = 0.9083$ in.

$t_2 = \frac{1}{8}$ in.

$d = 0.462$ in.

In many situations, stresses resulting from various types of loading will be combined to form a two-dimensional or three-dimensional state of stress. It is important to visualize loading situations that produce the more general states of stress.

Appropriate criteria of failure must be used. The von Mises criteria of energy distortion are commonly used for static load situations. Other modes of failure are fatigue and corrosion or any combination of static, fatigue, and corrosion. All possible modes of failure should be considered.

Beam Analysis

Beams are often used to support subsystems. They can experience various types of loading as they provide a structural frame for designs. Bending is the most significant load because it can generate relatively large stresses in comparison with direct pull or shear. For example, it is easier to break a stick by bending than by direct pull. That's because bending stresses are higher than axial stresses for the same load. Consider a solid rectangular beam that is 2″ by 4″. If 1000 lb is applied at the end of a foot-long cantilever beam, the bending stress is 22 500 psi (stiffest axis) and 90 000 psi (flexible axis). On the other hand, if the 1000 lb force is applied in tension, the tensile stress in the same member is only 125 psi.

The notion of equilibrium has been applied in numerous scientific fields, especially engineering. The acceptance and application of this science was slow to develop. As late as the early 1800s, many engineering designs were still based on empirical and historical approaches. Even the first metal bridge (cast iron) across the Severn River (1779) was developed empirically based on the shape of stone arch bridges.

In the early days, many felt science was too mathematical for practical use. By mid-1800s, the demand for longer bridges and larger (and safer) buildings brought about the scientific-based design approach. The Eads Bridge (designed and constructed by James Eads) spans the Mississippi River at St. Louis (Missouri) [1]. It was based on truss design. It was completed in 1867 and was the first scientific-based engineering bridge. It is still in use today. Eiffel, the French engineer, applied structural engineering principles to several bridges (1884) and structures, including the Eiffel Tower (1889) and the Statue of Liberty.

The Raftsundet Bridge (Norway) [1], completed in 1998, has the longest (978 ft) beam span ever for a bridge.

Shear and Bending Moment Diagrams

Beams are defined as structural members that carry transverse forces. Beams can be mounted in a variety of ways, such as simple supports and fixed supports.

Shear and bending moment diagrams show how internal shear (V) and internal bending (M) are distributed within beams, caused by external loading. Boundary conditions must also be considered. These diagrams show the location and magnitude of maximum shear and maximum bending.

The signs given to internal shear forces (V) and moments (M) are important because they help establish the direction of normal and shear stresses. While somewhat arbitrary, the sign conventions shown in Figure 2.2 are used here.

Example Consider the simply supported beam carrying a concentrated force at midpoint (Figure 2.3). The size of the beam is not a factor except for its distributed weight, which may or may not be considered. It is not considered in this example.

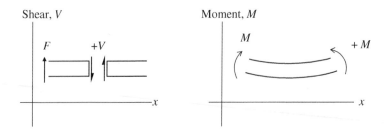

Figure 2.2 Sign convention for internal shear and bending.

Figure 2.3   Shear and bending moment diagrams.

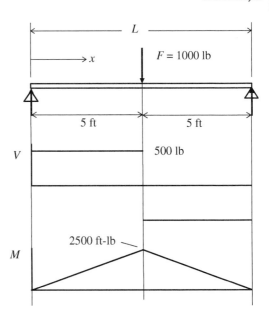

Mathematical expressions for shear and bending moments are

$V = +500 \text{ lb } (0–5 \text{ ft})$
$V = -500 \text{ lb } (5–10 \text{ ft})$
$M(x) = 500x \text{ ft-lb } (0–5 \text{ ft})$
$M(x) = (L - x)500 \text{ ft lb } (5–10 \text{ ft})$

with

$M_{max} = 2500 \text{ ft lb at } x = 5 \text{ ft}$

Example   Consider a beam supporting a distributed load (Figure 2.4). The total load distribution, w, includes the weight of the beam along with a live load.

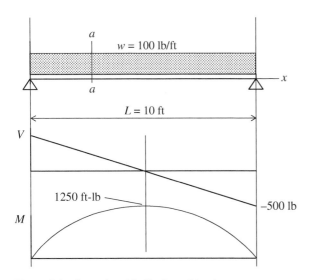

Figure 2.4   Example with distributed load.

Using a freebody diagram of the beam from the left end to section a–a, internal shear and bending are expressed mathematically as

$$V(x) = 500 - 100x \tag{2.3}$$

$$M(x) = 500x - 50x^2 \tag{2.4}$$

Maximum shear is located at $x = 0$ and $x = 10$ ft. Maximum bending moment is located by

$$\frac{dM}{dx} = 500 - 100x = 0 \tag{2.5}$$

$$x = 5\,\text{ft}$$

By substitution into Eq. (2.4)

$$M_{\max} = 500(5) - 50(5)^2$$

$$M_{\max} = 1250\,\text{ft-lb}$$

Shear and bending moment diagrams can also be constructed graphically. The method allows both diagrams to be sketched while identifying key aspects, such as maximum and minimum values, of each diagram. Equations that are useful in constructing shear and bending moment diagrams are developed from the diagram given in Figure 2.5.

Useful equations from this diagram are

$$\frac{dV}{dx} = w \text{ (slope of the shear diagram equals the local load rate)} \tag{2.6}$$

$$V_2 - V_1 = \int_1^2 w(x)\,dx \text{ (area under the load curve)} \tag{2.7}$$

$$\frac{dM}{dx} = V \text{ (slope of the moment diagram equals the local shear)} \tag{2.8}$$

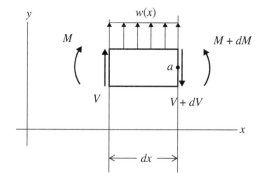

Figure 2.5 Freebody diagram of a beam element.

$$M_2 - M_1 = \int_1^2 V(x)dx \text{ (area under the shear curve)} \tag{2.9}$$

Applying these equations to the problem in Figure 2.4 gives

$$\frac{dV}{dx} = -100 \text{ lb/ft (slope of the shear diagram)}$$

The freebody diagram of the total beam shows the shear at both supports is 500 lb. Taking point 1 at the left boundary, the shear at the midpoint ($x = 5$ ft) is

$$V_2 - 500 = -5(100)$$
$$V_2 = 0 \text{ at } x = 5 \text{ ft}$$

Taking point 1 at the left end and point 2 at the right end and applying Eq. (2.7)

$$V_2 - 500 = -100 (10)$$
$$V_2 = -500 \text{ lb}$$

which agrees with the results from the freebody diagram.

Since the ends of the beam are assumed to be simply supported, moments at both ends are zero. Noting that the slope of the moment diagram is equal to local shear force, and shear force is zero at $x = 5$ ft, the maximum bending moment occurs at this point. Its magnitude is the area under the shear diagram between locations 1 and 2.

$$M_{max} = \frac{1}{2}5(500) = 1250 \text{ ft-lb}$$

Note the moment is (+) plus causing the beam to bend upward.

If a concentrated force or moment is applied anywhere on the beam, the forces and moments in Figure 2.6a and b apply.

If for example a concentrated force ($F = -1000$ lb) is applied at the center of the beam shown in Figure 2.7, internal shear is reduced by 1000 lb across the concentrated force as shown in the shear diagram. The maximum internal bending moment is determined graphically by

$$M_{max} = M(5) = \text{area 1} + \text{area 2}$$
$$M(5) = 500(5) + \frac{1}{2}500(5) = 3750 \text{ ft-lb}$$

Figure 2.6 Stepping across concentrated force and moment.

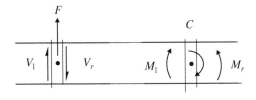

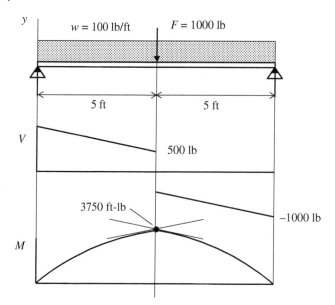

Figure 2.7 Concentrated force with distributed force.

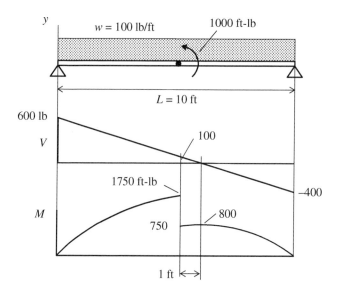

Figure 2.8 Local moment added to distributed loading.

The slope of the moment diagram is not zero at the midpoint because the shear is no longer zero. There is a discontinuity in the moment diagram at $x = 5$ ft.

If a 1000 ft lb couple is applied at the center point, shear and bending moment diagrams would be modified as shown in Figure 2.8.

A good check on each shear and moment diagram is reaction forces determined from a freebody diagram of the loaded beam. They should match the shear forces and moments at the ends of the two diagrams. The slope of the shear diagram is equal to the load rate, w, while the slope of the moment diagram is equal to the local internal shear, which is equal to zero at $x = 6$ ft.

Bending Stresses

Bending stresses and deflections in beams are based on Euler's model [2, 3] assuming a transverse plane remains flat after bending, i.e. transverse planes do not warp (Figure 2.9). From this, normal strain is linear with distance from the neutral axis of bending.

$$\varepsilon = -\frac{y\,d\theta}{dx} \tag{2.10}$$

Note, $dx = \rho\,d\theta$, so

$$\varepsilon = -\frac{y}{\rho} \quad \text{(normal strain)} \tag{2.11}$$

$$\sigma = -\frac{y}{\rho}E \quad \text{(normal stress)} \tag{2.12}$$

Summing the differential forces ($\sigma\,dA$) across the beams cross section gives

$$\int_A \sigma\,dA = \frac{1}{\rho}\int_A y\,dA = 0 \tag{2.13}$$

The integral on the right side defines the centroid of the area; therefore, the neutral axis of bending is the same as the centroid of the cross-sectional area.

The summation of the differential internal moments across the beam gives

$$M = \int_A y(\sigma\,dA) = \int \frac{E}{\rho}y^2\,dA$$

$$M = \frac{EI}{\rho} \tag{2.14}$$

Two important equations emerge from this derivation.

$$EI\frac{d^2y}{dx^2} = M \quad \text{(for small deflections)} \tag{2.15}$$

and by combining Eqs. (2.12) and (2.14)

$$\sigma = \frac{My}{I} \quad \text{(normal stress is compressive with $+y$)} \tag{2.16}$$

Figure 2.9 Euler's theory of beam stress.

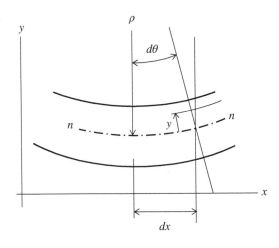

This equation is used along with the moment diagram to determine maximum bending stress in a beam. The bending diagram gives location and magnitude of the maximum *bending moment* in the beam. Equation (2.16) defines the *bending stress* at this location.

The objective in structural design is to select the lightest beam that is strong enough to support the applied loads. Material cost is directly related to weight of structure. The location and magnitude of the maximum bending moment is determined from the moment diagram. Cross-section properties of many standard beams (H beam, I beam, T beams, channel beam, and angle iron) are given by the American Institute of Steel Construction (AISC) [4]. Composite beams can also be formed by combinations of standard beams, such as channel beams and angle iron beams.

Beam selection depends on the required section modulus based on the allowable stress. Bending stress is the greatest at the point of maximum bending, as established by the bending moment diagram.

$$\sigma_{max} = \frac{M_{max}\, c}{I} = \frac{M_{max}}{S} \tag{2.17}$$

The allowable stress is found by

$$\sigma_a = \frac{\sigma_{yld}}{FS} \tag{2.18}$$

where FS is factor of safety. The required section modulus is determined from

$$S = \frac{M_{max}}{\sigma_a}\ \text{in.}^3 \tag{2.19}$$

It is the basis for beam selection.

Example Assume, for example, the required section modulus of a beam is $S = 55$ in.^{3} The lightest Standard I beam satisfying this requirement is (S15 × 42.9). This beam is 15 in. high, weighs 42.9 lb/ft, and has a cross-section area of $A = 12.57$ in.^{2} Since cost is directly related to beam weight, it is also important to choose a beam having the lightest weight.

Note, weight per unit length and cross-sectional area are related by

$$w\ell = A\ell\gamma_{stl} \tag{2.20}$$

$$w = A\left|\frac{1\ \text{ft}^2}{144\ \text{in.}^2}\right|(492\ \text{lb/ft}^3) = 3.417A$$

where $A(\text{in.}^2)$ and $w(\text{lb/ft})$.

Table 2.1 compares the relative strengths of different cross sections having the same area or weight per foot. The section moduli are compared to those of a solid rod having a diameter of 4 in., whose section moduli is 6.28 in.^{3} The section moduli of the H beam and Standard I beam

Table 2.1 Comparison of section moduli for different shapes. $A = 12.57$ in.^{2}, $w = 42.95$ lb/ft.

Shape	Section modulus (S, in.^{3})	Strength ratio
Circular rod (4 in dia)	6.28	1
Pipe (¼ in. wall)	49.5	7.88
S15 × 42.9	59.6	9.49
W14 × 43	62.7	9.98

are significantly higher than those for the solid bar. The diameter of the pipe is 16.23 in. with a ¼-in. wall thickness. While the weight in each case is the same, there is a wide difference in the section moduli.

Beam Deflection and Boundary Conditions

Develop the mathematical deflection function for the cantilevered beam shown. What is the maximum deflection? Consider a uniformly loaded cantilevered beam with a concentrated force at its free end (Figure 2.10).

The deflection function for this beam can be approached in two ways. First by starting with

$$EI\frac{d^2y}{dx^2} = M \tag{2.21}$$

The local internal bending moment is

$$M(x) = -\left[(L-x)F + \frac{(L-x)^2}{2}w\right] \tag{2.22}$$

$$EI\frac{d^2y}{dx^2} = -\left[(L-x)F + \frac{(L-x)^2}{2}w\right] \tag{2.23}$$

Both F and w create negative curvature in the beam. By integration and applying boundary conditions

$$EI\frac{dy}{dx} = -\left[-\frac{1}{2}(L-x)^2F - \frac{1}{6}(L-x)^3w + C_1\right] \tag{2.24}$$

$$EIy = -\left[\frac{1}{6}(L-x)^3F + \frac{1}{24}(L-x)^4w + C_1x + C_2\right] \tag{2.25}$$

Boundary conditions are

$$y(0) = 0 \quad \text{and} \quad \frac{dy}{dx}\bigg|_{x=0} = 0 \tag{2.26}$$

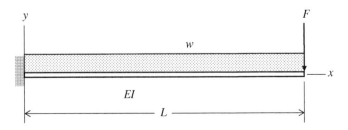

Figure 2.10 Deflection of cantilevered beam.

Giving

$$C_1 = \frac{L^2}{2}F + \frac{L^3 w}{6} \quad \text{and} \quad C_2 = -\frac{L^3}{6}F - \frac{L^4 w}{24} \tag{2.27}$$

The maximum deflection occurs at $x = L$

$$EIy_{max} = -\left(\frac{L^3}{3}F + \frac{L^4 w}{8}\right) \tag{2.28}$$

Note the total deflection at $x = L$ is the effects of the two loads superimposed as one expression. The minus sign is the downward direction. The maximum slope of the deflection occurs at $x = L$.

$$EI\theta_{max} = -\left(\frac{FL^2}{2} + \frac{wL^3}{6}\right)$$

The second approach starts with

$$EI\frac{d^4 y}{dx^4} = -w \tag{2.29}$$

By stepwise integration and applying boundary conditions

$$y(0) = 0 \tag{2.30a}$$

$$\left.\frac{dy}{dx}\right|_{x=0} = 0 \tag{2.30b}$$

$$\left.\frac{d^2 y}{dx^2}\right|_{x=L} = 0 \tag{2.30c}$$

$$EI\left.\frac{d^3 y}{dx^3}\right|_{x=L} = F \tag{2.30d}$$

This procedure leads to same expression for $y(x)$ and the same maximum deflection as before.

Shear Stress in Beams

The theory of shear stresses in rectangular beams was developed by D.J. Jourawski while designing wooden bridges (1844–1850) for the St. Petersburg–Moscow railroad line [5]. Jourawski recognized that wood beams are weak in shear along fibers, and these shear stresses cannot be disregarded in the design of wooden structures. He developed the mathematical formula that predicts shear stress magnitude across beams and showed the maximum shear stress occurs at the neutral axis.

Shear stresses in most beam applications are small compared with bending stresses and are usually ignored. However, shear stress in wood structures can be critical. Trees often fail by shear in a severe storm as manifested by gaps along center planes of the trunk or limbs. The theory behind stress magnitudes and distribution can be found in Refs. [2, 5]. This theory is based on the early work of Jourawski.

Shear stress at any location in the cross section of a beam is determined by

$$\tau = \frac{VQ}{It} \tag{2.31}$$

where

V	–	shear force, lb
I	–	cross-sectional area moment of inertia, in.^{4}
Q	–	moment of the area outside of the plane of interest, in.^{3}

For a rectangular cross section (Figure 2.11)

$$Q = \bar{y}A = \frac{1}{2}(c + y)(c - y)t \tag{2.32}$$

This equation shows that the maximum shear stress is located at the neutral axis.
After substitution, shear stress distribution in a rectangular cross section is

$$\tau = \frac{V}{2I}(c^2 - y^2) \tag{2.33}$$

The direction of this shear is shown in the left drawing; one component is horizontal, the other transverse. Since

$$I = \frac{t(2c)^2}{12}$$

$$\tau_{max} = \frac{Vc^2}{2I} = \frac{3}{2A} = 1.5\,\tau_{ave} \tag{2.34}$$

Consider a 10 ft beam, simply supported with a 1000 lb load applied in the middle; its cross section is 2″ by 4″ with the 4″ side vertical.

$$\sigma_{max} = \frac{Mc}{I} = 5623 \text{ psi}$$

$$\tau_{max} = \frac{3}{2}\frac{V}{A} = \frac{3}{2}\frac{500}{8} = 93.7 \text{ psi}$$

Jourawski's equation will now be applied to the "H" beam (not a standard beam) shown in Figure 2.12. Numbers were selection for convenience of calculation.

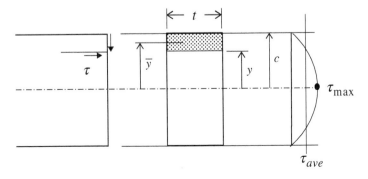

Figure 2.11 Shear stress distribution.

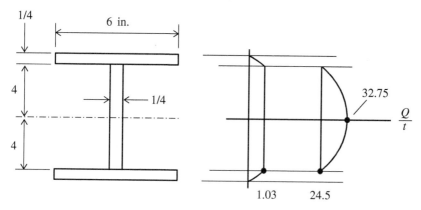

Figure 2.12 Shear distribution across an H beam.

Flange	Web
$\tau_1 = \frac{VQ}{It} = \left(\frac{V}{I}\right)\frac{Q_1}{t_1}$	$\tau_2 = \frac{VQ}{It} = \left(\frac{V}{I}\right)\frac{Q_2}{t_2}$
$Q_1 = 4.125\left(\frac{1}{4} \times 6\right) = 6.1875$ in.^{3}	$Q_2 = 6.1875$ in.^{3}
$t_1 = 6$ in.	$t_2 = 0.25$ in.
$\frac{Q_1}{t_1} = \frac{6.1875}{6} = 1.03$	$\frac{Q_2}{t_2} = 24.5$

Neutral Axis

$$Q_c = 4.125(6 \times 0.25) + 2(4 \times 0.25) = 8.187$$

$$t = 0.25$$

$$\frac{Q_c}{t_c} = 32.75$$

A direct way to determine the cross-sectional moment of inertia is illustrated in Figure 2.13.

$$I_c = \frac{6(8.5)^3}{12} - 2\frac{2.875(8)^3}{12} = 61.73 \text{ in.}^4$$

The cross-sectional moment of inertia with respect to the neutral axis is $I = 61.73$ in.^{4} Assuming the shear load, $V = 1000$ lb, the maximum shear stress is

$$\tau_{max} = 32.75\frac{1000}{61.73} = 530.5 \text{ psi}$$

Figure 2.13 Format of cross-sectional moment of inertia.

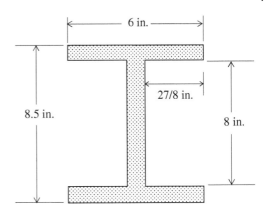

Example Consider a beam fabricated of two metal bars welded together as shown. Assume this beam is to be used to support the load as shown in Figure 2.14. If plate material has a yield strength of 75 000 psi and the weld material has a weld strength of 80 000 psi, determine the factor of safety for the beam. Which dictates the overall FS, shear in the weld or bending in the beam?

Bending Stress

$$I = \frac{bh^3}{12} = \frac{4(4)^3}{12} = 21.33 \text{ in.}^4$$

$$\sigma = \frac{Mc}{I} = \frac{48\,000(2)}{21.33} = 4571 \text{ psi}$$

$$FS = \frac{75\,000}{4571} = 16.4$$

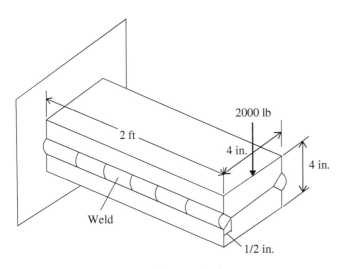

Figure 2.14 Welded beam of rectangular bars.

Shear Stress

$$Q = (1)2(4) = 8 \text{ in.}^3$$

$$\tau = \frac{VQ}{It} = \frac{2000(8)}{21.33(1)} = 750 \text{ psi}$$

Shear strength of weld material is

$$\tau_{yld} = 0.577\sigma_{yld} = 0.577(80\,000) = 46\,160 \text{ psi}$$

$$FS = \frac{46\,160}{750} = 61.5$$

Composite Cross Sections

A special application of the beam shear formula is in the design of attachments such as flanges and straps. These additions may be added to increase stiffness or to construct a symmetrical cross section, such as the composite channel beam. These additions can be attached by nails, bolts, glue, welds, etc. as illustrated in Figure 2.15.

The design issue in each case is how they are fastened (nail, bolt, weld) and how they are spaced to withstand the horizontal shear force. Buckling of straps may also be a consideration. The steps for designing each fastener type are:

1) Determine the shear force in the joint or connection. This is done by establishing the maximum shear stress at the fastener interface based on the maximum shear force as from the shear diagram.

$$\tau_J = \frac{V_{max}Q_J}{Ib_J} \tag{2.35}$$

2) Determine the joint force, F_J, to be transmitted by the fastener.

$$V_J = \tau_J A_J$$

$$A_J = b_J L$$

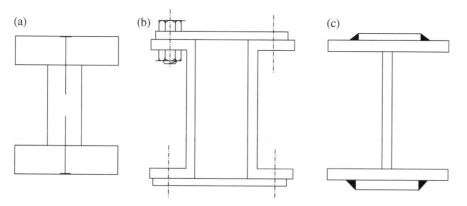

(a)   (b)   (c)

Figure 2.15   Composite beam cross sections.

$$F_J = b_J \tau_J L \tag{2.36}$$

3) Design of connectors

   a) Nail fastener (Figure 2.15a)

       F_n – shear force limit per nail

   $$F_N N = F_J L \tag{2.37}$$

   $$\frac{N}{L} = \frac{F_J}{F_N} \quad \text{(number of nails per length)}$$

   b) Bolted connection (Figure 2.15b)

       F_B – limit force of bolt in shear

   $$2 F_B N = F_J L \tag{2.38}$$

   $$\frac{2N}{L} = \frac{F_J}{F_B} \quad \text{(double bolt spacing)}$$

   c) Welded connection assuming continuous weld along strip (Figure 2.15c)

       τ_W – shear strength of weld

   $$2 \tau_{weld} t L = F_J L$$

   $$t = \frac{F_J}{2 \tau_{weld}} \text{(weld thickness)} \tag{2.39}$$

Example   Consider a special beam made by bolting two 2×6 in. boards onto a 2×4 in. board to form a special wood H beam (Figure 2.16). Assume this beam is simply supported and carries a 1000 lb at the center of a 12 ft long span. Determine the spacing for lag bolts if each bolt can support a 500 lb in shear.   Cross-section moment of inertia of the beam is

Figure 2.16   Wood H-beam.

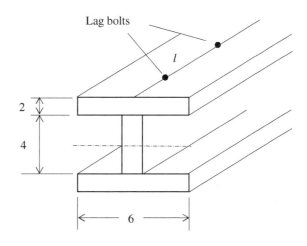

$$I = \frac{6(8)^3}{12} - 2\frac{2(4)^3}{12} = 256 - 21.3 = 234.7 \text{ in.}^4$$

$$Q_J = 3(12) = 36 \text{ in.}^3$$

$$\tau_J = \frac{VQ_J}{It} = \frac{500(36)}{234.7(2)} = 38.3 \text{ psi}$$

Joint force over length L is

$$F_J = \tau bL$$

$$F_J = 38.3(2)L = 76.6L$$

Required number of lag bolts is

$$500N = 76.7L$$

$$\frac{L}{N} = \frac{500}{76.7} = 6.52 \text{ in.spacing per bolt}$$

or one bolt per 6.5 in.

A check on the maximum bending stress under the 1000 lb load is

$$\sigma_{max} = \frac{M_{max}c}{I} = \frac{500(72)4}{234.7} = 614 \text{ psi}$$

which is within the limits of most wood.

Material Selection

One might ask, why not use the highest grade of material to minimize weight and cost? There are two reasons: (i) brittle fracture during service and (ii) manufacturing. High grades of steel are susceptible to embrittlement at low temperatures and stress corrosion cracking. They are also more difficult to machine and weld. As a rule, 80 000 psi yield is assumed to be the transition between mild steel and high strength steel. Structural steel has yield strength of about 55 000 psi. This level of strength allows for easier machining (drilling) and welding.

Mechanical Properties of Steel

Early forms of iron were weak in tension but strong in compression. This is due to a high level of carbon and impurities during the smelting process. Charcoal (a derivative of hard wood) was first used to reduce impurities. Coke (a derivative of coal) was developed by Abraham Darby in 1709. This was significant because there is a greater supply of coal than hard wood. His coke also produced a good quality of cast iron.

One of the early uses of cast iron was in the construction of the Severn Bridge in 1779. The bridge was designed to follow the shape of masonry arch bridges. The supporting arches were made of cast

Figure 2.17 Stress–strain relation.

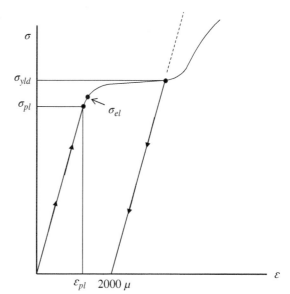

iron. The arched supporting frames were shaped by castings and required no machining or welding. This structure marked the end of masonry arch bridges. Cast iron components performed well because they were put in compression. This bridge is still in use.

Over time, steel became a higher quality of iron containing controlled amounts of carbon and other chemicals to achieve a wide range of material strengths able to perform under high levels of stress both in tension and compression. Low-strength steel or mild steel is ductile as shown in the stress–strain test diagram in Figure 2.17. The key parameters in this diagram are:

σ_{pl} – proportional limit
σ_{el} – elastic limit
σ_{yld} – yield strength
σ_{pl} – strain at proportional limit

The relation between stress and strain is linear up to a point. Stress and strain are related by the modulus of elasticity, E, defined by the slope of the straight line. Key points in the stress–strain diagram are plastic limit, elastic limit, yield strength, ultimate strength, and fracture strength. The plastic limit, elastic limit, and yield strength are approximately the same number. Yield strength is often considered the point of material failure.

Strain is an important parameter and strain levels matter, too. Consider steel having a proportional limit of 60 000 psi. At this stress level, corresponding strain is

$$\varepsilon = \frac{\sigma_{pl}}{E} = \frac{60\,000}{30\,000\,000} = 0.002 \text{ in./in.}$$

or 2000 μ strain, a good reference number for a limit strain. Actual operating strains should be much lower. If stress increases past the proportional limit, plastic elongation occurs leaving a permanent set in the material after the load is removed. The removal path is parallel to, but offset from, the loading path.

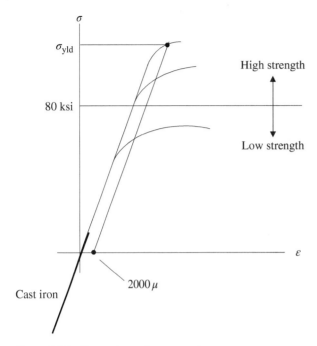

Figure 2.18 Comparison of stress–stain.

High-strength steels do not display significant elongation prior to separation. In this case, the yield strength is defined as the stress level producing a plastic set of 0.002 strain (2000 μ) in the material after the load has been removed (Figure 2.18).

High-strength steel exhibits little or no pronounced elongation prior to failure. There is no distinct stress plateau at the yield point. Yield strength for high strength is arbitrarily defined by the stress level that produces a 0.002 (or 2000 μ) permanent or plastic set after the load has been removed.

The relation between stress and strain for uniaxial loading is

$$\sigma = E\varepsilon \tag{2.40}$$

$$\tau = G\gamma \tag{2.41}$$

where the modulus of elasticity, E, is determined experimentally. The modulus of rigidity, G, is a calculated value.

$$G = \frac{E}{2(1 + \nu)} \tag{2.42}$$

Assuming a material has a modulus of elasticity of $E = 30 \times 10^6$ psi and Poisson's ratio of $\nu = 0.28$, its modulus of rigidity is $G = 11.72 \times 10^6$ psi (Table 2.2).

Shear stress yield is also a calculated value. It is not a measured value.

$$\tau_{yld} = 0.577\sigma_{yld} \tag{2.43}$$

This formula is based on the Mohr's energy of distortion criteria of yielding. Calculated shear yield values using this equation agree with test results.

Table 2.2 Material properties.

	Yield strength (ksi)		Ultimate strength (ksi)	
	Axial	Shear[a]	Axial	Shear
Wrought iron	30	17.3	48	25
Structural steel	36	20.8		66
Cast iron (gray)			110 (C), 20(T)	41
Cast iron (malleable)	32	18.5	50	
Stainless steel (304)	31	17.9	73	
Steel, SAE 4340, heat treated	132	76.2	150	95
Aluminum (wrought)	41	24	62	38
Douglas fir (air dry)	8.1			1.1
Red oak (air dry)	8.4			1.8

[a] The shear yield stress values are determined by $\tau_{yld} = 0.577\sigma_{yld}$, based on von Mises criteria for yielding.

Use of Stress–Strain Relationship in a Simple Truss

Example   Consider a simple truss made of three members supported at fixed points A and B (Figure 2.19). A side load of 10 kip is applied horizontally to point C. It is desired to find the defection of point C.

Inputs to this problem are the material physical and dimensional properties of members AC and BC.

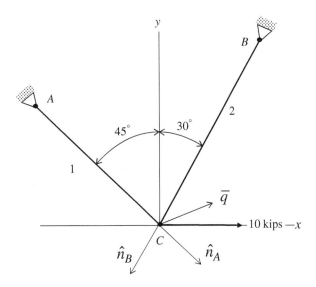

Figure 2.19   Deflections in simple truss.

Member AC	Member BC
Aluminum (1)	Steel (2)
$E_1 = 10\,000$ ksi	$E_2 = 29\,000$ ksi
$L_1 = 10$ ft	$L_2 = 15$ ft
$A_1 = 0.326$ in.^{2}	$A_2 = 0.508$ in.^{2}

The concurrent force system at point C gives

$$\sum F_y = 0 \tag{2.44}$$

$$F_{BC} \cos 30 + F_{AC} \cos 45 = 0 \tag{2.45}$$

$$F_{AC} = -1.22 F_{BC} \tag{2.46}$$

$$\sum F_x = 0 \tag{2.47}$$

$$F_{BC} \sin 30 + 10 - F_{AC} \sin 45 = 0 \tag{2.48}$$

By substitution

$$F_{BC} = -7.34 \text{ kip (compression)}$$

Therefore, $F_{AC} = 8.95$ kip (tension). These forces produce elongations in members AC and BC.

$$\delta_{AC} = \frac{F_{AC} L_{AC}}{E_{AC} A_{AC}} = \frac{8.95(10)(12)}{10\,000(0.326)} = 0.329 \text{ in (extension)} \tag{2.49}$$

$$\delta_{BC} = \frac{F_{BC} L_{BC}}{E_{BC} A_{BC}} = \frac{7.34(15)(12)}{29\,000(0.508)} = 0.0898 \text{ in.(compression)} \tag{2.50}$$

These axial deflections are compatible with internal forces of each member.

The movement of point C is defined by the displacement vector, q, as shown in the figure. Unit vectors in line with both members are also shown.

$$\hat{n}_A = 0.707i - 0.707j \tag{2.51}$$

$$\hat{n}_B = -0.5i - 0.866j \tag{2.52}$$

The elongation of each member is related to the displacement vector, q, by

$$\delta_B = \hat{n}_B \cdot \bar{q} = (-0.5i - 0.866j) \cdot (ui + vj)$$

$$-0.0898 = -0.5u - 0.866v$$

and

$$\delta_A = \hat{n}_A \cdot \bar{q} = (0.707i - 0.707j) \cdot (ui + vj)$$

$$0.329 = 0.707u - 0.707v$$

Algebraic solution is

$$0.0635 = 0.3535u + 0.6123v$$

$$-0.1645 = -0.3535u + 0.353v$$

$$\overline{}$$

$$-0.101 = 0.966v$$

$$v = \frac{-0.101}{0.966} = -0.105 \text{ in.}$$

$$u = \frac{0.0898 - 0.866(-0.105)}{0.5} = 0.362 \text{ in.}$$

$$q = 0.362i - 0.105j$$

Point C moves downward 0.105 in. and to the right by 0.362 in.

Statically Indeterminate Member

Example Consider a rigid bar pinned at the left end and supported by two flexible cables attached at points A and B (Figure 2.20). The problem is to find the tension forces in both cables when a force, F, is applied at the right end. There are three unknown forces involved, the vertical reaction at point 0 and tension forces in both cables, with only two equations of equilibrium. Therefore, cable deflections must be considered.

The equations of statics give

$$\sum M_0 = 0 \tag{2.53}$$

$$aT_A + 2aT_B - 3aF = 0 \tag{2.54}$$

$$T_A + 2T_B = 3F \tag{2.55}$$

and

$$\sum F_y = 0$$

$$T_A + T_B = F + R_0 \text{(assumes } R_0 \text{ pulls downward)} \tag{2.56}$$

Deflection equations yield

$$\delta_A = \frac{T_A L_A}{EA_A} \quad \delta_B = \frac{T_B L_B}{EA_B} \tag{2.57}$$

Also

$$\delta_A = a\theta \quad \delta_B = 2a\theta \text{ so } \delta_B = 2\delta_A$$

Giving

$$\frac{T_B L_B}{EA_B} = 2\frac{T_A L_A}{EA_A} \tag{2.58}$$

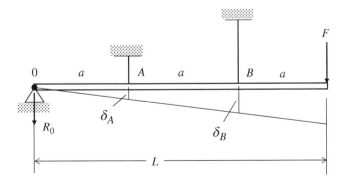

Figure 2.20 Pivoted rod with elastic cables.

We now have two statics equations and one deflection equation to solve for T_A, T_B, R_0. By substitution, Eq. (2.55) gives

$$T_A = 3F - 2\left(\frac{A_B L_A}{A_A L_B}\right) T_A \tag{2.59}$$

$$T_A\left[1 + 2\frac{A_B L_A}{A_A L_B}\right] = 3F$$

and by Eq. (2.58)

$$T_B = 2\frac{A_B L_A}{A_A L_B} T_A$$

Finally, the reaction at point 0 is

$$R_0 = T_A + T_B - F \tag{2.60}$$

Example Consider a pressurized tube with caps held in place by a tight wire with diameter of 0.1 in. (Figure 2.21). The pressure tube has an inside diameter of 2 in. and a wall thickness of 0.04 in. The wire is tightened to 500 lb before internal pressure is applied. The 500 lb is captured by the wedge. The problem is to determine the internal pressure that would cause leakage (contact force between the cap and tube is zero).

The equilibrium equation for the freebody before the applied pressure is

$$F = T \text{ (before pressure is applied)}$$

Since it is given that initial tension in the wire is 500 lb, the initial compression in the tube is also 500 lb.

When internal pressure is applied, the equation of statics becomes

$$T' = P + F' \tag{2.61}$$

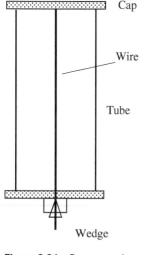

Cap

Wire

Tube

Wedge

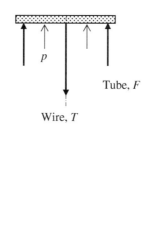

p

Tube, F

Wire, T

Figure 2.21 Pressure tube.

where

$T' = T + \Delta T$ (new wire tension)
$P = pA_{cap}$ (force against cap)
$F' = F - \Delta F$ (new compression in tube)

The problem now is statically indeterminate. By substitution

$$T + \Delta T = pA_{cap} + F - \Delta F \tag{2.62}$$

Since $T = F = 500$ lb

$$\Delta T = pA_{cap} - \Delta F \tag{2.63}$$

When pressure is applied, the change in position of both wire and tube is the same,

$$\Delta \delta_T = \Delta \delta_w \text{ (subscript } T \text{ refers to wire tension, } w \text{ refers to tube wall compression)}$$

or

$$\frac{\Delta TL}{EA_T} = \frac{\Delta FL}{EA_w}, \quad \Delta T = \frac{A_T}{A_w} \Delta F, \quad \text{and} \quad \Delta F = \frac{A_w}{A_T} \Delta T$$

So

$$\frac{A_T}{A_w} \Delta F = pA_{cap} - \Delta F$$

$$\Delta F \left(1 + \frac{A_T}{A_w} \right) = pA_{cap} \quad \text{and} \quad \Delta T \left(1 + \frac{A_w}{A_T} \right) = pA_{cap} \tag{2.64}$$

The areas are

$$A_T = \frac{\pi}{4}(0.1)^2 = 0.0079 \text{ in.}^2 \text{ (wire)}$$

$$A_W = \pi 2(0.04) = 0.2513 \text{ in.}^2 \text{ (tube wall)}$$

$$A_{cap} = \frac{\pi}{4}(2^2) = 3.1416 \text{ in.}^2 \text{ (cap)}$$

The ΔF and ΔT values are

$$\Delta F = \frac{pA_{cap}}{1.0314} = p\frac{3.1416}{1.0314} = 3.046p \tag{2.65}$$

$$\Delta T = \frac{pA_{cap}}{32.81} = \frac{3.1416}{32.81}p = 0.0958p \tag{2.66}$$

Therefore, the forces in the tube and wire after pressure is applied are

$F' = 500 - 3.046p$
$T' = 500 + 0.0958p$

At separation, $F' = 0$, so

$3.046p = 500$

$$p_{cr} = 164.15 \text{ psi} \tag{2.67}$$

and

$$T' = 500 + 15.7 = 515.7\,\text{lb}$$

As a check, the tension in the wire should be due only to the pressure against the cap or

$$T' = pA_{cap} = 164.15(3.1416) = 515.7\,\text{lb} \tag{2.68}$$

Modes of Failure

Material Yielding

Material yielding is a state of stress–strain where the material stretches with a small increase in load leading to permanent distortion in a material. Yield strength is a common failure criterion in design. Once yielding is reached, plastic deformation occurs, and low cycle fatigue is possible. The criteria of failure commonly used to determine material yielding is the von Mises criteria based on energy of distortion [3, 6]. For a biaxial state of stress this criterion shows, yielding occurs when

$$\sigma_A^2 - \sigma_A \sigma_B + \sigma_B^2 \geq \sigma_{yld}^2 \tag{2.69}$$

where σ_A and σ_B are local principal stresses. Brittle materials behave differently. Failure occurs suddenly without yield. The Coulomb–Mohr Criteria [2, 6] is commonly used in this case.

Stress Concentration

Classical means of calculating stress (bending, torsion, pressure, axial, and combined stress) are adequate in most cases to configure a reliable design considering factors of safety. However, in consideration of safety and cost, such as in aerospace and drilling equipment, it is desirable to refine stress calculations in critical areas of a design using finite element software or stress concentration factors that have been established analytically or experimentally.

Consider, for example, a small hole in the center of a long strap or plate (Figure 2.22). Using the theory of elasticity, Seely and Smith [7] show that local stress across a strap are defined by

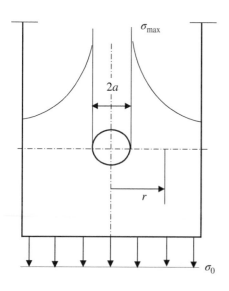

Figure 2.22 Plate with hole.

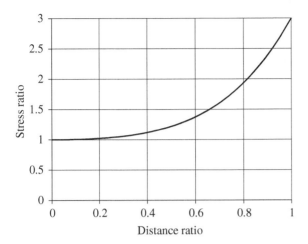

Figure 2.23 Stress distribution away from the hole.

$$\sigma_x = \frac{\sigma_0}{2}\left(2 + \frac{a^2}{r^2} + 3\frac{a^4}{r^4}\right) \tag{2.70}$$

This formula has been confirmed by strain measurements and photoelastic method.

Equation (2.70) is plotted in Figure 2.23 to show how local stress varies across the width of the plate starting at the edge of the hole and moving to the outside edge of the plate. The parameters in the drawing are

$$\text{Stress ratio}: \frac{\sigma_x}{\sigma_0} \tag{2.71}$$

$$\text{Distance ratio}: \frac{a}{r} \tag{2.72}$$

When $r = a$, the distance ratio is one and the stress ratio is 3, meaning that local stress at the hole is three times the nominal stress, σ_0. Also, the distance ratio reduces to zero, meaning distance, r, is very large and local stress approaches the nominal stress. The maximum stress is located at the edge of the hole where $\frac{\sigma_{\max}}{\sigma_0} = 3$. The stress concentration factor is 3 as $\sigma_{\max} = 3\sigma_0$.

Stress concentration factors for other geometric shapes have been established theoretically and experimentally [7]. Always avoid sharp geometric changes, such as fillets, openings, and threads, as they are points of stress concentration.

Wear
Mechanical wear is the result of friction between two sliding surfaces. Wear can be minimized or even eliminated by proper bearing design and lubrication.

Fatigue
Metal fatigue results from cyclic stresses defined by

$$\sigma = \sigma_m + \sigma_A \cos \omega t \tag{2.73}$$

where

σ_m	–	mean stress
σ_A	–	stress amplitude
ω	–	frequency of stress cycle

Material failure depends on magnitude of both σ_m and σ_A. When $\sigma_m = 0$, there is complete reversal of stress. Endurance limits are determined experimentally under this condition. Tests are conducted on multiple specimens, each being cycles at different stress levels of σ_A until failure occurs. The number of cycles to failure is documented at each stress level. The endurance limit is typically defined as the stress level at which complete reversals do not cause failure beyond 10^6 cycles.

Test data for steel show that endurance limits depend on the ultimate strength. As a guide endurance limit for steel having ultimate strength less than 200 ksi, assuming complete reversal, is 0.4–0.55 times ultimate strength.

In general, time-dependent stresses cycle about a mean or average stress level. In this case, the Goodman diagram (Figure 2.24) is useful for determining fatigue limits for stress levels having a nonzero mean stress. Mean strength is limited by yield strength, σ_{yld}, of the material.

The Goodman diagram is constructed by marking the endurance limit on both the plus (*a*) and minus (*b*) directions on the vertical axis. The mean stress is zero and consistent with complete stress reversal. The ultimate strength sets the maximum possible value for the mean stress and amplitude stress coordinates. This point is represented by point (*c*). Straight lines are drawn from *a* to *c* and *b* to *c*. The lines establish the fatigue limit for the material a different mean stress. Yield strength limits fatigue stress. The two lines *a*–*c* and *c*–*b* establish the fatigue limits for different mean stress levels. Amplitudes and mean stress levels near yield strength levels usually lead to low cycle fatigue.

Cyclic fatigue limits in the compression range are insensitive to mean stress [6].

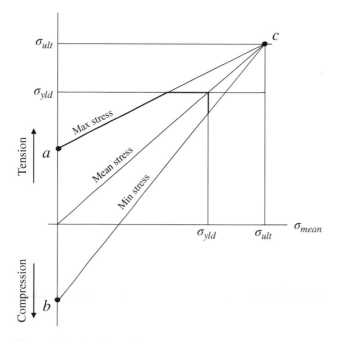

Figure 2.24 Goodman diagram.

Example Two 10-in. steel bars (1.5 in. diameter) are connected by a special 90° coupling (Figure 2.25). The steel has an ultimate strength of 120 ksi and a yield strength of 100 ksi. Equal and opposite forces, F, are applied as shown but vary periodically from a maximum value of 300 lb to a minimum value of 100 lb.

(a) Determine the factor of safety in member A from failure by fatigue.
(b) Determine the factor of safety in member B from failure by fatigue.
(c) Determine the factor of safety if force, F, at the elbow is removed.

For this example, we assume

$$S_e \sim 0.5\, S_{ult} \quad \text{and} \quad S_{ult} < 200 \text{ ksi}$$

Accordingly, the endurance limit is $\sigma_{end} = 60$ ksi for this material based on complete reversal of stress.

Cross-sectional properties of the bent rod are

$$I = \frac{\pi}{4}(0.75)^4 = 0.2485 \text{ in.}^4$$

$$J = 2I = 2(0.2485) = 0.497 \text{ in.}^4$$

Case a

The critical location in member A is at the elbow, which supports bending. Since the force, F, varies with time, bending stress will also vary between the following two stress levels. For the stated load condition, the stress varies between

$$\sigma_{max} = \frac{M_{max}c}{I} = \frac{300(10)0.75}{0.2485} = 9054 \text{ psi} \tag{2.74}$$

and a minimum value of

$$\sigma_{min} = \frac{M_{min}c}{I} = \frac{100(10)0.75}{0.2485} = 3018 \text{ psi} \tag{2.75}$$

over one cycle of periodic loading. The mean stress for this cycle is

$$\sigma_{mean} = \frac{1}{2}(\sigma_{max} + \sigma_{min}) = 6036 \text{ psi}$$

The amplitude of the stress variation is

$$\sigma_{amp} = \frac{\sigma_{max} - \sigma_{min}}{2} = 3018 \text{ psi} \tag{2.76}$$

Figure 2.25 Fatigue of analysis of bent rod.

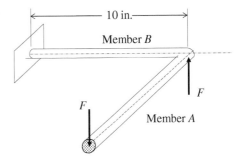

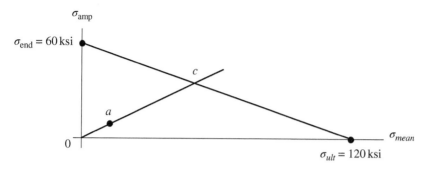

Figure 2.26 Modified Goodman diagram for normal stress.

These stresses are shown in the modified Goodman diagram (Figure 2.26) by the heavy dot on the positive sloped line.

The factor of safety is $FS = \frac{0c}{0a}$. This ratio is established as follows. Let the positive sloped line be defined by

$$y = m_0 x \tag{2.77}$$

and the negative slope line be defined by

$$y = -mx + b \tag{2.78}$$

The intersection of these lines defines the coordinates of point "c." The "y" coordinate of the intersection point is

$$y_c = \frac{b}{1 + \frac{m}{m_0}} \tag{2.79}$$

In terms of the modified Goodman diagram for member A

$$b = 60 \text{ ksi}$$

$$m = \frac{60}{120} = 0.5$$

$$m_0 = \frac{3.018}{6.036} = 0.5$$

Giving

$$y_c = \frac{60}{(1 + 1)} = 30$$

and a factor of safety of

$$FS_A = \frac{30}{3.018} = 9.94$$

Following the same procedure for member B

$$FS_B = 11.47$$

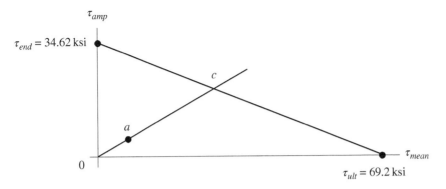

Figure 2.27 Modified Goodman diagram for shear fatigue.

Case b

Von Mises energy criteria of failure show that in the static stress case

$$\tau_{yld} = 0.577\sigma_{yld} \text{ (static load criteria of failure)}$$

$$\tau_{yld} = 57.7 \text{ ksi}$$

Experiments show that endurance limits in complete stress reversal in shear also follow the same criteria [8].

$$\tau_{end} = 0.577\sigma_{end} \tag{2.80}$$

For our example, the endurance limit in shear is

$$\tau_{end} = 0.577(60) = 34.62 \text{ ksi (complete reversal of stress)}$$

The modified Goodman diagram for shear is shown in Figure 2.27.
The cyclic shear stress levels in member B vary between

$$\tau_{max} = \frac{T_{max}c}{J} = \frac{3000(0.75)}{0.497} = 4527 \text{ psi} \tag{2.81}$$

$$\tau_{min} = \frac{T_{min}c}{J} = \frac{1000(0.75)}{0.497} = 1509 \text{ psi} \tag{2.82}$$

Using these numbers

$$\tau_{mean} = 3018 \text{ psi}$$

$$\tau_{amp} = 1509 \text{ psi}$$

These numbers define the location of the shear cycle by the heavy dot on the positive sloped line in Figure 2.27.

Case c

A two-dimensional stress condition arises at the support if force, F, at the elbow is removed. The bending stress in member A stays the same; however, member B experiences bending as well as torsion creating normal and shear stresses at the support.

We return to von Mises criteria of static failure, which for the static case, yielding occurs when

$$\sigma_{von} \succ \sigma_{yld} \tag{2.83}$$

where

$$\sigma_{von}^2 = \sigma_1^2 - \sigma_1\sigma_2 + \sigma_2^2$$

and σ_1 and σ_2 are principal stresses.

$$\sigma_{1,2} = \frac{1}{2}\sigma \pm \sqrt{\left(\frac{\sigma}{2}\right)^2 + \tau^2} \tag{2.84}$$

Applying von Mises's criteria of failure to the particular state of stress shown above, material yielding occurs when

$$\sigma_{yld}^2 = \sigma^2 + 3\tau^2 \tag{2.85}$$

Experiments show this formula also applies to this particular two-dimensional stress state under *completely reversed* cyclic stress levels provided [8]

$$\sigma_{end}^2 = \sigma^2 + 3\tau^2 \tag{2.86}$$

where σ_{end} is the endurance limit under uniaxial cyclic stress condition. This equation is rewritten in the form of

$$\left(\frac{\sigma}{\sigma_{end}}\right)^2 + 3\left(\frac{\tau}{\sigma_{end}}\right)^2 = 1$$

which is then potted as shown in Figure 2.28, where $\text{Ratio}(\sigma) = \frac{\sigma}{\sigma_{end}}$ and $\text{Ratio}(\tau) = \frac{\tau}{\sigma_{end}}$.

If shear stress is zero, the limiting normal stress is the endurance limit. If the normal stress is zero, the limiting shear stress level is

$$\tau = \frac{1}{\sqrt{3}}\sigma_{end} = 0.577\sigma_{end} \tag{2.87}$$

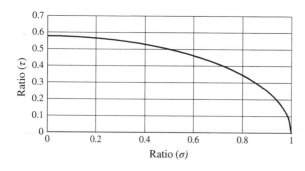

Figure 2.28 Endurance limits for biaxial stress state.

Stress Corrosion Cracking

Three factors must exist simultaneously for stress corrosion cracking to occur: (i) corrosive environment (such as the sea), (ii) local tension stress (could be residual stress caused by forming), and (iii) hard metal (base metal or produced by quenching). The corrosive environment causes irregularity in the surface to form and deepen a crack. Localized tension encourages crack growth. Hard metal cannot blunt a crack and therefore is susceptible to sudden crack growth and brittle fracture.

This type of failure can occur in offshore pipelines. In one case, a failure analysis showed that an elbow in the pipeline had been formed by heating it to high temperatures and then quenched for convenience of handling. In the process of forming the elbow, residual tension stresses were left in the hardened material. Local tension stresses could also have been generated during pipeline installation.

Brittle Fracture

Fracture mechanics deals with stresses at the tip of a crack. The crack could be visible at the surface or internal to the structural member. The major concern is a sudden, catastrophic failure – one that occurs without warning.

The Griffin theory [9] of brittle fracture states there are paths of weakness within the matrix of materials where cracks propagate. His work was based on experimental testing of glass. He showed that cracks form even though crystals within glass are much stronger than the nominal strength of glass.

There are two approaches for predicting brittle failure:

1) Experimental – Charpy V-Notch tests are conducted on several test pieces, having the same an initial crack, to determine impact energy levels required to cause the crack to suddenly extend and separate the material. Tests are conducted at different temperatures to determine a material's ability to support a crack. There is typically a temperature at which impact energy for failure substantially drops (Figure 2.29).

   This is the transition temperature, an important property of materials expected to operate at low temperatures. High-strength materials ($\sigma_{yld} \geq 80\,000$ psi) are particularly vulnerable to this type of failure, especially at low temperatures.

2) Theoretical – Stress predictions are based on the theory of elasticity, which predicts stresses at the end of an elliptical opening [6]. The elliptical opening can achieve the appearance of a crack

Figure 2.29 Charpy V-notch response.

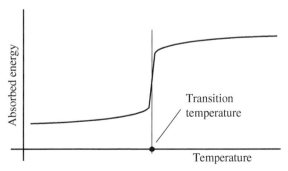

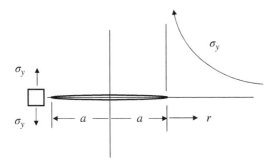

Figure 2.30   Critical crack length.

(see Figure 2.30). Stress predictions show that normal stresses away from the tip of a crack are defined by

$$\sigma_y = \frac{\sigma\sqrt{a}}{\sqrt{2r}} \qquad (2.88)$$

This equation shows the normal stress is ∞ when $r = 0$. In many cases, failure does not occur in the form of a crack propagation. Ductile materials can create a plastic region at the tip of the crack, and thus blunt the crack inhibiting sudden propagation. Even in high-strength materials, a crack may be stable. To quantify this phenomenon, the flowing equation is commonly used to determine critical crack length.

$$\sigma_y = \frac{K_0}{\sqrt{2\pi r}} \left(K_0 = \sigma\sqrt{\pi a}\right) \qquad (2.89)$$

where K_0 is the stress intensity factor. A crack propagates suddenly when $K_0 = K_{cr}$, a material constant determined experimentally.

Fluid Flow Through Pipe

Fluid mechanics is fundamental to many machine components and production systems. The building blocks for fluid analysis are:

- Continuity of fluid flow
- Energy equation commonly known as Bernoulli's equation
- Force analysis, including impulse momentum

Continuity of Fluid Flow

Continuity of flow is based on the conservation of mass, which requires that mass entering a control volume is equal to mass leaving the control volume (Figure 2.31).

$$(Q\rho)_{in} = (Q\rho)_{out} \qquad (2.90)$$

where

Q – flow rate (volume/time)
ρ – mass per volume

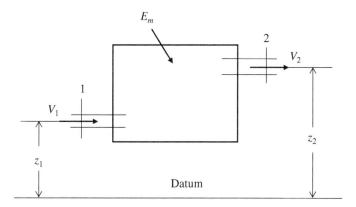

Figure 2.31 Control volume – open flow.

If fluid is assumed incompressible, the volume flow rates are the same and

$$(Q)_{in} = (Q)_{out} = \text{constant} \tag{2.91}$$

$$(VA)_{in} = (VA)_{out}$$

where

V – average velocity
A – cross-sectional area of flow

Bernoulli's Energy Equation (First Law)

The energy equation of fluid flow, assuming frictionless flow, is defined by the Bernoulli equation

$$\frac{V_1^2}{2g} + \frac{p_1}{\gamma} + y_1 + E_{in} = \frac{V_2^2}{2g} + \frac{p_2}{\gamma} + y_2 + h_f \tag{2.92}$$

where each term has units of feet. Fluid friction losses are defined by the friction head, h_f.

Each term can also be viewed as having units of ft-lb/lb. E_{in} is energy per lb (ft-lb/lb) being put into the control volume between locations 1 and 2. This formula is used to determine pump power required to move fluid from one location to another.

Reynolds Number

Reynolds number is a dimensionless number defined by

$$N_R = \frac{VD\rho}{\mu} = \frac{VD\gamma}{\mu g} \tag{2.93}$$

It can be viewed as the ratio of fluid kinetic energy divided by shear stress at the tubular surface. This number helps identify whether flow is laminar or turbulent. If kinetic energy is high and wall friction is low, flow is turbulent. On the other hand, if kinetic energy is low and wall friction is high, flow is laminar. Experiments on Newtonian fluids show that flow ceases to be laminar when Reynolds number is around 2000 and turbulent flow is fully developed when Reynolds number is around 4000.

Friction Head for Laminar Flow

The friction head for laminar flow in pipe is determined experimentally using the Bernoulli equation assuming, $V_1 = V_2$ and the pipe is horizontal.

$$h_f = \frac{\Delta p}{\gamma} \tag{2.94}$$

From Hagen–Poiseuille

$$\Delta p = L \frac{32\,\mu}{D^2} V \tag{2.95}$$

By substitution

$$h_f = \frac{L}{D} \frac{32\,\mu}{\gamma D} V \tag{2.96}$$

Noting that

$$N_R = \frac{VD\rho}{\mu}$$

and rearranging terms defines friction head for laminar pipe flow in terms of Reynolds gives

$$h_f = \left(\frac{64}{N_R}\right)\left(\frac{L}{D}\right)\frac{V^2}{2g} \tag{2.97}$$

Friction head is typically expressed by the Darcy–Weisbach empirical equation

$$h_f = f\left(\frac{L}{D}\right)\left(\frac{V^2}{2g}\right) \tag{2.98}$$

This formulation shows the friction factor for laminar flow is

$$f = \frac{64}{N_R}$$

Note that this equation was derived mathematically.

Turbulent Flow Through Pipe

In the case of turbulent flow, the factor, f, depends on Reynolds number and pipe surface roughness.

$$f = f\left(N_R, \frac{e}{D}\right) \tag{2.99}$$

This factor is determined experimentally and is given in the Moody diagram [10], which shows that the friction factor follows laminar flow theory up to a Reynolds number of 2000. There is a transition region between Reynolds numbers of 2000 and 4000. Above 4000 surface roughness is a factor. Data show that at high values of e/D, friction factor becomes less dependent on the Reynolds number.

Example Determine the pressure gradient along a pipe with the following flow conditions.

Pipe size	–	4½ in. (16.6 lb/ft) (drill pipe)
ID	–	3.826 in. (area = 11.497 in.²)
Flow rate	–	400 gpm
Viscosity	–	1 cp (water)
Density (γ)	–	62.4 lb/ft³ (water)

Converting to uniform units:

$$Q\left(\frac{ft^3}{min}\right) = 400\left(\frac{gal}{min}\right)\left|\frac{1\,ft^3}{7.48\,gal}\right| = 53.48\,ft^3/min$$

$$V\left(\frac{ft}{s}\right) = 53.48\left(\frac{ft^3}{min}\right)\left|\frac{1\,min}{60\,s}\right|\frac{1}{11.5\,in.^2}\left|\frac{144\,in.^2}{1\,ft^2}\right| = 11.16\,fps$$

Since

$$1\,cp = 1.45 \times 10^{-7}\,reyn$$

The viscosity of water in units of $\frac{lb\text{-}s}{ft^2}$ is

$$\mu = 1.45 \times 10^{-7}\left(\frac{lb\text{-}s}{in.^2}\right)\left|\frac{144\,in.^2}{ft^2}\right| = 208.8\,\frac{lb\text{-}s}{ft^2}$$

Pressure drop predictions for turbulent flow (Newtonian fluid) are determined as follows.

$$N_R = \frac{VD\rho}{\mu} = \frac{VD\gamma}{\mu g} = 3.43 \times 10^5$$

$$f = 0.038(\text{Moody diagram, assuming } e/D = 0.01)$$

$$h_f = f\left(\frac{L}{D}\right)\left(\frac{V^2}{2g}\right)(\text{Darcy – Weisbach})$$

$$\frac{p_1}{\gamma} - \frac{p_2}{\gamma} = h_f\,(\text{Bernoulli equation})$$

Combining last two equations gives

$$\frac{\Delta p}{L} = f\left(\frac{\gamma}{D}\right)\left(\frac{V^2}{2g}\right) \tag{2.100}$$

Substitution of all input data gives

$$\frac{\Delta p}{L} = 0.038\left(\frac{62.4}{3.826}12\right)\left[\frac{11.16^2}{2(32.2)}\right] = 14.383\,\frac{lb/ft^2}{ft}$$

$$\frac{\Delta p}{L} = 14.383\,\frac{lb/ft^2}{ft}\left|\frac{1\,ft^2}{144\,in.^2}\right| = 0.1\,psi/ft$$

For a 10 000 ft pipe length, the pressure loss due to fluid flow would be 1000 psi.

System losses, though important, are only part of the design of hydraulic systems. In some cases, the goal is to maximize the use of available mechanical/pump power available, such as in the hydraulics of drilling fluids in oil well drilling. Positive displacement-type pumps may deliver as much as 800 hydraulic horsepower (HHP) to the drilling fluid system. Here, drill pipe may be two to three miles in length (5000–15 000 ft) and friction losses are significant. Drill bit nozzles are selected to use up all HHP available at the drill bit for cleaning and removing cuttings. The hydraulic system may need to power downhole drilling motors, another demand on available HHP.

Senior Capstone Design Project

A senior capstone design project at the University of Tennessee required a hydraulic system designed to settle dust in an indoor equestrian arena. The size of the area is 140 ft by 240 ft. Specifications required that runoff of rainwater from the roof be stored and used instead of city water. The student team identified four subsystems.

1) Collection – gutter, return line, storage tank
2) Delivery system – piping, pumps, and nozzles
3) Controls – valves, supplemental water well

Central to the final design was pump selection.

Pump Selection

The basis for choosing a pump is: (i) pressure at the nozzle to create enough velocity to propel a jet of water capable of reaching across the length of the zone to be watered and (ii) enough flow rate to saturate the zone with an appropriate amount of water per time (gpm). Both pump pressure (p, psi) and flow rate (Q, gpm) define the performance requirements of the pump.

The two pump types considered were: positive displacement and impeller. As the following analysis shows, positive displacement-type pumps are best suited for this application because of better flow rate control. However, because of costs, an impeller-type pump was less expensive and became the focus of the analysis.

Required Nozzle Velocity

The initial step is to determine the exit velocity required from the nozzle to reach the full extent of the zone. The path of a fluid projectile is defined by

$$y(x) = x \tan \theta - \frac{gx^2}{2(V \cos \theta)^2} \tag{2.101}$$

Two locations where $y(x)$ is zero are $x = 0$ and $x = L$.

$$L = \frac{V^2}{g} \sin 2\theta \tag{2.102}$$

Assuming $L = 81$ ft and $\theta = 23°$, then the required exit velocity from the nozzle is

$$V^2 = \frac{Lg}{\sin 2\theta} = \frac{81(32.2)}{\sin 46} = 3626$$

$$V = 60.2 \text{ ft/s}$$

Nozzle Pressure

Now turn to Bernoulli's equation to determine nozzle pressure required to produce the require exit velocity.

Applying Bernoulli's energy equation between locations 1 and 2 gives

$$\frac{p_1}{\gamma} + \frac{V_1^2}{2g} = \frac{p_2}{\gamma} + \frac{V_2^2}{2g} \text{ (units are in ft)} \tag{2.103}$$

Since V_1 is small by comparison with V_2, it is dropped leaving

$$p_1 - p_2 = \Delta p = \frac{\gamma}{2g} V_2^2 \tag{2.104}$$

If the following units are chosen

p	–	psi
V	–	fps
γ	–	ppg

then Eq. (2.104) becomes

$$\Delta p \left(\text{lb/in.}^2\right) = \frac{\gamma(\text{lb/gal})}{2g(32.2\,\text{ft/s}^2)} \left|\frac{7.48\,\text{gal}}{\text{ft}^3}\right|\left|\frac{1\,\text{ft}^2}{144\,\text{in.}^2}\right| (V\,\text{ft/s})^2$$

$$\Delta p = \frac{7.48\gamma}{2(32.2)(144)} V_2^2 \tag{2.105}$$

$$\Delta p = p_1 = p = \frac{\gamma V^2}{1240} \tag{2.106}$$

Applying a nozzle coefficient of 0.95, V_2 is replaced according to $V = 0.95\,V_2$, giving

$$p = \frac{\gamma V^2}{1120} \tag{2.107}$$

This equation shows that a nozzle pressure of

$$p = \frac{8.342(60.2)^2}{1120} = 27\,\text{psi (density of fresh water is 8.342 ppg)}$$

is required to produce the 60.2 fps nozzle velocity.

Pump Flow Rate Requirement
Flow rate through the system, including the nozzle is constant. Exit velocity from the nozzle is expressed by

$$VA_n = Q \tag{2.108}$$

Accounting for units

$$V(\text{ft/s})A\left(\text{in.}^2\right)\left|\frac{1\,\text{ft}^2}{144\,\text{in.}^2}\right| = Q(\text{gal/min})\left|\frac{1\,\text{ft}^3}{7.48\,\text{gal}}\right|\left|\frac{1\,\text{min}}{60\,\text{s}}\right|$$

$$V = \frac{0.32Q}{A_n}\,\text{fps}$$

By substitution into Eq. (2.107)

$$p = \frac{\gamma V^2}{1120} = \frac{\gamma}{1120}\left(\frac{0.32Q}{A_n}\right)^2$$

giving

$$p = \frac{\gamma Q^2}{10938 A_n^2} \text{ (plotted in Figure 2.32)}$$

where

Δp	–	psi
Q	–	gpm
γ	–	ppg (8.342 ppg, fresh water)
A_n	–	nozzle area (total flow area, TFA), in.^{2}

which relates flow rate to the required pressure. Using the designated parameters,

A_n	=	0.093 in.^{2} (diameter is $11/32$ in.)
p	=	27 psi
γ	=	8.342 ppg

Then

$$Q^2 = \frac{10\,938 A_n^2 p}{\gamma} = \frac{10\,938 (0.093)^2 27}{8.342} = 306.2$$

$$Q = 17.5 \text{ gpm}$$

The required flow rate for two nozzles is $Q = 35$ gpm with nozzle pressure of still 27 psi. In summary, a pump must deliver:

One zone:	$p = 27$ psi at a flow rate of $Q = 17.5$ gpm
Two zones:	$p = 27$ psi at a flow rate of $Q = 35$ gpm

Figure 2.32 shows the broader response of the $11/32$ nozzle to different flow rates.

The diagram shows that the selected 1.5 hp pump was adequate to supply water to two nozzles at the required delivery pressure and flow rate.

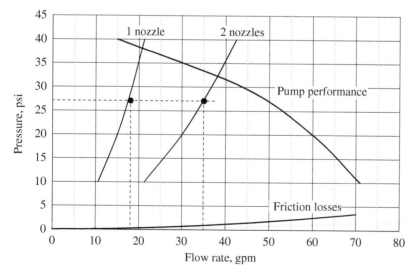

Figure 2.32 Performance of impeller-type pump.

Vibration Considerations

There is much to be learned from the theory of undamped single degree-of-freedom (SDOF) systems. Many mechanical vibration problems can be explained and resolved from this model. This is a good place to start. However, in high-speed machinery, multi-degrees of freedom may need to be considered.

Consider the single degree of freedom system of Figure 2.33. The elastic restoring force is the horizontal beam, the mass is a combination of the discrete mass and part of the mass of the beam. A simple harmonic is applied to the discrete mass to cause a forced vibration.

$$F(t) = F_0 \cos \omega t \tag{2.109}$$

The source of excitation is usually related to rotating components, such as the power source or moving parts in the power transmission sequence. The frequency of the excitation is represented by ω, which is different from the natural frequency ω_n.

The equation of motion for the mass, m, is

$$m\ddot{x} + kx = F_0 \cos t \tag{2.110}$$

The general solution to this equation has particular and complementary solutions. Considering only the particular solution (the complementary solution decays with time), the response is

$$x(t) = X \cos \omega t \tag{2.111}$$

By substitution

$$X = \frac{F_0}{k - m \omega^2} \tag{2.112}$$

$$X = \frac{\delta_{st}}{1 - r^2} \tag{2.113}$$

$$r = \frac{\omega}{\omega_n} \text{(frequency ratio)} \tag{2.114}$$

$$\delta_{st} = \frac{F_0}{k} \tag{2.115}$$

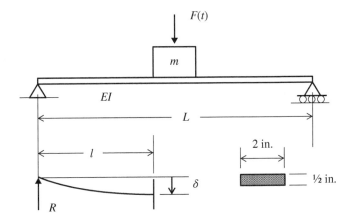

Figure 2.33   Forced vibration of SDF system.

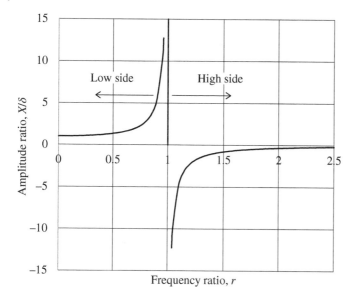

Figure 2.34 Frequency response of SDOF systems.

The response has the same frequency as the forcing function; however, the vibration amplitude depends on the frequency ratio, r. The response amplitude is shown in Figure 2.34.

Two possible frequencies giving an amplitude of one (1) for $\frac{X}{\delta_{st}}$ are $r = 0$ and $r = \sqrt{2}$.

Several practical observations can be made from this diagram. First and most important, large vibration amplitudes occur when $\omega = \omega_n$ ($r = 1$). Under this condition, vibration displacements and stress levels are largest; noise levels are usually high, too. When vibrations are excessive in machinery, the cause is nearly always, frequency tuning or resonance. The solution to this problem is simple – detune the system. This can be done by changing either the natural frequency of the system or changing the driving frequency.

1) Change omega
2) Change mass (usually increase)
3) Change stiffness (usually decrease)

Note that on the *low side* of resonance, the amplitude ratio, $\frac{X}{\delta_{st}}$ reaches 1.26 at a frequency ratio of $r = 0.455$. This means that the dynamic effects could be analyzed by a quasi-static approach with $FS = 1.26$.

On the high side of resonance, vibration amplitudes diminish rapidly to near zero with high values of the frequency ratio. This indicates the best operating conditions are at high-frequency ratios. It is shown that when $r = \sqrt{2}$, the amplitude ratio is 1 and the dynamic response is the same as the static displacement cause by F_0. The greatest reduction in vibration amplitude comes beyond a frequency ratio of $r = \sqrt{2}$.

Another aspect of the high side of resonance is the phase change, which is reflected in the negative amplitude. Physically, the force and amplitude move in opposite directions, i.e. when the force is applied downward, the mass is moving upward. When systems operate on the *low* side of resonance, force and displacement are in phase.

Example Assume the numerical details of this vibrating structure (Figure 2.33) are

$L = 40$ in.

$I = \dfrac{bh^3}{12} = \dfrac{1}{48}$ in.^{4}

$E = 30 \times 10^6$ psi

$W = 20$ lb

The spring constant is determined from the lower diagram noting that the deflection of a cantilever beam is

$$\delta = \frac{R\ell^3}{3EI}, \quad \text{with } R = \frac{F}{2} \tag{2.116}$$

The string constant of the total beam with a centrally applied force is

$$k = \frac{F}{\delta} = \frac{6EI}{\ell^3} \tag{2.117}$$

Taking a simplistic view of the system by letting $\ell = \frac{L}{2}$ leads to a spring constant of

$$k = \frac{48EI}{L^3} = \frac{48(30 \times 10^6)\frac{1}{48}}{40^3} = 468.7 \text{ lb/in.}$$

and a natural frequency of

$$f_n = \frac{1}{2\pi}\sqrt{\frac{468.7(386)}{20}} = 15.14 \text{ cps}$$

If we assume the mass is clamped over a 6-in. interval in the center, the elastic flexibility changes. In this case $\ell = 20 - 3 = 17$ in. and the spring constant changes as follows.

$$k = \frac{6EI}{17^3} = \frac{6(30 \times 10^6)\frac{1}{48}}{17^3} = 763.28 \text{ lb/in.}$$

The mass of the beam should also be considered. The total weight of the beam is

$Wt_{beam} = \text{volume} \times \text{density}$

$Wt_{beam} = 40 \text{ in.}^3 \left|\dfrac{1 \text{ ft}^3}{12(144)\text{in.}^3}\right| 490 \text{ lb/ft}^3 = 11.34 \text{ lb}$

Assuming half of the beam weight contributes to total vibrating mass, natural frequency is predicted as

$$f_n = \frac{1}{2\pi}\sqrt{\frac{763.28(386)}{25.67}} = 11.48 \text{ cps} \tag{2.118}$$

A somewhat lower value, but more realistic, than the first prediction.

Several worthwhile points are made from this example:

1) Assuming a dynamic force amplitude of $F_0 = 5$ lb, $\delta_{st} = \frac{5}{763.3} = 0.0066$ in. Inertia effects amplify this static displacement.
2) Reaction forces at the supports are also amplified through

$$X = \frac{R_A \ell^3}{3EI}, \quad R_A = \frac{3EI}{\ell^3} X \tag{2.119}$$

At a dynamic amplitude of $X = 0.25$ in., $R_A = 95$ lb.

3) Bending moment is greatest at the clamp.

$$\sigma = \frac{Mc}{I}, \quad M = R_A \ell = 95(17) = 1615 \text{ in.-lb}$$

$$\sigma = \frac{1615(0.25)}{1/48} = 19\,380 \text{ psi}$$

4) Time to reach 10^6 cycles

$$f_n = 11.4 \text{ cps} \quad 11.48 = \frac{10^6}{T}, T = 87\,108 \text{ seconds}$$

$$T = \frac{87\,108}{3600} = 24.2 \text{ hours}$$

5) Vibration affects both noise level and fatigue.
6) If the driving force, F_0, is generated by a rotating mass, then

$$f = \frac{N}{60} \text{ cps}, \quad N = 60(11.48) = 689 \text{ rpm}$$

The physical appearance of SDOF systems in an engineering setting may be different from the one shown in Figure 2.33. It is up to the designer to develop the appropriate mathematical model.

Natural Frequency of SDOF Systems

Per the above analysis, frequency tuning (resonance) defines whether there will be a vibration problem. Possible resonance can be established by considering only two factors in a design: natural frequency and frequency of excitation. Frequency of excitation is often related to rotary speed. Natural frequency can be determined analytically for simple structures, but for complex structures, it is best determined experimentally.

Engineering structures have many vibrations modes, and any one mode can be excited. The ability to calculate and measure natural frequencies becomes an important consideration in design. Natural frequencies are determined by consider free vibration of undamped systems. To have a vibration, there must be a spring and a mass. A simple spring–mass system (Figure 2.35) is often used to represent and illustrate the natural frequency of SDOF systems even though the actual mechanical arrangement may look completely different.

The spring can represent any elastic support to a mass. The equation of motion is

$$m\ddot{x} = W - k(x + \delta_{st}) \tag{2.120}$$

The static displacement due to W is represented by δ_{st}. It does not appear in the dynamic analysis because $W = k\delta_{st}$. Both terms cancel out of the above equation leaving

$$m\ddot{x} + kx = 0 \tag{2.121}$$

$$\ddot{x} + \frac{k}{m}x = 0$$

$$\ddot{x} + \omega_n^2 x = 0 \tag{2.122}$$

Figure 2.35 Freebody diagram for a SDOF system.

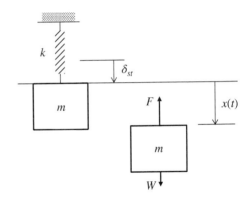

where

$$\omega_n = \sqrt{\frac{k}{m}}\ \text{rad/s}$$

The solution to this differential equation is

$$x(t) = A\sin\omega_n t + B\cos\omega_n t \tag{2.123}$$

Applying initial conditions, $x(0) = B$ and $\dot{x}(0) = 0$ gives

$$x(t) = B\cos\omega_n t \tag{2.124}$$

Over one cycle of motion having a period of T seconds

$$\omega_n T = 2\pi \tag{2.125}$$

Therefore, the period of one oscillation of the vibration is

$$T = \frac{2\pi}{\omega_n} \tag{2.126}$$

The natural frequency of the vibration is

$$f_n = \frac{1}{T} = \frac{\omega_n}{2\pi} = \frac{1}{2\pi}\sqrt{\frac{k}{m}}\ \text{cps} \tag{2.127}$$

The natural frequency increases with structural stiffness but reduces with increase in mass. This equation defines the natural frequency of any SDOF system regardless of its appearance. Several different SDOF systems are shown in Figure 2.36. Natural frequencies can be determined directly from the differential equation of motion. The coefficient to the dependent variable defines the natural circular frequency ω_n^2.

$$\ddot{\theta} + \frac{g}{\ell}\theta = 0 \quad \text{(pendulum)} \tag{2.128}$$

$$\ddot{\theta} + \frac{ga}{r^2}\theta = 0 \quad (r - \text{radius of gyration}) \tag{2.129}$$

$$\ddot{x} + \frac{2T}{ma}x = 0 \quad \text{(mass on a tight wire)} \tag{2.130}$$

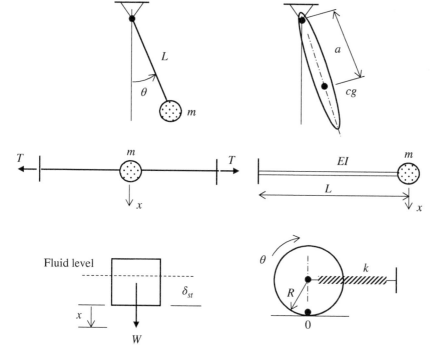

Figure 2.36 Single degree of freedom systems.

$$\ddot{x} + \frac{T}{m}\left(\frac{a+b}{ab}\right)x = 0 \quad \text{(tight wire with uneven spacing)} \tag{2.131}$$

$$\ddot{y} + \frac{3EI}{mL^3}x = 0 \quad \text{(cantilevered beam)} \tag{2.132}$$

$$\ddot{x} + \frac{g}{\delta_{st}}x = 0 \quad \text{(floating object)} \tag{2.133}$$

$$\ddot{\theta} + \frac{2}{3}\frac{k}{M}\theta = 0 \quad \text{(disk)} \tag{2.134}$$

The reader is encouraged to verify each equation of motion. Note that in each case, the motion is simple harmonic and fits the elementary definition: acceleration is proportional and opposite of the displacement. For example, the equation of motion for the rolling disk is

$$I_0\ddot{\theta} = -Rkx \tag{2.135}$$

Noting that

$$I_0 = \frac{3}{2}MR^2 \quad \text{and} \quad x = R\theta$$

Then

$$\ddot{\theta} + \frac{2}{3}\frac{k}{M}\theta = 0$$

and

$$\omega^2 = \frac{2}{3}\frac{k}{M} \tag{2.136}$$

On another note, natural frequency of the spring/mass system in Figure 2.35 can also be expressed in terms of static deflection δ_{st}. Since

$$m = \frac{W}{g} \quad \text{and} \quad k\delta_{st} = W$$

$$\omega_n = \sqrt{\frac{g}{\delta_{st}}} \tag{2.137}$$

This equation is useful if δ_{st} is known or specified. For example, if $\delta_{st} = 0.25$ in., when a structure is fully loaded, then

$$\omega_n = \sqrt{\frac{386 \text{ in.}/\text{s}^2}{0.25 \text{ in.}}} = 39.29 \text{ rad}/\text{s}$$

and

$$f_n = 6.25 \text{ cps}$$

In some structural designs, such as balconies, platforms, and walkways, a maximum static deflection is specified. This specification also sets the natural frequency of the structure.

In many cases, natural frequencies are determined by setting up the differential equation of motion. In other cases, natural frequencies are best determined experimentally.

Example   Next, consider a semicircular rim pivoted at point O as shown in Figure 2.37. The approach is to first determine its center of gravity and mass moment of inertia about the pivot point, O. The equation of motion is the same as for a rigid body pendulum (second case in Figure 2.36).

$$I_A\ddot{\theta} + aMg\theta = 0$$

$$\omega_n{}^2 = \frac{aMg}{I_A}$$

The objective is to relate I_A and "a" to the variables describing the geometry of the semicircle.

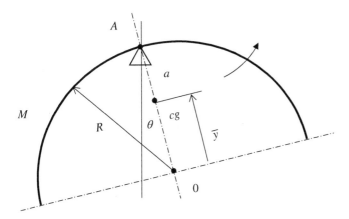

Figure 2.37a   Natural frequency of semicircular rim.

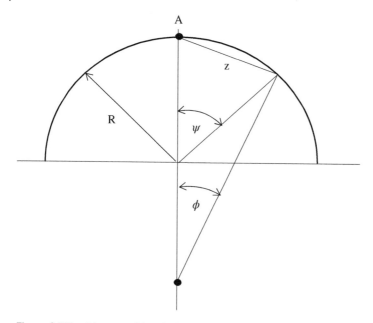

Figure 2.37b Moment of Inertia by Integration.

Location of Center of Gravity

$$\bar{y}M = 2\int_0^{\frac{\pi}{2}} R\sin\phi(R\,d\phi m) = 2R^2m\int_0^{\frac{\pi}{2}}\sin\phi\,d\phi \tag{2.138}$$

where

M – total mass
m – mass per length

$$\bar{y}M = 2R^2m(-\cos\phi)_0^{\frac{\pi}{2}} = 2R^2m(0-1)(-1) = 2R^2m \tag{2.139}$$

$$\bar{y} = \frac{2R^2m}{0.5(2\pi Rm)} = \frac{2R}{\pi} \tag{2.140}$$

From geometry of the semicircular rim
$$a = R - \bar{y}$$

Moment of Inertia with Respect to Point A
Two methods will be given. The first is based on the transfer function. The second on direct integration.
Method #1

$$I_0 = R^2M$$

$$I_{cg} = I_0 - \bar{y}^2M$$

$$I_A = I_{cg} + (R-\bar{y})^2M \tag{2.141}$$

$$I_A = 2RMR\left(1 - \frac{2}{\pi}\right) = 2RMa$$

Method #2 (see Figure 2.37b)

$$dI_A = z^2 \mu R d\psi = z^2 \mu R(2\phi)$$

$$I_A = (2)2R\mu \int_0^{\frac{\pi}{4}} (2R\sin\phi)^2 d\phi$$

$$I_A = 16R^3\mu \left[\frac{\phi}{2} - \frac{\sin 2\phi}{4}\right]_0^{\frac{\pi}{4}}$$

$$I_A = 2RMR\left(1 - \frac{2}{\pi}\right) = 2RMa$$

The natural frequency of the half ring is

$$\omega^2 = \frac{aMg}{2RaM} = \frac{g}{2R}$$

Springs in Series, Parallel

Mechanical springs may be arranged in series or in parallel (Figure 2.38). In the first case, the equivalent spring constant is

$$k_{eq} = k_1 + k_2 \quad \text{(springs in parallel)} \tag{2.142}$$

In the second case

$$k_{eq} = \frac{k_1 k_2}{k_1 + k_2} \quad \text{(springs in series)} \tag{2.143}$$

The spring constant of the beam

$$k_1 = \frac{F}{\delta} = \frac{3EI}{L^3} \quad \text{(beam)} \tag{2.144}$$

$$k_2 = \frac{EA}{L} \quad \text{(rod)} \tag{2.145}$$

$$k_\theta = \frac{GJ}{L} \quad \text{(torsion bar)} \tag{2.146}$$

Coil springs have much lower spring constants than rods in extension or twist.

(a)   (b)

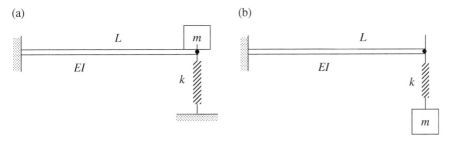

Figure 2.38 Springs in parallel and series. (a) Parallel. (b) Series.

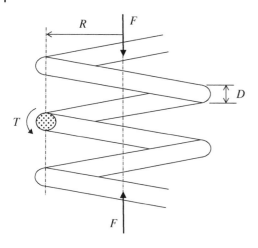

Figure 2.39 Deflection of coiled springs.

Deflection of Coiled Springs

The linear deflection across the coiled spring is mainly to twisting in the coiled rod (Figure 2.39). Twisting per unit circumferential length is

$$d\theta = \frac{T\,ds}{GJ} \tag{2.147}$$

Vertical movement of force, F, due to $d\theta$ is $d\delta = R\,d\theta$. Total axial displacement of force, F, is

$$\delta = \frac{RT}{GJ} \int_0^{2\pi} R\,d\theta = \frac{2\pi R^3 F}{GJ} \quad \text{(checks with Castigliano method)} \tag{2.148}$$

Spring constants can be expressed as

$$k = \frac{F}{\delta} = \frac{GJ}{2\pi R^3} \text{ per one coil} \tag{2.149}$$

$$k = \frac{F}{\delta} = \frac{Gr^4}{4R^3} \text{ per one coil}$$

$$k = \frac{F}{\delta} = \frac{GJ}{2n\pi R^3} \text{ per } n \text{ coils} \tag{2.150}$$

$$k = \frac{F}{\delta} = \frac{Gr^4}{4nR^3} \text{ per } n \text{ coils}$$

For example, if a coil is 4 in. across, made of ¼ in. wire and has 10 coils, its spring constant is calculated to be

$$k = \frac{12 \times 10^6 (0.125)^4}{4(10)(2)^3} = 9.16\,\text{lb/in.} \tag{2.151}$$

Free Vibration with Damping

Spring constants and masses can usually be determined directly through calculations. Damping levels are difficult to quantify except in dampers containing fluids in laminar flow. Therefore, damping coefficients are best determined experimentally.

Every vibrating system has a certain amount of friction, which affects response amplitude. In the case of free vibration, dynamic behavior is predicted by

$$m\frac{d^2x}{dt^2} + c\frac{dx}{dt} + kx = 0 \tag{2.152}$$

The damping coefficient, c, becomes an important parameter when predicting responses to free or force vibrations. It is often difficult to calculate except in special cases. It is a parameter that is usually determined experimentally. As in most cases, friction forces are modeled with linear damping or its equivalent. This assumption greatly simplifies the mathematical solution while giving reasonable engineering results.

The solution to Eq. (2.152), assuming an underdamped system, is

$$x(t) = e^{-\varsigma\omega_n t}\left[A\sin\sqrt{1-\varsigma^2}\omega_n t + B\cos\sqrt{1-\varsigma^2}\omega_n t\right] \tag{2.153}$$

where

$\varsigma = \frac{c}{c_{cr}}$, damping factor assuming under damping; $\varsigma < 1$

$c_{cr} = 2\sqrt{km}$, critical damping coefficient

The first term, $e^{-\varsigma\omega_n t}$, defines the decay of the free vibration. The other terms define the cyclic motion.

Quantifying Damping

This section gives background equations for extracting damping factors from experimental free vibration data.

The free vibration response is

$$x = e^{-\varsigma\omega t}(A\sin\omega_d t + B\cos\omega_d t) \tag{2.154}$$

where

$$\omega_d = \sqrt{1-\varsigma^2}\omega_n$$

$$\omega = \omega_n = \sqrt{\frac{k}{m}}$$

The amplitude of successive vibrations is

$$\frac{X_1}{X_2} = \frac{Xe^{-\varsigma\omega t}}{Ce^{-\varsigma\omega(t+T)}} = e^{\varsigma\omega T_d} \tag{2.155}$$

giving

$$\varsigma\omega T_d = \ln\frac{X_1}{X_2} \tag{2.156}$$

Log decrement is defined by

$$\delta = \ln\frac{X_1}{X_2} \tag{2.157}$$

or

$$\delta = \varsigma\omega T_d \tag{2.158}$$

But

$$\omega_d T_d = 2\pi \tag{2.159}$$

By substitution

$$\delta = \frac{2\pi\varsigma\omega}{\omega_d} = \frac{2\pi\varsigma}{\sqrt{1-\varsigma^2}} \tag{2.160}$$

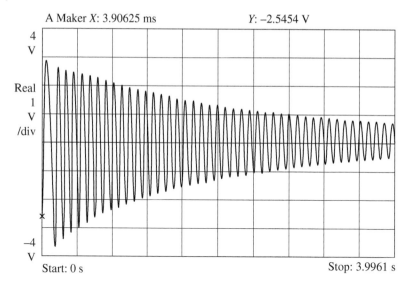

Figure 2.40 Damped free vibration response of SDOF.

The damping factor, ζ, is typically small (≈ 0.004), so a reasonable approximation for δ is

$$\delta = 2\pi\zeta \qquad (2.161)$$

The damping, ζ, is determined once the log decrement has been established experimentally. The log decrement is established from successive amplitude using Eq. (2.157).

A more accurate approach for determining the log decrement (δ) and the damping factor (ζ) is

$$\delta = \frac{1}{n} \ln \frac{X_0}{X_n} \qquad (2.162)$$

where

X_0	–	amplitude at point "0"
X_n	–	amplitude at point "n"
n	–	number of cycles between points 0 and n

Using data from Figure 2.40

$$\delta = \frac{1}{40} \ln \frac{2.7}{0.7} = 0.0337$$

$$\varsigma = \frac{\delta}{2\pi} = 0.0054$$

$$\varsigma = \frac{c}{c_{cr}}$$

Keep in mind that the experimental value of the damping factor accounts for all energy dissipating mechanisms in the systems, such as material hysteresis, support friction, and air drag, to mention a few. These friction effects have been collectively modeled as linear damping.

Critical Damping in Vibrating Bar System

In general, SDOF systems may look completely different from a simple spring mass system. In each case, the equation of motion of free vibration system will be of the same form. Consider, for example, the vibrating bar of Figure 2.41.

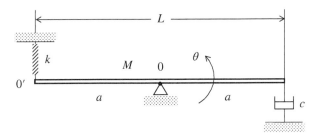

Figure 2.41 Critical damping in a rotational model.

Equation of motion is

$$I_0\ddot{\theta} + ca^2\dot{\theta} + ka^2\theta = 0 \tag{2.163}$$

where

$$I_0 = \frac{1}{12}ML^2 = \frac{1}{12}M(2a)^2 = \frac{1}{3}Ma^2$$

From the equation of motion, critical damping is

$$\left(ca^2\right)_{cr} = 2\sqrt{ka^2I_0} = 2a^2\sqrt{k\frac{M}{3}}$$

and the critical damping factor is

$$\varsigma = \frac{ca^2}{(ca)_{cr}} = \frac{c}{2\sqrt{k\dfrac{M}{3}}} \tag{2.164}$$

Forced Vibration of SDOF Systems with Damping

The differential equation of motion is similar to Eq. (2.110) except a damping term has been added (Figure 2.42).

$$m\frac{d^2x}{dt^2} + c\frac{dx}{dt} + kx = F_0 \cos \omega t \tag{2.165}$$

The solution to this equation has particular and complementary components. We are interested in only the particular solution as the complementary solution is transient and decays with time.

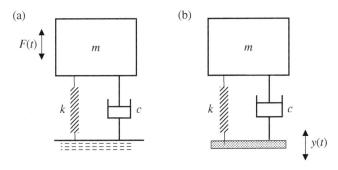

Figure 2.42 Single degree of freedom systems.

One approach is to assume the solution has the form of

$$x(t) = A \sin \omega t + B \cos \omega t$$

Substituting this expression into Eq. (2.165) yields two algebraic equations with unknowns, A and B.

$$A = \frac{2\zeta r \delta_{st}}{(1-r^2)^2 + (2\zeta r)^2} \tag{2.166}$$

$$B = \frac{(1-r^2)\delta_{st}}{(1-r^2)^2 + (2\zeta r)^2} \tag{2.167}$$

where $\delta_{st} = \frac{F_0}{k}$, and

r – frequency ratio
ζ – damping factor

Combining these equations gives

$$x(t) = X \cos(\omega t - \phi) \tag{2.168}$$

where

$$X = \frac{\delta_{st}}{[(1-r^2)^2 + (2\zeta r)^2]^{\frac{1}{2}}}$$

$$\tan \phi = \frac{2\zeta r}{1 - r^2}$$

These equations are plotted in Figures 2.43 and 2.44. Notice that the maximum amplitude for each level of damping occurs near $r = 1$ and is quantified by

$$\left(\frac{X}{\delta_{st}}\right)_{max} = \frac{1}{2\zeta} \tag{2.169}$$

This equation shows that the damping factor (not damping coefficient) is a good indicator of the effects of damping on response.

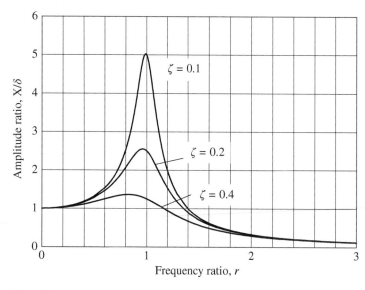

Figure 2.43 Frequency response for damped – forced vibrations.

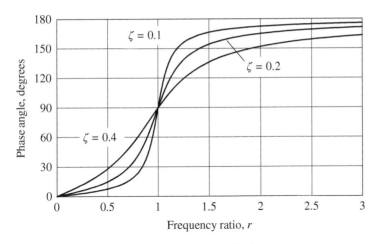

Figure 2.44 Phase angle between force and displacement.

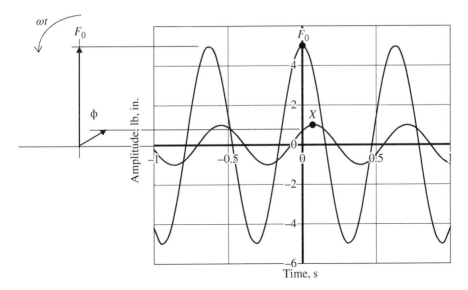

Figure 2.45 Phase angle between force and response.

Several conclusions can be made from this analysis. The amplitude of vibration is greatest when the driving frequency, ω, is the same as the natural frequency of the SDOF system. Under resonant conditions bad things, such as material fatigue and noise, can happen. One way to alleviate vibration amplitude at resonance is to operate the system to the far right of resonance. Another approach is to apply damping, but this adds cost.

The relation between force and displacement functions is represented in a vector diagram (Figure 2.45). This figure is based on the following data.

$\omega = 10\,\text{rad/s}$ ($T = 0.629$ seconds)
$\phi = 45°$ (phase angle)
$F_0 = 5\,\text{lb}$
$X = 1\,\text{in.}$

The two vectors, F_0 and X, on the left side are separated by the fixed angle, ϕ. As the vector pair rotate with angular velocity, ω, their projections onto the vertical line define amplitudes vs. time.

$$F(t) = F_0 \cos \omega t$$

$$x(t) = X \cos (\omega t - \phi)$$

The phase lag of X is explained in terms of energy balance between energy input by the force and the energy dissipated by the damper. The differential work done by F over distance dx is

$$dW = F dx = F \frac{dx}{dt} dt = F(t)V(t)dt \tag{2.170}$$

The velocity of the mass is defined by

$$\dot{x}(t) = V(t) = -X\omega \sin (\omega t - \phi)$$

Equation (2.170) shows that energy is put into a vibration when the applied force is in phase with the velocity of the vibration. When pushing a child in a swing, a force applied at the bottom of the motion, where the velocity is the greatest, puts the most energy into the back-and-forth motion. By substitution

$$W = -F_0 X \int_0^T \sin (\omega t - \phi) \cos \omega t (\omega\, dt) \tag{2.171}$$

The integration over time, T, gives the work per cycle done by $F(t)$. The amount of work required to overcome the energy dissipated by the damper is expressed by

$$W_{cycle} = \pi F_0 X \sin \phi \tag{2.172}$$

At resonance ($r = 1$), the phase angle becomes $90°$ so maximum work per cycle becomes

$$W_{cycle} = \pi F_0 X \tag{2.173}$$

This is the amount of energy required to sustain the vibration at resonance.

From the frequency response diagram (Figure 2.43 and Eq. (2.169)), the vibration amplitude, X, depends only on the damping factor and δ_{st}.

$$\frac{X}{\delta_{st}} = \frac{1}{2\zeta} \tag{2.174}$$

which converts to

$$X = \frac{F_0}{c\omega_n} \tag{2.175}$$

Using this expression in Eq. (2.173) gives the work per cycle at resonance for a linearly damped system.

$$W = \pi c \omega_n X^2 \tag{2.176}$$

Nonlinear Damping

This equation is commonly used to determine the equivalent damping of other types of damping, which may be nonlinear. For example, if a damping force is defined by

$$F_d = \pm a\dot{x}^2 \tag{2.177}$$

and $x(t)$ is approximated by $x(t) = X \cos \omega t$, $\dot{x}(t) = -\omega X \sin \omega t$, then the work per cycle of this damping force is

$$W = 2\int_0^T a\dot{x}^2 \frac{dx}{dt} dt \tag{2.178}$$

$$W = 2\int_0^T aX^3\omega^3 \sin^3\omega t \, dt$$

$$W = \frac{8}{3}a\omega^2 X^3 \tag{2.179}$$

The equivalent linear damping coefficient is determined by equating work per cycles.

$$\pi c_{eq}\omega X^2 = \frac{8}{3}a\omega^2 X^3 \tag{2.180}$$

Giving

$$c_{eq} = \frac{8}{3\pi}a\omega X \tag{2.181}$$

The amplitude at resonance is predicted by

$$X = \frac{F_0}{\omega}\frac{3\pi}{8a\omega X}$$

$$X = \sqrt{\frac{3\pi F_0}{8a\omega^2}} \tag{2.182}$$

Vibration Control

Basic steps that can be taken to reduce severe vibrations are:

1) Eliminate the source of excitation.
2) Change the frequency of the force of excitation (usually linked to rotary speed).
3) Reduce the magnitude of the driving force.
4) Change the natural frequency by altering k or m. It is preferable to reduce natural frequency by increasing m or reducing stiffness. This puts the frequency ratio, r, on the high side of resonance where response amplitudes are lowest.
5) Apply damping

a) Some damping is inherent.
b) Apply dashpots (mechanical dampers). Damping is the costliest and least desirable approach.

Other Vibration Considerations

Transmissibility

Transmissibility is defined as the ratio of force transmitted to a foundation to the driving force. Total transmitted force is the sum of force transmitted through the spring and dampener. One is displacement dependent; the other is velocity dependent. Both vary with time. The total force transmitted to the support (foundation, floor, etc.) can be greater than the magnitude of the driving force. This is of practical concern in situations where nearby equipment may be affected.

The equation of motion for this case is generated from the freebody diagram given in Figure 2.46a.

The transmitted force varies with time according to

$$F_T(t) = kx + c\dot{x} \tag{2.183}$$

Transmissibility is expressed as [11]

$$T_r = \frac{|F_T(t)|}{F_0}$$

$$T_r = \frac{\left[1 + (2\zeta r)^2\right]^{\frac{1}{2}}}{\left[(1 - r^2)^2 + (2\zeta r)^2\right]^{\frac{1}{2}}} \tag{2.184}$$

Transmissibility defines how much of the driving force, F_0, gets transmitted to the foundation. It is significantly affected by damping and frequency ratio. A plot of this function is given in Figure 2.47. Notice that damping reduces transmissibility up to a frequency ratio of $\sqrt{2}$. Beyond this point, damping increases magnitude of the transmitted force.

In an industrial setting, it is common practice to install air conditioning units on roofs. Flat space is available. Reciprocating and rotating components in these units may generate forces that can cause structural-borne noise, which can be eliminated by properly mounting these units on soft springs. This greatly reduces the transmitted force.

Example   Consider a compressor unit weighing 200 lb mounted on a support frame, which weighs 50 lb. An unbalanced rotating mass produces a vertical periodic force on the frame. Its rotating speed is 250 rpm producing a variable force of 20 lb. Determine the stiffness of the four spring

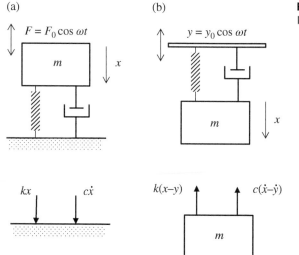

(a)

$F = F_0 \cos \omega t$

m

x

kx $c\dot{x}$

(b)

$y = y_0 \cos \omega t$

m

x

$k(x-y)$ $c(\dot{x}-\dot{y})$

m

Figure 2.46   (a) Transmissibility. (b) Vibration Isolation.

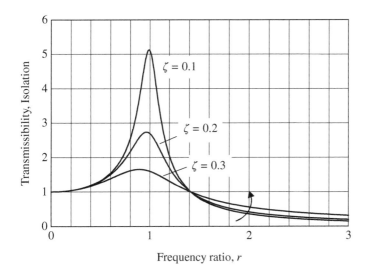

Figure 2.47 Transmissibility vs. frequency.

mounts such that the magnitude of the dynamic force transmitted to the roof is 10% of the magnitude of the driving force. Assume damping is negligible.

Applying Eq. (2.184) with zero damping in the springs gives

$$- T_r = \frac{1}{1-r^2} \quad \text{(minus because vibration on high side of resonance)} \tag{2.185}$$

$$(0.1)\left(1-r^2\right) = 1$$

$$r = 3.32 = \frac{\omega}{\omega_n}$$

$$\omega = \frac{2\pi N}{60}$$

$$\omega_n = \frac{2\pi}{3.32}\frac{250}{60} = 7.89 \text{ rad/s}$$

Recall

$$\omega_n = \sqrt{\frac{K}{M}}, \text{ where } M = \frac{250}{386}\left|\frac{\text{lb-s}^2}{\text{in.}}\right|$$

giving, $K = 40.3$ lb/in. Therefore, each spring would have a spring constant of $k = 10.1$ lb/in. The initial compression of each spring would be 6.2 in.

Vibration Isolation

Base motion is illustrated in Figure 2.46b. Response is similar to the transmissibility theory except in this case the base is moving, and force is applied to the mass through the motions of the spring and damper.

The differential equation of motion for base motion is

$$m\ddot{x} = - c(\dot{x}-\dot{y}) - k(x-y) \tag{2.186}$$

$$m\frac{d^2x}{dt^2} + c\frac{dx}{dt} + kx = c\frac{dy}{dt} + ky \tag{2.187}$$

The response ratio is

$$\frac{X}{y_0} = \frac{\left[1 + (2\zeta r)^2\right]^{\frac{1}{2}}}{\left[(1 - r^2)^2 + (2\zeta r^2)\right]^{\frac{1}{2}}} \tag{2.188}$$

This equation defines how much of base motion gets transmitted to the mass, m.

Commonality of Responses

Notice that the equations for transmissibility and vibration isolation are the same. Therefore, Figure 2.47 applies to both. Resonance still occurs when the two frequencies are tuned. Damping has a slightly different effect. It reduces vibration amplitude up to $\frac{\omega}{\omega_n} = \sqrt{2}$. Beyond this point, damping is detrimental to the system.

Instruments are sometimes isolated from support frames, which may be subjected to severe vibrations or shock forces. Electronic sensors within downhole drilling tools are good examples.

Application of Vibration Absorbers in Drill Collars

Vibration analyses show how drill collars respond to bit motion [12]. Drill pipe responds too, but it is their higher modes that respond to typical rotary speeds. It is easier to resonant the drill collar section than to develop the higher drill pipe modes. Therefore, the drill collar section is a critical area in the drillstring to monitor [13].

Drill collar length is dictated by required weight on bit (WOB). In hard rock drilling areas, more drill collars are carried to generate a high bit force. The irony of this practice is the natural frequency of axial modes of vibration becomes close to frequencies of excitation associated with normal drilling speeds. This situation creates frequency tuning almost automatically. The numbers for this scenario are given below.

The natural frequencies of a drill collars (pinned at the bottom and free at the top) are determined from

$$f_{na} = \frac{n16\,850}{4L}, \quad n = 1, 3, 5, \ldots \quad \text{(axial mode)} \tag{2.189}$$

Note that collar size is not a factor, only length. For the first mode, $n = 1$ and

$$f_{1a} = \frac{4212}{L} \tag{2.190}$$

If, for example, $L = 600$ ft, $f_{1a} = 7.02$ cps.

Now compare this frequency to that associated with a three-roller cone bit rotating at 140 rpm. The drill bit frequency in this case would be

$$f = \frac{3N}{60} = \frac{140}{20} = 7\,\text{cps} \tag{2.191}$$

Actual drilling cases may vary, but the reality remains. Both natural frequency and driving frequency are close enough to excite the first mode of the drill collars. The same condition could exist for higher drill collar modes when positive displacement motors (PDMs) and turbines are used. Drill collar response under this assumption, i.e. ignoring drill pipe participation, is shown in Figure 2.48.

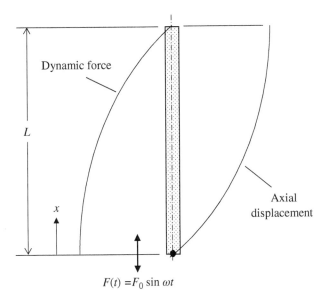

Figure 2.48 Vibration mode shape of drill collars.

Natural Frequencies with Vibration Absorbers

Vibration absorbers offer a practical solution for changing natural frequency of drill collars. Drill collar weight (length) can remain the same for a required WOB, while lowering natural frequency.

The natural frequencies of drill collars with vibration absorber is developed as follows. Starting with the basic equation of motion for the distributed model.

$$EA\frac{\partial^2 u}{\partial x^2} = m\frac{\partial^2 u}{\partial t^2}, \text{ noting that } c^2 = \frac{AE}{m} = \frac{E}{\rho}$$

$$(2.192)$$

where m is mass per unit length, ρ is mass density, and c is acoustic velocity of a compression in the collars (16 850 ft/s).

The formulation for determining natural frequencies of drill collars with a vibration absorber is given below. The math model is shown in Figure 2.49.

$$\frac{\partial^2 u}{\partial x^2} = \frac{1}{c^2}\frac{\partial^2 u}{\partial t^2}$$

$$(2.193)$$

For the solution, assume

$$u(x,t) = X(x)\sin \omega t$$

$$(2.194)$$

By substitution

$$\frac{d^2 X}{dx^2}\sin \omega t = -\frac{\omega^2}{c^2}X\sin \omega t$$

$$(2.195)$$

$$\frac{d^2 X}{dx^2} + \left(\frac{\omega}{c}\right)^2 X = 0$$

$$(2.196)$$

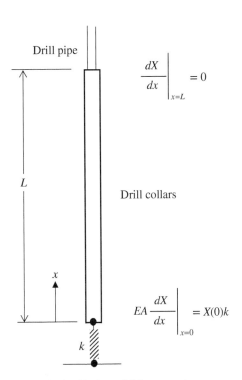

Figure 2.49 Math model for natural frequencies with vibration absorber.

General solution is

$$X(x) = A_1 \sin \frac{\omega}{c}x + B_1 \cos \frac{\omega}{c}x \tag{2.197}$$

$$\frac{dX}{dx} = A_1 \frac{\omega}{c} \cos \frac{\omega}{c}x - B_1 \frac{\omega}{c} \sin \frac{\omega}{c}x \tag{2.198}$$

Imposing boundary conditions

$$\frac{dX}{dx}\bigg|_{x=L} = 0 \qquad 0 = A_1 \cos \frac{\omega}{c}L - B_2 \sin \frac{\omega}{c}L$$

and

$$EA\frac{dX}{dx}\bigg|_{x=0} = kX(0) \qquad EA\left[A_1\frac{\omega}{c}\right] = kB_1$$

Combining these two equations gives

$$\left(\frac{\omega}{c}L\right) \tan \left(\frac{\omega}{c}L\right) = \frac{kL}{EA} \tag{2.199}$$

Let

$$\alpha = \frac{kL}{EA} \quad \text{and} \quad \beta = \frac{\omega}{c}L$$

Then

$$\beta \tan \beta = \alpha \tag{2.200}$$

The eigenvalue we seek is $\beta = \frac{\omega}{c}L$, which is implicit in the characteristic equation. This equation was programed to obtain the relation between the two dimensionless numbers. The relationship α and β is shown in Figure 2.50 and Table 2.3.

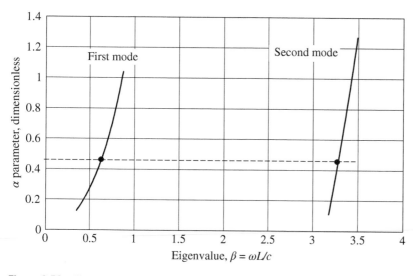

Figure 2.50 Eigenvalue of first mode with vibration absorber.

Table 2.3 Eigenvalues for first and second modes.

First mode		Second mode	
β	α	β	α
0.453 785	0.221 326		
0.488 692	0.259 842		
0.523 598	0.302 299		
0.558 505	0.348 992	3.176 497	0.110 917
0.593 411	0.400 261	3.211 403	0.224 554
0.628 318	0.456 499	3.246 31	0.341 192
0.663 225	0.518 167	3.281 216	0.461 136
0.698 131	0.585 801	3.316 123	0.584 712
0.733 038	0.660 029	3.351 029	0.712 273
0.767 944	0.741 594	3.385 936	0.844 198
0.802 851	0.831 375	3.420 842	0.980 9
0.837 757	0.930 422	3.455 749	1.122 83
0.872 664	1.039 999	3.490 656	1.270 483

Example Determine the natural frequency of the first and second modes (axial mode) of a drill collar $6\frac{1}{2} \times 2$ (102 lb/ft), 700 ft long for the following two cases:

a) No vibration absorber

b) A vibration absorber having a spring constant of 50 000 lb/in.

c) No vibration absorber

$$f_n = \frac{n16\,800}{4L}, n = 1, 3, 5$$

$$f_1 = \frac{16\,850}{4(700)} = 6 \text{ cps(first mode)} \qquad (2.201)$$

$$f_2 = \frac{3(16\,850)}{4L} = \frac{12\,638}{700} = 18.05 \text{ cps (second mode)} \qquad (2.202)$$

d) With vibration absorber ($k = 50\,000$ lb/in.)

$$\left(\frac{\omega}{c}L\right) \tan \left(\frac{\omega}{c}L\right) = \frac{kL}{EA}$$

$$\alpha = \frac{kL}{EA} \quad \text{and} \quad \beta = \frac{\omega}{c}L$$

$$\beta \tan \beta = \alpha$$

By substitution

$$\alpha = \frac{kL}{EA} = \frac{50\,000(700)12}{30 \times 10^6(30)} = 0.467 \text{ (dimensionless)}$$

From tables

First mode: $\beta \sim 0.628 \quad \dfrac{\omega}{c}L = 0.628 \quad \omega_1 = 0.628 \dfrac{16\,850}{700} = 15.12\,\text{rad/s}$

$f_1 = 2.41\,\text{cps}$

Second mode: $\beta \sim 3.28 \quad \dfrac{\omega}{c}L = 3.28 \quad \omega_2 = 3.28 \dfrac{16\,850}{700} = 78.95\,\text{rad/s}$

$f_{n2} = \dfrac{78.95}{2\pi} = 12.57\,\text{cps}$

These natural frequencies are in the range of frequencies generated by roller cone rock bits associated with rotational speeds drill collars and drilling motors. Frequencies from this type of drill bit relate to rotary speed by

$$f_{PDM} = \frac{3N_{PDM}}{60} = \frac{N_{PDM}}{20}\ \text{cps} \tag{2.203}$$

$$N_{PDM} = 20(12.57) = 251.4\ \text{rpm} \tag{2.204}$$

This rotational speed is within the range of PDM motors.

It is worth noting that the natural frequency of the first mode using the lumped mass model is

$$f_{n1} = \frac{1}{2\pi}\sqrt{\frac{k}{M}} = 2.61\ \text{cps (drill collars plus vibration absorber)}$$

where

$k = 50\,000\ \text{lb/in.}$

$W = 700(102) = 71\,400\ \text{lb}$

$M = \dfrac{71\,400}{385}\dfrac{\text{lb-s}^2}{\text{in.}} = 185.45\dfrac{\text{lb-s}^2}{\text{in.}}$

The distributed model gave a natural frequency of 2.4 cps.

The vibration absorber reduced the natural frequency of the first mode from 6 to 2.6 cps.

This means typical rotary speeds are on the high side of resonance when vibration dampers where vibration response is lowest.

Responses to Nonperiodic Forces

So far, forces of excitation have been periodic. However, there are many cases where applied forces are not periodic. In this case, the solution is based on the principle of impulse momentum.

Consider the mass in Figure 2.51 is impacted by a force, F, over time, Δt.

This impulsive force brings about a sudden change in momentum according to

$$F\Delta t = m(v_2 - v_1) \tag{2.205}$$

At $t = 0$, the initial conditions are $x = x_0$ and $v = v_0$. After the force impact, the velocity at point 2 becomes

$$v_2 = \frac{Im}{m} + v_0 \tag{2.206}$$

where "Im" is impulse ($F\Delta t$). This increase in velocity is achieved without movement of the mass. The response can now be viewed as a free vibration having an initial velocity of v_2. After incorporating the initial conditions, the mass responds by

Figure 2.51 Impulse of an impact force.

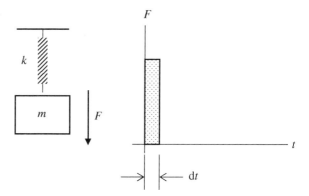

Figure 2.52 Response to impulse.

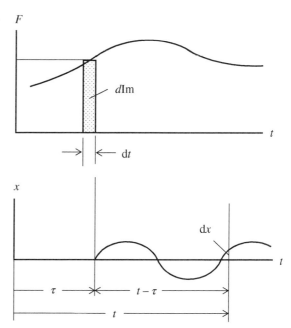

$$x(t) = x_0 \cos \omega_n t + \frac{v_0}{\omega_n} \sin \omega_n t + \frac{I}{m \omega_n} \sin \omega_n t \qquad (2.207)$$

"I" in this case represents the differential impulse, Im. The first two terms show the effects of initial conditions. The third term represents the effect of the impulse. In the case of damping

$$x(t) = e^{-\zeta \omega_n t} \left(x_0 \cos \omega_d t + \frac{v_0 + \zeta \omega_n x_0}{\omega_d} \sin \omega_d t \right) + e^{-\zeta \omega_n t} \frac{I}{m \omega_d} \sin \omega_d t \qquad (2.208)$$

The effects of the initial conditions decay with time, leaving the effect of the impulse (last term).

The impulse–momentum concept is useful for obtaining the response of SDOF systems to continuous force such as shown in Figure 2.52. The response at any time, t, is the result of consecutive differential impulses. A displacement response due to impulse, "I," is also shown in the figure.

The total displacement at time t is the sum of all differential responses twisted together or convoluted together.

Assuming there is no damping, the total effect of each differential impulse, $d\,\text{Im} = F(\tau)d\tau$ becomes

$$dx = \frac{d\text{Im}}{m\omega_n}\sin\omega_n(t-\tau) \tag{2.209}$$

$$dx = \frac{Fd\tau}{m\,\omega_n}\sin\omega_n(t-\tau) \tag{2.210}$$

$$x(t) = \frac{1}{m\,\omega_n}\int_0^t F(\tau)\sin\omega_n(t-\tau)d\tau \tag{2.211}$$

The integral can be viewed as the sum of several sine waves stacked side by side in a convoluted fashion, thus the term convolution integral. It is also called Duhamel's integral.

Dynamic Load Factor

One useful application is the response to a suddenly applied force, which remains constant with time as shown in Figure 2.53.

Duhamel's integral for this case gives

$$x(t) = \frac{1}{m\omega_n}\frac{F_0}{\omega_n}[\cos\omega_n(t-\tau)]_0^t \tag{2.212}$$

or

$$x(t) = \frac{F_0}{k}[1-\cos\omega_n t] \tag{2.213}$$

This equation can also be developed directly from Newton's second law by solving

$$m\ddot{x} + kx = F_0$$

The maximum displacement of the mass occurs when $\omega_n t = 2\pi$ at which

$$x_{max} = \frac{2F_0}{k} \tag{2.214}$$

Figure 2.53 Suddenly applied constant force.

This result implies the maximum displacement in the spring (or structure) is *twice* the displacement produced by F_0 acting as a static force. This gives justification for a factor of safety of 2, which is often used to account for dynamic effects on structures.

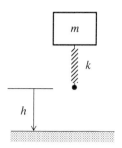

Packaging

Another practical example of this theory is packaging delicate items. Assume that mass, m, is the item to be protected and is packed inside a box containing cushion material having spring constant of k (Figure 2.54).

Figure 2.54 Packaging model.

When the box strikes the floor, the cushion material begins to compress. The velocity of the box and contents at point of contact is

$$\dot{x}_0 = \sqrt{2hg} \tag{2.215}$$

The constant force, F_0, becomes, mg, the weight of the packaged item. The equation of motion becomes

$$m\ddot{x} + kx = W \tag{2.216}$$

The solution to the problem can be found by solving the equation directly or using Duhamel's integral. Solving Eq. (2.216) directly and using initial conditions at first point of contact of the spring

$$x(0) = 0$$

$$\dot{x}(0) = \sqrt{2hg}$$

The complementary solution is

$$x_c = A\cos\omega_n t + B\sin\omega_n t$$

The particular solution is

$$x_p = \frac{W}{k}$$

Applying the first initial condition

$$A = -\frac{W}{k}$$

The second initial condition gives

$$B = \frac{\sqrt{2hg}}{\omega_n}$$

Bringing the terms together gives

$$x(t) = \frac{\sqrt{2hg}}{\omega_n}\sin\omega_n t + \frac{W}{k}(1 - \cos\omega_n t) \tag{2.217}$$

The first term is the response caused by the velocity of the mass during impact. The second term is the response due to sudden application of the weight of the mass. This term is small compared with the first term.

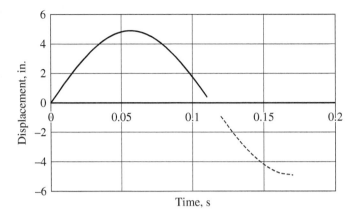

Figure 2.55 Time response of mass into packing material.

Example Assume the following conditions.

$h = 24$ in.
$W = 0.5$ lb
$k = 1$ lb/in.
$m = \frac{W}{g} = \frac{0.5}{386} = 0.001\ 295$ lb-s^2/in.
$\omega_n = \sqrt{\frac{k}{m}} = \sqrt{\frac{1}{0.001\ 295}} = 27.78$ rad/s
$f = 4.422$ cps
$T = 0.226$ seconds

The displacement of the mass (m) into the packing (k), due only to the velocity at impact, reaches a maximum value of 4.88 in. The corresponding impact force on the instrument is therefore 4.88 lb. The dashed line in Figure 2.55 is a continuation of the sine function. It has no physical meaning except to indicate separation of the mass from the packaging material.

Duhamel's integral applies to any mechanical system. Consider a compound wheel given a torque, which varies with time (Figure 2.56). One can determine the angular response of the wheel in terms of the variables (M, I_g, T, t_1, t_2, k) after the couple (T) is removed at t_2.

Figure 2.56 Application to a rolling disk.

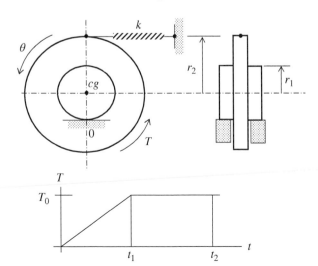

Since the acceleration of point 0 passes through *cg*, the equation of motion is

$$\sum M_0 = I_0 \alpha \tag{2.218}$$

$$T - k(r_1 + r_2)^2 \theta = I_0 \ddot{\theta}$$

$$I_0 \ddot{\theta} + k(r_1 + r_2)^2 \theta = T(t) \tag{2.219}$$

with

$$\omega_n^2 = \frac{k(r_1 - r_2)^2}{I_0} \tag{2.220}$$

The mathematical form is the same as for a simple spring–mass system. In terms of angular parameters

$$\theta(t) = \frac{1}{I_0 \omega_n} \left[\int_0^t T(\tau) \sin \omega_n(t - \tau) d\tau \right] \tag{2.221}$$

Specific to this example

$$\theta(t) = \frac{1}{I_0 \omega_n} \left[\int_0^{t_1} \frac{T_0}{t_1} \tau \sin \omega_n(t - \tau) d\tau + T_0 \int_{t_1}^{t_2} \sin \omega_n(t - \tau) d\tau \right] \tag{2.222}$$

where *T(t)* beyond t_2 is zero.

Vibrations Caused by Rotor Imbalance

Response to an Imbalanced Rotating Mass

A common source of excitation in machinery is rotational imbalance. The diagram of Figure 2.57 shows a model of an SDOF system driven by imbalance, *em*.

The acceleration of the rotating mass is

$$a_m = a_o + a_{m/o} \tag{2.223}$$

The *x* component of the acceleration of the rotating mass is

$$a_{mx} = \frac{d^2 x}{dt^2} - e \omega^2 \cos \omega t \tag{2.224}$$

The *x* component of the force required to accelerate the mass, *m*, is

$$f_x = m \left(\frac{d^2 x}{dt^2} - e \omega^2 \cos \omega t \right) \tag{2.225}$$

There is an equal but opposite force on the block. From the freebody diagram, the equation of motion for the block is

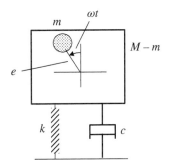

Figure 2.57 Rotating imbalance.

$$(M - m)\frac{d^2x}{dt^2} = -f_x - c\frac{dx}{dt} - kx \tag{2.226}$$

From which

$$M\frac{d^2x}{dt^2} + c\frac{dx}{dt} + kx = me\,\omega^2 \cos \omega t \tag{2.227}$$

where M is the total mass of the system.

$$M\frac{d^2x}{dt^2} + c\frac{dx}{dt} + kx = me\,\omega^2\,e^{i\,\omega t} \tag{2.228}$$

Using the complex variable approach, we assume the solution to be

$$x = Xe^{i\omega t} \tag{2.229}$$

Substituting this assumed solution into the equation of motion gives

$$\frac{\overline{X}M}{em} = \frac{r^2}{(1 - r^2) + i2\varsigma r} \tag{2.230}$$

and

$$\frac{XM}{em} = \frac{r^2}{\left[(1 - r^2)^2 + (2\varsigma r)^2\right]^{1/2}} \tag{2.231}$$

$$\tan \varphi = \frac{2\varsigma r}{1 - r^2} \tag{2.232}$$

where

$\varsigma = \frac{c}{c_{cr}}$

$c_{cr} = 2M\omega_n$

A plot of Eq. (2.231) is shown in Figure 2.58.
The time history of the response is described by

$$x = X \cos (\omega t - \varphi) \tag{2.233}$$

The amplitude ratio depends on frequency ratio and damping factor. These effects are illustrated in the frequency response curves. These results show the largest amplitudes at a frequency ratio of one (1) as expected. The damping factor, ζ, limits the maximum response amplitude per

$$\frac{XM}{em} = \frac{1}{2\zeta} \text{(at } r = 1) \tag{2.234}$$

Synchronous Whirl of an Imbalanced Rotating Disk

The physical situation is one in which an unbalanced disk is mounted on a shaft (Figure 2.59). The center of rotation of the shaft is indicated by axis a–a. The geometric center of the disk is indicated by line 0–0, and the location of the center of gravity is indicated by line c–c. Center distances are shown in the left diagram.

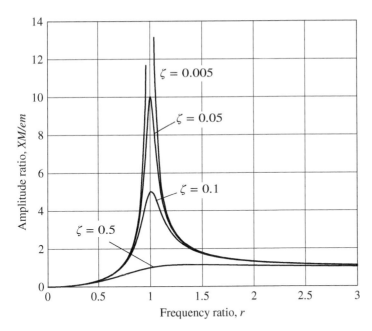

Figure 2.58 Frequency response of imbalanced mass system.

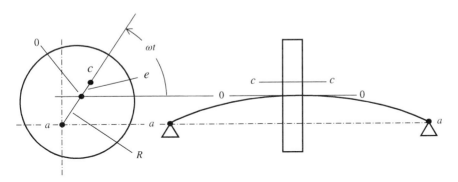

Figure 2.59 Synchronous whirl of disk on a rotating shaft.

Synchronous whirl occurs when a given point on the outer edge of a disk always points inward. At any point in time, the radius, R, and eccentricity, e, are fixed and each line rotates with angular velocity, ω. The objective is to locate radial displacements R and e.

Following the equation of motion for rigid bodies

$$M\bar{a}_{cg} = \sum \bar{F} \tag{2.235}$$

$$\bar{a}_c = \bar{a}_0 + \bar{a}_{c/0} \tag{2.236}$$

where

$$\bar{a}_0 = \ddot{x}i + \ddot{y}j$$

$$\bar{a}_{c/0} = -\omega^2 e(i\cos\omega t + j\sin\omega t)$$

Giving

$$M\ddot{x} + kx = M\omega^2 e\cos\omega t \qquad (2.237)$$

$$M\ddot{y} + ky = M\omega^2 e\sin\omega t \qquad (2.238)$$

and

$$\bar{R} = xi + yj$$

k – spring constant of shaft at location of the disc

Combining the solutions gives

$$R = \frac{er^2}{1-r^2}, \quad \text{where } r = \frac{\omega}{\omega_n} \text{ (frequency ratio)} \qquad (2.239)$$

This equation is interpreted as follows (Figure 2.60). On the low side of resonance, lines R and e are inline, with the center of gravity, c, located outside of point O. Displacement, R, grows with increase in the frequency ratio, r. On the high side of resonance, the position of c moves inside arm, R as the distance R diminishes with increased frequency ratio. At very high frequency ratios, the center of gravity, c, moves closer to axis a–a.

By comparison, the imbalance rotating mass of Figure 2.57, if $\zeta = 0$ is

$$\frac{XM}{m} = \frac{er^2}{1-r^2} \qquad (2.240)$$

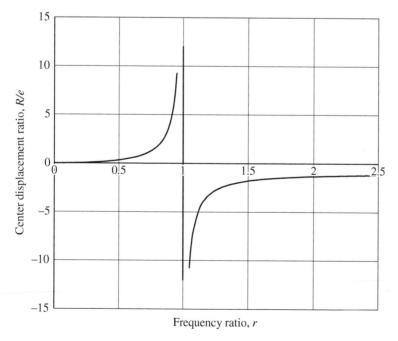

Frequency ratio, r

Figure 2.60 Response to rotation.

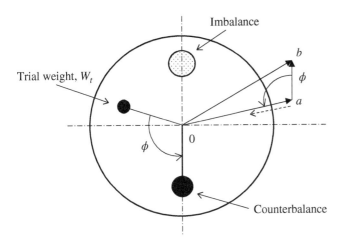

Figure 2.61 Imbalanced rotating disk.

Balancing a Single Disk

Consider a thin disk with an unknown imbalance, i.e. the location and magnitude of the mass are unknown. It is desirable to add a single mass at a location, which balances the disk. Regarding Figure 2.61, the shaded circle represents the unknown imbalance.

The steps to balance the disk are [11]:

Step 1 – Run the disk at any speed resulting in a measurable amplitude. The direction and amplitude are indicated by vector 0–a.

Step 2 – Attach a trial weight, W_t, at any location on the disk and measure the new displacement vector, 0–b, at the same speed as before. The amplitude 0–b is the result of both the unknown imbalance and the effect of the added weight, W_t. The difference between the two vectors, a–b, is the effect of the trial weight, W_t alone.

Step 3 – Move the trial weight counterclockwise by the angle ϕ. In this new location, the displacement vector, a–b, will be parallel and opposite to vector, 0–a, as indicated by the dashed line. Next, increase its weight (W_t) by the ratio of $\frac{0a}{ab}$. The disk will then be balanced.

Due to damping, the phase angle of vector (0a) will lag the unknown position of the imbalance. This is the basis of auto-tire balancing machines.

Synchronous Whirl of Rotating Pipe

Synchronous whirl of rotating pipe refers to motion where the inside surface of the deflected pipe is always pointing inward toward the axis of the support bearings. The outside surface is always pointing outward away from the bearing axis.

Assuming there is no tension in the pipe, the equation of motion is

$$EI\frac{d^4y}{dx^4} - m\,\omega^2 y = 0 \tag{2.241}$$

$$\frac{d^4y}{dx^4} - \alpha^4 y = 0 \tag{2.242}$$

where

$$\alpha^4 = \frac{m\,\omega^4}{EI}$$

The solution to Eq. (2.241) is

$$y = A\sin\alpha x + B\cos\alpha x + C\sinh\alpha x + D\cosh\alpha x \tag{2.243}$$

Boundary conditions for a simply supported beam are:

$$y(0) = 0 \tag{2.244a}$$

$$\frac{d^2y}{dx^2}\bigg|_{x=0} = 0 \tag{2.244b}$$

$$y(L) = 0 \tag{2.244c}$$

$$\frac{d^2y}{dx^2}\bigg|_{x=L} = 0 \tag{2.244d}$$

Constants B, C, D are zero leaving

$$y = A\sin\alpha L \tag{2.245}$$

For a nontrivial solution

$$\alpha L = n\pi, \quad n = 1, 2, 3, K$$

$$\frac{m\,\omega^2}{EI} = \left(\frac{n\pi}{L}\right)^4 \tag{2.246}$$

The critical rotary speed for the first mode ($n = 1$) is

$$\omega^2_{crit} = \left(\frac{\pi}{L}\right)^4 \frac{EI}{m} \tag{2.247}$$

Stability of Rotating Pipe under Axial Load

A similar solution is given by Den Hartog [14] for a rotating beam (pipe) subjected to compressive end loading, Q. If the shaft rotates slowly and jarred sidewise, the elastic stiffness will restore the shaft to its original straight position. At some higher rotational speed, the shaft, if disturbed laterally, will not return. Under this condition, the shaft has buckled from the rotation. This scenario is different from the whirling of unbalanced rotating shafts, which experience radial displacement once the shaft begins to rotate.

The differential equation of bending for this case is similar to the previous case except that the applied end force is compressive and the side-loading inertia is due to centrifugal forces based on synchronous whirl.

$$\frac{d^4y}{dx^4} + \frac{Q}{EI}\frac{d^2y}{dx^2} - \frac{m\omega^2}{EI}y = 0 \tag{2.248}$$

The end force, Q, is compressive and carries a (+) sign. It can also be viewed as $(Q_{eff})_{ave}$. The notation for the speed of rotation is ω rad/s.

Lateral buckling is imminent when

$$\frac{Q}{2EI} + \sqrt{\left(\frac{Q}{2EI}\right)^2 + \frac{m\omega^2}{EI}} = \frac{\pi^2}{L^2} \tag{2.249}$$

Equation (2.249) is put into the form

$$\frac{Q}{Q_{cr}} + \left(\frac{\omega}{\omega_{cr}}\right)^2 \geq 1 \tag{2.250}$$

where

$$Q_{cr} = \frac{\pi^2 EI}{L^2}$$

$$\omega_{cr} = \left(\frac{\pi}{L}\right)^2 \sqrt{\frac{EI}{m}}$$

If Q is tension instead of compression, Q changes signs and the rotary speed required to create whirling motion by the centrifugal force increases as a result. If the left side of Eq. (2.250) is greater than one (1), the rotating column (pipe) is dynamically unstable.

Example Consider, for example, a 60 ft span of drill collars between two stabilizers. Assuming pinned support at each stabilizer and

OD = 7 in.
ID = 2 in.
$I = 117.07$ in.^{4}
$E = 29\,(10)^6$ psi
$w = 120$ lb/ft
$m = 3.73$ slugs/ft
$N = 100$ rpm
$\omega = 10.47$ rad/s; $\omega = \frac{2\pi N}{60}$
$WOB = 50\,000$ lb

then

$$Q_{cr} = \frac{\pi^2 29(10^6)117.07}{[12(60)]^2} = 64\,636\ \text{lb}$$

and

$$\Omega_{cr} = \left(\frac{\pi}{60}\right)^2 \sqrt{\frac{29(10^6)117.07}{3.73} \left|\frac{1\ \text{ft}^2}{144\ \text{in.}^2}\right|} = 6.78\ \text{rad/s}$$

Applying Eq. (2.250)

$$\frac{50\,000}{64\,636} + \left(\frac{10.47}{6.78}\right)^2 = 0.774 + 2.38 = 3.15$$

Based on these numbers, the collar section between the two stabilizers is unstable and will experience synchronous whirl. The compressive force contributes to this instability, but collar rotary speed is the main culprit.

Balancing Rotating Masses in Two Planes

Consider the rotation of the two masses mounted on a shaft (Figure 2.62). The objective here is to place one mass in plane 0 and a second mass in plane N such that the four masses will be balanced.

Equilibrium of the four masses, including the balancing masses, must satisfy the following two equations

$$\sum_{0}^{N} \bar{a}_i x \bar{a}_i \left(M_i \omega^2 R_i \right) = 0 \tag{2.251}$$

$$\sum_{0}^{N} \bar{a}_i \left(M_i \omega^2 R_i \right) = 0 \tag{2.252}$$

The input data for this case are given in Table 2.4.

Substituting the numbers in Table 2.4 into Eqs. (2.251) and (2.252) gives

$$- \left[6m_1 R_1 + 16m_N R_N \sin \theta_N \right] \hat{i} + \left[12m_2 R_2 + 16m_N R_N \cos \theta_N \right] \hat{j} = 0 \,(\text{moment}) \tag{2.253}$$

$$\left[m_0 R_0 \cos \theta_0 + m_2 R_2 + m_N R_N \cos \theta_N \right] \hat{i} + \left[m_0 R_0 \sin \theta_0 + m_1 R_1 + m_N R_N \sin \theta_N \right] \hat{j} = 0 \,(\text{force}) \tag{2.254}$$

From the moment equation

$$\tan \theta_N = \frac{1}{2} \frac{m_1 R_1}{m_2 R_2} = 0.375 \tag{2.255}$$

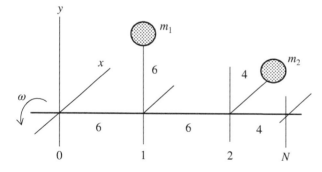

Figure 2.62 Unbalanced rotating mass. system.

Table 2.4 Balancing rotating masses.

	Planes containing the masses			
	0	1	2	N
a_i	0	6	12	16
$\hat{a}_i$	$\cos\theta_0 \hat{i} + \sin\theta_0 \hat{j}$	$\hat{j}$	$\hat{i}$	$\cos\theta_N \hat{i} + \sin\theta_N \hat{j}$
m_i	m_0	10	20	m_N
R_i	6	6	4	6

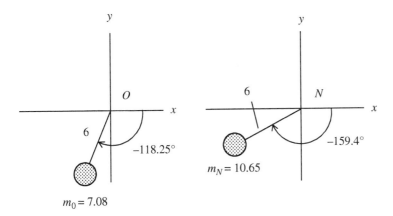

Figure 2.63 Location of masses in balancing planes. (*z* axis is out of paper, right-hand rule).

The arc tan of 0.375 has multiple angles. Only $\theta_N = -159.4°$ satisfies Eq. (2.252), the moment equation. By substitution, $m_N R_N = 63.94$ leading to $m_N = 10.65$.

From the force equation

$$\tan \theta_0 = \frac{m_1 R_1 + m_N R_N \sin \theta_N}{m_2 R_2 + m_N R_N \cos \theta_N} = 1.86 \qquad (2.256)$$

The arctan has multiple units; however, only $\theta_0 = -118.25$ satisfies the force equation. By substitution, $m_0 R_0 = 42.56$ leading to $m_0 = 7.08$. The solution is shown in Figure 2.63. The units of the balancing masses are the same as the unbalanced masses.

Refining the Design

Once the design has been configured in terms of its subsystems, it is desirable to identify critical areas for further analysis. This may require computer software to establish stress distributions and points of high stress, temperature, etc. Testing components with instrumentation, such as strain gages may be useful as check on computer models, especially if there are safety concerns. Testing of certain subsystems in isolation may also be useful.

Final steps to a final design and manufacturing are shown in Figure 1.1. Fabrication drawings communicate the final design and how it is to be fabricated.

Manufacturing

The fabrication of each component should be considered during the early stages of design. Consideration should be given to how each part is to be made and the sequencing of fabrication. Parts may be machined from a casting, which can reduce machining time and waste, or it may be machined from bulk stock.

The machine tool cutting of metal began with tools powered by water wheels. One of the problems that James Watt faced in manufacturing his steam engines was boring large holes in an engine block. The difficulty in manufacturing large cylinders with tightly fitted pistons was solved by John Wilkinson, an ironmaster in Staffordshire. In 1774, he had invented a machine, powered by water wheels, for boring cylinders with extreme accuracy. He used this method for boring canons. This machining technology greatly improved the quality of the Boulton–Watt steam engine. The steam engine, coupled with the development of electric motors during the 1800s, led to better and more accurate technology for cutting metal.

These early milling and boring machines led to stepwise improvements in metal-cutting devices, such as milling and boring machines capable of high-cutting speeds and quality surface finishes.

Manufacturing Drawings

Communication of the design is done through group discussions, oral presentations, and mechanical drawings. It is critical that mechanical drawings are accurate and explain every dimension, including tolerances.

The geometric configuration of a design, including all subsystems and components, is established following the procedure discussed earlier. In many cases, components, such as bearings, motors, and springs, can be obtained from numerous vendors. On the other hand, many design components must be fabricated. The overall assembly usually contains a combination of both. Design information is communicated to manufacturing through fabrication drawings, which convey specific information about dimensions, tolerances, and surface finish as well as how the various parts are to be assembled and fastened together. Fabrication drawings may be a part of a broader engineering report giving back-up material, engineering analysis, quality control, quality testing, and feedback process.

Dimensioning

Dimensioning standards have been established to clarify and unify how components are to be communicated from design to manufacturing. Dimensions on a drawing define the linear size of a part in terms of length, width, and thickness and various cutouts that shape the part. The American National Standards Institute (ANSI) defines dimensioning as "a numerical value expressed in appropriate units of measure and indicated on a drawing and in other documents along with lines, symbols, and notes to define the size or geometric characteristic of a part or part feature." Proper dimensioning conveys the intent of the designer.

Dimensioning practices are established by ANSI and emphasized by ASME [15] and other authors [16–18]. Dimensioning practices are summarized below.

1) A dimension line is a thin line that shows the extent of a dimension.
2) Extension lines provide end limits for dimension lines. They should start about 1/16 in. from the object and extend past the last dimension line about 1/8 in.
3) Dimension lines can be stacked with the first-dimension line about 3/8 in. from the object and the remaining lines ¼ in. apart. Overall, dimensions are placed outside smaller dimensions.
4) Dimension lines are typically not placed inside object lines.
5) The dimension value is placed at a break in the dimension line. Numbers are always read from the bottom.
6) Dimension numbers in the US are expressed in inches. There is no need to follow the numbers by the (") symbol. For decimal numbers, less than 1 in., the zero is omitted, for example, 0.25 in.
7) Dimension lines should be lighter than object lines in the drawing but visible. Use narrow arrow heads.
8) Dimensions should not be given to hidden lines, if possible.
9) Place dimension lines on the view that best conveys the shape or contour of the part to the machinist.

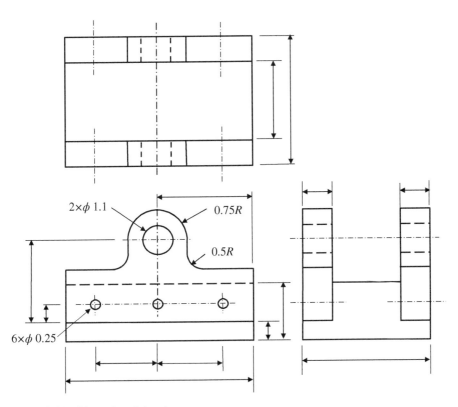

Figure 2.64 Dimensioned drawing.

10) Critical dimensions should give tolerances for that dimension.
11) Never have a dimension line cross another dimension line.
12) Dimension lines are drawn from center lines where necessary. Never use a center line as a dimension line.
13) Always give the diameter of a hole, not its radius. The symbol, , precedes hole dimensions. For example, 0.75, indicates a hole diameter of ¾ in. Drill sizes are expressed in decimals.
14) The symbol, *R*, should be placed after the value. For example, a filet radius of ⅛ in. is indicated by 0.125*R*.

The drawing in Figure 2.64 illustrates a few of these dimensioning principles.

Tolerances

Design may require components to operate with relative motion in areas of contact. The quality of both geometries could affect friction and simply how parts fit together. The performance of slider bearings and journal bearings depends on clearance between the moving parts. Assembly of sub-parts may also be affected by clearances.

It is impossible for a part to be manufactured exactly to a prescribed dimension. This could be due to machine tool flexibility, tool face interaction with work piece, interface dynamics, thermal expansion, etc. It is therefore necessary to define, in the drawing, acceptable variations from the

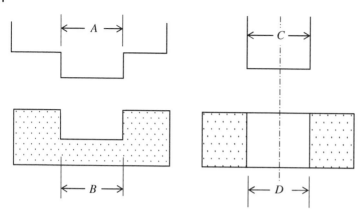

Figure 2.65 Situations requiring dimensions with tolerance.

base dimension. This allowable variation is call tolerance. Large tolerances may affect functionality. Small tolerances will affect cost thru precise manufacturing and parts inspection (and rejection).

Acceptable tolerances depend on the functionality of the mating surfaces (Figure 2.65). They control the manufacturing process, control variation between mating parts, and allow parts to be interchangeable. Smaller tolerances increase manufacturing cost exponentially.

The ANSI standard [16–18] defines *tolerance* as "the total amount by which a specific dimension is permitted to vary. The tolerance is the difference between the maximum and minimum limits." Tolerance on a dimension can be expressed in three ways.

Bilateral tolerance – Tolerance is expressed by using allowable variation in both plus (+) and minus (−) directions, for example, 1.125 ± 0.005. In this case, the upper limit is reached when the measured value is 1.13. The lower limit is reached when the measured value is 1.25. These two limits must allow the proper function between mating parts, such as a sliding guide or a shaft and collar fit.

Unilateral tolerance – In this case, a tolerance is allowed in one direction, either plus or minus. For example

$$A = 2.5\,{}^{0.005}_{0} \quad \text{or} \quad A = 2.5\,{}^{0}_{-0.005}$$

Limits of size – A dimension in this case is set between to limits, such as

$$A = {}^{2.505}_{2.495}$$

Three Types of Fits

The fit between mating parts, as illustrated in Figure 2.65, can be classified into three types:

a) Clearance fit – One part easily fits into another part with a clearance gap. The shaft is always smaller than the hole.

$$C = {}^{0.595}_{0.593} \quad D = {}^{0.600}_{0.602}$$

Tolerance on shaft is 0.002
Tolerance on hole is 0.002
Minimum clearance is $0.600 - 595 = 0.005$
Maximum clearance is $0.602 - 0.593 = 0.009$ in.
Tightest fit is 0.005 in.

b) Force (interference) fit – One part is force fitted into the other.

$$C = \frac{0.503}{0.502} \qquad D = \frac{0.500}{0.501}$$

Tolerance on shaft is 0.001
Tolerance on hole is 0.001
Minimum clearance is $0.500 - 503 = -0.003$
Tightest fit is 0.003 in. interference
Maximum clearance is $0.501 - 0.502 = -0.001$ (loosest fit)

c) Transition fit – The loosest case provides clearance fit and the tightest case gives an interference fit.

$$C = \frac{0.507}{0.502} \qquad D = \frac{0.500}{0.505}$$

Tolerance on shaft is 0.005 in.
Tolerance on hole is 0.005 in.
Minimum clearance is $0.500 - 0.507 = -0.007$
Tightest fit is 0.007 in. interference
Maximum clearance is $0.505 - 0.002 = 0.503$ in.
Loosest fit is 0.003 in. clearance
Used where accuracy is important but either a clearance or interference is permitted.

Surface Finishes

Surface finish can affect the performance of certain aspects of a design. Mechanical friction and the performance of nearly every type of bearings (as will be shown later) depend directly on surface finish. Rolling contact bearings (ball and cylindrical) require a surface finish of 20 µin. for long-term performance. Slider and journal bearings require a surface finish of 0.003–0.005 in. for useful service, too.

The texture of a machine surface can be important for several reasons. The finish affects the appearance of the part, stress concentration, friction, and bonding. Surface finishes are quantified in terms of surface roughness, the average of the surface undulations with respect to the nominal surface plane. Typical surface roughness for the three machining operations mentioned above are:

Turning 0.5–6 µm (15–250 µin.)
Drilling 1.5–6 µm (60–250 µin.)
Milling 1.0–6 µm (30–250 µin.)

Per profilometer measurements, the rms values of highly polished surfaces are about 0.5–1 millionth of an inch (0.0127–0.254 µm). The peak-to-valley distance is even higher.

Nanosurface Undulations

Surface undulations (Figure 2.66) have been manufactured at the nanolevel using special tools. These undulations have amplitudes of ± 5 nm with a wavelength of ~200 nm. They were manufactured in a silicon surface over a $5\,\mu m \times 5\,\mu m$ area.

In considering nanosize dimensions, remember

Nanoscale – atomic to 100 nm
Microscale – 100 nm to $500\,\mu m$
Macroscale – $500\,\mu m$ and above
$1\,\text{Å} = 0.1$ nm
1 nm $= 10^{-9}$ m
1000 nm $= 1\,\mu m = 10^{-6}$ m

Measurements of these undulations (Figure 2.67) show and quantify the geometry of the manufactured surface pattern [19]. Keep in mind that the range of atomic attractions is less than a nanometer (see Figures 11.18 and 11.19). Friction between contacting nanoundulations is probably due to a combination of molecular and mechanical effects.

Figure 2.66 Manufactured undulations in a silicon surface. Source: Courtesy of Oak Ridge National Laboratory.

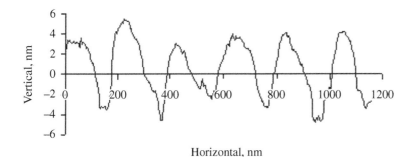

Figure 2.67 Image of surface topography. Source: Courtesy of Oak Ridge National Laboratory.

Machining Tools

Three principal machining processes are turning, drilling, and milling.

Lathes

Lathes are used to machine circular parts and drill holes. In this case, the work piece rotates and the cutting tool is stationary and fixed to a platform that can be translated longitudinally and transversely. Some of the cutting operations are:

- Facing
- Taper turning
- Contour turning
- Form turning
- Chamfering
- Cutoff
- Threading (internal and external)
- Boring (enlarging an existing hole)
- Drilling
- Knurling

These operations are conducted manually.

Drill Press

Several operations are made with drill presses. A hole is typically drill first, then it can be modified by:

- Reaming
- Tapping
- Counter boring
- Counter sinking
- Centering
- Spot facing

Each modification requires a special cutter head. In this operation, the cutter head rotates, and the work piece is stationary.

Milling Machines

These machines use a spindle, which rotates a cutting tool against a translating work piece. The rotating cutting tool can move vertically in z-direction. The work piece can move horizontally in either x or y direction. It is manually operated.

Milling can be peripheral (cutting takes place on the side of the cutting tool) or face (cutting takes place on the end of the cutting tool). In either case, the main purpose of milling is to produce a flat surface on the work piece.

A few operations of peripheral milling are:

- Slab milling
- Slotting
- Side milling
- Straddle milling

A few operations of face milling are:

- Conventional face milling
- Partial face milling
- End milling
- Profile milling
- Pocket milling
- Surface contouring

Machining Centers

Machine centers are milling machines capable of conducting many automated sequential operations, including tool changes. The spindle rotates the cutting tool and can move vertically as well as laterally. The work piece can move horizontally in both x and y directions. The motions of the cutting tool and work piece are controlled by a computer code (CNC). Once set up, these centers are fast and accurate in producing a single part with complex geometries.

Turning Centers

Turning centers operate similar to milling centers except that the work piece turns at a given cutting speed. The cutting tool attachment is fixed but can rotate as in a terete to change cutters. Programmed computer codes control tool interchange.

Turn centers can turn a cylindrical surface, facing the surface, and drilling a transverse hole. Computer-controlled machines produce accurately machined parts consistently and at a high rate. They can machine complex geometries faster than can be done by hand-operated machines. Setup time should be considered, but this cost is often offset by reduction of cost per part, when large quantities are to be made.

References

1 Ressler, S. (2011). Understanding the World's Greatest Structures. DVD Video. The Great Courses.

2 Timoshenko, S. (1955). *Strength of Materials (Parts I and II)*, 3e. NY: D. Van Nostrand Co. Inc.

3 Hibbler, R.C. (2011). *Mechanics of Materials*, 8e. Prentice Hall.

4 (1989). *Manual of Steel Construction*, 9e. NY: American Institute of Steel Construction.

5 Timoshenko, S.P. (1953). *History of Strength of Materials*. New York: McGraw-Hill.

6 Shigley, J.E. and Mitchell, L.D. (1983). *Mechanical Engineering Design*. McGraw-Hill Book Co.

7 Seely, F.B. and Smith, J.O. (1956). *Advanced Mechanics of Materials*, 2e. New York: McGraw-Hill.

8 Timoshenko, S. and Young, D.H. (1968). *Elements of Strength of Materials*. D. Van Nostrand Co. Inc. (see p 330).

9 Griffith, A.A. (1920). The phenomena of rupture and flow in solids. *Philos. Trans. Royal Soc.* 221A: 163.

10 Moody, L.F. (1944). Friction factors for pipe flow. *Trans. ASME* 66: 671–684.

11 Den Hartog, J.P. (1956). *Mechanical Vibrations*, 4e. New York: McGraw-Hill.

12 Dareing, D.W. and Livesay, B.J. (1968). Longitudinal and angular drill-string vibration with damping. *Trans. ASME J. Eng. Ind.* 90, Series B (4): 671–679.

13 Dareing, D.W. (1984). Drill collar length is a major factor in vibration control. *J. Pet. Technol.*: 637–644.

14 Den Hartog, J.P. (1952). *Advanced Strength of Materials*, 297. McGraw-Hill Book Company.

15 ASME Y14.5 and Y14.100 (2018). Dimensioning, Tolerancing, and Engineering Drawing Practice Package.

16 Puncochar, D.E. (2011). *Interpretation of Geometric Dimensioning and Tolerancing*, 3e. Industrial Press.

17 Lowell W. Foster; *Geometric Dimensioning and Tolerancing Per ASME Y14.5*, Self-Published, 2019.

18 Drake, P.J. (1999). *Dimensioning and Tolerancing Handbook*. McGraw-Hill Education.

19 Binning, G. and Rohrer, H. (1985). Scanning tunneling microscopy. *Physica* 127B: 37–45.

Part II

Power Generation, Transmission, Consumption

Every machine has three functional aspects: (i) Power supply, (ii) power transmission and (iii) end use. Machines are designed to perform to a given set of specifications as discussed in Part I. During the process of achieving an end use, energy is consumed as indicated in the drawing.

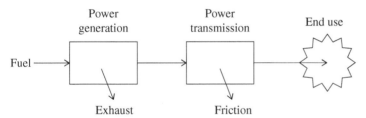

There are several options for a power source (electric motors, gasoline engines, diesel engines, gas turbines). Output performance of each dictates which is best for a given application. Power is transmitted by means of any one of several mechanisms (gears, pulleys, linkages, power screw, hydraulics) to achieve a specified end effect. Part of input energy will be lost to friction or other inefficiencies. Part II covers these aspects of machine design. The oil well drilling rig and its five subsystems is used to illustrate this process.

Engineering Practice with Oilfield and Drilling Applications, First Edition. Donald W. Dareing.
© 2022 John Wiley & Sons, Inc. Published 2022 by John Wiley & Sons, Inc.

3

Power Generation

Water Wheels

The water wheel was used as a source of power for many years. The evolution of the water wheel was driven by population growth and the need for greater food output. The Egyptian Nora (~700 BC) was used to lift water for irrigation. The Romans milled grain during the fourth century AD and at the time of William the Conqueror, England had about 5000 grist mills. By 1790, there were about 2000 grist mills in colonel America. By the time of the Civil War, there were some 55 000 water wheels in use, many powered manufacturing facilities. Power generated by the water wheel (~10 hp) was transmitted through gear trains to achieve a required output torque and speed. Grist mills typically have a gear ratio of 25 : 1 with the millstone having the higher speed and lower torque. Wooden gear teeth were common.

Early water wheels were constructed by empirical methods which evolved and were improved with experience (Figure 3.1). They worked and that is what mattered, first as a pump and a later as a milling machine. Performance of water wheels is explained in terms of current engineering mechanic.

Fluid Mechanics of Water Wheels

With reference to Figure 3.2 and Bernoulli's energy equation:

$$y_1 = y_2 + E_m \tag{3.1}$$

where

y_1	–	elevation of water entering the wheel, ft
y_2	–	elevation of water leaving the wheel, ft
E_m	–	mechanical energy imparted to the wheel shaft, ft or (ft-lb per lb of fluid)

The energy equation reduces to

$$E_m = H(\text{ft}) = H\left(\frac{\text{ft-lb}}{\text{lb}}\right) \tag{3.2}$$

where H is the drop in elevation.

The power equation is

$$P = H\left(\frac{\text{fl-lb}}{\text{lb}}\right)Q\left(\frac{\text{ft}^3}{\text{s}}\right)\gamma\left(\frac{\text{lb}}{\text{ft}^3}\right) = HQ\gamma\left(\frac{\text{ft-lb}}{\text{s}}\right) \tag{3.3}$$

Engineering Practice with Oilfield and Drilling Applications, First Edition. Donald W. Dareing.
© 2022 John Wiley & Sons, Inc. Published 2022 by John Wiley & Sons, Inc.

Figure 3.1 Lenoir Museum water wheel (Norris, Tennessee).

Expressed in terms of horsepower

$$P = \frac{HQ\gamma}{550}\, \text{hp}$$

Assuming

$Q = 2\,\text{ft}^3/\text{s}$
$\gamma = 62.4\,\text{lb/ft}^3$
$H = 12\,\text{ft}$

Then power delivered by the water wheel is

$$P = \frac{12(62.4)2}{550} = 2.7\,\text{hp (assuming 100\% hydraulic efficiency)}$$

In terms of rotary speed (rpm) and torque (ft-lb)

$$P = \frac{TN}{5252}\, \text{hp}$$

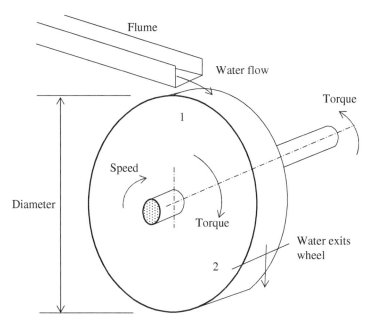

Figure 3.2 Water wheel mechanics.

If the wheel turns a $N = 1$ revolution per 8 seconds or 7.5 rpm, then output torque is

$$T = \frac{2.7(5252)}{7.5} = 1891 \text{ft} - \text{lb} \ (N = 7.5 \text{ rpm})$$

Power output is adjusted by flow rate, Q, with a side gate in the flume.

Steam Engines

The steam engines of Newcomen and Watt were based on steam vacuum, created by reducing the temperature of steam trapped in an enclosed volume. These early engines were named "atmospheric engines" because atmospheric pressure pushed the piston into a vacuum. James Watt was aware of the advantages of using positive steam pressure but was concerned about the capability of boilers to withstand positive pressure. With time, especially during the application of steam engines to river boats, positive steam pressure became common. Steam pressure level was kept somewhat below 200 psi to minimize the risk of explosion.

Early steam engines were limited in their ability to contain high steam pressure because of questionable quality of design, manufacturing, and materials. However, to increase power output, steam pressure had to be elevated to much higher levels. The fear of boiler explosions was a concern, especially when steam engines were used on steamboats. This concern was realized in the Sultana steamboat disaster of 1865.

The American Society of Mechanical Engineers (ASME) was founded in 1880 and began to address some of these issues. The first uniform code for testing boilers was presented in 1884. The code evolved over the next several years and was eventually published in 1915. It included all aspects of design, construction, and testing of boiler and pressure vessels. These standards are found in the current ASME Boiler and Pressure Vessel Code, which has been adapted across North America and in 60 countries around the world. As a result, steam boilers and pressure vessels in power stations safely operate near 1000 psia.

Steam Locomotives

Steam in locomotive engines reach pressures of 200 psi and temperatures near 800 °F. Water boils at 382 °F at this pressure (200 psia). Heat is added at constant temperature to the water/steam combination until all the water has been turned to steam. Beyond this point, the steam becomes super-saturated and the temperature of the steam increases.

The bulk of a steam-driven locomotive is the boiler. This is where the fuel, usually coal, is burned and steam is generated. Steam energy within the boiler is conveyed to the piston/cylinder chamber to generate the driving force. This force is converted to torque on the wheels through a slider crank arrangement. The steam chamber is relatively small compared to the boiler, but this is where mechanical power is generated and transmitted (Figure 3.3). Energy is transmitted to a wheel by means of a slider crank mechanism, which converts linear motion to rotational motion.

A steam locomotive includes a pump, a furnace and boiler, steam cylinder and a condenser. The furnace and boiler occupy most of the volume of the total *steam engine*. This is where fuel (coal) is burned and the heat gets transferred into the water/steam, which is moving through a long session of pipes (boiler). Steam comes out of the boiler and enters a *steam cylinder*, a relatively small device near the wheels.

In 1880, Ephraim Shay built a prototype of a geared steam engine with steam cylinders oriented vertically. Reciprocating motion of the piston is converted to rotary motion by a slider crank arrangement. This orientation allowed the use of bevel gears to increase torque to the wheels. Two bevel gears are used. The larger bevel gear is attached to the axial of the train wheels. The smaller gear is attached to the crankshaft (Figure 3.4). This gear arrangement reduced speed but increased torque to the wheels. This design feature allowed the train to pull a heavier load, which

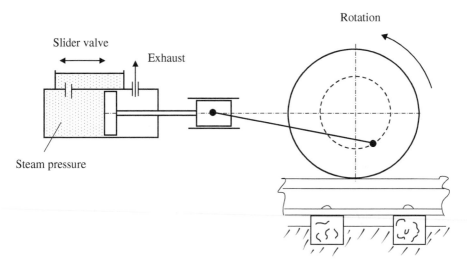

Figure 3.3 Steam cylinder and transfer mechanism.

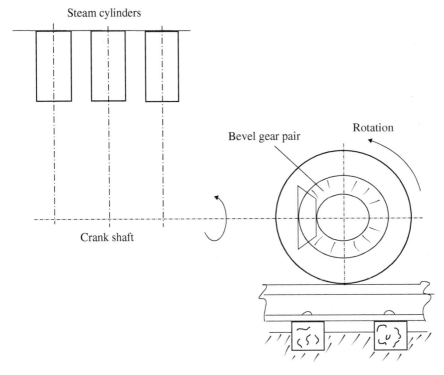

Figure 3.4 Shay locomotive power system.

was especially important in removing logs in a mountainous terrain. Another feature of this engine is the use of three steam cylinders side by side.

Shay engines were used by the Little River Railroad Lumber Company in timbering the Smoky Mountains near Townsend, Tennessee during the early 1900s (Figure 3.5). Note the steam loader in the rear car.

The miracle of the steam engine (and steam turbine) lies in the transformation of raw fuel (wood, coal, oil, uranium) to mechanical torque and speed. Steam is the linkage that allows this to happen. Steam is generated external to the cylinder.

Output power from the various electric motors or mechanical engines is delivered in terms of torque and speed. Power is expressed in units of horsepower or Watts. In horsepower units

$$P = \frac{2\pi TN}{550(60)} = \frac{2\pi TN}{33\ 000} = \frac{TN}{5252} \ \text{hp} \tag{3.4}$$

where

P	–	power, hp
T	–	torque, ft-lb
N	–	rotational speed, rpm
1 hp	–	550 ft-lb/s

In SI units,

$$P(watts) = T(N-m)\omega(\text{rad/s}) \tag{3.5}$$

$$P(watts) = T\omega(\text{Joules/s})$$

Figure 3.5 Shay engine designed for steep terrain. *Source:* Courtesy of Little River Railroad and Lumber Company Museum, Townsend, Tennessee.

where

P	–	power, W
T	–	torque, N-m
ω	–	rotational speed, rad/s
1 hp	–	745.7 W

Engine performance, which is determined experimentally, varies with torque and speed. Engines are rated at maximum output conditions. Actual operating output may be different than the rated output.

Power Units in Isolated Locations

Electrical power is sometimes needed in remote places away from stationary power plants. Examples are drilling operations, cruises ships, and locomotives. In each case, diesel engines drive electric generators which produce electrical power. Electrical power is then delivered through electric cables to various points of operation.

For example, in early drilling operations power was taken directly from internal combustion engines (ICE) and transmitted through chain and gear drives to an end use. This meant the ICE was close to operations often creating safety issues. Currently, ICEs generate electric power through generators. In recent years, power units are located away from operations, and electric power is delivered through cables, which are simple, reliable, safer, and cheaper. Isolated power units of this type are found on drilling rigs, cruise lines, and diesel locomotives.

The amount of deliverable power (kilowatts) is limited by space. Typically, portable power units are mounted on skids for easy set up and positioning. The engine room in cruise ships contains diesel engines and electric generators, while diesel engines and electric generator occupy the front car in railroad locomotives.

Regional Power Stations

Regional power stations are fixed and usually located near rivers. Electrical output is measured in terms of megawatts. Output from local power plants supplies electricity regionally and can be input to a national power grid. The source of power is steam. Since space is usually not a problem, solid fuels, such as coal, natural gas, and uranium are options.

Physical Properties of Steam

Thermodynamic properties of steam have been tabulated in steam tables [1]. The state of the mixture of water and steam is illustrated in the following figure. The bell curve shows where water begins to boil and where steam is completely saturated (Figure 3.6).

For example, consider a certain volume of water at room temperature (70 °F) and atmospheric pressure (14.7 psia). When heat is added to water, temperature begins to increase, and the volume expands until the water starts to boil. Water has reached its saturation temperature or boiling temperature of 212 °F ($p = 14.7$ psia). If heat is continually added at a constant pressure, more and more steam is formed, until all the water is converted to steam (saturated vapor). This expansion takes place without an increase in temperature. Beyond the saturated vapor point, steam temperature increases with added heat and becomes superheated steam.

If the heat is added at a higher pressure, say 100 psia, the saturation temperature increases. On the other hand, if the pressure is lower than 14.7 psia, the saturation temperature reduces;

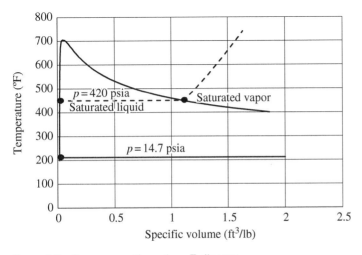

Figure 3.6 Steam properties – (p, v, T) diagram.

therefore, it is easier to boil water at higher elevations. Desalination systems boil seawater in a vacuum to distill fresh water at a lower boiling temperature.

Energy Extraction from Steam

Liquid/steam changes take place continuously throughout the steam cycle of a steam engine. The mechanical components of a steam engine, such as, boiler, piston/cylinder (or turbine), condenser, and pump control the state of the steam. The whole process is one of energy conversions, starting with combustion of raw fuel (coal), and ultimately converting heat into mechanical work through steam.

First Law of Thermodynamics – Enthalpy

For a non-flow system, the first law is expressed as

$$Q - \frac{W}{J} = u_2 - u_1 \tag{3.6}$$

The energy equation for an open flow system is

$$Q + u_1 + \frac{p_1 v_1}{J} + \frac{V_1^2}{2gJ} + \frac{z_1}{J} = \frac{W}{J} + u_2 + \frac{p_2 v_2}{J} + \frac{V_2^2}{2gJ} + \frac{z_2}{J} \tag{3.7}$$

Enthalpy is a state variable that includes internal energy and pressure and is useful in tracking energy conversions throughout a thermodynamic process. It is a component of the energy equation, a formulation of the first law.

Ignoring the elevation change and combining internal energy and pressure energy gives

$$Q + h_1 + \frac{V_1^2}{2gJ} = \frac{W}{J} + h_2 + \frac{V_2^2}{2gJ} \tag{3.8}$$

where
Q – heat supplied between station 1 and 2 Btu/lb
$h = u + \frac{pv}{J}$ Btu/lb (enthalpy, a state variable)
W – shaft work leaving apparatus, ft-lb/lb
J – 778 ft-lb/Btu
V – velocity, fps
g – acceleration due to gravity, 32.2 ft/s^2

Entropy – Second Law

Entropy is a state variable used to determine the amount of heat transferred into and out of a thermodynamic process. Since heat and work are related, entropy is used to determine energy taken from heat engines.

For a reversible process [2]:

$$\sum \frac{dQ}{T} = \text{Constant} \tag{3.9}$$

The constant is defined as the difference between the entropy between points 2 and 1 in the process:

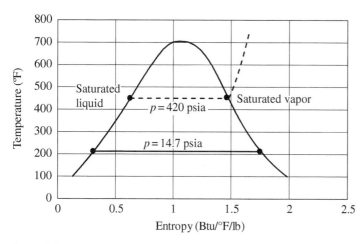

Figure 3.7 Entropy chart for steam.

$$s_2 - s_1 = \int_1^2 \frac{dQ}{T} \tag{3.10}$$

Expressing steam properties in terms of (T, p, s), allows heat flow in a thermodynamic process to be determined by

$$Q_{1-2} = \int_1^2 T\,ds \tag{3.11}$$

This is a huge advantage in determining heat flow into and out of a steam cycle. The state of steam expressed in terms of temperature and entropy is shown in Figure 3.7.

The efficiency of a thermodynamic steam cycle becomes

$$\text{efficency} = \frac{Q_{in} - Q_{out}}{Q_{in}} \tag{3.12}$$

Thermodynamics of Heat Engines

A thermodynamic cycle is one in which water/steam is taken through various physical states and returned to its original state. An example is a thermodynamic cycle (Rankin heat cycle) shown in Figure 3.8. The key mechanical components in this cycle are pump, boiler, turbine, and condenser. Its efficiency is determined by output work divided by heat going into the cycle. The letters on the rectangular flow diagram indicate the phase of the water/steam as depicted on the above steam properties diagram. Unlike the steam locomotive, water is recycled. This process requires a con- denser, which lowers pressure and temperature and converts steam to water for pumping. In an ideal condenser, pressure and temperature remain constant.

Utility power stations are based on a thermodynamic cycle; they are typically located near rivers. Water from the rivers cool the condensers and remove heat from the steam.

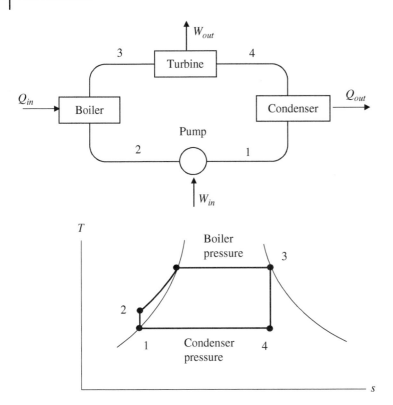

Figure 3.8 Rankins heat cycle with *T–s* diagram.

This process is, perhaps, the only remaining external combustion engine in use today. The external combustion generates steam, which drives steam turbines. This process allows the use of variety of fuels, such as coal, natural gas, and uranium.

Utility power stations are usually located near rivers, which serve as heat sinks for James Watt's condensers. The thermodynamic cycle in this case is the same regardless of the fuel source (coal, oil, or uranium). Coal and oil generated heat by combustion. Nuclear power stations generate heat by nuclear fission, caused by splitting atoms. Fission produces a huge amount of heat that is used to heat water/steam. Combustion is not involved.

The engines or power units are steam turbines. Steam is expanded through these turbines to develop mechanical power in terms of torque and speed. This mechanical power drives an electric generator, which supplies electric power to the cities. The energy conversions are raw fuel, steam heat, mechanical turbine power, and electric power. Once the electrical power gets to your home it goes through other power transformations depending on household needs.

These liquid/steam changes are taking place continuously throughout the steam cycle. The basic mechanical components of a steam power plant (boiler, turbine, condenser, and pump) control the state of the steam. The whole process is one of energy conversions, starting with combustion of raw fuel (coal), and ultimately converting heat into mechanical work through steam.

Mechanical output comes off the steam turbine shaft in the form of torque and rotary speed. Typical output power is 50 MW at about 3500 rpm.

Steam Turbines

Steam turbines are mechanically simpler than steam engines. Steam turbines simply rotate, while the motions of piston, connecting rod and crankshaft are much more complicated. The rotor moves in response to momentum changes of the steam through its blades. Turbine shaft speed ranges between 3000 and 15 000 rpm with an output power of 20–100 MW. Typical operating conditions are 500 psia and 300 °C (572 °F).

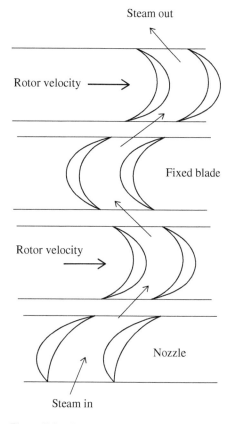

Figure 3.9 Principle of the steam turbine (isentropic process).

The physical arrangement of steam turbines is given in Figure 3.9. Properties of steam continuously change as energy is removed from the steam as manifested in pressure drop, temperature, and density. Steam density directly affects momentum as steam moves across each moving turbine blade. The thermodynamic process of steam through the turbines is assumed to be isentropic.

Steam turbines generate power from steam, which inter turbines under very high pressure (~1500 psia). High steam pressure is converted into velocity by means of nozzles. Nozzles can be converging-diverging type (Laval nozzle) or simply fixed blades which direct the pressurized steam onto rotor blades. To create a gradual extraction of energy from the steam, multiple stages of rotors are used to bring steam down from high pressure to a lower pressure. Pressure reduction is stepped down over each stage in a controlled and calculated manner until the entire steam pressure has been used. This staged reduction in pressure means that there is a progressive increase in volume requiring gradual increase in turbine blade diameter.

Mechanical power delivered at the output shaft of steam turbines is generated by the change of momentum change across the rotor blades [3]. Stator blades or fixed blades redirect steam onto the rotor blades. Steam engines of the type used on locomotives rely on steam pressure to push against a reciprocating piston. Rotary motion is achieved by a slider crank mechanism, made up of a connecting rod, piston, and crankshaft. Dynamic forces in turbines are the result of rotary motion, while dynamic forces in reciprocating engines are caused by the combination of translation and rotation of moving parts. Therefore, the steam turbine can operate at much higher rotary speeds than reciprocating engines.

The main driver of the turbine is energy from steam, which enters the turbine at temperatures around 1000 °F and at pressures around 900 psig. By comparison, these operating temperatures and pressures are much higher than those (near atmospheric pressure and 220 °F) used in the early steam engines by James Watt. Steam turbines are the primary generators of electric power.

Large electric generators used by the electric utility industry today have a stationary conductor. A permanent magnet is attached to a rotating shaft and positioned inside a stationary conducting

(a)

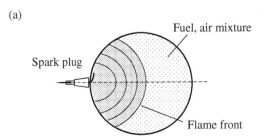

Spark plug

Fuel, air mixture

Flame front

Figure 3.10 Combustion and pressure cycle of otto ICE engines.

(b)

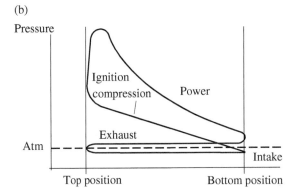

Pressure

Ignition
compression   Power

Exhaust

Atm

Intake

Top position   Bottom position

coil made up of a long, continuous piece of a wire conductor. When the magnet rotates, it generates a small electric current in each section of wire. Every small current of individual sections adds up to one current of considerable size. This is the basis of the generation of electric power whether its electricity comes from a steam power utility plant or in a hydroelectric power station.

Electric Motors

Power transfer from engines was initially made by mechanical devices, such as chain drives, pulleys, and gears. This often created a hazardous working environment. In recent years, engine power is transformed into electrically. Electrical power is very transportable and safe. As a result, electric motors are used throughout industry.

During the early 1830s, Michael Faraday made an important discovery while studying magnetism. Faraday discovered that electric current and an electrical potential are generated between the ends of a conductor when the conductor is moved perpendicular across magnetic lines. His discovery laid the foundation for modern-day electrical power used in homes and industry today. The electric generator is a vital link in the energy chain. Electricity is very transportable and much safer than mechanical linkages. Electric power can be applied locally to machinery or household equipment and is easily conveyed throughout a city or country.

Following this discovery, Faraday built the first electromagnetic generator, which produced a small direct current (DC) voltage and a large amount of current. At first, water wheels drove electric generators. During the second half of the 1800s, electric generators were driven by steam engines. By the turn of the century, steam turbines became the choice of prime movers for electric generators and still are today. The steam turbine can produce high torque and horsepower because of the simplicity of motion.

There are two general types of electric motors: alternating current (AC) and DC. AC electric motors are divided further into single phase and three phase motors.

Single-phase motors are adequate for applications requiring up to about 5 hp. They draw more current than three-phase motors, thus making three-phase motors a more efficient choice for industrial applications. Single-phase motors are typically found in home products. AC motors are simple in design, low cost, reliable, and come in numerous sizes and performance characteristics.

Three-phase motors require smaller wiring and less voltage making them safer and less expensive to operate. Three phase motors are lighter in weight and more efficient than comparable single-phase motors. More power is supplied to them than to a single-phase motor over the same period. For this reason, they are commonly used in industry and manufacturing. However, three-phase motors and controls are more complex and expensive.

There are three types of DC motors: brush motors, brushless motors, and stepper motors. Brush motors are the most common because they are easy to build and cost effective. Carbon brushes, which are used to transfer electric current to the rotating part, wear over time. DC brushless motors overcome this wear problem but are costlier and require complicated drive electronics to operate. In applications where speed and torque need to be controlled with high accuracy, brushed DC motors are good choices. If higher performance and reliability are required, brushless DC motors may be preferred.

Selection of the right motor depends on application. There are multiple choices; however, the right choice has to match design requirements. In many cases, motor power is delivered at high speed and low torque.

Internal Combustion Engines

The difference in the steam engine and ICEs is in the location of the combustion of the fuel or energy source. Combustion of fuel to energize steam engines, occur external to the piston/cylinder chamber, while combustion for gasoline engines takes place inside the cylinder. The internal combustion process allows for a much higher power to weight ratio in engines. The external combustion engine allows for the use of bulk fuels, such as coal, wood, and uranium.

Four Stroke Engine

The first ICE was built and tested by Nicolaus Otto (German) in 1876. The basic cycle of his engine (called the Otto Cycle Engine) has four strokes: air intake, fuel injection and compression, spark ignition and expansion of burned gas, and exhaust. The injected fuel was liquid (gasoline), which was another product of the rock oil produced by Rockefeller. The Otto Cycle Engine (internal combustion) was the first alternative to the steam engine, an external combustion engine.

The internal combustion has a much lower weight to power ratio (~2 lb/hp) than do steam engines and is more adaptable to automobiles and other vehicles. They use petroleum fuels having higher energy content than coal. Thermal energy is released and converted into mechanical energy during the combustion process. The liquid fuel is mixed with air in a carburetor to form a combustible mixture, which is compressed by the piston within the cylinder before being ignited by an electric spark [4]. The explosion and expansion of the air-fuel mixture moves the piston downward-generating torque through a connecting rod and crankshaft (Figure 3.10).

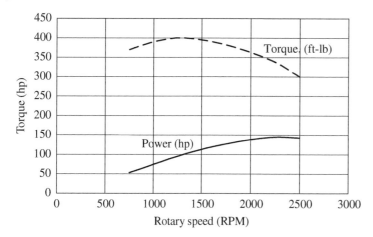

Figure 3.11 Performance diagram of six cylinder SI engine.

In 1885, Gottlieb Daimler and Wilhelm Mayback improved on the Otto engine. Their patented engine is recognized as the foundation for the modern-day gasoline engine. The same year they adapted their engine to a stagecoach and produced the first four-wheel automobile. Also, in 1885, Karl Benz designed and built the first complete automobile. By 1900 Benz & Company became the world's largest maker of cars.

Several pistons and cylinders are arranged to provide uniform torque. Proper balancing of each group of cylinders is also necessary for smooth operation of the engine.

Figure 3.11 gives the performance data for a spark ignition engine (4⅜ in. diameter, 5¾ in. stroke) having a compression ratio of 5.5 : 1. Tests were made with a wide-open throttle. The reduction of output torque at high speeds is due to internal friction. Gear transmissions are used to increase torque at low speed.

In a gasoline engine, the fuel–air mixture, which is premixed in a carburetor, is drawn into a cylinder, then, compressed to a ratio ranging from 4 : 1 to 10 : 1 before being ignited by a spark (Spark Ignition, SI). The spark is sufficient to produce smooth combustion and thus the engine operates smoothly. Gasoline engines require a lighter engine frame, making it preferable for light vehicles and other mobile applications. High compression ratios are not possible because of pre-ignition.

Two Stroke Engines

The two-stroke engine accomplishes these same four events over two strokes instead of four. Starting with the downward or first stroke, ignited gasoline (the power stroke) expands, and drives the piston downward. The piston reaches a position where an exhaust valve is exposed and allows the expended gas to escape. Still moving downward, the piston reaches another opening where a mixture of fuel and air is blown into the cylinder by a scavenging fan, removing any remaining exhaust gases while filling the cylinder with a new gas mixture.

When the piston moves upward (beginning of the second stroke), the intake valve closes and subsequently the piston blocks the exhaust opening. Over the remaining part of the second stroke, the gas mixture is compressed and ignited at the top position by means of a spark plug.

Alternatively, a crankcase-scavenged two stroke engine does not have a scavenging fan. The crankcase is sealed to allow the intake fuel mixture to be compresses by the bottom part of the piston. Once the fuel mixture is trapped, it is compressed within the crankcase by the downward travel

(a)

(b)

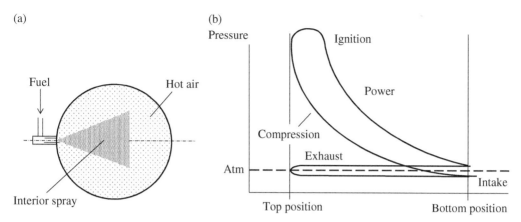

Figure 3.12 Combustion and pressure cycle of diesel engines.

of the piston. When the piston moves upward to the intake valve, the gas mixture is forced into the cylinder for further compression and ignition. The power stroke (downward stroke) is the same as the first engine configuration. The crankcase scavenger replaces the fan scavenger.

Diesel Engines

The diesel engine carries the name of its inventor and developer, Rudolf Diesel (German). His idea (1890) was to achieve a much higher compression than the Otto engine such that fuel ignition is accomplished by temperature caused by the higher compression. The four strokes are the same, except fuel is injected at the peak of compression, not during the compression cycle.

In a diesel engine, air is drawn into the cylinder alone and then compressed to a ratio ranging from 14 : 1 to 25 : 1. The air at these compression ratios reach temperatures ranging between 700 and 900 °C (compression ignition, CI). Since diesel fuels do not mix with air, they are injected through an atomizer into the highly compressed air where mixing takes place [4]. Fuel droplets must be mixed with heated air before ignition (Figure 3.12). There is a time lag between fuel injection and ignition measured terms of fractions of a second. Combustion could begin anywhere in the combustion chamber; it is an uncontrolled process.

Because of the high compression ratio and uncontrollable combustion, diesel engines require a more rugged frame.

Engines with high compression ratios are more efficient, meaning that diesel engines are more efficient than gasoline engine. High compression means higher torque especially at low speeds. Because of this diesel engines are used in heavy industries, such as trucking, shipping, construction, oil drilling, and production. A key advantage in the diesel engine or gasoline engine is the low maintenance cost.

Gas Turbine Engines

In some ways, gas turbines are similar steam turbines in that energy is extract from high temperature and high-pressure gases to rotate stages of turbine blades. Gases are caused to impinge on rotors to bring about mechanical power in the form of torque and speed. Gas turbines have three key components: (i) compressor, (ii) combustion chamber, and (iii) turbine. The compressor brings in outside air and compresses it to a pressure of about 50–75 psia. The compressed air inters a combustion chamber, where liquid fuel is injected and ignited (~650 °C). This mixture then passed through a turbine at high velocity and generates mechanical power (torque and rotational speed) [5]. This power is often used to drive electric generators. In this configuration, the turbine, compressor, and generator are mounted on one shaft. Part of the power generated by the turbine is used to drive the compressor and part to drive the electric generator.

John Barber, an Englishman, developed the technical basis for the current gas turbine engine around 1790. He invented and patented a gas turbine engine in 1791. It was not until 1903 that Ægidius Elling, a Norwegian, built a gas turbine that generated more power (11 hp) than needed to run the engine. Sir Frank Whittle later used his work to design and build the first jet engine for aircraft propulsion (1930). The first successful use of his jet engine was in April 1937.

An axial gas pressure gradient toward the exhaust end of the engine creates the momentum change between the gas entering and leaving the engine. This pressure gradient is amplified in the compression chamber by combustion of the fuel. Per Newton's third law, there is an equal and opposite force applied to the engine structure which pushed the airplane forward.

Gas turbines are typically used to power aircraft and in this application, they are rated by how much thrust they produce. Velocity to lift an aircraft is a key factor. Per to the law of linear impulse-momentum:

$$T\Delta t = m(v_2 - v_1) \tag{3.13}$$

Starting from rest at the end of a runway, the thrust required to reach a lift-off velocity is

$$v_2 = \frac{\Delta t}{m} T \tag{3.14}$$

Gas turbines are also used with propellers to form turbo-prop engines. In this case, the propeller is attached to the power shaft in front of the compressor. Power is also taken off the main shaft to drive mechanical devices, such as pumps.

Basically, gas turbine engines pull in air and discharge a heavier gaseous mixture made up of air and burnt jet fuel at high velocities (Figure 3.13). The result is a thrust force somewhat like the force propelling a pressurized toy balloon that has been released. The theory behind the thrust is based on momentum change of the gases coming into and out of the engine.

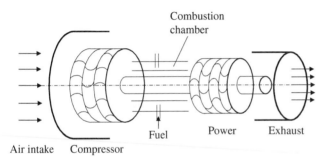

Figure 3.13 Components of gas turbine engines.

The forward thrust to the jet housing is equal and opposite of the resultant force, F, applied to the gases inside the engine (Newton's third law). The higher the mass flow rate and exit velocity the greater the thrust. To elevate both requires higher combustion rates, which create higher gas temperatures within these engines.

The basic activities inside the engine are air intake, compression by the entry turbine blades, injection of jet fuel with simultaneous ignitions, and escape of the gas mixture at high velocities. The temperature in the combustion chamber rises to about 3 500 °F in the hottest part of the flame. These hot gases pass through the power turbine and impart power to the main shaft, driving the compressor blades up front. The compression power is about 60% of total power leaving the compression/combustion chamber. The remaining power is used to create the thrust of propulsion or torque on a separate drive shaft.

The compressor turbine blades and the power turbine blades near the exit are on the same shaft and rotate at the same revolutions per minute. Rotary speeds are in the range of 20 000 rpm. By comparison typical rotary speeds of automobile engines are 2 000 rpm.

Gas turbines are also used to power electric generators using mechanical power off the main turbine shaft. Gear trains reduce shaft speed and increase output torque.

The Boeing Company developed the first commercial jet airplane in the early 1950s. The Boeing 707 was certified for commercial service in 1958. This airplane was the first of subsequent 7 × 7 model airplanes and ushered in modern commercial jet travel.

Jet engines have greatly improved over the past 50 years. The goal of higher and higher thrust to weight ratios, drives this technology.

General Electric recently presented their new high thrust jet engine (90115 B) capable of generating thrusts forces between 115 000 and 126 000 lb[1] [6]. For comparison, this level of thrust is equivalent to three 727 airplanes using all three engines. This new engine is so powerful that just one of them could fly a Boeing 747 jetliner. The power capability of this engine is over 100 000 hp a quantum leap in power from James Watt's first steam engine.

Impulse/Momentum

The thrust produced by a gas turbine is determined from impulse–momentum. Consider the control volume in Figure 3.14

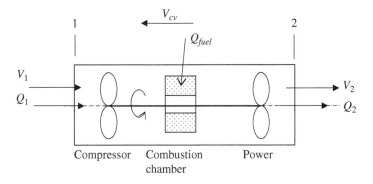

Figure 3.14 Control volume within a gas turbine.

Assuming the gas in the exit nozzle expands to ambient pressure and subsonic flight and the inlet pressure is ambient, total internal forces applied to the gases within the control volume is

$$\sum \overline{F}_i = \overline{V}_2 Q_2 - \overline{V}_1 Q_1 \tag{3.15}$$

where the mass flow rate of the exhaust is

$$Q_2 = Q_1 + Q_{fuel}$$

Since the mass flow rate of the fuel is small by comparison with air flow, it is neglected. Also

$$V_1 = v_i + V_{CV}$$

$$V_2 = v_e + V_{CV}$$

where

V_{CV}	–	velocity of control volume (engine)
V_1	–	absolute velocity of air entering the control volume
V_2	–	absolute velocity of air leaving the control volume
v_i	–	intake velocity relative to control volume
v_e	–	exhaust velocity relative to control volume

By substitution

$$F = Q(v_e - v_i) \text{ (total force on gas in control volume)} \tag{3.16}$$

where $v_i = V_{CV}$ and thrust is opposite to F [5].

$$\text{Thrust} = Q(v_e - V_{CV})$$

Energy Considerations

The power generated by the turbine near the exit is determined from the energy equation. This energy is converted to torque and rotation speed of the shaft which becomes the drive shaft for power take-off:

$$Q_{fuel} + h_1 + \frac{V_1^2}{2gJ} = \frac{W_{power}}{J} + h_2 + \frac{V_2^2}{2gJ} \tag{3.17}$$

The energy process through gas turbines is usually assumed to be isentropic.

Engine Configurations

Turboprop engines use power off the power shaft to rotate a propeller in front of the gas turbine engine. An initial application of the gas turbine is in jet aircraft propulsion. Gas turbines can also be

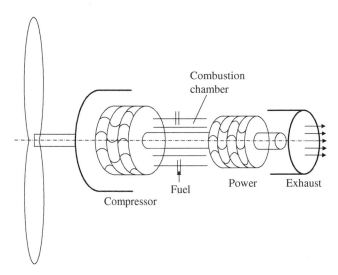

Figure 3.15 Turboprop engine configuration.

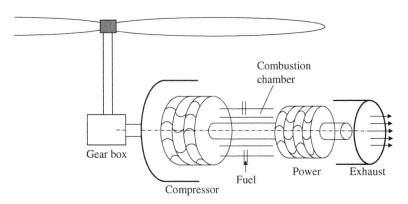

Figure 3.16 Helicopter configuration.

configured to produce torque on the main shaft for driving propellers (Figure 3.15). The turboprop engine is a slight variation of the thruster concept. The turboprop engine is used in commercial aircraft and in helicopters (Figure 3.16). This same configuration is also used to power electric generators.

Gas turbines are often used for pumping oil or gas from offshore production facilities, as the turbine can be set to operate at optimum impeller pump flow rate speeds (Figure 3.17). Linkage between the power drive and impeller shaft is a direct connection.

A gas turbine is started with an electric motor which caused the compressor to create pressure within the combustion chamber. Fuel is brought in to start the turbine, which in turn begins to power the compressor.

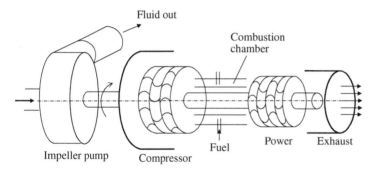

Figure 3.17 Impeller pump configuration.

Rocket Engines

The power system includes two solid busters and pivotal main engines that use liquid hydrogen as a fuel. Liquid oxygen accelerated the combustion process in these engines. Thrust is like the jet propulsion of airplanes. In both cases, thrust is created by momentum changes into and out a combustion chamber. With the jet engine, air is brought into the combustion chamber from the environment, compressed and combined with ignited fuel to increase the momentum of the exhaust. With the space rocket, fuel is ignited by liquid oxygen, which is carried with the engine during and after lift-off. Since the magnitude of thrust depends on rate of change of momentum, higher burning rates mean higher thrusts. Higher burning rates also mean higher temperatures.

Rocketdyne F-1 Engine

High-speed turbo pumps are essential to a successful lift off. In the case of the main engine of the space shuttle, the turbo pumps are used to delivering liquid hydrogen and oxygen to the combustion chamber at extremely high rates (Figure 3.18). To do this, the turbo pumps must rotate at speeds near 80 000 rpm. These speeds usher in a whole host of mechanical issues from friction to vibrations. This engine is considered the most powerful engine every made.

Rocket and jet engines were just immerging as new weapons during World War II. But there were peaceful uses too. Space scientist, Werner von Braun, and his Redstone Arsenal team were already at work, when on 4 October 1957, the Soviet Union successfully launched Sputnik I.

On 31 January 1958, the United States launched Explorer I, which carried a scientific instrument that led to the discovery of the James van Allen magnetic radiation belts. In July 1958, Congress passed the Space Act, which established the National Aeronautics and Space Administration (NASA). The space race was on, and Congress was ready to pour money into research and education.

Atlas Booster Engine

Figure 3.19 is a photo of an Atlas Booster Engine, one of three engines used to boost the Atlas Missile. It had a thrust of 165 000 lb with a 130 seconds burn time. The first Atlas launch was 19 July

Figure 3.18 Rocketdyne F-1 engine. *Source:* NASA.

Figure 3.19 Atlas F series booster engine. *Source:* Courtesy of Cape Canaveral Air Force Station Museum.

1958. These engines were made by Rocketdyne. These and other rocket engines are mounted on an adjustable linkage (center of photo) which allow for navigation control.

Gas Dynamics Within Rocket Engines

Thrust produced by a rocket engine is determined from impulse momentum principles. Following the procedure outlined above, total impulse required to bring about total change in momentum of all gas particles across the control volume is (See Figure 3.20)

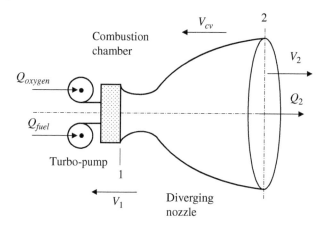

Figure 3.20 de Laval nozzle.

$$\text{Thrust} = Q_{fuel}v_e + (p_e - p_a)A_e \tag{3.18}$$

where

Q_{fuel}	–	total mass flow rate of oxidizer and fuel
v_e	–	velocity of exhaust relative to engine
p_e	–	exhaust pressure
p_a	–	ambient pressure

The de Laval nozzle expands and accelerates combustion gases causing exhaust gases to exit the nozzle at hypersonic velocities.

A measure of performance of these engines is a parameter called Specific Impulse having units of "pounds of thrust per pound of propellant burning per second." The Specific Impulse of hydrogen/oxygen rocket engines (~415) is as high as anything we know (Albert C. Martin, personal communication). Basic to these engines is the storage of the fuel at cryogenic temperatures and the turbo pump which delivered the fuel at rotary speeds up to 80 000 rpm. Testing before each launch was critical to the reliability of these engines. Lyle C. Bjorn, manager of testing for North American-Rockwell at Cape Canaveral, managed some 400 engineers and technicians in preparing these engines for each launch. In addition to fuel storage at cryogenic temperatures, special attention was given to turbo pump especially, the contact bearings wear at these high speeds and vibration modes in the turbine blades. There were no launch failures during the Saturn/Apollo project during the 1960s.

The total weight (dead weight) of the Saturn V rocket was about 6.5 million lb. The first stage contained five (5) Rocketdyne engines each capable of generating 1.5 million lb of thrust. These engines were fueled by jet fuel which was oxidized by liquid oxygen. The second stage also contained five (5) Rocketdyne engines each capable of generating 2 million lb of thrust. These engines were fueled by liquid hydrogen which was oxidized by liquid oxygen.

Rocket Dynamics

The equation of motion a rocket for rockets ignoring drag is

$$T = Q_{fuel}v_e \tag{3.19}$$

where

Q_{fuel}	–	rate at which mass is being exhausted from nozzle
v_e	–	velocity of exhaust relative to rocket

Newton's second law becomes [7]

$$\frac{dm}{dt}v_e = M\frac{dV}{dt} \tag{3.20}$$

where

M	–	mass of rocket
V	–	velocity of rocket

The mass of the rocket is continuously changing as fuel is burned.

$$M(t) = M_0 - \mu t \tag{3.21}$$

$$\mu = \frac{dM}{dt}$$

$$\frac{dM}{dt} = \frac{dm}{dt} = \mu$$

Combining everything gives

$$(M_0 - \mu t)\frac{dV}{dt} = \mu v_e \tag{3.22}$$

The first stage launched the rocket down range about 500 mi and achieved an altitude of about 45 mi and a speed of 7000 mph. The second stage launched the remaining part of the rocket down range another 500 mi to an altitude of 80 mi and a speed of 17 000 mph; great enough to put the command module in earth orbit.

It is noteworthy that the Apollo 11 (1969) occurred 200 years after James Watt patented his steam engine (1769). This launch also occurred 300 years after Sir Isaac Newton developed calculus and established his laws of dynamics. Both are essential to propel, navigate, and monitor space flight.

Energy Consumption in US

In conclusion, consider the dependents of the US economy on fossils fuels. Figure 3.21 shows sources and amount of energy consumed in the United States since 1800.[2] Coal was the main energy source for smelting iron and making steel during the eighteenth and nineteenth centuries. Even during the twentieth century, coal remained a major fuel source for steam power plants.

Oil overcame coal as a major energy source during the mid-twentieth century due mainly to the rapid growth in automotive and trucking industries. Around 1950 steam locomotives through my small hometown of Miami, Oklahoma disappeared. They were replaced by diesel engines. At this time oil began to outpace coal as indicated.

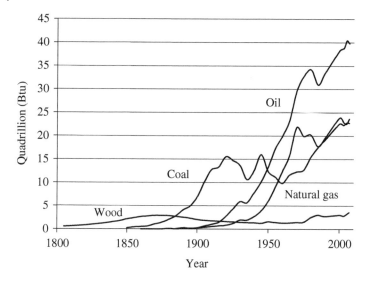

Figure 3.21 Energy consumption in the US since 1800. *Source:* Data from Energy Consumption in the US by Source, 1800-2000 (Quadrillion Btu), U.S. Energy Information Administration.

The petroleum industry responded during the 1960s with research in drilling and production at the same time the space program was also growing. Oil exploration and production expanded into remote areas, especially offshore. Two technologies accelerated this expansion.

1) Geophysics greatly improved seismic mapping of promising formations through digital computers. Improvement came in quality and speed of mapping especially offshore.
2) Directional drilling allowed accurate navigation through multiple formations to reach complex reservoir structures deep within the earth.

Lateral horizontal drilling into production zones along with hydraulic fracking greatly increased oil production and oil recovery.

Industry also responded to the growing population by expanding production of goods and services. The technology for engines and electric motors was in place along with mechanisms and controls to improve manufacturing.

Currently nearly all electricity is generated by steam turbines. The technology and thermo sciences are fully developed to optimize power output. Steam is the linkage between fuel (such as coal and uranium) and electric power.

Unfortunately, both external and internal combustion engines produce carbon dioxide which threatens the earth's environment and climate. New energy sources will be needed to supplement and perhaps eventually replace fossil fuels.

Solar Energy

Photovoltaic solar energy cells convert sunlight directly into electricity through the use of solar induced electron–hole pairs. These solar cells are non-mechanical and are usually made from silicon alloys. Photons carried by sunlight contain various amounts of energy corresponding to the different wavelengths of the solar spectrum. Electrons are dislodged from the semiconductor's atoms and migrate to one surface, when solar energy is adsorbed. This migration of electrons, which are negatively charged, leaves positive charges behind creating a voltage potential between positive and negative surfaces. When these two surfaces (positive and negative) are connected to electrical conductors, electricity flows. One cell produces about 1–2 W of power. Many photovoltaic cells are typically linked together to generate power levels great enough to run most applications.

Hydrogen as a Fuel

Hydrogen fuel cells produce electrical power by bringing hydrogen and oxygen together in chemical reaction. The process produces water and electricity. The mechanical power in hydrogen-fueled vehicles is derived from this process. The only exhaust is water and heat. The process involves no combustion or carbon exhaust and therefore is environmentally friendly. These energy cells are efficient but expensive to build.

There is, however, a price to pay for the hydrogen. Hydrogen does not occur naturally in the atmosphere. It must be separated from heavier molecules, such as water. Energy is required to bring about this separation through the process of electrolysis. However, if energy is produced by hydroelectric power stations, wind, or solar, hydrogen production and its use as an energy carrier is environmentally friendly. Hydrogen has been called the perfect energy carrier.

Hydroelectric Power

Hydroelectric energy is generated from turbine-driven electric generators. The turbines are driven by water flowing from a high level to a lower level. The difference in elevation represents the energy potential to do work on the turbine. If the water is at the same level on both sides of the dam, there is no energy potential, and the water can do no work on the turbine.

The elevation difference of the water across a dam is maintained by rain up stream. Rain is caused by the sun, so the indirect energy from the sun is driving the turbine. Water wheels and ocean tidal wave entrapment are also based on elevation potential.

Wind Turbines

There is enormous energy in hurricanes and tornadoes, and we have all witnessed their destructive power. We have yet to harness this energy for constructive purposes. In fact, the potential of wind-generated energy is much greater than the current consumption of energy worldwide. This is theoretical of course. The energy still has to be captured. It is interesting to know that there is much energy potential all around us in the form of wind and solar. There has to be a means of economically capturing this energy for useful purposes.

Geothermal Energy

The earth's crust is about 5 mi thick. Beyond that the rock becomes extremely hot. The hot core of the earth is residual heat left from when it was formed some five billion years ago. There is plenty of thermal energy within this core. So far the process of making this energy useful relies on drilling into this very hot formation and pumping water into the hot rock. Supersaturated steam is produce out of an adjacent well bore.

The closest distances to hot rock are in the caldera of old volcanoes. Some oil companies have been very successful in producing thermal energy this way, especially on the west coast.

Atomic Energy

Atomic energy refers to the generation of high levels of heat through one of two types of reactions, fission, and fusion.

Nuclear fission refers to a process of splitting atoms into small fragments. Heat is given off as a source of energy, which is used in atomic power plants, to generate steam. The steam cycle to generate mechanical and electrical power is the same as for other types of fuels, such as coal and oil. This process is also the basis of an atomic bomb. The use of nuclear fission for peaceful or destructive purposes depends on the rate of the atomic splitting process. In atomic power plants, the splitting process is highly controlled.

Nuclear fusion is the process of fusing two atomic nuclei together to form a heavier nucleus. Heat energy is also generated in this process.

While there are no toxic gases exhausted to the atmosphere, the by-products of nuclear fission are highly radioactive, creating a nuclear waste problem.

Biofuels

Biofuels are derived from near term "organisms." Unlike fossil fuels, which have taken millions of years to form, biomass has about a two gestation period. It is derived typically from plants such as corn, soybeans, wheat, sugar beets, and sugar cane. A common type of biofuels is ethanol. It is made through fermentation of certain biomasses such as sugar cane and corn. It must be blended with standard fuels up to a maximum of around 10% to operate in standard cars. Engines have to be modified to greater concentrations of biofuels with fossil fuels. Some cars are being manufactured to run on any combination of fossil fuel and ethanol, even 100% ethanol.

Even though carbon dioxide is produced during combustion, biofuels are carbon neutral. This means that carbon dioxide produced during combustion is adsorbed by new plant growth at the same rate, and there is no net increase in carbon dioxide levels. Biofuels can help reduce the world's dependence on fossil fuel and carbon dioxide production.

Notes

1 www.boeing.com/commercial/787family/background.
2 http://www.eia.doe.gov/emeu/aer/eh/intro.html.

References

1 Keenan, J.H. and Keyes, F.B. (1938). *Thermodynamic Properties of Steam*, 1e. Wiley.
2 Stoever, H.J. (1953). *Essentials of Engineering Thermodynamics*. Wiley.
3 (1977). *How Things Work*. Granada Publishing Limited.
4 Obert, E.F. (ed.) (1950). Internal combustion engines. In: *Analysis and Practice*, 2e. Scranton, PA: International Textbook Company.
5 John, J.E.A. and Haberman, W.L. (1988). *Introduction to Fluid Mechanics*. Englewood Cliffs, NJ: Prentice Hall.
6 Eisenstein, P. (2004). Biggest jet engine. *Popular Mechanics*.
7 Housner, G.W. and Hudson, D.E. (1959). *Applied Mechanics – Dynamics*, 2e. Princeton, NJ: D. Van Nostrand Co.

4

Power Transmission

Throughout antiquity stone was the primary building material. Stone was used to build structures, such as the Pyramids of Misa (2600 BC), Mycenae (1600 BC), Parthenon (600 BC), and the famous arches of Rome. Considering the density of stone (~160 lb/ft^3), a 3-ft cube of stone weighs

$$W = 160(27) = 4320 \, \text{lb}$$

or about 2 tons. This is the approximate weight of each stone used to build the pyramids.

Archimedes (287 to 212 BC) identified three devices from antiquity that were used to transmit force and motion. These devices were: (i) lever, (ii) pulley, and (iii) the screw. Later, three other devices were added: (iv) wedge, (v) inclined plane, and (vi) wheel and axle (see Figure 4.1).

These devices were used to remove, shape, and transport large blocks of stones from quarries to a construction site. The wedge was a significant tool used by early builders to remove cubic shaped stones. The sides were cut with chisels. Cracks underneath the stone's base were created by wedges and large hammers. Heavy stones were lifted in place by human effort using these devices. Each represents a means for transmitting force to move huge blocks of stone or other objects.

The wheel and axle were used for battle and transport of goods. Wheels can be attached to the axle in two ways: rigidly fixed to the axle or freewheeling on bearings. The latter case eliminates skidding on turns.

Even today, we find these devices used in machinery for transmitting force and power.

Settlers, during early days, made fences of split rails created by splitting logs with wooden wedges and a wooden mallet. The wedge equation

$$N = \frac{F}{2(\mu \cos \theta + \sin \theta)} \tag{4.1}$$

shows that if the half angle of the wedge is, say, 10° and the coefficient of friction is 0.3 then

$$N = 1.066F$$

Engineering Practice with Oilfield and Drilling Applications, First Edition. Donald W. Dareing.
© 2022 John Wiley & Sons, Inc. Published 2022 by John Wiley & Sons, Inc.

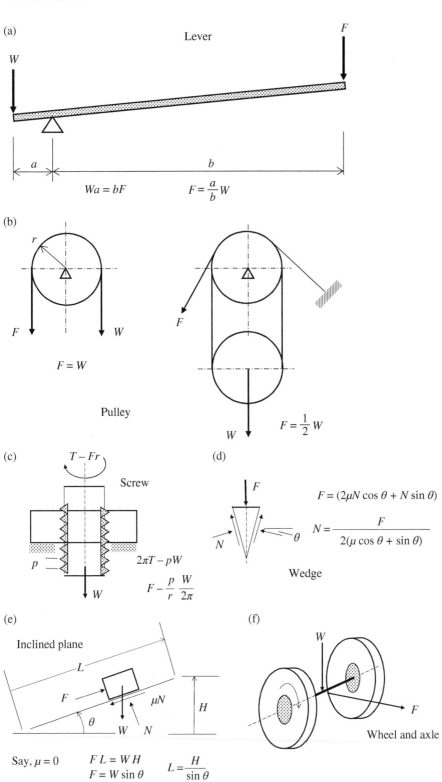

Figure 4.1 (a–f) Devices of antiquity.

Figure 4.2 Horse collar – a major breakthrough. *Source:* Courtesy of Little River Railroad and Lumber Company Museum, Townsend, Tennessee.

In this example, the wedge generates an opening force, N, having a magnitude approximately equal to the impact force, F, caused by the hammer.

For many years, oxen and horses were used to pull plows and wagons. Force generated by oxen was transferred through a yoke. Earlier attempts to harness the horse included throat-girth harness and breast-collar harness. The throat-girth harness (used as far back as ancient Chaldea, third Millennium BC) restricted breathing of the horse. It was not improved until the breast-strap harness was developed in China (481 to 221 BC). In this harness, a strap or cinch was place around the belly of the horse from which load was applied.

A breakthrough came with the horse collar, which puts pressure on the sternum and transmits the pull force through the skeletal system of the horse without restricting breathing. The horse collar was developed in China during fifth century AD and greatly improved the capability of the horse. Studies show that a horse with a collar can apply 50% more force than an ox and at greater speed. The horse collar is essentially unchanged since its inception.

Both horse and ox pull along straight lines and lack maneuverability, a limitation in the transmission of animal power. Figure 4.2 shows two horses skidding a log into position for loading onto a rail car. Force is transmitted to the log by means of horse collar and chain.

Gear Train Transmission

Water Wheel Transmission

In engineering practice, power is transformed into torque and speed to accomplish an end use. This is done through a gear transmission or a pulley drive. Early designers of water wheels used gear reduction drives (~1–25) to increase rotary speed of the millstones or saw blade.

Gears convert power from one set of torque and speed to another set. Often rotary speed is reduced, and torque increased across a gear pair. This section gives basic features of gears and reviews considerations in choosing gear tooth size for specific applications.

The shape of early gear teeth was not critical to the overall performance of a machine, such as the water wheel, because gear speeds were low. Gear teeth were often made of wood having a tapered flat surface of engagement. Gear teeth were made of apple wood, which is less likely to split during operation. The flat contacting surface smooths out with use and performed quite well under water wheel loads and speed.

As machinery began to run at higher and higher speeds, gear tooth profile became important. An important feature of gear tooth geometry is that speed ratio between mating gears remains constant. If not, the output speed of the driven gear would vary with time and become a source of machine vibration.

Fundamental Gear Tooth Law

For the speed ratio to be constant, gear tooth profiles must follow the fundamental gear tooth law, which states that the common normal to two mating surfaces must pass through the line of centers at a fixed point, called the pitch point. This is essential to assure angular accelerations are zero throughout the contact region. The proof of this law is well documented [1].

The speed ratio of two mating gears is expressed as

$$\frac{\omega_3}{\omega_2} = \frac{r_2}{r_3} \tag{4.2}$$

where r_2 and r_3 are the radii of the pitch circles for the two gears.

Many tooth profiles can satisfy the fundamental gear tooth law, but most have operational and manufacturing limitations. Three profiles that have found practical application over the years are involute [2], cycloid [1], and Wildhaber and Novikov [3, 4] profiles. Involute gears are by far the most used and a brief summary is given below.

Other gear tooth geometries are possible provided they satisfy the fundamental gear tooth law [5]. Almost any geometry can be arbitrarily selected for one of the matching tooth pair. The challenge is to find the tooth geometry that will mesh with the assumed profile while satisfying the fundamental gear tooth law.

Involute Gear Features

The involute profile gets its name from how it is generated. An involute curve is generated by tracing the path of the end of a tight string as it unwinds from a disc (Figure 4.3). The tip of the string scribes an involute. Visualize a string wrapped around the base circle of gears 2 and 3. As the string is unwound from the base circle on gear 2 onto the base circle on gear 3, any point (such as a knot) would scribe involute curves on both gear pair. The normal to contacting tooth pairs is tangent to both base circles and intersects the line of centers at the pitch point. The fundamental gear tooth law is, therefore, satisfied. The radius of each base circle depends on the angle of the line of action.

$$r_B = r \cos \phi \tag{4.3}$$

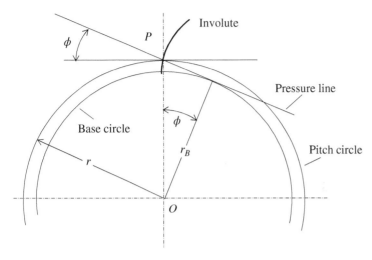

Figure 4.3 Involute gear tooth profile.

where r is the pitch radius and ϕ is the pressure angle. Pressure angles usually range between 20° and 25°, even though 14½° was once used. The pitch circle is different from the base circle. Its radius is the distance to the pitch point from gear center.

By definition, diametral pitch (P) of a gear set is the number of teeth per inch of diameter of the pitch circle.

$$P = \frac{N}{d} \tag{4.4}$$

where d is diameter of the pitch circle and N is number of teeth on the gear.

It is a measure of the size of gear teeth in a gear set. For example, if a diametral pitch of 2 is specified for a gear pair for which the speed ratio is 1 : 2 and the pitch diameter of the pinion is 12 in., then the number of teeth on the pinion is

$$N_p = P\, d_p$$
$$N_p = 2(12) = 24 \text{ teeth}$$

The pitch diameter of the gear is 24 in. and

$$N_g = 2d_g$$
$$N_g = 2(24) = 48 \text{ teeth}$$

The addendum circle defines the outer edge of gear teeth, while the dedendum circle defines the depth of tooth profile, both measured from the pitch circle.

The standard addendum and dedendum distances for interchangeable teeth are $\dfrac{1}{P}$ and $\dfrac{1.25}{P}$, respectively [1].

Circular pitch (p) is the arc length of a tooth plus tooth space.

$$p = \frac{\pi d}{N} = \frac{\pi}{P} \tag{4.5}$$

$$pP = \pi \tag{4.6}$$

Since p is the same for mating gears, P is the same.

The line of action, ϕ, is oriented from the pitch circle tangent line passing through the pitch point at P.

Gear Tooth Size – Spur Gears

Pitch diameters establish the velocity and torque ratios in gear pairs regardless of tooth size. Tooth size depends on force transmitted and geometric factors. Gear teeth must be large enough to transmit the maximum force while keeping bending stress within allowable limits. Wilfred Lewis [6] formulated a bending stress equation in terms of tooth form and load. His stress model is based on treating gear teeth as cantilever beams (Figure 4.4).

The transmitted forced, W, is assumed to occur at the initial engagement point and with no other tooth pairs in contact. Bending stress formula is developed below, using Lewis nomenclature.

$$\sigma = \frac{Mc}{I} = \frac{6W_t l}{Ft^2} \tag{4.7}$$

This formula has been modified to automatically account for standard gear tooth forms. The resulting Lewis formula thus is written

$$\sigma = \frac{W_t P}{K_v FY} \quad \text{(Lewis equation, } Y \text{ is the form factor)} \tag{4.8}$$

where

P	–	diameter pitch, teeth per diameter of pitch circle
W_t	–	tangent force
F	–	width of tooth
Y	–	form factor (tabulated by AGMA, also accounts for stress concentration)
K_v	–	dynamic factor

Various formulas are used to determine the dynamic factor. The Barth formula is stated below.

$$K_v = \frac{600}{600 + V} \quad (V \text{ is pitch line velocity in fpm}) \tag{4.9}$$

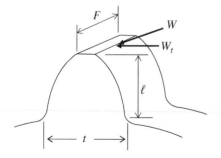

Figure 4.4 Lewis gear tooth model.

If gears are of high precision and there is no appreciable dynamic load, AGMA recommends $K_v = 1$.

The AGMA also propose a modified version of the Lewis equation.

$$\sigma = \frac{W_t P}{K_v F J} \tag{4.10}$$

The parameter, J, is a geometric factor, which includes a modified form factor Y, a fatigue stress concentration factor, K_f, and a load-sharing ratio, m_N. J factors are also tabulated for various gear forms.

Gear teeth are sized to transmit a specified force and maintain bending stress within specified limits. Two controlling parameters are tooth thickness, which relates to diametral pitch, P, and tooth width, F. As a rule, face width should be within the range $3p \leq F \leq 5p$. If face width is too long, the transmitted force may not be uniform across the tooth. This becomes a guide to be considered when determining cross-sectional geometry of gear teeth. In general, small gear teeth have a large diametral pitch and are used to transmit power at high speeds and low torque. Larger gear teeth are needed to transmit larger forces and thus have a smaller diametral pitch (low number of teeth per inch of the pitch circle diameter).

Example   Consider a simple speed reducer having the following design specifications.

Power transmitted – 50 hp
Input speed – 1000 rpm
Output speed – 500 rpm
Gear/Pinion material – UNS G10400 treated and drawn to 1000 °F
Gear teeth – 20° full depth
Safety factor–3

The challenge is to determine the pitch diameter of both gear and pinion, the diametral pitch, and tooth face width using the Lewis formula.

$$\sigma_a = \frac{W_t P}{K_v F Y} \tag{4.11}$$

The torque coming into the pinion is

$$\text{Power} = \frac{TN}{5252} \quad T = \frac{50(5252)}{1000} = 262.6 \ \text{ft-lb}$$

Following AGMA for coarse pitch, we choose $P = 10$ recommended for general use. Also, for 20° pressure line, the recommended minimum number of teeth is 18. Our choice for number of teeth for the pinion is 60.

$$P = \frac{N}{d} \quad d = \frac{N}{P} = \frac{60}{10} = 6 \ \text{in.} \quad \left(p = \frac{\pi}{P} = \frac{\pi}{10} = 0.314\,16 \ \text{in.} \right)$$

This fixes tooth size and diameter for the pinion. Based on this diameter,

$$W_t = \frac{262.6}{3} \left| \frac{12 \ \text{in.}}{1 \ \text{ft}} \right| = 1050 \ \text{lb} \tag{4.12}$$

Numerical values in the Lewis formula are

$$\sigma_a = \frac{\sigma_{yld}}{FS} = \frac{80\,000}{3} = 26\,667 \text{ psi}$$

$P = 10$ (arbitrarily chosen)

$W_t = 1050 \text{ lb}$

$Y = 0.494$

Using these numbers and $K_v = 1$

$$F = \frac{W_t P}{\sigma_a Y} = \frac{1050(10)}{26667(0.494)} = 0.797 \text{ in.(close to guideline)} \tag{4.13}$$

AGMA recommends face width, F, between $3p \leq F \leq 5p$.

Simple Gear Train

Kinematics

The diagram in Figure 4.5 shows pitch circles of a gear set. Assume the pinion (driving gear) is centered at O_1 and the driven gear is centered at O_2. From the left diagram, the tangent velocity of both circles at the contact point is the same giving

$$\omega_2 = \frac{r_1}{r_2} \omega_1 \tag{4.14}$$

From the right diagram, the tangent force (action reaction) is the same giving

$$T_2 = \frac{r_2}{r_1} T_1 \tag{4.15}$$

Since power in is $T_1 \omega_1$ and power out is $T_2 \omega_2$

$$P_{out} = T_2 \omega_2 = \frac{r_2}{r_1} T_1 \frac{r_1}{r_2} \omega_1 = T_1 \omega_1 \tag{4.16}$$

$$P_{out} = P_{in} \tag{4.17}$$

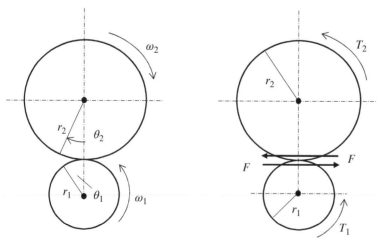

Figure 4.5 Gear pair.

Gear pairs of this type are often used with electric motors to reduce rotary speed and increase torque. Power is coming into the smaller gear and taken off the larger gear.

The starter motor and gear drive, found in every vehicle, are good examples of a simple gear train. DC electric motors deliver moderate torque at high speeds and are ideal for this application. The motor shaft, containing the pinion, drives the large gear, which elevates the torque and lowers the speed. The higher torque to the engine drive shaft is necessary to overcome friction in the engine and start-up inertia forces in each of the piston/connecting rods as well as the crank shaft. It is also needed to initiate the compression and ignition strokes.

Worm Gear Train

A worm gear drive is useful when a large torque is required on the pinion gear shaft. In this case, the worm gear rotates at a high speed and low torque, which is typical of electric motors. Figure 4.6 depicts a worm gear with input speed, ω_1, and torque, T_1. The torque demand, T_2, on the pinion gear is much greater than T_1.

The speed and torque relationships are determined as follows.

The pitch line velocity of the pinion gear is $V = r\,\omega_2$. The pitch line velocity is due to the advancement caused by the threads on the worm gear,

$$\frac{\Delta\theta_1}{2\pi} = \frac{\Delta x}{p} \tag{4.18}$$

where

$\Delta\theta_1$	– angular displacement of worm gear
Δx	– pitch line displacement of pinion

Dividing by Δt gives

$$V = \frac{p}{2\pi}\omega_1 \tag{4.19}$$

So

$$\omega_2 = \frac{p}{2\pi r}\omega_1 \tag{4.20}$$

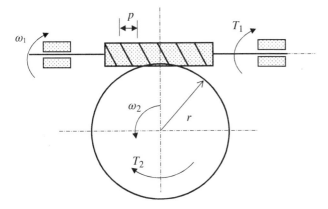

Figure 4.6 Worm gear drive.

Since the number of teeth on the pinion gear is $n = \dfrac{2\pi r}{p}$

$$\omega_2 = \frac{\omega_1}{n} \tag{4.21}$$

showing that the output speed of the gear is a small fraction of the input speed of the worm gear. Equating power-in to power-out gives

$$T_2 = \frac{\omega_1}{\omega_2} T_1 = nT_1 \tag{4.22}$$

Worm gears are useful in cases where input power is working against a relatively large torque. They are especially useful with electric motors, which produce power at high speed and low torque. Worm gear drives can achieve speed and torque ratios in the order of 20 : 1. The axes of rotation of the worm and gear are at 90° as shown. Worm gears can be self-locking, which is important in many cases. Musical instruments are a good example. They occupy a smaller space than compound gear trains.

Planetary Gear Trains

Planetary gear trains involve a gear (or gears) moving around another gear (or gears) as shown in Figure 4.7. The outside gears are planet gears. The gears rotating about a fixed axis are sun gears. This gear arrangement can produce a large step up (or down) in speed and torque.

Power can be supplied to planetary gear trains through gear a (link 2) and the rotating arm (link 5) separately or simultaneously. Both inputs affect output speed according to

$$\omega_4 = k_1 \, \omega_5 + k_2 \, \omega_2 \tag{4.23}$$

The output speed of link 4 or gear d is the superposition of two inputs: links 5 (arm) and 2 (gear a). Consider first the input from the arm (link 5).

The first gear constant (k_1) in Eq. (4.23) is determined as follows. In this case, assume link 2 is fixed. The driver becomes the arm (link 5). Assume the entire gear train is rotated clockwise, including the arm (link 5). The motion of links 5, 2, and 4 is the same as shown in Table 4.1. Next imagine that link 2 (gear a) rotates counterclockwise one turn back to its initial fixed position; arm 5 is fixed during this motion. The resulting motions of the three links (2, 5, 4) are shown in the second line. Adding each column gives the result of rotating the arm, link 5, one turn.

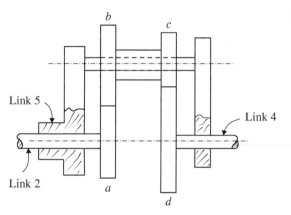

Figure 4.7   Planetary gear train.

Table 4.1 Relative motions in planetary gear trains.

Member	5	2	4
Motion with 5	1	1	1
Motion relative to 5	0	−1	$-\dfrac{ac}{bd}$
Total	1	0	$1 - \dfrac{ac}{bd}$

For one (1) rotation of the arm (link 5), gear 4 rotates

$$k_1 = \left(1 - \frac{ac}{bd}\right) \tag{4.24}$$

Fixing arm (link 5) and rotating gear a (link 2) gives

$$k_2 = \frac{ac}{bd} \tag{4.25}$$

Planetary gear trains reduce speed (increase torque) at high ratios and occupy less space than compound gear trains. They can be mounted in line with input and output shafts. They have better service life and higher stability than compound gear trains. Speed and torque ratios of 10 : 1 are common.

Compound Gear Trains

Now consider a typical application of a gear drive with input power being supplied by a 5 hp electric motor rated at a speed of 1000 rpm (Figure 4.8). Using these two numbers establishes the output torque of the motor.

$$P = \frac{2\pi TN}{550(60)} = \frac{2\pi TN}{33000} \text{ hp} \tag{4.26}$$

$$P = \frac{TN}{5252} \text{ hp}$$

where

P	–	hp
T	–	ft-lb
N	–	rpm

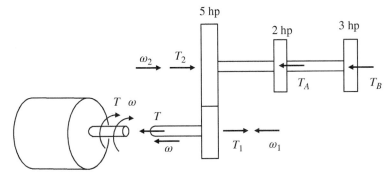

Figure 4.8 Gear train drive.

By substitution $T = 26.26$ ft-lb. Note the low-output torque and high-output speed of the electric motor.

Electric motors typically deliver power at high speed and low torque, so the first set of gears transform power to a higher torque and lower speed (gear 2). This is done with a $1 : 4$ gear pair ($T_2 = 4$ $T_1 = 105$ ft-lb and $N_2 = 250$ rpm). The transmitted power can be consumed in several ways, such as two tear drives A (2 hp) and B (3 hp). Gear 2 sets the rotary speed ($N_A = N_B = N_2 = 250$ rpm). The torque at gears A and B is determined from the power equation, $P = T\omega$. $T_2 = 105$ ft-lb and $N_2 = 250$ rpm

$$T_A = \frac{33\,000(2)}{2\pi 250} = 42 \text{ ft-lb}$$

$$T_B = \frac{33\,000(3)}{2\pi 250} = 63 \text{ ft-lb}$$

These torques drive something else.

The shaft section with the greatest torque is between gears 2 and A ($T = 105$ ft-lb). Assuming the allowable shear stress in the shafting is $\tau_a = 20\,000$ psi, shaft diameter is

$$\tau_a = \frac{Tr}{J} = \frac{2T_2}{\pi r^3} \tag{4.27}$$

$$r^3 = \frac{2(105)12}{\pi 20000} = 0.0401$$

$$r = 0.3423 \text{ in}$$

$$d = 0.6847 \text{ in}$$

Substituting and solving for r gives, $r = 0.3423$ in and a shaft diameter of $d = 0.6847$ in. The closest standard shaft size would be $d = 0.75$ in.

Pulley Drives

Rope and Friction Pulleys

Another means of transferring power is through pulleys. We start with a simple rope and friction pulley (Figure 4.9). This type of pulley is often found on fishing boats to pull in traps. The operator pulls on one end of the rope to create a larger pull-in force, F_0.

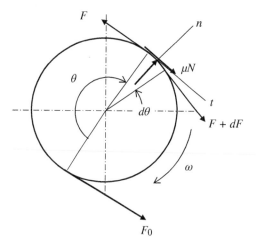

Figure 4.9 Rope and friction pulley.

Consider the freebody diagram of a differential section of the rope against a pulley.

$$\sum F_n = 0 \quad \sum F_t = 0 \tag{4.28}$$

$$N = (F + F + dF)\cos\frac{d\theta}{2} \quad \cos\frac{d\theta}{2}F = \mu N + (F + dF)\cos\frac{d\theta}{2}$$

$$N = F\,d\theta \quad dF = -\mu N \tag{4.29}$$

Combining gives

$$\frac{dF}{F} = -\mu\,d\theta \tag{4.30}$$

In this case, μ is the kinetic coefficient of friction as slippage is assumed between rope and pulley. By direct integration and applying the boundary condition at $\theta = 0$

$$F_0 = Fe^{\mu\theta} \quad (\theta = n2\pi + \theta_0) \tag{4.31}$$

The rope usually wraps around the pulley several times to create high pulling force, F_0. The operators end provides a back pull, which affects the magnitude of the pulling force.

If, for example, the required pull-in force is $F_0 = 500$ lb and assuming two complete wraps around a pulley, i.e. $\theta = 2(2\pi) = 4\pi$ and $\mu = 0.2$ then the operator applies

$$F = \frac{F_0}{e^{\mu\theta}} = \frac{500}{e^{0.2(4\pi)}} = 40.5\,\text{lb} \tag{4.32}$$

Power delivered to the incoming rope is $P_{out} = $ rope velocity times force in rope. Total power consumed by the drum is

$$P = P_{out} + \text{friction losses} \tag{4.33}$$

A cat head is an attachment (Figure 4.10) on a power shaft used to pull objects into a workspace. Such a device is used around drilling rigs to lift and pull pipe or other pieces of equipment onto the rig floor. Cat heads are also used to pull in and lift fishing traps onto boats. The shaft typically rotates at a low rotary speed (rpm). A rope is wrapped around the cat head as shown and pulled

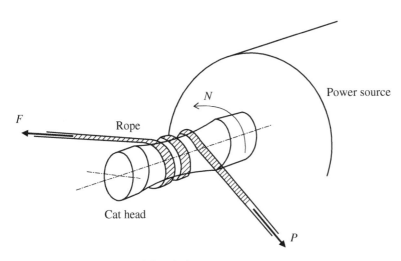

Figure 4.10 Pulling and lifting device.

with force F to move against a force P. A problem is to determine the magnitude of force, F to generate a force of $P = 300$ lb. Assume the rope makes two complete loops (plus 45°) round the cat head and the coefficient of kinetic friction between rope and pulley is $\mu = 0.2$.

$$F = 300e^{-(0.2)2.125(2\pi)} = 20.8 \text{ lb} \tag{4.34}$$

This example illustrates the unique advantage in the use of cathead pulleys on oil rigs and fishing boats.

Belted Connections Between Pulley Drives

Flexible connections, such as belts and chains, transmit power over a relatively long distance and allow power transmission between shafts through pulleys (Figure 4.11). They are simpler and cheaper than gear trains. Transmission ratios for speed and torque are the same as for gears except for the possibility of slippage between belt and pulley. Since belts are long and stretch, they have the added advantage of absorbing shock and damping torsional vibrations. Three common types of belts are:

1) Flat belts – Flat belts were initially made of oak tanned leather. Modern flat belts are made of a strong elastic core, such as steel or nylon, for strength in transmitting force (or torque). The core is coated with special coating to enhance friction, while maintaining belt flexibility. Belt drives are efficient and can be used over long distances between pulleys.

   The application of the previous equation to belt drives assumes slippage is impending, so coefficient of friction in this case is static coefficient of friction. To prevent slipping, the belt must be pretensioned. With this in mind

$$T_1 = r_1(F - Q) \tag{4.35}$$

where force F is the transmitted force and Q is the pretension in the belt ($Q < F$). Since

$$F = Qe^{\mu\theta}$$

The allowable torque and pretension are related by

$$T_1 = r_1 Q(e^{\mu\theta} - 1)$$

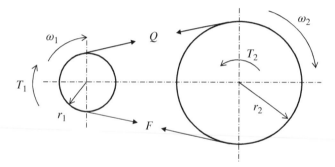

Figure 4.11 Power transmission through pulleys.

where θ is the angle of contact between belt and pulley. The required pretension for a given driving torque, T_1, is

$$Q = \frac{T_1}{r_1(e^{\mu\theta} - 1)} \tag{4.36}$$

where μ is the static coefficient of friction. Belt slippage occurs when

$$Q < \frac{T_1}{r_1(e^{\mu\theta} - 1)} \tag{4.37}$$

This equation also applies to the driven pulley since $\dfrac{T_1}{r_1} = \dfrac{T_2}{r_2}$. Note that $\theta_2 > \theta_1$.

Example Using the following numbers determine the minimum preload, Q, required to transmit 3 hp at 800 rpm so the belt does not slip.

$r_1 = 2$ in.
$r_2 = 4$ in.
$\mu = 0.3$
$d = 14$ in.
$N_1 = 800$ rpm

The torque, T_1, applied to pulley 1 is

$$T_1 = 5252\frac{P}{N} = 5252\frac{3}{800} = 19.7 \text{ ft-lb}$$

$$T_1 = 19.7(12) = 236.3 \text{ in.-lb}$$

The belt angle, ϕ, is approximately

$$\tan\phi = \frac{r_2 - r_1}{d} = \frac{2}{14} = 0.143$$

$$\phi = 8.13°$$

The angle of contact on pulley 1 is $\theta_1 = 180 - 16.26 = 163.7°$. The angle of contact on pulley 2 is $\theta_2 = 180 + 16.26 = 196.26°$

$$\theta_1 = \frac{163.7}{180}\pi = 2.86 \text{ rad}$$

$$\theta_2 = \frac{196.26}{180}\pi = 3.43 \text{ rad}$$

By substitution, the minimum pretension is

$$Q = \frac{236.3}{2(e^{0.3(2.86)} - 1)} = 87 \text{ lb}$$

This calculation is based on impending slippage because we used the coefficient of static friction. The corresponding force, F is

$$F = \frac{T_1}{r_1} + Q = 205.3 \text{ lb}$$

Slippage between the belt and pulley 2 is

$$Q = \frac{472.6}{4(e^{0.3(3.43)} - 1)} = 65.7 \, \text{lb}$$

The smaller pulley dictates the required pretension because the angle of contact is smaller.

2) V-belts – V belts are made of fabric and cord, typically cotton, rayon, and nylon, which are impregnated with rubber. The cord gives it strength, while the rubber gives flexibility and enhances friction. They are used for shorter distances. They require grooved pulleys (sheaves). While that are somewhat less efficient than blat belts, they can be used in a multiple drive. They are also continuous and do not require a joining connection.
3) Timing belts – Timing belts contain rubber "teeth," which provide a positive transfer of motion and torque between pulleys. The surface of pulleys has recesses to engage the timing belt teeth. They are made up of a wire core with a rubberized fabric bonded to it. The teeth made it possible to run at any speed, fast or slow. Tooth engagement and disengagement may cause dynamic fluctuations across the drive.

The relationships between torque and speed are the same as for gear pairs. However, in this case, both pulleys rotate in the same direction.

Fundamentals of Shaft Design

Shafts are an important element in power transmission. As indicated previously, pulleys, gears, and other linkages are attached to them, and it is through the shafts that torque gets transferred. Shafts must be sized so that stress is kept within strength limits of the material. Torque in shafts produces shear stress and shear strain.

When torque is applied (Figure 4.12), shear stress is generated on a small element as shown. One component lies in the cross section, another also lies in a cross section some small distance to the left. Two other shear stress components are also developed but in horizontal planes. The directions of each stress component are consistent with the applied torque.

The sign convention for applied torque is plus (+) as shown, with the torque vector pointing outward for the cross-sectional surface. The direction of the torque vector, T, follows the right-hand rule, with fingers pointing in the direction of twist and the thumb in the direction of the vector. If the torque vector points inward, then torque is considered negative (−).

Figure 4.12 Torsional shear stress.

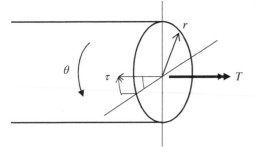

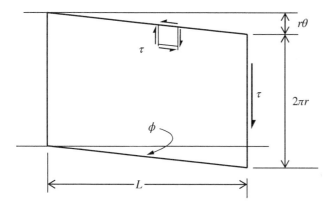

Figure 4.13 Shear strain due to torque.

Shear Stress

Consider an imaginary cylinder within the shaft with radius, r. Let the outer surface be unrolled into a flat surface (Figure 4.13)
where

L	–	length of shaft
r	–	radius of imaginary cylinder within the shaft

The twist in the shaft, due to torque, distorts this surface as indicated by ϕ. Simple shear strain is developed from this diagram.

$$\gamma = \phi = \frac{r\theta}{L} \tag{4.38}$$

It follows that

$$\tau = G\gamma = G\frac{r}{L}\theta \tag{4.39}$$

Integrating the shear stress across the face of the shaft

$$T = \int_0^R r\tau \, dA = G\frac{\theta}{L}\int_0^R r^2 \, dA \tag{4.40}$$

$$T = G\frac{\theta}{L}J$$

$$\theta = \frac{TL}{GJ} \tag{4.41}$$

This equation defines the amount of twist over the length of a shaft and is used to find torsion spring constants. The parameters are

θ – rotational twist of one end relative to the other
T – applied torque
L – length of shaft
G – modulus of rigidity
J – cross-section polar moment of inertia

Units must be compatible

By substitution, the magnitude of shear stress at any radial distance, r, is given by

$$\tau = \frac{Tr}{J} \tag{4.42}$$

Equation (4.42) shows the maximum shear stress is located at the outer surface, $r = R$. The direction of the shear strain and stress is consistent with the transmitted torque.

Example Consider a shaft with the following parameters

$T = 1500$ ft-lb

$\sigma_{yld} = 60\,000$ psi

$\tau_{yld} = 0.577(60\,000) = 34\,620$ psi

$G = 12 \times 10^6$ psi

$FS = 1.5$

$L = 12$ in.

$\tau_{allowable} = \dfrac{\tau_{yld}}{FS} = 23\,080$ psi

The shaft size under these conditions is

$$\tau_a = \frac{TR}{J} = \frac{2T}{\pi R^3}$$
$$R^3 = \frac{2T}{\pi \tau_a} = \frac{2(1500)12}{\pi 23\,080} = 0.4965 \tag{4.43}$$

Giving $R = 0.792$ in. Shaft diameter (1.584 in.) would be rounded up to the lowest standard shaft sizes, say $D = 1\frac{5}{8}$ in diameter.

The expected twist over the active length of the shaft (12 in.) is

$$\theta = \frac{TL}{GJ}, \quad J = \frac{\pi}{2}\left(\frac{1.625}{2}\right)^4 = 0.6846 \text{ in.}^4$$

$$\theta = \frac{1500(12)12}{12 \times 10^6(0.6846)} = 0.026\,29 \text{ rad} = 1.51° \tag{4.44}$$

There are no normal stresses in the stress element shown in Figure 4.13. However, there could be normal stresses developed simultaneously by other loads.

This derivation applies to circular shafts. Stresses in noncircular shafts, such as elliptical, triangular, and other shapes, are discussed in Ref. [7].

Example Consider a case where bending and torque are applied simultaneous (Figure 4.14). In this example, it is required to determine

a) stress condition at point A showing the stresses on an element;

b) principal stresses and maximum shear stress;

c) safety factor based on the energy of distortion criteria of failure. Yield strength of the material is 100 ksi.

The cross-sectional moment of inertia of the 1″ diameter shaft is

$$I = \frac{\pi}{4}R^4 = \frac{\pi}{4}(0.5)^4 = 0.05 \text{ in.}^4$$

Its polar moment of inertia is $J = 0.1$ in.^{4}

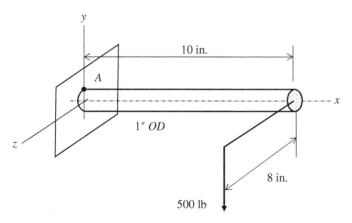

Figure 4.14 Shaft under bending and torque.

The bending moment at location A is 5000 in.-lb. The torque applied to cross section A is 4000 in.-lb. Corresponding bending and torsion stresses are

$$\sigma = \frac{Mc}{I} = \frac{5000(0.5)}{0.05} = 50\,000 \text{ psi}$$

$$\tau = \frac{Tc}{J} = \frac{4000(0.5)}{0.1} = 20\,000 \text{ psi}$$

The stress element for point A is shown in Figure 4.15.
From Mohr's stress circle

$$R = \left[25^2 + 20^2\right]^{\frac{1}{2}} = 32 \tag{4.45}$$

There

$$\sigma_1 = 25 + 32 = 57 \text{ ksi} \tag{4.46}$$

$$\sigma_2 = 25 - 32 = -7 \text{ ksi} \tag{4.47}$$

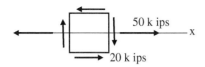

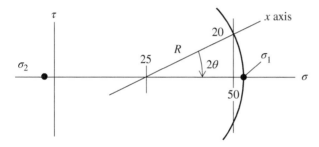

Figure 4.15 Mohr's circle for normal and shear stress.

Applying the energy of distortion criteria of failure

$$\sigma' = \left[\sigma_1^2 - \sigma_1\sigma_2 + \sigma_2^2\right]^{\frac{1}{2}} \tag{4.48}$$

$$\sigma' = \left[57^2 + 57(7) + (-7)^2\right]^{\frac{1}{2}} = 60.8 \text{ ksi}$$

Factor of Safety is

$$FS = \frac{100}{60.6} = 1.64 \tag{4.49}$$

Stress Analysis of Shafts

Example Consider the power train shown in Figure 4.16. An electric motor delivers power to pulley B by way of a V belt. The rotary speed of pulley B is 500 rpm. The magnitude of the force in the belt on one side of the pulley is 400 lb and the magnitude of the force on the other side is 200 lb. Determine the maximum normal stress in the shaft at location "a" during each revolution. Assume the shaft is 1 in. in diameter and simply supported at the bearings.

Torque generate at pulley B is

$$T_B = 5(400 - 200) = 1000 \text{ in.-lb}$$

This torque is opposed by output torque, T, which does useful work on something. Shear stress produced in the shaft is

$$\tau = \frac{Tc}{J} \tag{4.50}$$

where

$T = 1000 \text{ in.-lb}$
$c = 0.5 \text{ in.}$
$J = \dfrac{\pi}{2}(0.5)^4 = 0.0982 \text{ in.}^4$

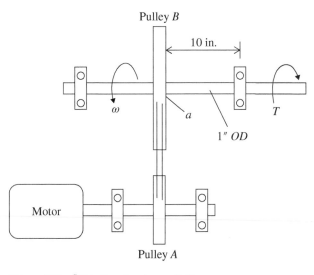

Figure 4.16 "V" belt pulley transmission.

So

$$\tau = \frac{1000(0.5\pi)}{0.0982} = 5092 \text{ psi} \tag{4.51}$$

Bending stress at point "a" is

$$M_a = 300(10) = 3000 \text{ in.-lb}$$

$$\sigma_a = \frac{Mc}{I} = \frac{3000(0.5)}{0.0491} = 30\,550 \text{ psi} \tag{4.52}$$

Principal stress

$$\sigma_{p1} = \frac{\sigma_x}{2} + \sqrt{\left(\frac{\sigma_x}{2}\right)^2 + \tau_{xy}^2} \tag{4.53}$$

$$\sigma_{p1} = \frac{30\,550}{2} + \sqrt{\left(\frac{30\,550}{2}\right)^2 + 5092^2} = 31\,376 \text{ psi} \tag{4.54}$$

Twisting in Shafts Having Multiple Gears

When a shaft contains several gears (Figure 4.17), internal torque between each gear is constant, but different in magnitude. The torque coming into each gear is indicated by the torque vectors. Internal torque between each gear is also shown along with the angular twist of each gear relative to gear "a."

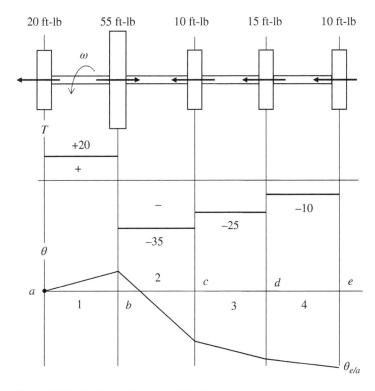

Figure 4.17 Shaft rotation of shaft between gears.

The twist of gear b relative to gear a is

$$\theta_{a/b} = \frac{TL}{GJ} = \frac{20\,L}{GJ} \tag{4.55}$$

The twist of gear "j" relative to gear a is

$$\theta_{j/a} = \sum_{1}^{j} \frac{T_i L_i}{GJ} \tag{4.56}$$

Transverse loads generated by gears and pulleys create bending stresses in addition to shear stress. All applied loads have to be considered in designing shafts as well as bearing supports.

Keyway Design

Keys, pins, and types of retainers are used to transmit torque from gears and pulleys. Examples of such fasteners are:

- Square keys
- Round keys
- Round pins passing through a hub and shaft
- Tapered pins

Square keys are the most common means of transmitting torque to shafts. As a rule, the sides of square keys are ¼ shaft diameter. Length is adjusted according hub length. Two keys may be used if necessary.

Example Consider a gear-transmitting torque of 4200 lb-in. (Figure 4.18). The steel shaft has yield strength of 75 ksi and a diameter of 1.5 in. A square key having side dimensions of 0.375 in. and yield strength of 65 ksi is chosen. Determine the factor of safety for the key if it is 1.5 in. long.

The shear force in the key is

$$V = \frac{T}{0.75} = \frac{4200}{0.75} = 5600\,\text{lb} \tag{4.57}$$

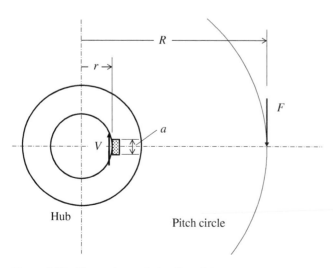

Figure 4.18 Torque transmission though keys.

The shear strength of the key material, per the von Mises energy of distortion criteria, is

$$\tau_{yld} = 0.577\sigma_{yld} = 37.5 \, \text{ksi} \tag{4.58}$$

The applied shear stress in the $\frac{3}{8}$ in key is

$$\tau = \frac{5600}{0.375(1.5)} = 9955 \, \text{psi}$$

The factor of safety of the key is

$$FS = \frac{37\,500}{9955} = 3.77 \tag{4.59}$$

Mechanical Linkages

The study of mechanical linkages requires understanding of position, velocity, and acceleration of key points as well as angular velocities and accelerations. Two factors affect the design and application of linkages; (i) desired motion (kinematics) and (ii) inertia forces (dynamics) generated by achieving the desired motion. Inertia forces radiate throughout machinery affecting bearings, shafts, and frame.

Relative Motion Between Two Points

Consider two particles A and B having motions defined in relation to coordinates XY (fixed) and xy (translating, not rotating). In the *first case*, assume the two points are not connected (Figure 4.19a). The position vectors locating points A and B are

$$\overline{R}_A = X_A i + Y_A j \quad \text{Point } A \tag{4.60}$$
$$\overline{R}_B = X_B i + Y_B j \quad \text{Point } B \tag{4.61}$$
$$\overline{r}_{A/B} = xi + yj \tag{4.62}$$

The position of point A is

$$\overline{R}_A = \overline{R}_B + \overline{r}_{A/B} \tag{4.63}$$

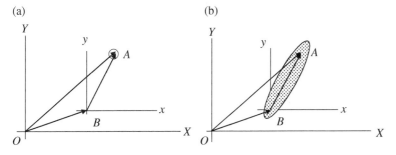

Figure 4.19 Relative velocity and acceleration. (a) Point A independent of B. (b) Points A and B on same link (special case of a).

By differentiation

$$\bar{v}_A = \frac{d\bar{R}_A}{dt} = \frac{d\bar{R}_B}{dt} + \frac{d\bar{r}_{A/B}}{dt} \tag{4.64}$$

$$v_A = v_B + v_{A/B} \tag{4.65}$$

In each case, the unit vectors do not change with time. Typically, the velocities and accelerations of both points are known, and the relative velocity and acceleration are to be determined. For the *first case* (Figure 4.19a), assume

$$\bar{v}_A = 200(-\cos 30i + \sin 30j)\,\text{mph}$$

$$\bar{v}_B = 400(-\cos 45i - \sin 45j)\,\text{mph}$$

Then the velocity of A with respect to B is

$$\bar{v}_{A/B} = \bar{v}_A - \bar{v}_B \tag{4.66}$$

$$\bar{v}_{A/B} = 109.64i + 382.8j\,\text{mph}$$

Relative acceleration is determined in a similarly starting with

$$a_A = a_B + a_{A/B}$$

When a_A and a_B are known

$$\bar{a}_{A/B} = \bar{a}_A - \bar{a}_B$$

The *second case* (Figure 4.19b) shows two points fixed on a rigid body. The velocity and acceleration of one point (say Point B) and angular velocity and acceleration of the rigid body are usually known or can be determined from the linkage configuration. The objective is to determine the velocity and acceleration of Point A. The following information is given for this example.

$$\bar{v}_B = 10(\cos 60i + \sin 60j)\,\text{fps} \tag{4.67}$$

$$\bar{a}_B = 5(\cos 45i - \sin 45j)\,\text{fps}^2 \tag{4.68}$$

The angular velocity of body AB is

$$\omega = 2\,\text{rad/s}\,(\text{CCW})$$

The angular acceleration of body AB is

$$\alpha = 4\,\text{rad/s}^2\,(\text{CCW})$$

$$L = 1\,\text{ft (fixed distance between A and B)}$$

$$\theta = 60°$$

The *velocity* of point A is determined as follows.

$$v_A = v_B + v_{A/B}\,(\text{velocity})$$

where

$$\bar{v}_{A/B} = L\omega(-i\sin\theta + j\cos\theta) \tag{4.69}$$

$$\bar{v}_{A/B} = 2(-0.866i + 0.5j)\,\text{fps}$$

By substitution

$$\bar{v}_A = 10(\hat{i}\cos 60 + \hat{j}\sin 60) - 1.732\hat{i} + \hat{j}$$

$$\bar{v}_A = 3.268\hat{i} + 9.66\hat{j}$$

The *acceleration* of point A is determined from

$$a_A = a_B + a_{A/B}$$

The acceleration of A relative to B has two components, normal and tangent.

$$\bar{a}_{A/B} = \left(\bar{a}_{A/B}\right)_n + \left(\bar{a}_{A/B}\right)_t \tag{4.70}$$

The normal component is

$$\left(\bar{a}_{A/B}\right)_n = \omega^2 L(-\cos\theta i - \sin\theta j) \tag{4.71}$$

$$\left(\bar{a}_{A/B}\right)_n = 4(-0.5i - 0.866j)$$

The tangent component is

$$\left(\bar{a}_{A/B}\right)_t = \alpha L(-\sin\theta i + \cos\theta j) \tag{4.72}$$

$$\left(\bar{a}_{A/B}\right)_t = 4(-0.866i + 0.5j)\,\text{fps}^2$$

Combining the terms gives

$$\bar{a}_A = 5(\hat{i}\cos 45 - \hat{j}\sin 45) + 4(-0.5\hat{i} - \hat{j}0.886) + 4(-0.866\hat{i} + 0.5\hat{j})$$

$$\bar{a}_A = -1.93i - 5j$$

Absolute Motion Within a Rotating Reference Frame

In this arrangement, a point A moves in space and this motion is observed from a rotating x,y reference frame as shown in Figure 4.20. The rotating reference frame (xy) has angular acceleration, α, as well as angular velocity, ω. Point B is the origin of the moving reference frame. Since the x,y reference frame rotates as well as translates the derivatives of the unit vectors, i and j, are not zero.

The position of point A relative to the fixed XY frame is

$$\bar{r}_A = \bar{r}_B + \bar{r}$$

By differentiation of the position vectors, the *absolute* velocity of point A is determined as follows.

$$\bar{v}_A = \bar{v}_B + \frac{d\bar{r}}{dt} \tag{4.73}$$

Since

$$\bar{r} = ix + jy$$

$$\frac{d\bar{r}}{dt} = x\frac{di}{dt} + \bar{i}\dot{x} + y\frac{dj}{dt} + \bar{j}\dot{y} \tag{4.74}$$

Figure 4.20 Point moving within a rotating reference frame.

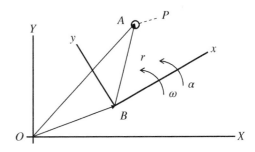

Also

$$\frac{d\bar{i}}{dt} = \omega j \quad \text{and} \quad \frac{d\bar{j}}{dt} = -\omega\bar{i}$$

By substitution

$$\bar{v}_A = \bar{v}_B + \omega(-\bar{i}y + \bar{j}x) + \bar{v} \tag{4.75}$$

$$\bar{v}_A = \bar{v}_B + \overline{\omega}x\bar{r} \tag{4.76}$$

The interpretation of each term in this equation is important. First note that point A is the point of interest. Point P is located within the xy frame and is coincident with point A at a given instant of time.

$\bar{v}_A$ – absolute velocity of A
$\bar{v}_B$ – absolute velocity of B (origin of the xy frame)
ω – angular velocity of the rotating reference frame, xy
$\bar{r}$ – position of point P within the xy reference frame coincident with point A
$\overline{\omega}x\bar{r}$ – absolute velocity of point P located within the xy frame and coincident with point A
$\bar{v}$ – velocity of point A relative to the xy reference frame $(\bar{v}_{A/P})$

If point B is fixed in space, the velocity of point A is

$$\bar{v}_A = \overline{\omega}x\bar{r} + \bar{v} \tag{4.77}$$

or

$$\bar{v}_A = \bar{v}_P + \bar{v}_{A/P} \tag{4.78}$$

Acceleration of point A is established by rate of change of velocity with time. By differentiation of Eq. (4.78) and in consideration of the derivatives of the unit vectors

$$a_A = a_B + \alpha x r + \omega x(\omega x r) + a + 2\omega x v \tag{4.79}$$

where

$\bar{a}_A$	–	absolute acceleration of A
$\bar{a}_B$	–	absolute acceleration of B
α	–	angular acceleration of the xy reference frame
ω	–	angular velocity of rotating reference frame
$\overline{\omega}x\bar{r}$	–	tangent component of acceleration of a coincident point P in xy frame
$\omega x(\omega x r)$	–	normal component of acceleration of coincident point P in xy frame
r	–	radial position of point P (and A) relative to xy frame
$\bar{a}$	–	acceleration of point A relative to the xy frame, $\bar{a}_{xy}$
$\bar{v}$	–	velocity of point A relative to the xy frame, $\bar{v}_{xy}$

The last term in Eq. (4.79) is the Coriolis component of acceleration. If point B is fixed, Eq. (4.79) becomes

$$a_A = (a_P)_t + (a_P)_n + a_{xy} + 2\omega v_{xy} \tag{4.80}$$

Example Consider a situation of a person (dot A) moving outwardly along the x axis, while a merry-go-round is rotating with ω (Figure 4.21).

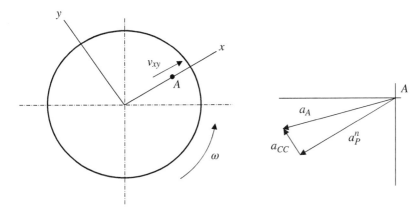

Figure 4.21 Coriolis component of point moving on a rotating frame.

Applying Eq. (4.81)

P – point on x axis coincident with dot A

A – person moving outward along x axis

N – 20 rpm

$$\omega = \frac{2\pi}{60} 20 = 2.09 \text{ rad/s}$$

$r = 2$ ft (radial distance from origin)

$v_{xy} = v_{A/P} = 0.5$ fps (assumed)

$\bar{v}_A = \bar{v}_P + \bar{v}_{A/P}$

The acceleration of point A is

$$\bar{a}_A = a_P^n + 2\omega v_{A/P} \tag{4.81}$$

$$a_P^n = \omega^2 r = (2.09)^2 2 = 8.74 \text{ fps}^2 (\text{directed toward the origin})$$

$$2\omega v_{A/P} = 2(2.09)0.5 = 2.09 \text{ fps}^2 \quad (\text{Coriolis component})$$

Mechanical linkages are used to impart specific motions to accomplish a given output or effect. There are many possible configurations depending on the required output. The following covers five common types of linkages. In practice, linkages may contain a combination of each [8, 9].

Scotch Yoke

The Scotch Yoke (Figure 4.22) is a mechanical analog for generating simple harmonic motion. The motion of point B is defined by

$$x = r - r\cos\theta \tag{4.82}$$

$$\dot{x} = r\sin\theta\dot{\theta} = v_A\sin\theta \tag{4.83}$$

$$\ddot{x} = r(\dot{\theta}\cos\theta\dot{\theta} + \sin\theta\ddot{\theta}) \tag{4.84}$$

$$\ddot{x} = r\omega^2\cos\theta + r\ddot{\theta}\sin\theta \tag{4.85}$$

If the crank rotates at a constant velocity, ω, then $\ddot{\theta} = 0$ and

$$\ddot{x} = r\omega^2\cos\theta$$

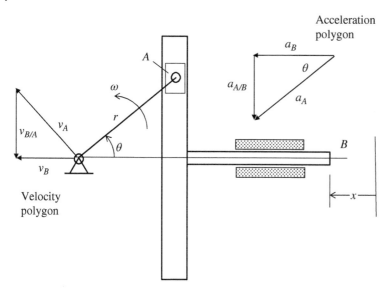

Figure 4.22 Scotch yoke mechanism.

The displacement of the yoke is the same throughout. Taking B in the yoke as a coincident point with point A and knowing the velocity of point A, the relative sliding velocity of point A within the grove is determined from the relative velocity equation.

$$v_A = v_B + v_{A/B} \tag{4.86}$$

The velocity polygon is shown with the center coincident with the center of rotation of the crank. Its construction starts with known velocity of point A ($v_A = r\omega$). The direction of the other two components is used to close the polygon.

$$v_{A/B} = v_A - v_B \tag{4.87}$$

The relation between these vectors is shown in Figure 4.22. Also

$$\bar{a}_B = \bar{a}_A + \bar{a}_{B/A} \tag{4.88}$$

Here the acceleration of point A is known ($a_A = \omega^2 r$). The direction of the other two components closes the polygon.

Slider Crank Mechanism

The history of the slider crank mechanism starts with James Watt's steam engine (1769) and the need to convert linear motion into rotary motion. Steam was used as the driving force to displace a piston, while the connecting rod provided the means to generate pure rotation to a crank shaft. The slider crank mechanism was used throughout the Industrial Revolution of the 1800s, most notably the steam locomotive.

The slider crank mechanism illustrates all three types of rigid body dynamics: translation (piston), rotation (crank shaft), and general (connecting rod) motion. With reference to Figure 4.23 and assuming the crank rotates with constant velocity, it is desirable to find the velocities and accelerations of points A and B as well as the angular velocity and acceleration of the connecting rod.

Knowing the angular velocity (ω) of the crank shaft, the angular velocity of the connecting rod and linear velocity of the piston are determined from a velocity polygon. This is needed to determine the angular acceleration of the connecting rod and the linear acceleration of the piston, point B. The velocity polygon is shown (Figure 4.23) along with the slider crank to show the true directions of each vector with respect to the alignment of each link. The angular velocity of the connecting rod is extracted from the relative velocity, $v_{B/A}$, or from the instantaneous center (IC).

Velocity Analysis

The velocity polygon is based on

$$\bar{v}_B = \bar{v}_A + \bar{v}_{B/A} \tag{4.89}$$

The velocity of point A ($v_A = r\omega$) is known. The velocity polygon is closed with the known directions of the other two components.

The x and y components of the velocity vectors in the polygon give

$$\mathbf{x} : v_B = v_A \sin\theta + v_{B/A} \sin\phi \tag{4.90}$$

$$\mathbf{y} : 0 = v_A \cos\theta - v_{B/A} \cos\phi \tag{4.91}$$

Solving them together gives

$$v_B = v_A(\sin\theta + \sin\phi) \tag{4.92}$$

$$v_{B/A} = \frac{\cos\theta}{\cos\phi} v_A \tag{4.93}$$

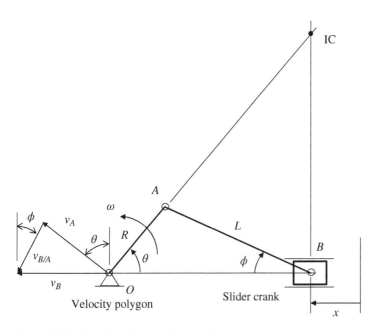

Figure 4.23 Angular velocity of connecting rod.

where angles ϕ and θ are related by

$$\sin \phi = \frac{R}{L} \sin \theta \tag{4.94}$$

Once $v_{B/A}$ is determined,

$$\omega_{AB} = \frac{v_{B/A}}{L} \text{ (clockwise)} \tag{4.95}$$

The velocity of point B and angular velocity of link AB can also be determined by use of the instantaneous center.

Acceleration Analysis

The acceleration polygon is drawn with origin at point O along with the slider crank to show the true direction of each acceleration component with respect each linkage (Figure 4.24). The acceleration polygon is based on

$$\bar{a}_B = \bar{a}_A + \bar{a}_{B/A} \tag{4.96}$$

The acceleration of point A is known ($a_A = \omega^2 R$). Only the direction of point B is known. The acceleration B relative to A has two components. The normal component is known $\left(a_{B/A}^n = \omega_{AB}^2 L \right)$. Only the direction of $a_{B/A}^t$ is known. Closure of the acceleration polygon is shown in Figure 4.24.

$$x : a_B = a_A \cos \theta + a_{B/A}^n \cos \phi - a_{B/A}^t \sin \phi \tag{4.97}$$

$$y : 0 = a_A \sin \theta - a_{B/A}^n \sin \phi - a_{B/A}^t \cos \phi \tag{4.98}$$

From Eq. (4.98)

$$a_{B/A}^t = a_A \frac{\sin \theta}{\cos \phi} - a_{B/A}^n \tan \phi \tag{4.99}$$

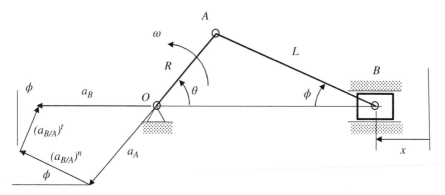

Figure 4.24 Angular acceleration of connecting rod.

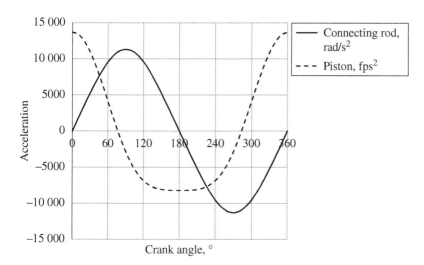

Figure 4.25 Acceleration of connecting rod and piston.

where

$$a^n_{B/A} = \omega^2_{AB}L \tag{4.100}$$

The tangent component, $a^t_{B/A} = \alpha_{AB}L$. The vector sum of both normal and tangents components yields α_{AB}.

The dynamic forces in the bearings at connections O, A, B are affected by the magnitude of the accelerations of each of the three components (crank, connecting rod, piston). Angular acceleration of the connecting rod and linear accelerations of the pistons are displayed in Figure 4.25 for one complete rotation of the crank. The numbers are based on a crank radius of $R = 3$ in., a connecting rod of $L = 12$ in., and a crank rotary speed of 2000 rpm.

The block represents a piston sliding inside a cylinder. The linear motion of the block is a function of angular position of the crank defined by

$$x(\theta) = R(1 - \cos\theta) + L(1 - \cos\phi) \tag{4.101}$$

where

$$r\sin\theta = L\sin\phi$$

These formulations are used to predict velocities and accelerations of the piston as continuous functions of θ.

Four-Bar Linkage

This mechanism contains three moving links plus a fixed link represented by 00* (Figure 4.26). The connecting rod, AB, is extended out to point C for generality. This linkage allows a multitude of motions for point C depending on requirements of a design. Movement of any point can be determined by stepping the driving link, OA, through different angular positions, θ, and tracking other points, including point C. Linear and angular velocities are determined from the velocity polygon as shown. Velocities are required to determine linear and angular accelerations. They are needed to determine normal acceleration components of the linkages.

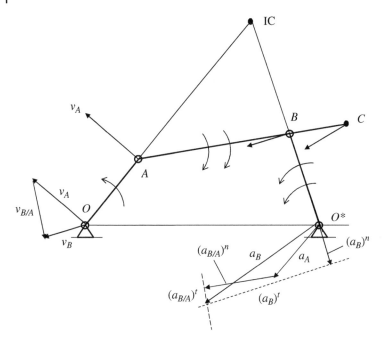

Figure 4.26 Parameters to define a four-bar linkage.

The motion of each member (a, b, c, d) and any point on each member can be established as follows. The relation between the lengths of each member and angular positions, defined by (θ, ψ, ϕ), is determined from

$$c \sin \phi - b \sin \psi = a \sin \theta \tag{4.102}$$

$$c \cos \phi + b \cos \psi = d - a \cos \theta \tag{4.103}$$

Squaring both sides of each equation and then adding the results of the two gives

$$\cos (\phi + \psi) = \frac{a^2 + d^2 - c^2 - b^2 - 2da \cos \theta}{2bc} \tag{4.104}$$

$$\phi + \psi = \cos^{-1} \left[\frac{1}{2bc} (a^2 + d^2 - c^2 - b^2 - 2da \cos \theta) \right] \tag{4.105}$$

where

$a = OA$	$b = AB$	$c = O^*B$	$d = OO^*$
θ = slope of OA	ϕ = interior slope of O^*B	ψ = slope of AB	

Typically, the angle θ is the independent variable. Angles ϕ and ψ are determined by solving Eqs. (4.103) and (4.105) simultaneously.

Example Consider the geometry of a four-bar linkage having the following parameters

$a = 4$ in.
$b = 8$ in.

$c = 10$ in.
$d = 12$ in.
$\theta = 60°$
$\omega = 2\pi$ rad/s ($N = 60$ rpm)

With $\theta = 60°$, we next establish values for and ψ by solving Eqs. (4.103) and (4.105) simultaneously. These values, determined by trial and error, are $\psi = 44°$ and $= 65°$. Note that the sum of and ψ is 109° according to Eq. (4.105).

Velocity Analysis

The unknowns are ω_{AB} and v_B. These parameters are determined from

$$\bar{v}_B = \bar{v}_A + \bar{v}_{B/A} \tag{4.106}$$

This vector equation yields two scalar equations for finding the magnitudes of v_B and $v_{B/A}$.

$$v_B\left(-\hat{i}\cos(90-\phi)-\hat{j}\sin(90-\phi)\right) = v_A\left(-\hat{i}\sin\theta+\hat{j}\cos\theta\right) + v_{B/A}\left(\hat{i}\sin\psi-\hat{j}\cos\psi\right) \tag{4.107}$$

$$x: \ -v_B\cos 25 = -v_A\sin\theta + v_{B/A}\sin\psi \tag{4.108}$$

$$y: \ -v_B\sin 25 = v_A\cos\theta - v_{B/A}\cos\psi \tag{4.109}$$

$$x: \ -v_B\cos 25 = -v_A\sin 60 + v_{B/A}\sin 44$$

$$y: \ -v_B\sin 25 = v_A\cos 60 - v_{B/A}\cos 44$$

$$x: \ -0.91v_B = -0.866v_A + 0.69v_{B/A}$$

$$y: \ -0.42v_B = 0.5v_A - 0.72v_{B/A}$$

Since

$$v_A = a\omega = 4(2\pi) = 25.13 \text{ ips}$$

The two simultaneous equations become

$$x: 0.9063\,v_B + 0.6947\,v_{B/A} = 21.7626 \tag{4.110}$$

$$y: -0.4226\,v_B + 0.7193\,v_{B/A} = 12.565 \tag{4.111}$$

which give $v_B = 7.33$ ips and $v_{B/A} = 21.77$ ips. The angular velocities of link AB and $0*B$ are

$$\omega_{AB} = \frac{v_{B/A}}{b} = \frac{21.77}{8} = 2.72 \text{ rad/s}$$

$$\omega_{0*B} = \frac{v_B}{c} = \frac{7.33}{10} = 0.733 \text{ rad/s}$$

The velocity polygon is shown in Figure 4.26.

Acceleration Analysis

Accelerations are important, especially in high-speed machinery, where inertia effects can generate large bearing loads and other reactions. It is a good idea to quantify the inertia effects and forces throughout any mechanism. The relative acceleration equations can be useful in this regard.

$$\bar{a}_B = \bar{a}_A + \bar{a}_{A/B} \tag{4.112}$$

$$(\bar{a}_B)^n + (\bar{a}_B)^t = (\bar{a}_A) + (\bar{a}_{B/A})^n + (\bar{a}_{B/A})^t \tag{4.113}$$

where

$$(a_B)^n = (\omega_{O*B})^2 c$$

$$(a_B)^t = \alpha_{O*B} c$$

$$a_A = \omega^2 a$$

$$(a_{B/A})^n = (\omega_{AB})^2 b$$

$$(a_{B/A})^t = \alpha_{AB} b$$

The acceleration polygon is also shown in Figure 4.26.

The acceleration equation yields two scalar equations. They are used to find the magnitude of $(a_B)^t$ and $(a_{B/A})^t$, which are then used to find angular accelerations. The directions of all vectors are known from the drawings. The velocity and acceleration of point C can be determined from similar polygons.

Three-Bar Linkage

A three-bar linkage is one in which two rotating members are connected by a sliding collar as shown in Figure 4.27. One end of each member is pin connected to a rigid frame: the frame being the third link. The link (OA) is connected to a collar and typically rotates at a constant angular velocity, Ω. It is desired to determine the angular velocity (ω) and acceleration (α) of the second link (y axis), on which the collar slides.

The first step is to determine the angular velocity (ω) of the x,y reference frame and relative velocity of the sliding collar. Velocity information is needed to complete the acceleration analysis. Keep in mind that point A is located on the collar, while point P is coincident with A, but located on the y axis.

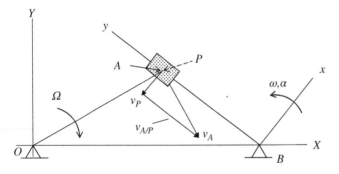

Figure 4.27   Three-bar mechanism.

Velocity Equation

The velocity of point A on the collar is related to ω by

$$\bar{v}_A = \bar{v}_P + \bar{v}_{A/P} \tag{4.114}$$

where

$\bar{v}_A$	–	absolute velocity of point A
$\bar{v}_P$	–	velocity of point P on the xy rotating frame coincident with point A
$\bar{v}_{A/P}$	–	velocity of point A relative to the xy rotating frame

The velocity polygon, which is the vector representation of Eq. (4.114), is shown in Figure 4.27. This vector equation separates into two scalar equations, which yield the magnitude of the sliding velocity and the absolute velocity of P. The angular velocity of the y axis is then found from the velocity of point P.

Acceleration Equation

The acceleration of point A relates to α by

$$\bar{a}_A = (\bar{a}_P)_t + (\bar{a}_P)_n + \bar{a}_{xy} + 2\bar{\omega}x\bar{v}_{xy} \tag{4.115}$$

where

$\bar{v}_{xy}$	–	velocity of A relative to the y axis $(\bar{v}_{A/P})$ or the sliding velocity of the collar on the rotating y axis. It is determined from the velocity analysis
$\bar{a}_{xy}$	–	acceleration of A relative to the xy reference frame. In this case, it is the sliding acceleration of A on the y axis. Its direction is known, but not its magnitude
$(\bar{a}_P)_n$	–	normal component of the acceleration of P $(=\omega^2 r)$. It is directed toward point B
$(\bar{a}_P)_t$	–	tangent component of the acceleration of P. Its magnitude is αr
$2\bar{\omega}x\bar{v}_{xy}$	–	Coriolis acceleration component

This vector equation separates into two scalar equations whose solutions give the magnitude of the sliding acceleration of A on the y axis $(a_{A/P})$ and the tangent component of the absolute acceleration of point P. Angular acceleration, α, is determined from this tangent component.

Example Consider the example (Figure 4.28) with following parameters.

$a = 8$ in.
$c = 12$ in.

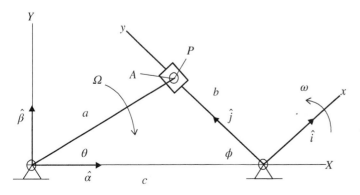

Figure 4.28 Rotating reference frame.

$\theta = 30°$

$\Omega = 10 \, \text{rad/s (constant)}$

Essential parameters used in solving this problem are shown in the diagram below. The *xy* reference frame rotates. The *XY* frame is fixed. Point *P* is coincident with point *A* and lies within to the rotating frame, while point *A* moves with the collar along the *y* axis.

The problem is to determine the angular velocity (ω) and acceleration (α) of the *xy* reference frame. Distance, *b* (or *r*), is determined by the law of cosine.

$$b^2 = a^2 + c^2 - 2\,ac\,\cos 30 \tag{4.116}$$

$$b = 6.46$$

Angle, ϕ, is determined from

$$b\cos\phi + 8\cos 30 = 12$$

to be $\phi = 38.2°$.

Velocity Analysis
The velocity polygon is based on Eq. (4.114) (also see Figure 4.28). The velocity vectors are

$$\bar{v}_A = v_A\left(\hat{\alpha}\sin\theta - \hat{\beta}\cos\theta\right) \tag{4.117}$$

where $v_A = \Omega\,a = 10\,(8) = 80$ ips

$$\bar{v}_P = -\hat{i}v_P$$

where

$$v_P = b\,\omega = 6.46\,\omega$$

$$\bar{v}_{A/P} = \bar{v}_A - \bar{v}_P \tag{4.118}$$

By substitution

$$\bar{v}_{A/P} = 80\left(\hat{\alpha}\sin\theta - \hat{\beta}\cos\theta\right) - \left(-6.46\omega\hat{i}\right) \tag{4.119}$$

In terms of $\hat{i}, \hat{j}$,

$$\hat{\alpha} = \hat{i}\sin\phi - \hat{j}\cos\phi$$
$$\hat{\beta} = \hat{i}\cos\phi + \hat{j}\sin\phi$$

By substitution

$$\bar{v}_{A/P} = \left(-29.7 + 6.46\omega\right)\hat{i} - 74.25\hat{j} \tag{4.120}$$

From the velocity polygon, we see that

$$\left(\bar{v}_{A/P}\right)_x = 0$$

$$-29.74 + 6.43\omega = 0$$

$$\omega = 4.62 \ \text{rad/s}$$

Also

$$v_{A/P} = \left(\bar{v}_{A/P}\right)_y = -74.24 \ \text{ips} \tag{4.121}$$

The minus sign means the collar is moving toward the origin. Bringing terms together

$$\bar{v}_A = -29.8\hat{i} - 74.3\hat{j}$$

Acceleration Analysis

The two unknowns at this point are α and the linear acceleration, $(a_A)_{xy}$, the sliding acceleration of the collar A on the y axis. Starting with the acceleration equation

$$\bar{a}_A = (\bar{a}_P)_t + (\bar{a}_P)_n + \bar{a}_{xy} + 2\bar{\omega}x\bar{v}_{xy} \qquad (4.122)$$

where

$$(a_P)_t = r\alpha$$

$$(a_P)_n = \omega^2 r$$

$$\bar{a}_{xy} = (\bar{a}_A)_{xy}$$

Components in the acceleration equation are

$$(\bar{a}_P)_t = b\alpha(-\hat{i}) = -\hat{i}6.46\alpha \qquad (4.123)$$

$$(\bar{a}_P)_n = \omega^2 b(-\hat{j}) = -4.62^2 6.46\hat{j} = -137.9\hat{j} \qquad (4.124)$$

$$(\bar{a}_{xy}) = a_{xy}(\hat{j}) \quad \text{(assumed direction)}$$

$$(\bar{a}_{xy})_n = 0$$

Coriolis component

$$2\vec{\omega}x\bar{v}_{xy} = 2\omega v_{xy}(\hat{i}) = 2\omega v_{A/P}(\hat{i})$$
$$= 2(4.62)(74.24) = 686\hat{i} \qquad (4.125)$$

By substitution

$$\bar{a}_A = (-6.46\alpha + 686)\hat{i} - (138 - a_{xy})\hat{j} \qquad (4.126)$$

The absolute acceleration of point A or the collar is known.

$$\bar{a}_A = 800(\hat{i}\sin(\phi + \theta) + \hat{j}\cos(\phi + \theta))$$

$$\bar{a}_A = -742.8\hat{i} + 297.1\hat{j} \qquad (4.127)$$

Bringing all terms into Eq. (4.122) gives

$$x:\ -742.8 = -6.46\alpha + 686 \quad \text{giving}\ \alpha = 221.2\ \text{rad/s}^2(\text{CCW})$$
$$y:\ 297.1 = -137.88 + a_{xy} \qquad \text{giving}\ a_{xy} = 435\ \text{ips}^2$$

The acceleration polygon is shown in Figure 4.29.

Figure 4.29 Acceleration polygon.

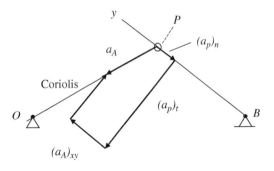

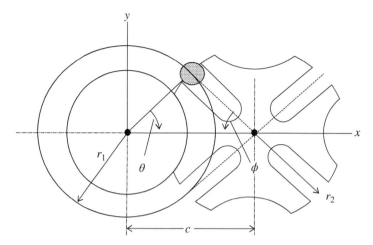

Figure 4.30 Geneva mechanism – intermittent motion.

The direction of the acceleration of point A should be toward point 0 (as shown). A good check on the polygon.

Geneva Mechanism

The Geneva Mechanism (Figure 4.30) is a special case of a three-bar mechanism. It is a mechanism that is commonly used in watches. Driver 1 rotates at a constant angular velocity and imparts rotation to the receiver 2, which experiences intermittent motion at a nonuniform speed.

The position of the Geneva wheel is determined by

$$\tan \phi = \frac{\sin \theta}{\dfrac{c}{r_1} - \cos \theta} \tag{4.128}$$

where c is the center distance. By differentiation

$$\omega_2 = \omega_1 \frac{\dfrac{c}{r_1} \cos \theta - 1}{1 + \left(\dfrac{c}{r_1}\right)^2 - 2\dfrac{c}{r_1} \cos \theta} \tag{4.129}$$

Differentiating again gives the expression for angular acceleration of the Geneva wheel.

$$\alpha_2 = \omega_1^2 \frac{\dfrac{c}{r_1} \sin \theta \left[1 - \left(\dfrac{c}{r_1}\right)^2\right]}{\left[1 + \left(\dfrac{c}{r_1}\right)^2 - 2\dfrac{c}{r_1} \cos \theta\right]^2} \tag{4.130}$$

The receiving member can have more than four slots. An important feature is that the pin enters the receiver tangent to the slot. These equations apply only when $45° \geq \theta \geq -45°$.

As a comparison with the vector analysis, assume ($c/r_1 = 12/8 = 1.5$ and $\theta = 30°$).

$\phi = 38.26°$	(using (4.128))	$\phi = 38.26°$	
$\omega = 4.59$ rad/s	(using (4.129))	$\omega = 4.62$ rad/s	(Eq. (4.120))
$\alpha = 220.58$ rad/s^2	(using (4.130))	$\alpha = 217.4$ rad/s^2	(Eq. (4.127))

Flat Gear Tooth and Mating Profile

Another interesting kinematic situation is the profile of novel gear teeth. In this case, the motion between two gears is given, i.e. the angular velocity ratio of two gears is constant. The question becomes, knowing the profile of one gear tooth, what the profile of the mating gear tooth. Both profiles must satisfy the fundamental gear tooth law. The attraction to such a gear set is reduced Hertz contact stresses.

Consider, for example, a case where one tooth profile is flat, and the other profile is to be determined. The setup to the problem is illustrated in Figure 4.31 along with variables and reference frames. The flat tooth profile is fixed to the rotating xy frame, while the profile to be determined is fixed to the rotating x_1y_1 rotating frame. The solution strategy is to define the path of the contact point, P, in terms of the x_1y_1 frame. Vector positions of point P in terms of each reference frame are

$$\bar{r} = \hat{i}x + \hat{j}y \quad \text{and} \quad \bar{r}_1 = \hat{\alpha}x_1 + \hat{\beta}y_1 \tag{4.131}$$

The relation between these vectors is ($d = a + b$)

$$\bar{r}_1 = \bar{r} - \bar{d} \tag{4.132}$$

$$\hat{\alpha}x_1 + \hat{\beta}y_1 = \hat{i}(x - d\cos\theta) + \hat{j}(y - d\sin\theta) \tag{4.133}$$

The unit vectors are related by

$$\hat{i} = \hat{\alpha}\cos(\theta + \phi) - \hat{\beta}\sin(\theta + \phi) \tag{4.134}$$

$$\hat{j} = \hat{\alpha}\sin(\theta + \phi) + \hat{\beta}\cos(\theta + \phi) \tag{4.135}$$

By substitution

$$x_1 = x\cos(\theta + \phi) + y\sin(\theta + \phi) - d\cos\phi \tag{4.136}$$

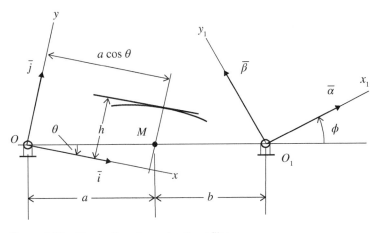

Figure 4.31 Kinematics of gear tooth profiles.

$$y_1 = -x\sin(\theta + \phi) + y\cos(\theta + \phi) + d\sin\phi \tag{4.137}$$

Note that θ and ϕ are related by $\theta = b/a\phi$. These two equations define the profile we seek. The first step is to define the parametric equation $x(\theta)$ and $y(\theta)$. With regard to the above figure,

$$x(\theta) = \cos\alpha\left[\cos(\alpha - \beta)(h^2 + a^2 - 2ha\sin\theta)^{\frac{1}{2}}\right] \tag{4.138}$$

$$y(\theta) = h + \sin\alpha\left[\cos(\alpha - \beta)(h^2 + a^2 - 2ha\sin\theta)^{\frac{1}{2}}\right] \tag{4.139}$$

where

$$\tan\beta = \frac{h - a\sin\theta}{a\cos\theta}$$

Alternatively,

$$y = (\tan\theta)x + h \tag{4.140}$$

Both coordinates define the location of P in the xy frame. The coordinates of point P in the x_1y_1 frame are determined by the transformation equations given above.

Example Consider the following parameters.

$a = 6$ in.
$b = 3$ in.
$d = 9$ in.
$\alpha = 0.17453$ rad ($10°$)
$h = 1$ in.

The above equations were programed to obtain the mating tooth profile as viewed from the x_1y_1 reference from. Both profiles are shown Figure 4.32. The portion of each profile used for form teeth on both gears depends on the diametral pitch of the gear system.

These equations were used to generate gear tooth pairs for a variety of flat surface orientations. One such gear pair is shown in Figure 4.33.

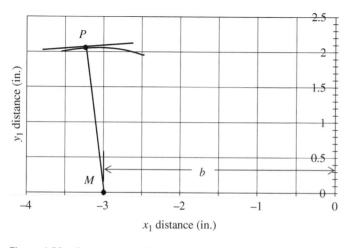

Figure 4.32 Gear tooth profiles.

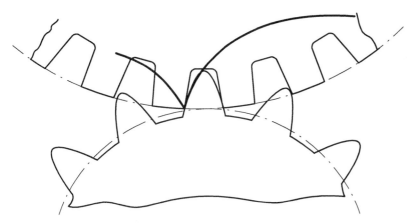

Figure 4.33 Gear pair with flat teeth.

A companion study defines flat tooth profiles in hypoid gears [10].

Cam Drives

The purpose of a cam is to convert rotary motion into linear motion with specified displacements over one complete rotation of the cam. The specified motion will normally have a low and a high point with intermediate positions related to the shape of the cam.

Velocity and acceleration of the follower are also important as they affect contact forces and possible separation. Cams of a given shape are keyed to a shaft and rotated at constant velocities. The shape of a cam is referenced from a rotating frame as shown. The challenge is to determine velocity and acceleration of the follower at different angular position of the cam.

Consider the velocity and acceleration of cam and follower. Three cases will be discussed: (i) linear displacement of a flat surface follower, (ii) linear displacement of a roller follower, and (iii) angular displacement of a rocker arm follower.

In each of these examples, point P is chosen as a point within a rotating xy reference frame.

Cam Drives – Linear Follower

In this case, the follower experiences linear motion, which is dictated by the geometry of the cam. Follower displacement is important as the purpose of cam drives is to convert rotary motion into a prescribed displacement to the follower. Once cam geometry has been set, based on the desired follower displacement, it is important to know the velocity and acceleration of the follower.

Example The first case is illustrated in Figure 4.34. The objective is to determine the velocity and acceleration of the follower. Calculations are based on the following input parameters.

$r = 4$ in.
$R = 6$ in. (radius of curvature of cam surface at point of contact)
$\omega = 2$ rad/s
$\theta = 30°$
$\phi = 25°$

Velocity Analysis
The equation used to find the vertical velocity of the linear follower is based on

$$\bar{v}_A = \bar{v}_P + \bar{v}_{A/P} \tag{4.141}$$

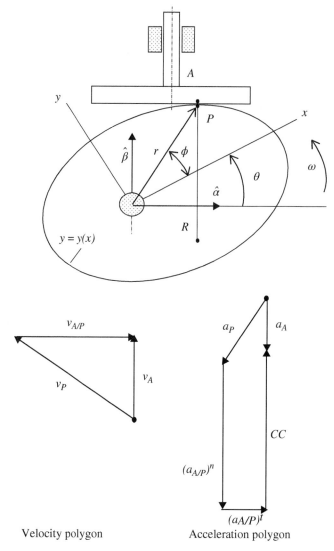

Velocity polygon   Acceleration polygon

Figure 4.34 Cam and linear follower.

The component velocities are

$$\bar{v}_A = v_A \hat{\beta}$$

The magnitude v_A is unknown, but its direction is known. The velocity of point P is

$$\bar{v}_P = \omega r \left[-\sin (\theta + \phi) \hat{\alpha} + \cos (\theta + \phi) \hat{\beta} \right] \tag{4.142}$$

$$\bar{v}_P = 2(4) \left[-\sin 55 \hat{\alpha} + \cos 55 \hat{\beta} \right] = -6.55 \hat{\alpha} + 4.59 \hat{\beta}$$

The angle ϕ depends on cam geometry, θ, is the independent variable.

The velocity of point A relative to point P can be visualized by the principle of kinematic inversion, i.e. the relative motion is the same regardless of which link is fixed. In this case, we choose to fix the cam and let the fixture rotate *CW*. The direction of point *A* relative to point *P* is tangent to the cam surface with direction to the right; therefore

$$\bar{v}_{A/P} = \hat{\alpha} v_{A/P} \tag{4.143}$$

The vector components are:

$$\bar{v}_P = 4(2)\left[-0.8192\hat{\alpha} + 0.5736\hat{\beta}\right] = -6.55\hat{\alpha} + 4.59\hat{\beta} \text{ ips} \tag{4.144}$$

$$\bar{v}_{A/P} = v_{A/P}\hat{\alpha} \text{ ips} \tag{4.145}$$

$$\bar{v}_A = v_A\hat{\beta} \tag{4.146}$$

By substitution

$$v_A\hat{\beta} = -6.55\hat{\alpha} + 4.59\hat{\beta} + v_{A/P}\hat{\alpha} \tag{4.147}$$

Therefore

$$v_A = 4.59 \text{ ips} \quad \text{and} \quad v_{A/P} = 6.55 \text{ ips}$$

The velocity polygon is shown in Figure 4.34. These velocities apply only for the given values of and θ.

Acceleration Polygon

The acceleration polygon is based on

$$\bar{a}_A = (\bar{a}_P)_t + (\bar{a}_P)_n + \bar{a}_{xy} + 2\bar{\omega}x\bar{v}_{xy} \tag{4.148}$$

where

$$\bar{v}_{xy} = \bar{v}_{A/P} = 6.55\hat{\alpha}$$

$$\bar{a}_{xy} = \bar{a}_{A/P}$$

The relative acceleration components can also be visualized by use of the principle of kinematic inversion where the cam is fixed and other linkages, including the frame, rotate clockwise. The normal component of point *A* relative to *P* results from the curvature of the cam. The tangent component of *A* relative to *P* is tangent to the cam surface at the contact point.

a_A – magnitude is unknown, but its direction is along the vertical center line
$(a_P)_t$ – zero, because ω is constant
$(a_P)_n = \omega^2 r = (2)^2 4 = 16 \text{ ips}^2$
$(a_{A/P})_n = \omega^2 R = (2)^2 6 = 24 \text{ ips}^2$ (relative to rotating *xy* reference frame)
$(a_{A/P})_t$ – magnitude unknown, direction tangent to contact surface
$2\omega x v_{A/P} = 2(2)6.55 = 26.2 \text{ ips}^2$ (Coriolis component)

By substitution

$$\bar{a}_A = (\bar{a}_P)^n + (\bar{a}_{A/P})^t + (\bar{a}_{A/P})^n + 2\bar{\omega}x\bar{v}_{A/P} \tag{4.149}$$

Giving

$$a_A\hat{\beta} = 16(-16\cos 55\hat{\alpha} - \sin 55\hat{\beta}) + a^t_{A/P}\hat{\alpha} - 24\hat{\beta} + 26.2\hat{\beta}$$

which in turn gives $a_A = -10.9\text{in.}/\text{sec}^2$ and $a^t_{A/P} = 9.18 \text{ in.}/\text{s}^2$

The acceleration polygon is shown in Figure 4.34.

Cam with Linear Follower, Roller Contact

The cam chosen for this discussion contains two circular arcs (Figure 4.35). The instant of time under consideration assumes the follower is engaged with the cam at the beginning of the circular arc having radius, R. Rotation of the cam is clockwise. We wish to determine velocity and acceleration of the follower using: (i) Coriolis' law and (ii) Ritterhaus model (shown as a slider crank), which is kinematically equivalent of the cam over the circular arc travel.

Example Numerical input to the example is

- $R = 1\frac{1}{2}$ in.
- $r = \frac{1}{2}$ in.

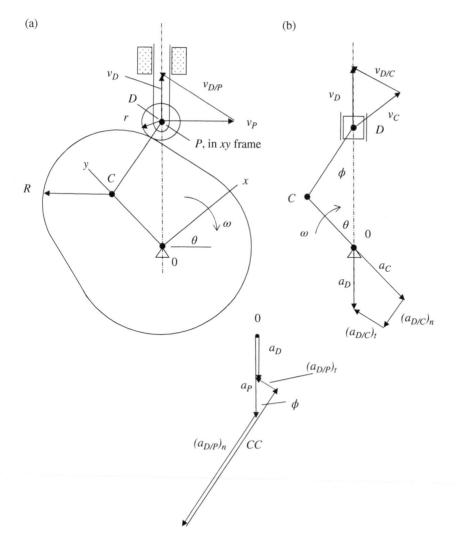

Figure 4.35 Cam with linear follower, roller contact. (a) (b) Ritterhaus model.

- $OC = 1.75$ in.
- $N = 10$ rpm $\quad \omega = \dfrac{2\pi N}{60} = \dfrac{2\pi 10}{60} = 1.05 \, \text{rad/s}$
- $\theta = 45°$
- From the law of cosines, $OD = 2.81$ in.

Velocity Analysis – Rotating Reference Frame
With reference to Figure 4.35a, the velocity of point A (on follower) is determined from

$$\bar{v}_D = \bar{v}_P + \bar{v}_{D/P}$$

We know the direction and magnitude of point P. Only the direction of the other two is known. The distance OD is determined by

$$OD = OC \cos\theta + CD \cos\phi$$

where

$$OC \sin\theta = CD \sin\phi$$

When $\theta = 45°$, then $\phi = 38.2°$ and $OD = 2.81$ in. The velocity of point P then becomes

$$\bar{v}_P = 2.81(1.05) = 2.95 \, \text{ips}$$

From the velocity polygon

$$v_D = v_P \tan\phi = 2.95 \tan 38.2 = 2.33 \, \text{ips}$$

$$v_{D/P} = \frac{v_P}{\cos\phi} = \frac{2.95}{0.786} = 3.75 \, \text{ips}$$

Acceleration Analysis – Rotating Reference Frame
The acceleration of the linear slider is determined by both the Coriolis approach and the Ritterhaus model. The acceleration polygon based on the Ritterhaus model is shown in the figure. The Coriolis approach is given below.

The controlling equation is

$$\bar{a}_D = (\bar{a}_P)_t + (\bar{a}_P)_n + \bar{a}_{xy} + 2\bar{\omega}x\bar{v}_{xy} \tag{4.150}$$

$$\bar{a}_D = (\bar{a}_P)_t + (\bar{a}_P)_n + (\bar{a}_{xy})_t + (a_{xy})_n + 2\bar{\omega}x\bar{v}_{xy} \tag{4.151}$$

where point P is located on the cam coincident with point D. The values of each component are listed below.

$\bar{a}_D$	–	direction is along a vertical line, *magnitude unknown*
$(\bar{a}_P)_t = 0$	–	cam has constant angular velocity
$(\bar{a}_P)_n = OD\omega^2$	–	direction known, from point D to point O
$(\bar{a}_{xy})_t$	–	direction known (perpendicular to CD), *magnitude unknown*
$(a_{xy})_n = \dfrac{v_{D/P}^2}{(R+r)}$	–	direction from point D toward C
$2\bar{\omega} \times \bar{v}_{xy} = 2\omega(CD\omega)$	–	magnitude known, direction along line DC but away from C

Using the input numbers

$$a_P = (a_P)_n = OD\omega^2 = 2.81(1.05)^2 = 3.1 \, \text{in./s}^2$$

$$\left(a_{xy}\right)_n = \left(a_{D/P}\right)_n = \frac{3.75^2}{2} = 7.03 \text{ in./s}^2$$

$$2\bar{\omega} \times \bar{v}_{xy} = CC = 2(1.05)(3.75) = 7.88 \text{ in./s}^2 \tag{4.152}$$

Bring the components together and using the acceleration polygon

$$a_D = a_P - \frac{\left(CC - a_{D/P}^n\right)}{\cos 38.2} = 3.1 - \frac{(7.88 - 7.03)}{0.786} = 2.01 \text{ ips}$$

$$a_{D/P}^t = 1.08 \sin 38.2 = 0.67 \text{ ips}^2$$

The solution is shown graphically in Figure 4.35.

Velocity Analysis – Ritterhaus Model

The velocity of the piston (point D) is determined from

$$\bar{v}_D = \bar{v}_C + \bar{v}_{D/C}$$

where

$$v_C = OC\omega = 1.75(1.05) = 1.84 \text{ ips}$$

$$v_C \cos\theta = v_{D/C} \cos\phi$$

$$v_{D/C} = \frac{\cos 45}{\cos 38.2} v_C = 1.66 \text{ ips}$$

$$v_D = v_{D/C} \sin\phi + v_C \sin\theta = 1.66(0.618) + 1.84(0.707) = 2.33 \text{ ips (close agreement)}$$

Acceleration Analysis – Ritterhaus Model

The acceleration of point D is determined from

$$\bar{a}_D = \bar{a}_C + \bar{a}_{D/C}$$

$$\bar{a}_D = \left(\bar{a}_C\right)_n + \left(\bar{a}_{D/C}\right)_n + \left(\bar{a}_{D/C}\right)_t$$

$$a_D\left(-\hat{\beta}\right) = OC\left(\omega^2\right)\left(\hat{\alpha}\sin\theta - \hat{\beta}\cos\theta\right) - (CD)\Omega^2\left(\hat{\alpha}\sin\phi + \hat{\beta}\cos\phi\right) + CD\alpha\left(-\hat{\alpha}\cos\phi + \hat{\beta}\sin\phi\right)$$

$$-a_D\hat{\beta} = 1.36\left(\hat{\alpha} - \hat{\beta}\right) - 0.85\hat{\alpha} - 1.08\hat{\beta} + \alpha\left(-1.57\hat{\alpha} + 1.24\hat{\beta}\right)$$

The solution comes from the two unit-vector components

$$-a_D = -1.36 - 1.08 + 1.24\alpha$$

$$0 = 1.36 - 0.85 - 1.57\alpha$$

which give

$$\alpha = 0.372 \text{ rad/s}^2 \quad \text{and} \quad a_D = 1.98 \text{ ips}^2 \text{ (close agreement)}$$

Cam with Pivoted Follower

Consider the cam arrangement shown in Figure 4.36. In this case, the coincident point P is the center of the disc follower. Point A is coincident with P but located on an imaginary extension (dashed line) of the cam; member 2. In this example, we assume the cam rotates with constant velocity, Ω

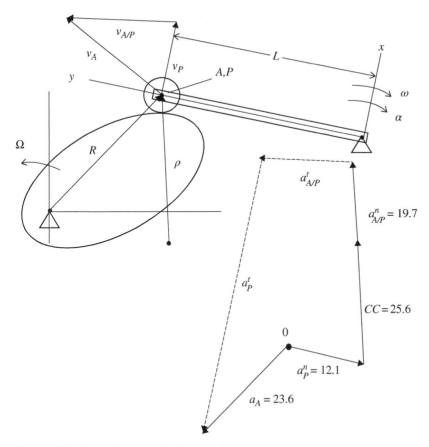

Figure 4.36 Cam with pivoted follower, roller contact.

counterclockwise. The problem is to determine both angular velocity (ω) and angular acceleration (α) of the follower arm for this position.

Example Parameters used in this example are

$R = 5.4$ in.
$L = 6$ in.
$\rho = 4.5$ in. (determined from the geometry of the cam)
$\Omega = 20$ rpm $= 2.09$ rad/s

With the rotation reference (xy) frame attached to cam follower

$$\bar{v}_A = \bar{v}_P + \bar{v}_{A/P} \tag{4.153}$$

$$v_A = 5.4(2.09) = 11.3 \text{ ips}$$

The direction of relative velocity, $v_{A/P}$, is tangent to the contact surface (see velocity polygon). The relative velocity can be visualized by use of the principle of inversion; fix the cam and imagine the remaining portion of the mechanism move clockwise about the cam giving $v_{P/A}$. The direction of the velocity of point P is perpendicular to the pivoting arm. For the sake of simplicity, velocities v_P and $v_{A/P}$ are scaled from the velocity polygon as

$$v_P \sim 8.5 \text{ ips}$$

$$v_{A/P} \sim 9.0 \text{ ips}$$

$$\omega = \frac{8.5}{6} = 1.42 \text{ rad/s}$$

The acceleration equation is

$$\bar{a}_A = \bar{a}_P^t + \bar{a}_P^n + \bar{a}_{A/P} + 2\varpi x v_{A/P} \tag{4.154}$$

where

$$\bar{a}_{A/P} = a_{A/P}^t + a_{A/P}^n \tag{4.155}$$

Equation (4.154) is the basis for the acceleration polygon shown in Figure 4.36. The known vector components are shown as solid vectors. The two unknown vectors are shown as dashed vectors. Their intersection closes the polygon producing the angular acceleration of the follower arm.

Note that point P is the coincident point located on the y axis and point A is located on the imaginary "extension" of the cam.

The magnitudes of the acceleration components are determined from

$$a_A = a_A^n = \Omega^2 R = 2.09^2(5.4) = 23.6 \text{ ips}^2 \tag{4.156}$$

$$a_P^n = L\omega^2 = 6\omega^2 = 6(1.42)^2 = 12.1 \text{ ips}^2 \tag{4.157}$$

a_P^t only direction is known

$$a_{A/P}^n = \Omega^2 \rho = 2.09^2(4.5) = 19.66 \text{ ips}^2 \tag{4.158}$$

$a_{A/P}^t$ only direction is known

$$a_{CC} = 2\omega v_{A/P} = 2(1.42)9 = 25.6 \text{ ips}^2 \tag{4.159}$$

The direction of $\bar{a}_{CC}$ is based on the cross product of $\varpi x \bar{v}_{A/P}$ following the right-hand rule. It is parallel to radius of curvature, ρ.

The angular velocity of the arm is determined by $\omega = \dfrac{v_P}{L}$. The unknowns in the polygon are a_P^t and $a_{A/P}^t$. These components are represented by the dashed lines in the polygon. The angular acceleration of the follower is determined by $\alpha = \dfrac{(a_P)^t}{L} \sim \dfrac{50}{6} = 8.3 \text{ rad/s}^2$. Its direction is opposite to the assumed direction in the drawing.

Power Screw

Power screws are used to convert rotary motion of a screw to linear motion of a nut. Threads are usually acme type for applications of high load. Force is also converted from torque to linear force on the nut. A machine component is typically attached to the nut to achieve a desired motion.

The relation between rotary torque and liner force is

$$2\pi T = pF \tag{4.160}$$

where

T	–	torque, in-lb
p	–	pitch of threads on screw drive, in
F	–	axial force to be moved, lb

The equation is based on the input energy supplied by torque, T, over on complete rotation of the screw and output energy of displacing the force, F, acting over one thread pitch. The power for motivating screw drives can come in through the nut or through the screw. Power transmission is often achieved by a stepping motor or by a direct drive to the screw itself. Stepping motors are commonly used when computers control platform motion. The mass of movable platforms affects input power according to

$$2\pi T(t) = p[Ma(t)] \tag{4.161}$$

where $T(t)$ varies with time. M is mass and $a(t)$ is its linear acceleration.

The efficiency of acme lead screws is 20–25%. Much of the input power is lost to friction in the threads. Also, there is backlash in the threads and this limits its application.

Hydraulic Transmission of Power

Downhole motors are used in oil well drilling to: (i) increase mechanical power to drill bits and (ii) control the direction of drilling. These motors are located within the drillstring near a drill bit (see Figure 4.37). They are powered by drilling fluid, which arrives at these motors under pressure and a rate of flow. These motors essentially convert hydraulic power to mechanical power in the form of bit torque and rotary speed.

The power in the circulating system goes through several transformations starting with mechanical power delivered to the mud pumps [11]. The power chain includes:

- Diesel engine – mechanical
- Generator – electrical
- Electric motor – electrical
- Mud pump – mechanical to hydraulic
- Positive displacement motor (PDM)/turbine – hydraulic to mechanical

The mechanical power output of downhole motors/turbines starts with mechanical power delivered by diesel engines.

$$P_M = \frac{2\pi TN}{33\,000} = \frac{TN}{5252}\,\text{hp} \tag{4.162}$$

where

P_M	–	mechanical power, hp
T	–	torque, ft-lb
N	–	rotary speed, rpm

Hydraulic power is related to fluid pressure and flow rate by

$$P_H = \frac{pQ}{1714}\,\text{hp} \tag{4.163}$$

Figure 4.37 Energy transformations in drilling operations.

Drill fluid pump

Fluid flow

Drill pipe

Rotation

Drill bit

Drilling motor

Bit torque

where

P_H	–	hydraulic power, hp
p	–	fluid pressure, psi
Q	–	flow rate, gpm

The exchange of power across a motor/turbine is illustrated in Figure 4.38. Applying Bernoulli's energy equation across the motor gives

$$\frac{p_1}{\gamma} = \frac{p_2}{\gamma} + E_{out} \frac{\text{ft-lb}}{\text{lb}} \tag{4.164}$$

The unit of each term is "ft" or ft-lb per lb of fluid flowing thought the control volume. To convert this equation to units of power

$$\frac{\Delta p}{\gamma}(\gamma Q) = E_{out}(\gamma Q) \quad \text{(energy per time)} \tag{4.165}$$

Figure 4.38 Hydraulic power converted to mechanical power.

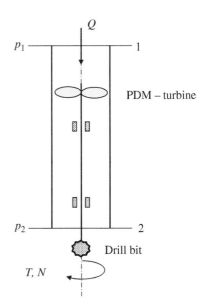

In terms of hydraulic and mechanical power

$$(P_H)_{in} = (P_M)_{out} \tag{4.166}$$

$$\frac{\Delta pQ}{1714} = \frac{TN}{5252} \tag{4.167}$$

assuming no friction losses.

The efficiency of a motor is determined by

$$\eta = \frac{(P_M)_{out}}{(P_H)_{in}} \tag{4.168}$$

The performance of PDMs is monitored through rig data. Speed (N) is directly related to flow rate (Q) and torque (T) is directly related to pressure drop (Δp). These data are available to the driller by means a flow meter and the standpipe pressure gauge.

It has long been recognized that only a small portion of total power (~8000 hp) at the surface is used at the drill bit for boring into a formation. For example, under normal rotary drilling operations ($N = 100$ rpm, $T = 1500$ ft-lb), horsepower delivered to drill bits for boring into a formation is

$$P_{bit} = \frac{2\pi TN}{33\,000} \tag{4.169}$$

$$P_{bit} = \frac{2\pi(1500)100}{33\,000} = 28.6 \text{ hp} \tag{4.170}$$

a very small portion of total rig power. By using PDMs (Figure 4.39), bit rotary speed increases to, say, 400 rpm, creating four times the power at the drill bit.

$$P_{bit} = \frac{2\pi(1500)400}{33\,000} = 114.2 \text{ hp} \tag{4.171}$$

This increase in drilling power comes at a cost, which must be weighed against rig rate ($/d) and increase in rate of penetration (ROP).

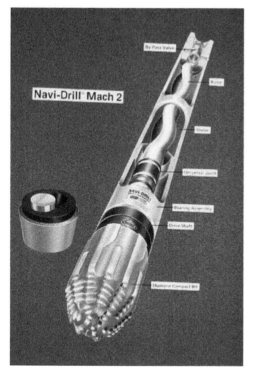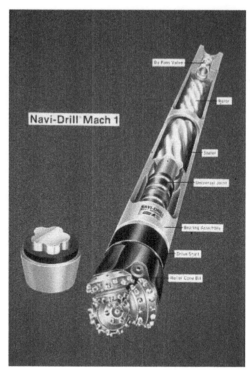

Figure 4.39 Downhole drilling motors. *Source:* Baker Hughes.

Kinematics of the Moineau Pump/Motor

The Moineau principle has been applied to pumping applications, especially in petrochemical industries. As a pump, the Moineau pump is a positive displacement tool delivering fluids at a rate proportional to input rotary speed. It is especially adapted to pipelines and flowlines because of its tubular length and diameter. Since the early 1970s, the Moineau principle has been used as a drilling motor, which is driven by hydraulic power. In this case, power is taken from fluid power and converted to mechanical power in terms of rotary speed and torque [12].

The kinematics of the rotor relative to the fixed stator can be analyzed as a planet pitch circle (rotor) and an annular pitch circle as shown in Figure 4.40. Of interest is the motion of a point on the rotor pitch circle. The motion of a point P on the pitch circle scribes a hypocycloid in space relative to the fixed stator. The path of the hypocycloid is defined by the following *xy* coordinates.

The vector position $(\overline{R})$ of point P is

$$\overline{R} = \left(a\hat{i}\cos\phi + a\hat{j}\sin\phi\right) + \left(r\hat{i}\cos\theta - r\hat{j}\sin\theta\right) \tag{4.172}$$

Gathering the x and y components gives

$$\overline{R} = (r\cos\theta + a\cos\phi)\hat{i} + (-r\sin\theta + a\sin\phi)\hat{j} \tag{4.173}$$

From which

$$x = r\cos\theta + a\cos\phi \tag{4.174}$$

$$y = -r\sin\theta + a\sin\phi \tag{4.175}$$

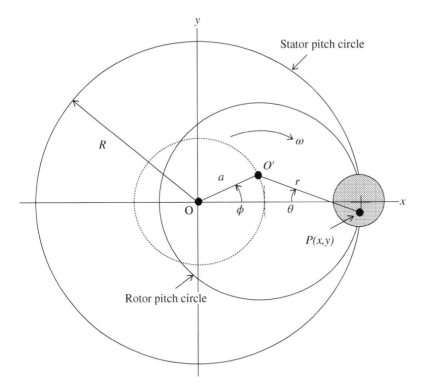

Figure 4.40 Hypocycloid geometry.

The velocity and acceleration of point P can easily be determined by differentiation. The angles, and θ, are related through the motion of the pitch circles and either or θ is known. If, for example, the ratio of the radii of the pitch circles is $\dfrac{r}{R} = \dfrac{3}{4}$ then the hypocycloid path is as shown in Figure 4.41. If a circle is attached to the rotor pitch circle with point P as its center, an outline of a stator geometry is generated representing the shape of a 3 : 4 stator "gear."

Using a planetary gear table, the relation between ϕ and θ is

$$\frac{\phi}{\theta} = \frac{1}{1 - \dfrac{R}{r}} = \frac{1}{1 - \dfrac{4}{3}} = -3 \tag{4.176}$$

Therefore, $\phi = 3\theta$.

Mechanics of Positive Displacement Motors

The geometry of rotors and stators can be configured to achieve a desired relation between flow rate Q and rotational speed N. Output speed and flow rate are related by $N = CQ$. The following correlates the design variables that affect the value of C.

$$C = \frac{n}{q} \tag{4.177}$$

where

n	–	turns of output shaft (drill bit)
q	–	corresponding volume through motor

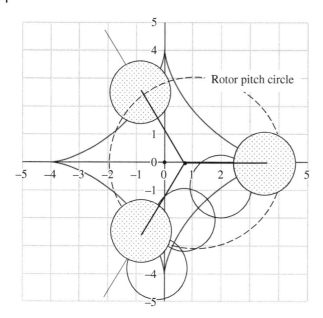

Figure 4.41 Hypocycloid path of point P for a 3 : 4 lobed motor.

Since the rotor and stator interact as a planetary gear train, the number of turns (n) of the output shaft (equal to number of turns of the rotor) per one complete circular travel of the rotor center is

$$n = \frac{a}{b} - 1 \tag{4.178}$$

where

a	–	number of lobes on the stator
b	–	number of lobes on the rotor

For example, if a stator has two lobes and the rotor one lobe, $n = 1$, then the output shaft makes one turn for one circular path of the rotor center. Or, if a stator has 10 lobes and the rotor 9, the output rotation would be 1/9 rotation for each circular path of the rotor, a much smaller rotation.

Consider the volume throughput (q) for one circular movement of the rotor center. Let the two cylinders in Figure 4.42 represent the pitch cylinders of a rotor and stator. Let the length of both pitch cylinders be the length of one stage of the stator, L_s. For the sake of visualization, assume:

- center lines O and O' are fixed in space
- both pitch cylinders are torsional flexible
- top ends of both rotor and stator are fixed
- lower end of the stator makes one complete turn causing the lower end of the rotor to make a/b turns
- rotation of both rotor and stator are linear with distance from the top fixed end.

Based on the principle of kinematic inversion, the volume between the rotor and stator over stator length, L_s, is displaced down the motor whether point O is viewed as making a circular path

Figure 4.42 Pitch cylinders of rotor and stator.

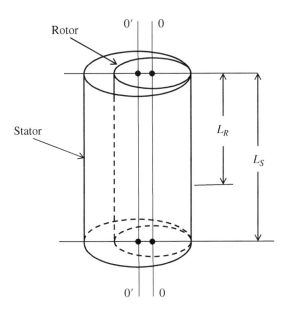

about point O', or both O and O' are viewed as being fixed and the two pitch circles mesh as a gear pair. This volume is

$$q = AL_S \tag{4.179}$$

where

A	–	cross-sectional area of the space between rotor and stator cavity
L_s	–	length of one stage as measured on the stator

Note, this volume is different for each isolated cavity entrapped between rotor and stator. It represents the total volume over one stage (L_S).

Combining Eqs. (4.178) and (4.179) with Eq. (4.177) gives

$$C = \frac{n}{q} = \frac{\frac{a}{b} - 1}{AL_s} \tag{4.180}$$

Since $a = b + 1$ for PDMs

$$C = \frac{1}{bAL_s}$$

and since $L_r = \frac{b}{a}L_s$

$$C = \frac{1}{aAL_r}$$

The relation between rotary speed and flow rate, $N = C \cdot Q$ becomes

$$N = \frac{1}{bAL_s}Q \tag{4.181}$$

and noting that $bL_s = aL_r$

$$N = \frac{1}{aAL_r}Q \tag{4.182}$$

AL_r is the total entrapped volume between rotor and stator over length L_r. In consideration of Eqs. (4.181) and (4.182), the smaller the denominator, the larger the output rotational speed for a given flow rate, Q. Also, the larger the number of lobes, the lower the output speed. That's why multilobed motors operate at low speeds and high torque. Note that N and Q are proportional, which is characteristic of PDMs and pumps.

The dimensional unit of the motor constant, C, is output revolution per gallon pumped through the motor. If, for example, 450 gpm flow rate through a one to two lobed motor produces an output rotational speed of 450 rpm,

$$Q = 450 \text{ gpm}$$
$$b = 1$$

the volume of fluid between rotor and stator over one stage of the stator is

$$AL_S = 1 \text{ gal}$$

This 1-gal volume can be achieved with either a large area and small stage length or a small area and large stage length. Almost any flow rate and rotational speed relationship can be established by various combinations of b, A, and L_S.

Consider a motor design where $AL_S = 1$ gal and the number of lobes on the rotor is $b = 3$. The motor constant in this case is

$$C = \frac{1}{bAL_S} = \frac{1}{3} \tag{4.183}$$

and

$$N = \frac{1}{3}Q$$

A flow rate of 450 gpm produces an output rotational speed of 150 rpm.

A high-speed motor could be designed by making the product bAL_S small. The smallest that "b" can be is one. A one to two lobed motor with either a small stage length or a small void area or both would be a high-speed motor. A low-speed motor could be designed by making the product bAL_S large. Therefore, low-speed/high-torque motors are multilobed.

If a motor is to be designed to deliver a certain output speed (RPM) at a given flow rate (gpm), the motor constant, C, is known and becomes a design specification. Motor geometry is determined as follows.

$$C = \frac{N}{Q} = \frac{1}{bAL_s} \tag{4.184}$$

To balance the units when AL_s has units of in^3 and Q (gal)

$$C = \frac{144(12)}{bAL_s(7.48)}$$

or

$$bAL_s = \frac{231}{C} \tag{4.185}$$

where

b	–	number of lobes on rotor
A	–	flow area through motor, in.^{2}
L_s	–	length of stator, in.

Motor dimensions can be selected once the motor constant, C, have been specified.

Example Assume that a motor is specified to have a rotational speed of 300 rpm at a flow rate of 400 gpm. In this case

$$C = \frac{N}{Q} = \frac{300}{400} = \frac{3}{4} \tag{4.186}$$

The motor has to be configured so that

$$bAL_s = \frac{4}{3}231 = 308$$

The selection of A and L_s depends on tool size and length. The final dimensions are made through a series of design trade-offs.

Note that by measuring A and L_s of a motor one can determine the motor constant, C, using Eq. (4.183) and estimate the relationship between output speed and flow rate through the motor by assuming no slipping.

The conversion of fluid pressure to mechanical torque is the result of the fluid wedge at the fluid entry or first stage. Fluid density and flow rate have no effect on torque generated by the motor. PDMs are typically made up of several stages, and each stage contains an entrapped volume of fluid, which travels through the motor. Even though these motors have multiple stages, no work is done by the constant volume cavities, and only the top stage converts hydraulic pressure to torque.

Multiple stages, usually about five, perform as dynamic seals, so reasonable pressure differentials across these motors can be developed. Performance tests show that even these back-up stages allow a certain amount of fluid leakage. The extra stages do, however, create friction and affect the overall efficiency.

Since the hydraulic horsepower consumed by the motor is equal to the output mechanical output power (assuming no losses)

$$\Delta pQ = TN \tag{4.187}$$

So, output torque is

$$T = \frac{Q}{N}\Delta p$$
$$T = (bAL_s)\Delta p \tag{4.188}$$

This equation shows that output torque increases with number of lobes on the rotor (and stator). Torque also increases with stage length, which increases the wedging effect of fluid entering the motor.

It has long been recognized that out of the total mechanical power (~8000 hp) required to operate a drilling rig, only a small fraction is used at drill bits. For example, under normal rotary drilling operations, bit horsepower is

$$P_{bit} = \frac{2\pi TN}{33\ 000} \tag{4.189}$$

$$P_{bit} = \frac{2\pi(1500)80}{33\ 000} = 22.8 \text{ hp} \tag{4.190}$$

The output speed of a 1 : 2 PDM is in the range of 400 rpm. Corresponding output power is

$$P_{bit} = \frac{2\pi(1500)400}{33\ 000} = 144 \text{ hp} \tag{4.191}$$

Bit power is increased by a factor of 4. This increase in bit power comes at a cost because motors cost (C_{pdm}) is added into footage cost calculations. However, PDM cost is offset by increased ROP.

Note that mud density is not a factor in torque or power generation in PDMs. This is not the case in power generation of turbines.

References

1 Ham, C.W., Crane, E.J., and Rogers, W.L. (1958). *Mechanics of Machinery*. NY: McGraw-Hill.

2 Shigley, J.E. and Mitchell, L.D. (1983). *Mechanical Engineering Design*, 4e. McGraw-Hill Inc.

3 Wildhaber, E. (1961). Gears with Circular Tooth Profile Similar to the Novikov System, VDI Berichte, No. 47, Germany (first introduced by Ernest Wildhaber in 1910).

4 Novikov, M. (1956). L. USSR Patent No. 109,750.

5 Stadtfeld, D.H.J. and Saewe, J.K. (2015). *Non-Involute Gearing, Function and Manufacturing Compared to Established Gear Designs*, 42–51. Gear Technology.

6 Mitchner, R.G. and Mabie, H.H. (1982). The determination of the Lewis form factor and the AGMA geometry factor J for external spur gear teeth. *ASME J. Mech. Des.* **104** (1): 148–158.

7 Timoshenko, S. and Goodier, J.N. (1951). *Theory of Elasticity*, 2e. McGraw-Hill Book Co. Inc. (see torque of non-circular cross sections).

8 Myszka, D.H. (2013). *Machines and Mechanisms, Applied Kinematic Analysis*, 4e. Pearson – Prentice Hall.

9 Shigley, J.E. (1961). *Theory of Machines*. McGraw-Hill Book Co, Inc.

10 Dareing, D. W. and Chung-Moon Chen, C.-M. (1973). *J. Eng. Ind. Trans. ASME*, **95**, B (4): 1171–1177.

11 Dareing, D.W. (2019). *Oilwell Drilling Engineering*. ASME Press.

12 R.J.L. Moineau (1932). Gear Mechanism, U.S. Patent 1 892 217.

5

Friction, Bearings, and Lubrication

In 1951 Vogelpohl [1] estimated that about $\frac{1}{3}$ to $\frac{1}{2}$ of energy produced is consumed by friction. Considering all automobiles, trucks, and other modes of transportation plus machinery used in manufacturing, it is not surprising the level of energy lost to friction is large. Also, equipment wears out because of friction and must be replaced; another economic loss. Because of this, mechanical friction and wear became a focus of research during the early 1960s. A study conducted (1966) in Great Britain by Jost [2] concluded that by following good design and lubrication procedures, about 500-million-pound sterling could be saved per year. Studies in Europe and America have reached similar conclusions. The general area of friction and wear became known the science of tribology, a Greek word meaning a study of friction, lubrication, and wear between moving surfaces.

Rolling Contact Bearings

Shafts must be supported on some type of bearing to accommodate applied loads and allow for shaft rotation. There are two broad categories of mechanical bearings: rolling contact bearings and thick film–lubricated bearings. Within these two bearing categories, there are various alternatives. In both cases, bearings are selected to support applied loads with minimum friction and wear.

Roller contact bearings are used in nearly every industry as well as jet engines and rocket engines. They are reliable and forgiving. Failure usually comes in the form of pitting, but even then, they still perform. Types of roller contact bearings are (Figure 5.1):

a) Cylindrical roller
b) Tapered roller
c) Deep groove ball
d) Angular contact ball

There is a great deal of history on performance and a comprehensive data bank for making statistical predictions on their performance.

Table 5.1 lists coefficients of friction for various bearings and materials.

Equipment literally "wears out" because of wear. In most cases, friction and wear produce heat, which accelerates wear and if left unchecked can destroy equipment. Temperature is a measure of heat and a good indicator of friction and wear. The objective in design is to minimize friction and thus minimize mechanical wear and heat.

Engineering Practice with Oilfield and Drilling Applications, First Edition. Donald W. Dareing.
© 2022 John Wiley & Sons, Inc. Published 2022 by John Wiley & Sons, Inc.

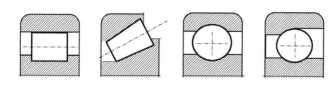

Figure 5.1 Rolling contact bearings.

Table 5.1 Comparison of friction.

	Coefficient of friction (typical)
Dry friction	
Generic	0.3
Steel on copper	0.15
Teflon on anything	0.1
Thick film bearings	
Journal bearings	0.001
Slider bearings	0.001
Roller contact bearings	
Cylindrical roller	0.001–0.003
Tapered roller	0.002–0.005
Deep groove ball	0.0015–0.003
Angular contact ball	0.0015–0.002

Rolling contact bearings normally fail by pitting caused by cyclic stresses beneath the surface of either the roller or race. The criterion for bearing failure is a 6 mm² (0.01 in.²) area pit in the surface of either bearing race. These bearings are forgiving in that they will still perform even with a pit in the surface, and their performance can be monitored easily with vibration and noise sensors.

Advantages of rolling contact bearing over thick film–lubricated bearings are simplicity and reliability. On the other hand, hydrostatic bearings can be designed to support roofs of sports arenas with minimal space in comparison to ball bearings.

Rated Load of Rolling Contact Bearings

In a typical industrial setting, rolling contact bearings are lubricated and, in many cases, sealed. The life of these bearings is related to a steady load by

$$L = \left(\frac{C}{P}\right)^q \tag{5.1}$$

where

L – expected bearing life in millions of revolutions
P – bearing load; assumed to be steady, lb
C – basic load rating of a given bearing size and type, lb (corresponding to 10^6 rev)
q – exponent (3.33 for roller bearings, 3.0 for ball bearings)

This empirical equation is based on laboratory testing and operating performance. Performance data are displayed on a log–log graph (Figure 5.2).

The exponent, q, is determined as follows:

$$\frac{L}{10^6} = \left(\frac{C}{P}\right)^q \tag{5.2}$$

$$\text{Log}\frac{L}{10^6} = \log\left(\frac{C}{P}\right)^q = q\log\frac{C}{P} \tag{5.3}$$

$$q = \frac{\log L - \log 10^6}{\log C - \log P} \tag{5.4}$$

Note that q is *not* the slope of the performance line in Figure 5.2, but it's reciprocal. This can be checked with the q values in Figure 5.5.

If "L" is defined as life in millions of revolutions, then Eq. (5.1) applies. The performance line is drawing through the data such that 90% of measured data is above the line and 10% is below the line. This means that bearing performance based on this line has a reliability of 0.9. Rolling contact bearing failure is defined as fatigue pitting on either the roller or race. Even under this condition, roller bearings can still function.

Commercial bearings are rated by the value of "C," which corresponds to a life of 10^6 cycles. This coordinate is for convenience only as bearing loads are not expected to perform as this load level. Equation (5.5) is used to select bearings for specific applications.

$$C = PL^{\frac{1}{q}} \tag{5.5}$$

where P and L are design requirements and C is the bearing "dynamic load rating" listed in catalogs.

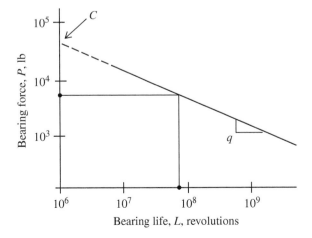

Figure 5.2 Bearing load rating.

There are many types of rolling contact bearings as indicated earlier, each having special features to satisfy specific operating requirements.

Catalog ratings are based on radial load. In general, axial loads may also be applied, in which case an equivalent radial load, F_e, is used to select an appropriate bearing. The Anti-Friction Bearing Manufactures Association (AFBMA) equation for finding the equivalent radial load for ball bearings is[1]

$$F_e = VF_r \tag{5.6}$$

or

$$F_e = XVF_r + YF_a \tag{5.7}$$

where

F_e	–	equivalent radial load
F_r	–	applied radial load
F_a	–	applied axial load
V	–	rotation factor (1 for inner race rotation, 1.2 for outer race rotation)
X	–	radial factor
Y	–	thrust factor

Values of X and Y depend on bearing geometry, size, and number of balls. Two columns are listed for each (Table 5.2). The pair that gives the highest value for F_e should be used.

Once a specific bearing model is chosen, the required load rating, C, is determined by Eq. (5.1) and used to select a bearing from supplier catalogs.

Example   A medium 300K Series Timken ball bearing:

Bearing number 305K (p. D22 Timken Catalog, see Footnote 1)
25 mm bore (0.9843 in., ID)
62 mm (2.4409 in., ID)
Dynamic load rating – 26 600 N (6000 lb)

Specifications requiring this bearing support a radial load of 600 lb. What is the expected life of this bearing whose inner race rotates at 1000 rpm? Using Eq. (5.1)

$$L = \left(\frac{C}{F}\right)^3 = \left(\frac{6000}{600}\right)^3 = 1000 \times 10^6 \text{ rev} \tag{5.8}$$

Table 5.2   Equivalent radial load factors.

	X_{10}	Y_1	X_2	Y_2
Radial contact ball bearings	1	0	0.5	1.4
Angular contact ball bearings (shallow angle)	1	1.25	0.45	1.2
Angular contact ball bearings (steep angle)	1	0.75	0.4	0.75

In terms of time, this bearing will last

$$T = \left| \frac{10^9 \text{ rev}}{1000 \text{ rpm}} \right| \left| \frac{1 \text{ h}}{60 \text{ min}} \right| \left| \frac{1 \text{ d}}{24 \text{ h}} \right| \left| \frac{1 \text{ yr}}{365 \text{ d}} \right| = 1.9 \text{ years}$$

Next assume the 600 lb radial load is supplemented with a 200-axial load. In this case,

$$F_e = XVF_r + YF_a \qquad (5.9)$$

$$F_e = XF_r + YF_a = 0.5(600) + 1.4(200) = 580 \text{ lb}$$

Since $580 < 600$ lb, the first solution applies. However, if the axial load is 300 lb

$$F_e = XF_r + YF_a = 0.5(600) + 1.4(300) = 720 \text{ lb}$$

then

$$L = \left(\frac{C}{F} \right)^3 = \left(\frac{6000}{720} \right)^3 = 578.7 \times 10^6 \text{ rev}$$

$$T = \left| \frac{578.7 \times 10^6 \text{ rev}}{1000 \text{ rpm}} \right| \left| \frac{1 \text{ h}}{60 \text{ min}} \right| \left| \frac{1 \text{ d}}{24 \text{ h}} \right| \left| \frac{1 \text{ yr}}{365 \text{ d}} \right| = 1.1 \text{ years}$$

The axial load reduces the bearing life by 9.6 months.

Effect of Vibrations on the Life of Rolling Contact Bearings

Radzimovsky and Dareing [3] predict the effect of dynamic forces on the life of rolling contact bearings. Dynamic forces are assumed to be of the form

$$P(t) = P_m + P_v \sin \frac{2\pi t}{T} \qquad (5.10)$$

where

P_m	–	mean value of load variation, lb
P_v	–	amplitude of variable component of load variation, lb
T	–	period of cyclic load, s
t	–	time, s

To derive the equivalent load, it may be assumed that if a certain load acts for a certain fraction of the life duration, which the bearing expects to have under this load, then the same fraction of the life of the bearing is consumed. For another load, having different magnitude, only the remaining part of the bearing capacity may be utilized. Experience supports this assumption. Therefore, over one load cycle

$$TP_e^q = \int_0^T P(t)^q dt \qquad (5.11)$$

$$P_e = P_m \left[\frac{1}{T} \int_0^T \left(1 - \frac{P_v}{P_m} \sin \frac{2\pi t}{T} \right)^q dt \right]^{\frac{1}{q}} \qquad (5.12)$$

From which

$$\beta = \left[\frac{1}{T} \int_0^T \left(1 - \alpha \sin \frac{2\pi t}{T} \right)^q dt \right]^{\frac{1}{q}} \tag{5.13}$$

where

$$\alpha = \frac{P_v}{P_m}$$

$$\beta = \frac{P_e}{P_m}$$

P_e is defined as a steady load that will cause fatigue failure in the bearing after the same number of revolutions (or hours, if speed is constant) as a given unsteady load defined by Eq. (5.10).

When contact roller bearings are properly lubricated, $q = 3.33$ for cylindrical roller bearings and $q = 3$ for ball bearings.

For ball bearings, the mathematical relation between these two parameters is

$$\beta = \left[1 + 1.5\alpha^2 \right]^{\frac{1}{3}} \tag{5.14}$$

For cylindrical bearings, the relationship must be determined numerically.

The effect of dynamic loading on bearing life is defined by the life reduction coefficient, K.

$$K = \frac{L_{unsteady}}{L_{steady}} = \frac{L_e}{L} \tag{5.15}$$

K and β are related as follows:

$$\frac{L_e}{L} = \left(\frac{\frac{C}{P_e}}{\frac{C}{P}} \right)^q = \frac{1}{\left(\frac{P_e}{P} \right)^q} \tag{5.16}$$

$$K = \frac{1}{\beta_q} \tag{5.17}$$

The relation between K and β (L_e and P_e) is illustrated in Figure 5.3.

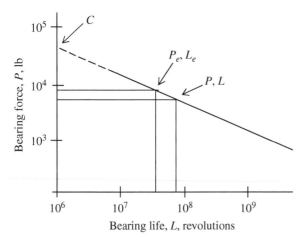

Figure 5.3 Effect of dynamic loading on bearing life.

Example Consider the following set of numbers:

$C = 10\ 700$ lb (load rating for deep groove radial ball bearing)
$P_{min} = 900$ lb
$P_{max} = 5100$ lb
$N = 1800$ rpm

Therefore

$P_m = 3000$ lb
$P_v = 2100$ lb

Expected bearing life under *steading* loading conditions is

$$L = \frac{(10\ 700/3000)^3 10^6 \text{ rev}}{1800(60)\text{rev/h}} = 421 \text{ hours}$$

From these input data,

$P_m = 3000$ lb
$P_v = 2100$ lb
$\alpha = \dfrac{2100}{3000} = 0.7$
$\beta = 1.202$ (Table 5.3)
$K = 0.576$ (Table 5.3)

The equivalent load is

$$P_e = 1.202\ (3000) = 3606 \text{ lb}$$

Table 5.3 Equivalent load and life reduction factors.

α	Roller bearings		Ball bearings	
	β_r	K_r	β_b	K_b
0.0	1.000 0	1.000	1.000	1.000
0.1	1.004 2	0.986	1.005	0.985
0.2	1.019 9	0.936	1.020	0.943
0.3	1.045 8	0.861	1.043	0.881
0.4	1.980 3	0.773	1.073	0.806
0.5	1.121 6	0.683	1.112	0.727
0.6	1.168 2	0.596	1.155	0.649
0.7	1.218 7	0.518	1.202	0.576
0.8	1.271 9	0.449	1.252	0.510
0.9	1.327 0	0.390	1.303	0.451
1.0	1.383 5	0.340	1.357	0.400

The bearing life under this dynamic load condition is

$$L_e = 421 \times 0.576 = 242\,\text{h}$$

Representing a $179/421 = 0.4252$ or a 42.52% reduction in bearing life.

Effect of Environment on Rolling Contact Bearing Life

One of the most severe environments for rolling contact bearings occurs in oil well drilling. Since the first roller bit patented by Hughes [4], the roller cone bit has gone through many improvements. The design challenges are bearing life and cutter wear. Figure 5.4 shows a typical bit cone cross section and the bearing arrangement. Initially, these bearings were open to drilling fluids, which contain barite and bentonite, additives for density and viscosity. Both additives are abrasive.

The ball bearing carries little or no load but is there to hold the cone in place. The balls are inserted through a hole that leads to the bearing cavity. Assuming weight-on-bit (WOB) is 45 000 lb, each of the three bearing legs supports 15 000 lb. This load generates lateral, thrust, and moment loads within the bearing arrangement. The small pin or sleeve bearing is required to support the thrust load with some help from the ball bearing. Lateral loads are carried by the cylindrical bearing and the sleeve bearing. This may be the best arrangement for the space, but in an industrial application, cylindrical and sleeve bearings would not be use in parallel. Sleeve bearings require radial space, which in this case causes nonuniform loading to the cylindrical rollers.

For many years, these bearings were unsealed and operated in a corrosive and abrasive environment caused by drilling fluid made up of: (i) a base fluid, usually water, (ii) barite to increase the fluids density, and (iii) bentonite to give the fluid gel strength. Under this condition, bearing life is about 8–10 hours.

In recent years, roller bearing life has increased to over 100 hours due to reliable dynamic seals and lubricant reservoirs and improvements in bearing design.

Bearings in roller bit bearings have typically been open to the drilling mud plus high loads required for acceptable rates of penetration (ROP). Also, the overall bearing assembly comprising

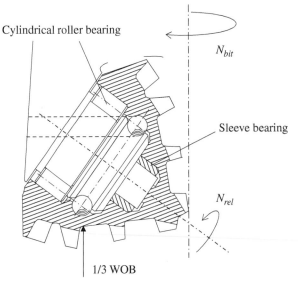

Cylindrical roller bearing

N_{bit}

Sleeve bearing

N_{rel}

1/3 WOB

Figure 5.4 Bearings mounted inside a roller cone.

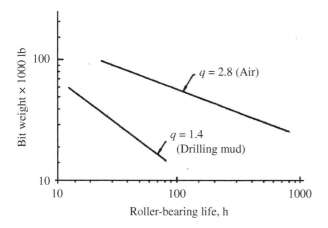

Figure 5.5 Life expectancy of unsealed bit bearing (*N* = 60 rpm).

cylindrical, ball and journal bearings within each bit cone is not conducive to prolonged bearing life. The cylindrical roller bearing usually fails first because of misalignment and abrasion.

Tests conducted by King [5] on 7 7/8 in. W7R drill bits in barite mud show that $q = 1.4$, while the same drill bit shows $q = 2.8$ in air (Figure 5.5). The values of q are strikingly different from those for well-lubricated ball and cylindrical bearings. These tests were conducted under steading external loading.

Effect of Vibration and Environment on Bearing Life

Dareing and Radzimovsky [6] expanded the analysis by King [5] to include other values of the exponent, q, to cover the effects of drilling mud. The definitions of the dimensionless parameters are still the same. Figure 5.6 relates, β, α, and q.

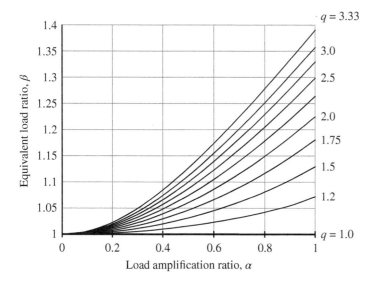

Figure 5.6 Equivalent steady loads yielding same bit life.

The results are also presented as life reduction due to dynamic loading. A life reduction coefficient, K, is used to show reduction in bearing life caused by unsteady loading. K is the ratio of bearing life corresponding loading to bearing life corresponding to steady loading.

$$K = \frac{L_{unsteady}}{L_{steady}} \tag{5.18}$$

This ratio can also be related to β through

$$K = \frac{1}{\beta^q} \tag{5.19}$$

Figure 5.7 and Table 5.4 relate K, α, and q.
As an example, consider the following case:

$\alpha = 0.25$ (variable force component is 25% of the average force component)
$P_m = 40\ 000\ \text{lb}$
$q = 1.4$ (bearing cavity open to drilling mud)

According to Figure 5.7,

$$\beta = \frac{P_e}{P_m} = 1.01$$

This means that the effect of the 25% vibration amplitude is the same as increasing mean bit force by 400 lb, an insignificant affect.
Alternatively, the life reduction coefficient, K, for this example is

$$K = \frac{L_{unsteady}}{L_{steady}} = 0.99 \qquad \text{(see Table 5.4)}$$

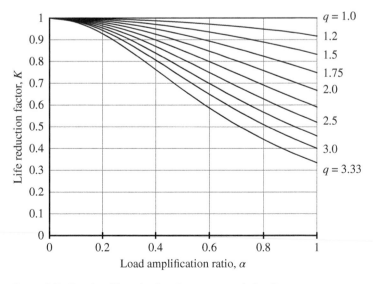

Figure 5.7 Bearing life reduction due to unsteady loading.

Table 5.4 Life reduction factors, K.

α	$q = 3.33$	3	2.75	2.5	2.25	2	1.75	1.5	1.25	$q = 1$
				K, life reduction factors						
0	1	1	1	1	1	1	1	1	1	1
0.1	0.980 968	0.985 222	0.988 113	0.990 713	0.993 018	0.995 025	0.996 729	0.998 128	0.999 219	1
0.2	0.927 925	0.943 396	0.954 105	0.963 878	0.972 657	0.980 393	0.987 035	0.992 542	0.996 874	1
0.3	0.851 058	0.881 058	0.902 392	0.922 291	0.940 545	0.956 939	0.971 265	0.983 334	0.992 965	1
0.4	0.762 356	0.806 452	0.838 831	0.869 853	0.899 051	0.925 925	0.949 965	0.970 653	0.987 486	1
0.5	0.671 915	0.727 273	0.769 301	0.810 744	0.850 884	0.888 889	0.923 827	0.954 689	0.980 427	1
0.6	0.586 415	0.649 351	0.698 694	0.748 77	0.798 723	0.847 458	0.893 634	0.935 668	0.971 763	1
0.7	0.509 306	0.576 369	0.630 517	0.686 978	0.744 934	0.803 212	0.860 209	0.913 826	0.961 448	1
0.8	0.441 753	0.510 204	0.566 93	0.627 543	0.691 435	0.757 575	0.824 35	0.889 389	0.949 387	1
0.9	0.383 566	0.451 467	0.509 022	0.571 849	0.639 658	0.711 744	0.786 777	0.862 508	0.935 369	1
1	0.333 895	0.4	0.457 135	0.520 65	0.590 592	0.666 663	0.748 039	0.833 038	0.916 869	1

Assuming the life expectancy for the steady case is 22 hours, then the life expectancy for the unsteady force case is

$$L_{unsteady} = 22(0.99) = 21.78 \text{ hours}$$

The effect of the unsteady bit force is a reduction in bit life of 13 minutes.

If the force variation is 100% instead of 25%, bit life would be reduced from 22 to 19 hours, a small but significant decrease in bit life.

Now assume that the bearing cavity is for which the sealed bearing cavity has a wear exponent of $q = 2.5$ and a 50 hours life under a steady bit load of 40 000 lb. A 40 000 lb-vibration variation gives $\alpha = 1$ and a K value of 0.52. Bit life under this condition is

$$K = \frac{L_{unsteady}}{L_{steady}} = 0.52$$

$$L_{unsteady} = 50(0.52) = 26 \text{ hours}$$

In this case, bearing life has been cut in about half.

These calculations show that the life of sealed bearings is more sensitive to dynamic forces than is the life of unsealed bearings. As bit bearing seals are improved, there is more incentive to eliminate bit force vibrations.

Hydrostatic Thrust Bearings

There are three basic types of thick film–lubricated bearings:

1) Hydrostatic thrust bearings
2) Hydrostatic squeeze films
3) Hydrodynamic lubricated bearings

If properly designed and maintained, all three types of bearings would experience no wear. Some bearings of these types have operated over a period of 30 years with no measurable wear. Friction-generated heat and temperature are important considerations though. Each of these three types of thick film–lubricated bearings are discussed in the remaining sections of this chapter.

Lubricant viscosity directly affects the performance of each bearing type. Viscosity is expressed in units of centipoise (Systems International [SI] system) or reyn (English). Viscosities of several fluids are given in Table 5.5.

Flow Between Parallel Plates

First, consider pressure-induced flow between two parallel plates (Figure 5.8). Pressure at point 1 is greater than at point 2. The pressure difference drives the fluid from left to right. The velocity profile across the film and flow rate through the film are determined as follows.

From the freebody of a differential fluid volume

$$p\,dyb - (p + dp)dyb - \tau\,dxb + (\tau + d\,\tau)dxb = 0$$

$$\frac{dp}{dx} = \frac{d\tau}{dy} \tag{5.20}$$

Table 5.5 Viscosities of common fluids.

	cp	reyn × 10^6
Honey	1500	218
Castor oil	1000	145
Glycerin	500	72
SAE 60	550	80
SAE 30	165	24
SAE 10	65	9.5
Kerosene	2	0.290
Mercury	1.5	0.217
Water	1.0	0.145
Gasoline	0.6	0.087
Air	0.018	0.0026

Note: 1 cp $= 1.45 \times 10^{-7}$ reyn.

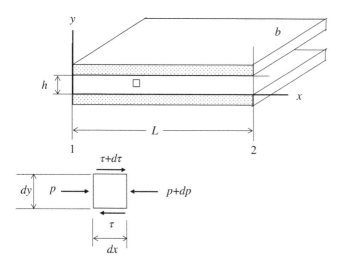

Figure 5.8 Flow between parallel plates.

Assuming the lubricant is Newtonian

$$\tau = \mu \frac{du}{dy}$$

$$\frac{dp}{dx} = \mu \frac{d^2 u}{dy^2} \tag{5.21}$$

By integration

$$\frac{du}{dy} = \frac{1}{\mu} \frac{dp}{dx} y + C_1 \tag{5.22}$$

$$u(y) = \frac{1}{2\mu} \frac{dp}{dx} y^2 + C_1 y + C_2 \tag{5.23}$$

Imposing boundary conditions, $u(0) = 0$, $u(h) = 0$ gives

$$u(y) = \frac{1}{2\mu} \frac{dp}{dx} (y^2 - yh) \tag{5.24}$$

If the pressure gradient is $(+)$, flow is to the left. If the pressure gradient is $(-)$, flow is to the right. Now consider the flow rate through the parallel film of width b.

$$Q = b \int_0^h u(y) dy \tag{5.25}$$

$$Q = \frac{b}{2\mu} \frac{dp}{dx} \int_0^h (y^2 - hy) dy \tag{5.26}$$

$$Q = -\frac{bh^3}{12\mu} \frac{dp}{dx} \tag{5.27}$$

Pressure gradient must be minus for positive flow to the right. This equation will now be adapted to radial flow.

Fluid Mechanics of Hydrostatic Bearings

Hydrostatic bearings rely on external pumps for fluid film pressure. Fluid is circulated through these bearings at specified pressure and flow rate to establish and maintain a specified operating film thickness, h. A simple hydrostatic bearing is shown in Figure 5.9.

Modifying Eq. (5.27), for this case $b = 2\pi r$ giving

$$Q = -\frac{\pi r h^3}{6\mu} \frac{dp}{dr} \tag{5.28}$$

Integrating Eq. (5.28) and noting that $p(r_2) = 0$, gives

$$p = \frac{6\mu Q}{\pi h^3} \ln \frac{r_2}{r} \tag{5.29}$$

Setting $r = r_1$ gives

$$p_0 = \frac{6\mu Q}{\pi h^3} \ln \frac{r_2}{r_1} \tag{5.30}$$

The flow rate, Q, required to create pressure, p_0 is

$$Q = \frac{p_0 \pi h^3}{6\mu \ln \frac{r_2}{r_1}} \tag{5.31}$$

By substitution

$$p(r) = p_0 \frac{\ln \frac{r_2}{r}}{\ln \frac{r_2}{r_1}} \tag{5.32}$$

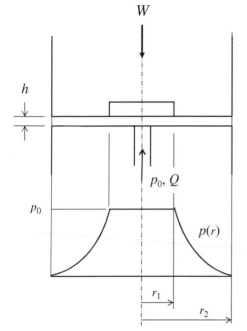

Figure 5.9 Hydrostatic thrust bearing.

W

h

p_0, Q

p_0

$p(r)$

r_1

r_2

The load-carrying capacity is determined by integrating the pressure profile.

$$W = p_0 \pi r_1^2 + \int_1^2 2\pi r p(r) \, dr \qquad (5.33)$$

By substitution for $p(r)$

$$W = p_0 \pi r_1^2 + 2\pi p_0 \int_1^2 r \frac{\ln \frac{r_2}{r}}{\ln \frac{r_2}{r_1}} \, dr \qquad (5.34)$$

$$W = \pi p_0 \left[r_1^2 + \frac{2}{\ln \frac{r_2}{r_1}} \int_1^2 r \ln \frac{r_2}{r} \, dr \right] \qquad (5.35)$$

$$W = \pi p_0 \left[r_1^2 + \frac{2}{\ln \frac{r_2}{r_1}} \left| \frac{r^2}{2} \ln \frac{r_2}{r} + \frac{r^2}{4} \right|_1^2 \right] \qquad (5.36)$$

$$W = \pi p_0 \left[r_1^2 + \frac{2}{\ln \frac{r_2}{r_1}} \left[\frac{r_2^2}{4} - \frac{r_1^2}{2} \ln \frac{r_2}{r_1} - \frac{r_1^2}{4} \right] \right] \qquad (5.37)$$

Bringing these terms together gives

$$W = \frac{\pi p_0}{2} \frac{(r_2^2 - r_1^2)}{\ln \frac{r_2}{r_1}} \qquad (5.38)$$

This equation relates supply pressure (recess pressure) to the load, W.

$$p_0 = \frac{2W \ln \frac{r_2}{r_1}}{\pi (r_2^2 - r_1^2)} \qquad (5.39)$$

Equations (5.31) and (5.39) are the basis of hydrostatic bearing design. The expected load, W, on the bearing would dictate the required pressure, p_0, from the external pump. The corresponding flow rate, Q, required to establish cavity pressure, p_0, is determined from Eq. (5.31). The hydraulic horsepower from the pump is determined from the product of p_0 and Q. This type of bearing finds application in the support and movement of heavy structures, such as roofs of sports arenas and space telescopes.

Bearing design requires consideration of space and load. Unknowns in the design are flow rate, delivery pressure, and viscosity.

Example Consider the following information

W	= 10 000 lb	(design load)
h	= 0.006 in.	(required minimum film thickness)
μ	= 15.63×10^{-7} reyn	(lubricant viscosity)
r_2	= 6 in.	(outer radius)
r_1	= 1.5 in.	(inner radius)

The problem is to determine the delivery pressure, p_0, and flow rate, Q, and horsepower, P, required to satisfy these design constraints. Using Eq. (5.39), $p_0 = 261.5$ psi.

$$p_0 = \frac{2(10\ 000)\ln\left(\frac{6}{1.5}\right)}{\pi(6^2 - 1.5^2)} = 261.5\ \text{psi}$$

Using Eq. (5.31),

$$Q = \frac{\pi(261.5)(0.006)^3}{6(15.63 \times 10^{-7})\ln\left(\frac{6}{1.5}\right)} = 13.65\ \text{in.}^3/\text{s}$$

Pump power required to deliver this pressure and flow rate is

$$P = p_0 Q = 261.5(13.65) = 3570\ \text{in.-lb/s}$$

$$P = 3570 \left|\frac{1\ \text{ft}}{12\ \text{in.}}\right|\left|\frac{1\ \text{hp}}{550\ \text{ft-lb/s}}\right| = 0.541\ \text{hp}$$

The power input to the pump depends on mechanical efficiency of the pumping unit.

$$\eta = \frac{\text{HHP(hydraulic)}}{\text{HP(mechanical)}} \tag{5.40}$$

$$\text{HP} = \frac{\text{HHP}}{\eta}$$

This example shows that a very large force (10 000 lb) can be supported by a relatively low pressure (261.5 psi). The required pumping power (0.541 hp) is relatively low. Analyses show that rolling contact bearings would be very large (and costly) by comparison with hydrostatic lubricating units.

We will now examine the best design for minimum power consumption.

Optimizing Hydrostatic Thrust Bearings

The above example addresses the pumping power required to support a 10 000 lb load while maintaining a film thickness of 0.006 in., which is normal for this type of bearing. Calculations show that a standard ¾ horsepower motor and pump (above the 0.541 hp mentioned above) would be required in this case.

Circular thrust bearings may also rotate and in this case, energy is lost in viscous friction. Both pumping requirements and rotational friction are considered in optimizing the design of thrust bearings.

In optimizing thrust bearings, there are two factors to consider [7]. The first is the radius of the recess (r_1) with respect to the bearing size (r_2). The second is the operating film thickness. Both affect energy consumption. The objective is to configure the bearing to minimize hydraulic power and thus the amount of mechanical power required to support a given load.

Pumping Requirements

The formula for hydraulic horse power is HHP (or P) = $p_0 Q$. By substitution

$$P = p_0 \left(\frac{p_0 \pi h^3}{6\mu \ln \frac{r_2}{r_1}}\right) \tag{5.41}$$

Making a further substitution for p_0,

$$P = \left(\frac{\pi h^3}{6\mu \ln \frac{r_2}{r_1}}\right)\left[\frac{2W \ln \frac{r_2}{r_1}}{\pi(r_2^2 - r_1^2)}\right]^2 \tag{5.42}$$

This expression defines the pump power output necessary to support the required load. It is rewritten as

$$P_p = \frac{8\pi \ln \frac{r_2}{r_1}}{\left[1 - \left(\frac{r_1}{r_2}\right)^2\right]^2}\left(\frac{W}{A}\right)^2 \frac{h^3}{12\mu} \tag{5.43}$$

$$P_p = C\left(\frac{W}{A}\right)^2 \frac{h^3}{12\mu} \tag{5.44}$$

where

$$A = \pi r_2^2$$

and

$$C = \frac{8\pi \ln \frac{r_2}{r_1}}{\left[1 - \left(\frac{r_1}{r_2}\right)^2\right]^2} \tag{5.45}$$

This equation is plotted to show how pumping power is affected by ratio of recess radius (r_1) to outside radius (r_2) (see Figure 5.10).

This plot shows that the coefficient, C, and pumping power are minimized when the radius ratio, r_1/r_2, is about 0.5.

Friction Losses Due to Rotation
In many cases, hydrostatic bearings are used as thrust bearings to support axial forces applied to a rotating shaft. The rotation produces friction in the film. Therefore, two energy components are required to sustain a given film thickness. Power is required to pump through the film from the

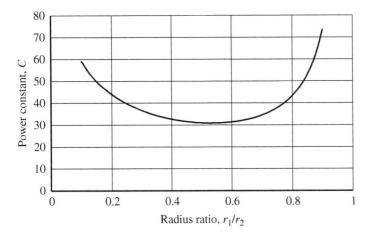

Figure 5.10 Best recess for minimum pumping power.

recess to the outside. Power is also required to overcome friction brought about by film shear caused by rotation. The total power required to operate a bearing at a given film thickness is the sum of both the pumping and film friction losses.

Friction losses are due to fluid shear across the film. Local shear force in a thin film is the shear force times a differential area, which in this case is a circular ring with radius, r, times width, dr.

$$dF = \tau \, dA$$

$$dF = \mu \frac{U}{h} dA, \quad dA = 2\pi r \, dr$$

$$dF = \mu 2\pi \, dr \frac{U}{h}, \quad \text{where } U = r\omega \text{ ips} \tag{5.46}$$

The moment of all forces due to local shear stress is the sum or integral of all shear forces over the thin-film area.

$$dM = r \, dF = \frac{2\pi\mu\omega}{h} r^2 \, dr$$

$$M = \frac{2\pi\mu\omega}{h} \int_1^2 r^3 \, dr$$

The resulting moment is

$$M = \frac{\pi\mu\omega}{2h} \left(r_2^4 - r_1^4 \right) \tag{5.47}$$

Friction power loss is

$$P_f = \omega M \tag{5.48}$$

$$P_f = \frac{\pi\mu\omega^2}{2h} \left(r_2^4 - r_1^4 \right)$$

$$P_f = \left(\frac{2\pi N}{60} \right)^2 \frac{\pi\mu}{2h} \left(r_2^4 - r_1^4 \right) \text{ in.-lb/s} \tag{5.49}$$

Total Energy Consumed

The total energy consumed by the bearing is the sum of pumping losses and rotational friction losses.

$$P_T = P_p + P_f \tag{5.50}$$

Example Consider the following example.

$W = 100\,000$ lb
$r_1 = 5$ in.
$r_2 = 8$ in.
$\mu = 29.24$ cp $= 42.4 \times 10^{-7}$ reyn
$N = 750$ rpm

Using Eq. (5.39), $p_0 = 774$ psi and from Eq. (5.31), $Q = 43.92$ in.^{3}/s or

$$Q = 43.92 \text{ in.}^3/\text{s} \left| \frac{1 \text{ ft}^3}{144(12) \text{ in.}^3} \right| \left| \frac{7.48 \text{ gal}}{1 \text{ ft}^3} \right| \left| \frac{60 \text{ s}}{1 \text{ min}} \right| = 11.41 \text{ gpm} \tag{5.51}$$

Power lost to pumping is (from Eq. (5.39))

$$P_p = 31.8 \left(\frac{100\,000}{201} \right)^2 \frac{h^3}{12(42.4 \times 10^{-7})} \text{ in.-lb/s} \tag{5.52}$$

$$P_p = \frac{(1000h)^3}{64.6} \left| \frac{1 \text{ ft}}{12 \text{ in.}} \right| \left| \frac{1 \text{ hp}}{550 \text{ ft-lb/s}} \right| \tag{5.53}$$

$$P_P = \frac{(1000h)^3}{42.6} \text{ hp} \tag{5.54}$$

Power lost to rotational friction is

$$P_f = \left(\frac{2\pi 750}{60} \right)^2 \frac{\pi 42.4 \times 10^{-7}}{2h} (8^4 - 5^4) = \frac{142}{h} \text{ in.-lb/s} \tag{5.55}$$

$$P_f = \frac{142}{h} \frac{1}{12(550)} = \frac{0.0216}{h} \text{ hp} \tag{5.56}$$

Bringing these numbers together into Eq. (5.50) gives the total power consumed by the bearing.

$$P_T = \frac{(1000h)^3}{42.6} + \frac{0.0216}{h} \text{ hp} \tag{5.57}$$

This equation is plotted in Figure 5.11, which shows that total energy required to operate this hydrostatic bearing is minimized with a film thickness of $h = 0.004$ in. Film thickness, h, is controlled by flow rate, Q, according to Eq. (5.31).

Coefficient of Friction

The coefficient of friction is determined as follows:

$$f = \frac{F}{W} \tag{5.58}$$

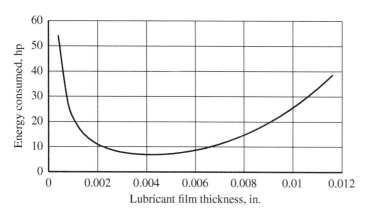

Figure 5.11 Optimizing film thickness against energy losses.

where

$$F = \frac{M}{r_{ave}} \quad \text{assuming } h = 0.004 \text{ in.} \tag{5.59}$$

$$M = \left(\frac{2\pi N}{60}\right) \frac{\pi \mu}{2h} \left(r_2^4 - r_1^4\right)$$

$$M = 45.41 \text{ in.-lb} \quad r_{ave} = 6.5 \text{ in.}$$

$$F = \frac{45.41}{6.5} = 6.99 \text{ lb}$$

Therefore

$$f = \frac{6.99}{10\ 000} \sim 0.0007$$

Squeeze Film Bearings

Hydrostatic squeeze films generate their film pressure from the motion of two approaching surfaces. Pressure builds in the film because the fluid cannot be squeezed out instantaneously. Pressure distribution in the film depends on shape and size of the bearing.

Pressure Distribution Under a Flat Disc

Pressure distribution in a circular squeeze film is derived from continuity of viscous flow between parallel surfaces (Figure 5.12). Flow within the squeeze film is also developed from Eq. (5.28). In this case, Q varies with r according to

$$Q(r) = V\pi r^2 \tag{5.60}$$

where V is velocity of approach.
By substitution into Eq. (5.28)

$$V\pi r^2 = -\frac{2\pi r h^3}{12\mu} \frac{dp}{dr} \tag{5.61}$$

The solution to Eq. (5.61) gives the pressure distribution within a circular squeeze film.

$$dp = -\frac{6\mu V}{h^3} r\, dr$$

$$p = -\frac{3\mu V}{h^3} r^2 + C_1$$

Applying the boundary condition, $p(R) = 0$,

$$p(r) = \frac{3\mu V}{h^3} \left[R^2 - r^2\right] \quad \text{(parabolic)} \tag{5.62}$$

The force–velocity relationship is determined by integrating the pressure across the circular face.

$$dW = p(r)2\pi r\, dr$$

Figure 5.12 Circular squeeze film.

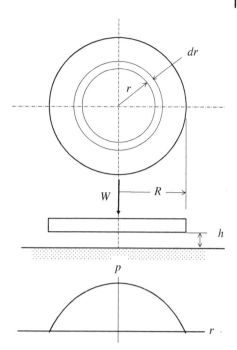

$$W = \frac{3\mu V}{h^3} \int_0^R (R^2 - r^2) 2\pi r \, dr$$

$$W = \frac{3\pi\mu R^4}{2h^3} V \tag{5.63}$$

The average fluid pressure is

$$p_{ave} = \frac{W}{\pi R^2}$$

If the applied force, W, is constant, the velocity of approach, V, changes with the inverse of film thickness, h^3.

In this case Eq. 5.63 is written as

$$V = \frac{dh}{dt} = \frac{2W}{3\pi\mu R} h^3 \tag{5.64}$$

and

$$dt = \frac{3\pi\mu R}{2W} \frac{dh}{h^3} \tag{5.65}$$

giving a means for tracking film thickness vs. time.

The time interval between two film thicknesses (h_1 to h_2) is

$$t = \frac{3\pi\mu R^4}{4W} \left[\frac{1}{h_2^2} - \frac{1}{h_1^2} \right] \tag{5.66}$$

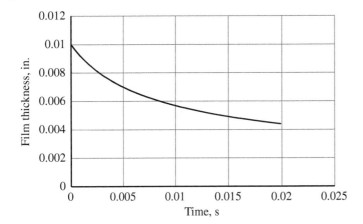

Figure 5.13 Film thickness change with time.

where h_1 is the reference film thickness. Figure 5.13 shows how film thickness changes with time assuming

$$
\begin{aligned}
W &= 10\,000\,\text{lb} \\
\mu &= 15.63 \times 10^{-7}\,\text{reyn} \\
R &= 6\,\text{in.} \\
h_1 &= 0.01\,\text{in.}
\end{aligned}
$$

The film change from 0.01 to 0.005 in. takes about 0.015 second (see Figure 5.13).

Comparison of Pressure Profiles

Pressure profiles for both hydrostatic and squeezed films are based on

$$
Q = \frac{bh^3}{12\mu}\left(-\frac{dp}{dr}\right) \tag{5.67}
$$

$$
\frac{dp}{dr} = -\frac{Q}{b}\frac{12\mu}{h^3} \tag{5.68}
$$

The flow rates for both are:

$$
\text{Hydrostatic film}: Q = \text{constant} \quad b = 2\pi r \quad \frac{dp}{dr} = -\frac{1}{r}C_1 \tag{5.69}
$$

$$
\text{Squeeze film}: Q = \pi r^2 V \quad b = 2\pi r \quad \frac{dp}{dr} = -rC_2 \tag{5.70}
$$

For the same value of p_0, the squeeze film carries the larger load. Both pressure profiles are shown in Figure 5.14.

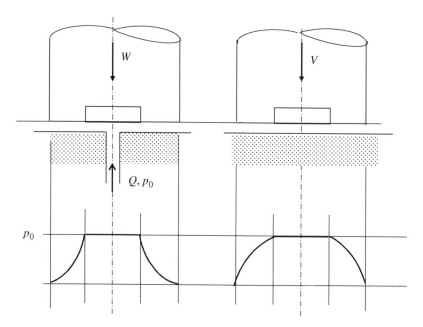

Figure 5.14 Comparison of hydrostatic and squeeze film.

Spring Constant of Hydrostatic Films

Hydrostatic films operate at a given film thickness, h, and have a certain rigidity with respect to h. Starting with

$$p_0 = \frac{6\mu Q}{\pi h^3} \ln \frac{r_2}{r_1} \tag{5.71}$$

$$W = \frac{\pi p_0}{2} \frac{(r_2^2 - r_1^2)}{\ln \frac{r_2}{r_1}} \tag{5.72}$$

Combining these two equations gives

$$W = 3\mu Q (r_2^2 - r_1^2) \frac{1}{h^3} \tag{5.73}$$

$$k = \frac{dW}{dh} = -9\mu Q (r_2^2 - r_1^2) \frac{1}{h^4} \tag{5.74}$$

The minus sign simply means that the slope of the function, $W(h)$, is negative. We only seek the magnitude of the slope. Equation (5.74) shows that $k \propto \frac{1}{h^4}$. Stiffness of the film is very high for small film thicknesses.

Damping Coefficient of Squeeze Films

Damping in squeeze films can be important for systems having high natural frequencies. From Eq. (5.64)

$$c = \frac{W}{V} = \frac{3\pi \mu R^4}{2h^3} \quad \text{(damping coefficient, force/velocity)} \tag{5.75}$$

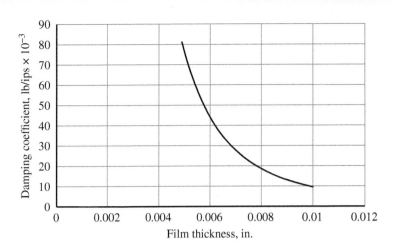

Figure 5.15 Damping coefficient vs. film thickness.

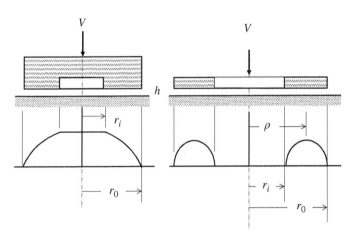

Figure 5.16 Squeeze film geometries.

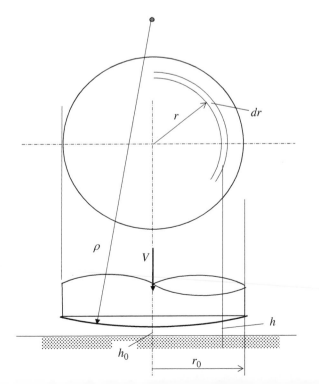

Using the previous numbers for a squeeze film

$$
\begin{aligned}
W &= 10\,000\,\text{lb} \\
\mu &= 15.63 \times 10^{-7}\,\text{reyn} \\
R &= 6\,\text{in.}
\end{aligned}
$$

Damping coefficients were determined and plotted against film thickness (Figure 5.15).

Damping coefficients are lowest at the thicker films. For this example, $c = 10\,000$ lb/ips at a film thickness of 0.01 in. Recall, critical damping for an single degree of freedom (SDOF) system is

$$c_{cr} = 2\sqrt{km} \tag{5.76}$$

For an underdamped system, $\zeta = \dfrac{c}{c_{cr}} < 1$. For the above case

$$\frac{10\,000}{2\sqrt{km}} < 1$$

The spring constant, k, and mass, m, would have to be large for an underdamped system.

Other Shapes of Squeeze Films

This section develops pressure distributions for three other shapes.

1) Flat plate with a cavity
2) Washer
3) Spherical film

In each case (Figures 5.16 and 5.17), pressure profile is established by the flow equation (Eq. (5.61)).

Squeeze Film with Recess

The pressure distribution in the thin film is

$$p(r) = \frac{3\mu V}{h^3}\left(r_0^2 - r^2\right) \tag{5.77}$$

The pressure in the cavity is assumed constant

$$p_i = \frac{3\mu V}{h^3}\left(r_0^2 - r_i^2\right) \tag{5.78}$$

The load-carrying capacity of this film is

$$W = W_1 + W_2 \tag{5.79}$$

where

$$W_1 = \pi r_i^2 p_i$$

$$W_2 = \int_{r_1}^{r_2} 2\pi r p(r) dr = \frac{3\pi\mu V}{h^3} \left[\frac{r_0^4}{2} - r_i^2 r_0^2 + \frac{r_i^4}{2} \right] \tag{5.80}$$

Combining expressions gives

$$W = \frac{3\pi\mu V}{2h^3} \left(r_0^4 - r_i^4 \right) \tag{5.81}$$

The average pressure is

$$p_{ave} = \frac{W}{\pi r_0^2} \tag{5.82}$$

Squeeze Film Under a Washer

The boundary conditions for this problem are zero pressure on the inside and outside of the washer. We anticipate a maximum pressure somewhere within the films. The location of this point is indicated by radius ρ. Fluid on the outside of ρ flows outward. Fluid on the inside of ρ flows inward.

Flow equation to the right of ρ is

$$\left(\pi r^2 - \pi\rho^2 \right) = \frac{\pi r h^3}{6\mu} \left(-\frac{dp}{dr} \right) \tag{5.83}$$

Flow equation to the left of ρ is

$$\left(\pi\rho^2 - \pi r^2 \right) = \frac{\pi r h^3}{6\mu} \left(\frac{dp}{dr} \right) \tag{5.84}$$

Note that the same equation applies to both sides of ρ. The equation is used for outside flow knowing that fluid pressure is zero at $r = r_0$. By direct integration

$$p(r) = \frac{6\mu V}{h^3} \left[\frac{r_0^2}{2} - \frac{r^2}{2} - \rho^2 \ln \frac{r}{r_0} \right] \tag{5.85}$$

Using this equation and knowing that $r = r_i$, $p = 0$ gives

$$\rho^2 = \frac{\left(r_0^2 - r_i^2 \right)}{2 \ln \dfrac{r_i}{r_0}} \tag{5.86}$$

and

$$p(r) = \frac{3\mu V}{h^3} \left[\left(r_0^2 - r^2 \right) - \left(r_0^2 - r_i^2 \right) \frac{\ln \dfrac{r}{r_0}}{\ln \dfrac{r_i}{r_0}} \right] \tag{5.87}$$

Note that maximum pressure occurs at $r = \rho$. The applied load, W, is determined from

$$dW = 2\pi r \, dr p(r)$$

By integration

$$W = \frac{3\pi\mu V}{2h^3}(r_0^2 - r_i^2)\left[1 + \frac{1}{\ln\frac{r_i}{r_0}} + \frac{2r_i^2}{(r_0^2 - r_i^2)}\right] \tag{5.88}$$

$$V = \frac{dh}{dt} \tag{5.89}$$

$$dt = \frac{3\pi\mu}{2W}(r_0^2 - r_i^2)\left[1 + \frac{1}{\ln\frac{r_i}{r_0}} + \frac{2r_i^2}{(r_0^2 - r_i^2)}\right]\frac{dh}{h^3} \tag{5.90}$$

Spherical Squeeze Film

This type of film symbolized cases, such as a hammer face, coming down in a viscous film. Experiments of dropping steel balls onto a film cause triangular shape indentation of the film surface indicating high local pressure at the center location (Figure 5.17).

The fluid flow equation is the same as in the previous cases; however, in the spherical dome example

$$h = h_0 + (\rho - \sqrt{\rho^2 - r^2}) \tag{5.91}$$

Using the binomial expansion with $\rho \gg r$ gives

$$h \sim h_0 + \frac{r^2}{2\rho} \tag{5.92}$$

By substitution

$$\pi r^2 V = \frac{2\pi r\left(h_0 + \frac{r^2}{2\rho}\right)^3}{12\mu}\left(-\frac{dp}{dr}\right) \tag{5.93}$$

$$dp = -6\mu V\int\frac{r\,dr}{\left(h_0 + \frac{r^2}{2\rho}\right)^3} \tag{5.94}$$

Integrating and applying the boundary condition at $r = r_0$, $(p = 0)$ gives

$$p(r) = 3\mu\rho V\left[\frac{1}{\left(h_0 + \frac{r^2}{2\rho}\right)^2} - \frac{1}{\left(h_0 + \frac{r_0^2}{2\rho}\right)^2}\right] \tag{5.95}$$

Load W relates to the design variables by

$$W = \int_0^{r_0} 2\pi r p(r)\,dr \tag{5.96}$$

Substituting for $p(t)$ and integrating

$$W = 6\pi\mu V\rho^2\left[\frac{1}{h_0} - \frac{h_0 + \frac{r_0^2}{\rho}}{\left(h_0 + \frac{r_0^2}{2\rho}\right)^2}\right] \tag{5.97}$$

Nonsymmetrical Boundaries

The general case of fluid flow in films is shown in Figure 5.18. The square element has dimensions of dx, dy, and film thickness, h.

Case 1 – First consider the trapped film is being squeezed by the top surface moving downward with velocity V. Fluid is entering and leaving the control volume (film) in the x and y directions. The downward movement of the top surface also causes fluid to enter the control volume from the top. To satisfy continuity of flow

$$Q_{in} = Q_{out}$$

$$Q_x + Q_y + V\,dx\,dy = \left(Q_x + \frac{dQ_x}{dx}dx\right) + \left(Q_y + \frac{dQ_y}{dy}dy\right) \tag{5.98}$$

Recall

$$Q_x = -dy\frac{h^3}{12\mu}\frac{dp}{dx} \quad \text{and} \quad Q_y = -dx\frac{h^3}{12\mu}\frac{dp}{dy} \tag{5.99}$$

By substitution, the continuity equation becomes

$$V = -\frac{h^3}{12\mu}\left[\frac{d^2p}{dx^2} + \frac{d^2p}{dy^2}\right] \tag{5.100}$$

or

$$-\frac{12\mu V}{h^3} = \left[\frac{d^2p}{dx^2} + \frac{d^2p}{dy^2}\right] \tag{5.101}$$

In polar coordinates, the Poisson equation becomes

$$\frac{\partial^2 p}{\partial r^2} + \frac{1}{r}\frac{\partial p}{\partial r} + \frac{1}{r^2}\frac{\partial^2 p}{\partial \theta^2} = -\frac{12\mu V}{h^3} \tag{5.102}$$

This form is useful in arc-shaped geometries. For the circular squeeze film, pressure is independent of θ, so the Poisson equation becomes

$$\frac{1}{r}\frac{d}{dr}\left(r\frac{dp}{dr}\right) = -\frac{12\mu V}{h^3} \tag{5.103}$$

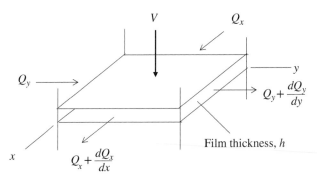

Figure 5.18 Two-dimensional fluid flow.

Figure 5.19 Squeeze film between journal and bearing.

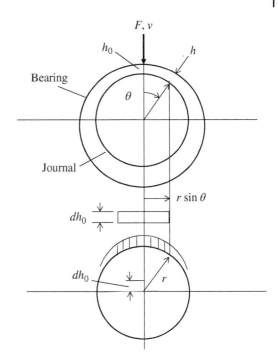

$$\frac{dp}{dr} = -\frac{r}{2}\frac{12\mu V}{h^3} = -\frac{6\mu V}{h^3}r \tag{5.104}$$

the same expression as Eq. (5.61).

Case 2 – There are situations where fluid enters a film having noncircular boundaries as previously discussed. Also, boundary pressures may not be uniform. The differential equation in this case is derived from the control volume of Figure 5.18 as well. The continuity equation (Eq. (5.101)) still applies except the film is assumed to be constant ($V = 0$).

$$\nabla^2 p = 0 \quad \text{(Laplace equation)} \tag{5.105}$$

There are many example solutions to both Poisson and Laplace equations [8]. Problems having noncircular boundaries and various internal shapes can also be solved by finite difference methods.

Application to Wrist Pins

An important application of squeeze films is the lubrication of wrist pins within pistons. A kinematic analysis of the slider crank mechanism shows that the relative angular motion of the pin at the end of the connecting rod is a rocking motion. The wrist pin does not make a complete revolution within the pistons; therefore, hydrodynamic lubrication is unlikely. However, the pressure force on the piston is cyclic allowing the pin to be lubricated with squeeze film action during the compression phase with the film recovering during the intake phase. The goal is to maintain a required minimum film thickness during the compression stroke.

During the power phase, the lubrication film is squeezed when the pin (journal) moves into the film displacing fluid (see Figure 5.19). The volume that is displaced through section h when the journal moves dh_0 is

$$d(\text{vol.}) = br \sin\theta \, dh_0 \tag{5.106}$$

Using

$$Q = -\frac{bh^3}{12\mu} \frac{dp}{r \, d\theta} \tag{5.107}$$

In this case, b is the length of the bearing. Film thickness is defined by

$$h = c - e\cos\theta$$

$$h = mr\,(1 - n\cos\theta); \quad m = \frac{c}{r}, \quad n = \frac{e}{c} \tag{5.108}$$

where

---

h – film thickness at position, θ
c – radial clearance between journal and bearing (concentric position)
e – center distance between bearing and journal (see Figure 5.30)
r – radius of journal
n – eccentricity ratio

$$Q = \frac{d(\text{vol.})}{dt} = br\sin\theta\frac{dh_0}{dt} = brV\sin\theta \text{ (see Figure 5.19)} \tag{5.109}$$

Substituting Eqs. (5.109) and (5.108) into Eq. (5.107) gives

$$dp = -\frac{12\mu V}{m^2 r} \frac{\sin\theta}{(1-n\cos\theta)^3}\,d\theta \tag{5.110}$$

giving

$$p(\theta) = \frac{12\mu V}{m^3 r}\left[\frac{1}{2n(1-n\cos\theta)^2} + C\right] \tag{5.111}$$

Assuming $p = 0$ at $\theta = \dfrac{\pi}{2}$, gives $C = -\dfrac{1}{2n}$

$$p(\theta) = \frac{6\mu V}{m^3 rn}\left[\frac{1}{(1-n\cos\theta)^2} - 1\right] \tag{5.112}$$

Local pressure in the film depends on V, n, and θ.

When $n = 0$ or the pin is concentric with the bearing, Eq. (5.111) is indeterminate. Applying L'Hopital's rule shows

$$p(\theta) = \frac{12\mu V\cos\theta}{m^3 r} \quad (n = 0) \tag{5.113}$$

Equation (5.112) is determinate for other values of n. Note that when $n = 1$, the film thickness is zero at $\theta = 0$ and pressure is ∞, as expected.

The relation between bearing load and these parameters is determined from equilibrium considerations.

$$dW = brd\,\theta p(\theta)\cos\theta \tag{5.114}$$

When $n = 0$ (concentric case)

$$W = \frac{24\mu Vb}{m^3} \int_0^{\frac{\pi}{2}} \cos^2\theta \, d\theta = 6\pi\left(\frac{\mu Vb}{m^3}\right) \quad (n = 0) \tag{5.115}$$

However, in general (using Eq. (5.113)) and being patient with the integration, we find that

$$W = \frac{\mu Vb}{m^3} K \tag{5.116}$$

where

$$K = 12\left[\frac{2}{(1-n^2)^{\frac{3}{2}}} \tan^{-1}\left(\frac{1+n}{1-n}\right)^{\frac{1}{2}} + \frac{n}{1-n^2}\right] \tag{5.117}$$

At $n = 0$, $K = 12[2\tan^{-1}1] = 6\pi$

The velocity of approach is

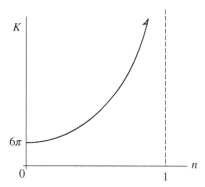

$$V = -\frac{dh}{dt}$$

But $h = mr(1 - n)$ and $dh = -mr\,dn$. By substitution

$$V = mr\frac{dn}{dt}$$

$$V = \frac{Wm^3}{\mu b}\frac{1}{K}$$

Substituting into Eq. (5.116) for V and separating variables gives

$$\int_{t_1}^{t_2} dt = \frac{12br\mu}{m^2W} \int_{n_1}^{n_2} \left[\tan^{-1}\left(\frac{1+n}{1-n}\right)^{\frac{1}{2}} \frac{2}{(1-n^2)^{\frac{3}{2}}} + \frac{n}{1-n^2}\right] dn \tag{5.118}$$

Integration gives

$$\Delta t = \frac{24br\mu}{m^2W}\left[\tan^{-1}\left(\frac{1+n}{1-n}\right)^{\frac{1}{2}} \frac{n}{(1-n^2)^{\frac{1}{2}}}\right]_{n_1}^{n_2} \tag{5.119}$$

Table 5.6 Time history of film thickness, inches.

h_1 (in.)	h_2 (in.)	Δh (in.)	h_{avg} (in.)	n_{avg}	K	V_{avg} (ips)	Δt (s)
0.000 30	0.000 25	0.000 05	0.000 275	0.084	20	0.008 6	0.005
0.000 25	0.000 20	0.000 05	0.000 225	0.25	25.4	0.006 7	0.007
0.000 20	0.000 15	0.000 05	0.000 175	0.417	40	0.004 3	0.011 7
0.000 15	0.000 10	0.000 05	0.000 125	0.584	58	0.002 9	0.016 9
0.000 10	0.000 05	0.000 05	0.000 075	0.74	125	0.002 1	0.036 5
0.000 05	0.000 03	0.000 02	0.000 040	0.866	330	0.000 52	0.038
							0.115

These equations were developed for a 180° bearing. They can be applied to 360° provided the negative pressures on the opposite side are small compared to the positive pressures.

Example Consider a piston having wrist pin parameters

Radial clearance = 0.0003 in.
Bearing length = 1¼ in.
Viscosity, $\mu = 15.63 \times 10^{-7}$ reyn
P (unit load) = 125 psi
Wrist pin diameter = 0.875 in.

The problem is to determine the time required to change film thickness from 0.0003 to 0.000 03 in. Compare this time with the time or the expansion stroke for the engine operating at 3000 rpm.

Actually $W = W(t)$ and is not a constant. The calculating procedure is as follows: W_1 will be known. Assume t_2 and find W_2 from $W(t)$. Calculate Δt and compare with t_2. Repeat until t_2 and Δt agree, etc.

For an engine running at 3000 rpm, the duration of the power stroke is 0.01 second. According to Table 5.6, the time to reduce film thickness from 0.0003 to 0.000 03 in. is 0.104 second. This time is greater by factor of 10 over the power stroke at 3000 rpm. The bearing should operate properly under the given assumptions.

Thick Film Slider Bearings

Slider Bearings with Fixed Shoe

Three factors are required for pressure to develop in hydrodynamic lubrication films: (i) viscous fluid, (ii) sliding velocity, and (iii) a fluid wedge. These three conditions are met with a slider bearing (Figure 5.20). In this example, the shoe is fixed with a slight taper. The runner has a velocity of U. The lubricant is a viscous fluid. The analysis in this case establishes the pressure profile over the active part of the bearing and the load-carrying capacity of the bearing.

Assuming a Newtonian fluid and no side leakage, the Navier–Stokes equation for this 2-D case is

$$\frac{dp}{dx} = \mu \frac{d^2 u}{dy^2} \qquad (5.120)$$

as before. However, in the slider-bearing case, the film is tapered.

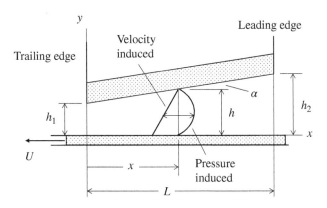

Figure 5.20 Flow within a slider-bearing film.

The velocity profile, $u(y)$, across the film is now affected by the pressure gradient of the runner. The velocity distribution across the film at any x location is modified by the boundary conditions; $u(0) = 0$ and $u(h) = -U$. These boundary conditions yield

$$u = \frac{1}{2\mu} \frac{dp}{dx} (y^2 - yh) - \left(1 - \frac{y}{h}\right) U \qquad (5.121)$$

This equation shows that velocity is the result of pressure-induced flow (first term) and velocity-induced flow (second term). The two flow components are shown in Figure 5.20.

For continuity of flow at any x location

$$Q = \int_0^h ub \, dy \qquad (5.122)$$

$$Q = b\left[-\frac{1}{2\mu} \frac{dp}{dx} \frac{h^3}{6} - \frac{1}{2} hU \right] \qquad (5.123)$$

Since flow, Q, is the same at every location, x,

$$\frac{dQ}{dx} = 0 \qquad (5.124)$$

A simplified form of Reynold's equation of lubrication is obtained by differentiating Eq. (5.203).

$$\frac{d}{dx}\left[\frac{h^3}{12\mu} \frac{dp}{dx} \right] = -\frac{d}{dx}\left[\frac{1}{2} hU \right] \qquad (5.125)$$

The solution to Eq. (5.125) defines the pressure distribution, $p(x)$, for a given slider-bearing geometry.

Pressure is assumed to be constant across the thickness of the film. Pressure distribution with x is found by direct integration.

$$\frac{h^3}{12\mu} \frac{dp}{dx} = -\frac{1}{2} Uh + C_1 \qquad (5.126)$$

At some point, x, the pressure gradient, $\frac{dp}{dx}$ is zero. The film thickness at this point will be defined as h_*. Therefore, $C_1 = \frac{1}{2}Uh_*$. Equation (5.126) can now be expressed as

$$\frac{dp}{dx} = 6\mu U \left(\frac{h_* - h}{h^3} \right) \tag{5.127}$$

The symbol h_* is a constant of integration. The other constant of integration comes from Eq. (5.127). Both are established from pressure boundary conditions. Once the bearing film shape has been defined, by $h(x)$, Eq. (5.127) can be integrated with respect to x to determine the pressure distribution, $p(x)$ along the length the film. This function is used to relate bearing load to other design parameters.

The expression for film thickness, $h(x)$, is typically defined by

$$h(x) = h_1 + \alpha x \tag{5.128}$$

where

$$\alpha = \frac{h_2 - h_1}{L}$$

The mathematical steps required to integrate Eq. (5.127) are a bit unwieldy. They can be found in references, such as Radzimovsky [9]. The resulting pressure function resolves into a simple expression.

$$p(x) = -\frac{\mu U}{L} \frac{6\alpha \left(1 - \frac{x}{L}\right) \frac{x}{L}}{(\alpha - 2a)\left(a - \alpha \frac{x}{L}\right)^2} \tag{5.129}$$

where $a = h_1/L$. In dimensionless terms, Eq (5.129) becomes

Note that Ref. [9] has the slider moving to the right and the shoe angle α with a negative slope. The equations apply directly to Figure 5.20 provided the angle α is given a minus sign and $a = h_1/L$. Location parameter, x, is consistent with Figure 5.20.

In dimensionless terms

$$\frac{pL}{\mu U} = -\frac{6\alpha \left(1 - \frac{x}{L}\right) \frac{x}{L}}{(\alpha - 2a)\left(a - \alpha \frac{x}{L}\right)^2} \tag{5.130}$$

A plot of Eq. (5.130) is shown in Figure 5.21.

The maximum value of film pressure occurs at $x/L \sim 0.32$ from the trailing edge, where $x = 0$.

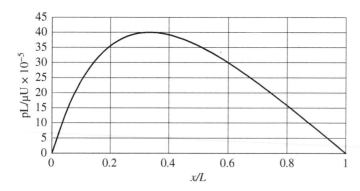

Figure 5.21 Pressure distribution in slider-bearing films.

Load-Carrying Capacity

The load-carrying capacity of slider bearings is determined by integrating under the pressure curve.

$$W = \int_0^L p(x)b\,dx \tag{5.131}$$

Again, the integration steps require patience but eventually lead to

$$W = 6\mu b U \frac{1}{\alpha^2}\left[\ln\frac{a-\alpha}{a} + \frac{2\alpha}{2a-\alpha}\right] \tag{5.132}$$

If the applied load is expressed in terms of load per area (P) of bearing

$$W = PLb \tag{5.133}$$

Then, Eq. (5.132) is expressed in dimensionless terms as

$$\frac{\mu U}{PL} = \frac{\alpha^2}{6}\left[\ln\frac{a-\alpha}{a} + \frac{2\alpha}{2a-\alpha}\right]^{-1} \tag{5.134}$$

This equation shows that the load capacity increases with a reduction in $a = h_1/L$ and a reduction in angle, α.

Friction in Slider Bearings

Friction in slider bearings is derived from

$$\tau = \mu\frac{du}{dy} \tag{5.135}$$

Starting with the velocity distribution across the oil film

$$\frac{du}{dy} = \frac{1}{2\mu}\frac{dp}{dx}(2y-h) + \frac{U}{h} \tag{5.136}$$

The differential shear force within the film is

$$dF = \mu dA\frac{du}{dx} \tag{5.137}$$

$$dF = b\left[\frac{1}{2}\frac{dp}{dx}(2y-h) + \frac{\mu U}{h}\right]dx \tag{5.138}$$

Friction force on the runner ($y = 0$) is

$$dF_R = b\left[-\frac{1}{2}\frac{dp}{dx}h + \frac{\mu U}{h}\right]dx \tag{5.139}$$

Using Eq. (5.130) and integrating from $x = 0$ to $x = L$ give

$$F_R = -\mu U b\left[\frac{4}{\alpha}\ln\left(\frac{a-\alpha}{a}\right) + \frac{6}{2a-\alpha}\right] \tag{5.140}$$

Similarly, friction on the fixed shoe is

$$F_h = \mu U b\left[\frac{2}{\alpha}\ln\left(\frac{a-\alpha}{a}\right) + \frac{6}{2a-\alpha}\right] \tag{5.141}$$

Coefficient of Friction

$$f = \frac{F_R}{W} = \frac{-2\alpha(2a - \alpha)\ln\left(\frac{a-\alpha}{a}\right) - 3\alpha^2}{3(2a - \alpha)\ln\left(\frac{a-\alpha}{a}\right) + 6\alpha} \tag{5.142}$$

where

$$\alpha = \frac{h_2 - h_1}{L}$$

$$a = \frac{h_1}{L}$$

Center of Pressure

The location of the center of pressure is determined from the first moment of the pressure curve.

$$AW = \int_0^L xbp(x)\,dx \tag{5.143}$$

"A" is x distance to the applied load, W, measured from the leading edge. Width of bearing is "b." Using the previous expression for $p(x)$ and integrating, we find that

$$\frac{A}{L} = \frac{(a - \alpha)(3a - \alpha)\ln\left(\frac{a - \alpha}{a}\right) - 2.5\alpha^2 + 3a\alpha}{a(a - 2a)\ln\left(\frac{a - \alpha}{a}\right) - 2\alpha^2} \tag{5.144}$$

The location of the center of pressure does not depend on W, U, or viscosity, μ. It depends on the angle of inclination, α, and the quantity, a, which contains the minimum film thickness, h_1.

Example   Consider a slider bearing having the following parameters.

Length (L)	– 3.0 in.
Width (b)	– 2.5 in.
Load (W)	– 3450 lb
Velocity (U)	– 100 in./s
Inclination (α)	– (–)0.000 266 7 rad
Viscosity (μ)	– 10×10^{-6} reyn

We wish to determine:

1) Minimum film thickness
2) Friction force on moving member
3) Coefficient of friction
4) Power loss in the bearing due to viscous friction

The unit pressure, P, is

$$P = \frac{W}{bL} = \frac{3450}{2.5(3)} = 460 \text{ psi}$$

Equation (5.134) is put in the form

$$\frac{PL\alpha^2}{6\mu U} = \ln\left(\frac{a - \alpha}{a}\right) + \frac{2\alpha}{2a - \alpha} \tag{5.145}$$

The left side is equal to

$$\frac{PL\alpha^2}{6\mu U} = \frac{460(3)(-0.000\ 266\ 7)^2}{6(10)10^{-6}(100)} = 0.0163$$

So

$$\ln\left(\frac{a-\alpha}{a}\right) + \frac{2\alpha}{2a-\alpha} = 0.0163$$

The shoe angle is given as $\alpha = -0.000\ 266\ 7$. The left side is implicit in "a." By trial and error, "a" is found to be 0.000 333.

$$h_1 = \alpha L = 0.000\ 333(3) = 0.001\ \text{in}.$$

The friction force on the runner under these conditions (using Eq. (5.140)) is 5.97 lb. The coefficient of friction is

$$f = \frac{F_R}{W} = \frac{5.97}{3450} = 0.001\ 73$$

Loss of power in friction is

$$H = \frac{F_R U}{550(12)} = 0.905\ \text{hp}$$

Slider Bearing with Pivoted Shoe

Equation (5.144) shows that the location of center of pressure depends only on minimum film thickness, h_1, and angle of inclination, α. It is independent of load (W), viscosity (μ), and speed (N). Analyses also show that load capacity is greatest and minimum film thickness is maximum when $h_2 \sim 2\ h_1$. Under this condition, $A \sim \tfrac{5}{9}L$ (Figure 5.22).

Therefore, if a bearing pad is designed to pivot on a point located $\tfrac{5}{9}$ of its length from the leading edge, we would expect the h_1 and h_2 (and α) to be established automatically giving greatest load and

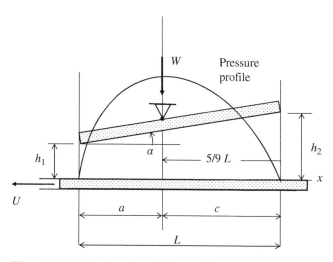

Figure 5.22 Slider bearing with pivoted shoe.

film thickness. This feature makes the pivoted slider bearing very attractive since it automatically sets the best angle, thus greatly reducing manufacturing costs.

It is useful in the analysis of pivoted slider bearing to use the following expression.

$$m = \frac{h_2}{h_1} - 1$$

Adjusting Eq. (5.104) to reflect this expression gives

$$W = \mu Ub \frac{6L^2}{m^2 h_1^2} \left[\ln(1 + m) - \frac{2m}{m + 2} \right] \tag{5.146}$$

$$W = \frac{6\mu UbL^2}{h_1^2} \left[\frac{1}{m^2} \ln(1 + m) - \frac{2m}{m(m + 2)} \right] \tag{5.147}$$

$$W = \frac{6\mu UbL^2}{h_1^2} K_W \tag{5.148}$$

where

$$K_W = \frac{1}{m^2} \ln(1 + m) - \frac{2}{m(m + 2)}$$

Because K_W is insensitive to values of m, some designers use an average value of $K_W = 0.025$ to simplify calculations.

Frictional Resistance

Expressing the friction force on the runner in terms of m gives

$$F_0 = \frac{\mu UbL}{h_1} \left[\frac{4}{m} \ln(1 + m) - \frac{6}{2 + m} \right] \tag{5.149}$$

$$F_0 = \frac{\mu UbL}{h_1} K_F \tag{5.150}$$

where

$$K_F = \frac{4}{m} \ln(1 + m) - \frac{6}{2 + m}$$

Coefficient of Friction

$$f = \frac{F_0}{W} \tag{5.151}$$

Combining terms gives

$$f = \frac{h_1}{L} \left[\frac{1}{6} \frac{K_F}{K_W} \right] = \frac{h_1}{L} K_f \tag{5.152}$$

where

$$K_f = \frac{K_F}{K_W}$$

Exponential Slider-Bearing Profiles

The shape of the film does not markedly influence pressure distribution. In some cases, the choice of film geometry greatly simplifies mathematical solution but still predicts useful film performance. The exponential film (Figure 5.23) is a good example [10].

The film geometry of an exponential slider is defined by

$$h(x) = h_1 e^{sx} \tag{5.153}$$

$$h_2 = h_1 e^{sL} \tag{5.154}$$

$$s = \frac{1}{L} \ln \left(\frac{h_2}{h_1} \right) \tag{5.155}$$

Note that h_1, h_2, and L define the exponential profile.

Pressure Distribution for Exponential Profile

Returning to Reynold's equation

$$\frac{d}{dx} \left[\frac{h^3}{12\mu} \frac{dp}{dx} \right] = - \frac{U}{2} \frac{dh}{dx} \tag{5.156}$$

The solution is determined directly by substituting for $h(x)$, integrating with respect to x, satisfying boundary pressure conditions, and solving for $p(x)$. An alternate approach is to define h^* as the film thickness where the film pressure is maximum, $dp/dx = 0$. In this case

$$\frac{h^3}{12\mu} \frac{dp}{dx} = \frac{U}{2}(h^* - h) \tag{5.157}$$

$$\frac{dp}{dx} = 6\mu U \left(\frac{h^* - h}{h^3} \right) \tag{5.158}$$

Substituting for $h(x)$, integration, and imposing $p(0) = 0$ boundary condition

$$p(x) = \frac{6\mu U}{h_1^2 s} \left[\frac{-1}{2} \left(1 - e^{-2sx} \right) + \frac{h^*}{3h_1} \left(1 - e^{-3sx} \right) \right] \tag{5.159}$$

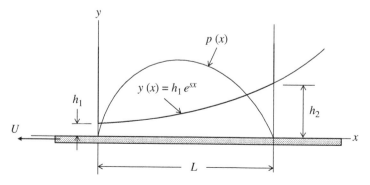

Figure 5.23 Exponential slider-bearing profile.

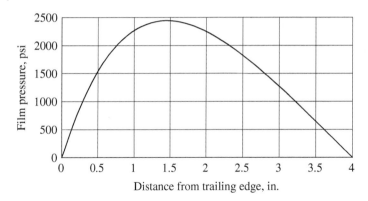

Figure 5.24 Pressure distribution in an exponential slider bearing.

Using the boundary condition, $p(L) = 0$ gives

$$\frac{h^*}{h_1} = \frac{3}{2} \frac{(1 - e^{-2sL})}{(1 - e^{-3sL})} \tag{5.160}$$

Consider an exponential slider bearing having the following dimensions.

h_1	=	0.001 in.
h_2	=	0.002 in.
L	=	4 in.
μ	=	24×10^{-6} reyn (165 cp – SAE 30 Oil)
U	=	100 ips

Figure 5.24 shows that maximum pressure is achieved at $x = 1.4$ in. The load-carrying capacity of the film is determined by integrating under the pressure curve.

Pressure Comparison with Straight Taper Profile

Pressure distribution for both straight and exponential profiles is given in Figure 5.25. The above input data were used for both. It is interesting that profile shape does not have a significant effect on film pressure distribution, a conclusion found by considering various profiles.

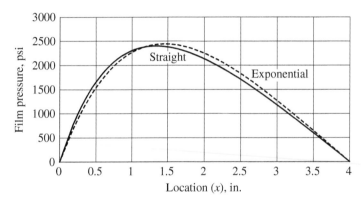

Figure 5.25 Comparison of pressure profiles.

Load-Carrying Capacity

The load-carrying capacity is defined by

$$W = \int_0^\ell p(x)b\,dx \tag{5.161}$$

Friction is

$$F = \int_0^\ell \tau(x)b\,dx \tag{5.162}$$

$$\tau = \mu\frac{du}{dx}\bigg|_{y=0} \tag{5.163}$$

$$\frac{du}{dy} = \frac{1}{2\mu}\frac{dp}{dx}(2y-h) + \frac{U}{h} \tag{5.164}$$

Coefficient of friction is

$$f = \frac{F}{W} \tag{5.165}$$

Pressure Distribution for Open Entry

In this case, the slider-bearing shoe extends well beyond the previous length, L. The exponential expression for the fixed shoe still applies

$$h(x) = h_1 e^{sx} \tag{5.166}$$

where (for the above example)

$$s = \frac{1}{L}\ln\left(\frac{h_2}{h_1}\right) = \frac{1}{4}\ln 2 = 0.1733 \tag{5.167}$$

However, in this case, the profile of the shoe extends beyond the distance, L (see Figure 5.26). Imposing $p(\infty) = 0$ boundary conditions on Eq. (5.160),

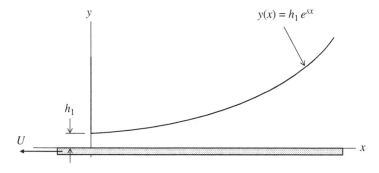

Figure 5.26 Exponential slider bearing with front opening.

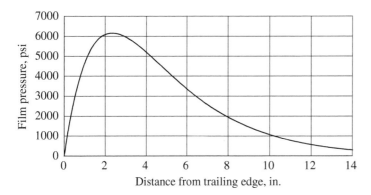

Figure 5.27 Pressure distribution in an exponential bearing with open leading edge.

$$\frac{h^*}{h_1} = \frac{3}{2} \tag{5.168}$$

This occurs at an x location of

$$\frac{h^*}{h_1} = e^{sx} \tag{5.169}$$

$$x = \frac{1}{s} \ln\left(\frac{h^*}{h_1}\right) = \frac{1}{0.1733}0.4055 = 2.34 \text{ in.(agrees with Figure 5.27)} \tag{5.170}$$

Replacing $\dfrac{h^*}{h_1} = \dfrac{3}{2}$ in Eq. (5.159) gives

$$p(x) = \frac{6\mu U}{h_1^2 s}\left[\frac{-1}{2}\left(1 - e^{-2sx}\right) + \frac{1}{2}\left(1 - e^{-3sx}\right)\right] \tag{5.171}$$

$$p(x) = \frac{3\mu U}{sh_1^2}\left[e^{-2sx} - e^{-3sx}\right] \tag{5.172}$$

Applying the same input data as before, except the leading edge is open and extends beyond 4 in., gives a pressure profile as shown in Figure 5.27. In this case, the maximum pressure (6100 psi) increases dramatically over the previous case (2400 psi).

Exponential Slider Bearing with Side Leakage

The exponential profile also provides a direct mathematical approach accounting for side leakage. A sketch of a 3-D control volume of a film is shown in (Figure 5.28). While the film geometry is the same as for a 2-D film, the pressure distribution and film velocities are different. The pressure profile varies as $p = p(x, z)$.

Fluid velocity components also vary with x and y.

$$u = u(x,y)$$

$$w = w(x,y)$$

The pressure distribution within the film is determined by solving Reynolds equation accounting for flow in the side (z) directions.

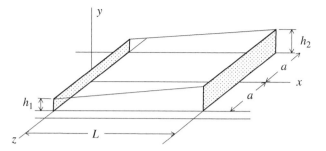

Figure 5.28 Control volume of 3-D film.

$$\frac{\partial}{\partial x}\left[\frac{h^3}{\mu}\frac{\partial p}{\partial x}\right] + \frac{\partial}{\partial z}\left[\frac{h^3}{\mu}\frac{\partial p}{\partial z}\right] = -6U\frac{dh}{dx} \tag{5.173}$$

Since fluid film pressure is symmetrical with respect to the x axis, it is reasonable to assume the pressure function, $p(x, z)$, can be represented by a series expansion containing even cosine terms.

$$p(x, z) = \sum_{n=1,3,5}^{\infty} X_n \cos \frac{n\pi z}{2a} \tag{5.174}$$

here $X_n(x)$ is an unknown function of x. Note that Eq. (5.174) satisfies two of the boundary conditions, i.e. zero pressure at $z = \pm a$.

After substitution, Reynolds equation becomes

$$\sum_{n=1,3,5}^{\infty}\left[h^3 X_n'' + 3h^2\frac{dh}{dx}X_n' - h^3\left(\frac{n\pi}{2a}\right)^2 X_n\right]\cos\frac{n\pi z}{2a} = -6\mu U\frac{dh}{dx} \tag{5.175}$$

If the right side of Eq. (5.175) is multiplied by one, which in turn is expanded into a Fourier series, the x and z variables can be separated.

$$\sum_{n=1,3,5}^{\infty}\left[h^3 X_n'' + 3h^2\frac{dh}{dx}X_n' - h^3\left(\frac{n\pi}{2a}\right)^2 X_n + 6\mu U\frac{dh}{dx}\frac{4}{n\pi}(-1)^{\frac{n-1}{2}}\right]\cos\frac{n\pi z}{2a} = 0 \tag{5.176}$$

From which

$$\frac{dh}{dx}[h^3 X_n'] - \left(\frac{n\pi}{2a}\right)^2 h^3 X_n = -6\mu U\frac{dh}{dx}\frac{4}{n\pi}(-1)^{\frac{n-1}{2}} \tag{5.177}$$

This is an ordinary differential equation with variable coefficients, which depend on the geometry of the lubricating film. Its solution is not simple. Hayes [11], however, solved this equation for the flat shoe.

The exponential shape of the bearing shoe, on the other hand, is amenable to a direct solution. The exponential shape is defined by

$$h(x) = h_1 e^{sx} \tag{5.178}$$

where

$$s = \frac{1}{L}\ln\left(\frac{h_2}{h_1}\right)$$

L – length of bearing

By substitution into Eq. (5.177)

$$X_n'' + \frac{3}{L}X_n' - \left(\frac{n\pi}{2a}\right)^2 X_n = -\frac{6\mu Us}{h_1^2}\frac{4}{n\pi}(-1)^{\frac{n-1}{2}}e^{-2sx} \tag{5.179}$$

or

$$X_n'' + 3sX_n' - \alpha^2 X_n = \beta e^{-2sx} \tag{5.180}$$

where

$$\alpha = \frac{n\pi}{2a}$$

$$\beta = -\frac{6\mu Us}{h_1^2}\frac{4}{n\pi}(-1)^{\frac{n-1}{2}}$$

The total solution to Eq. (5.180) (including both particular and complementary solutions) is

$$X_n(x) = e^{-\frac{3}{2}sx}[A_n \sinh \lambda x + B_n \cosh \lambda x] - \frac{\beta}{(2s^2 + \alpha^2)}e^{-2sx} \tag{5.181}$$

where

$$\lambda = \left[\frac{9}{4}s^2 + \alpha^2\right]^{\frac{1}{2}}$$

Imposing boundary conditions of $X_n(0) = 0$ and $X_n(L) = 0$

$$X_n(x) = \frac{\beta}{(2s^2 + \alpha^2)}e^{-\frac{3}{2}sx}\left[\frac{\sinh \lambda x}{\sinh \lambda L}\left(e^{-\frac{1}{2}Ls} - \cosh \lambda L\right) - \left(e^{-\frac{1}{2}sx} - \cosh \lambda x\right)\right] \tag{5.182}$$

This expression completes the pressure function defined by Eq. (5.174).

$$\frac{pL}{\mu U} = \sum_{1,3,5}\frac{\beta'}{(2s^2 + \alpha^2)}e^{-\frac{3}{2}sx}\left[\frac{\sinh \lambda x}{\sinh \lambda L}\left(e^{-\frac{1}{2}Ls} - \cosh \lambda L\right) - \left(e^{-\frac{1}{2}sx} - \cosh \lambda x\right)\right]\cos\frac{n\pi}{2}\frac{z}{a} \tag{5.183}$$

where

$$\alpha = \frac{n\pi}{2a}$$

$$\beta = -\frac{6\mu Us}{h_1^2}\frac{4}{n\pi}(-1)^{\frac{n-1}{2}} = \frac{\mu U}{L}\beta'$$

$$\beta' = -\frac{Ls}{h_1^2}\frac{24}{n\pi}(-1)^{\frac{n-1}{2}} \quad \text{where } Ls = \ln\left(\frac{h_2}{h_1}\right)$$

$$\lambda = \left[\frac{9}{4}s^2 + \alpha^2\right]^{\frac{1}{2}}$$

$$s = \frac{1}{L}\ln\left(\frac{h_2}{h_1}\right)$$

$$\lambda x = \frac{x}{L}\left[\frac{9}{4}\ln^2\left(\frac{h_2}{h_1}\right) + \left(\frac{n\pi}{2}\right)^2\left(\frac{L}{a}\right)^2\right]^{\frac{1}{2}} \tag{5.184}$$

Table 5.7 Pressure profile for an exponential bearing with side leakage.

x	$\frac{pL}{\mu U}$ (dimensionless)		p (psi)	
	z = 0	z = 0.8a	z = 0	z = 0.8a
0.0	0	0	0	0
0.5	1.748	0.792	1048	475
1.0	2.456	1.029	1473	618
1.5	2.558	1.033	1535	620
2.0	2.318	0.923	1391	554
2.5	1.886	0.755	1131	453
3.0	1.337	0.551	802	331
3.5	0.704	0.311	422	187
4.0	0.0	0.0	0	0

As an application of Eq. (5.183), consider a slider bearing operating under the following parameters.

h_1 = 0.001-in.
h_2 = 0.002-in.
L = 4-in. (bearing length)
$2a$ = 4-in. (bearing width)
μ = 165 cp or 24×10^{-6} reyn (SAE 30 lubricant)
U = 100 ips

The shape of the fixed shoe is of an exponential shape as defined above.

The pressure distribution is listed in Table 5.7 giving pressure along the spine of the pressure profile (z = 0) and for z = 0.8a.

For comparison, check the pressure distribution for the same exponential bearing without side leakage (Figure 5.24). We see the pressure distribution along the spine of Figure 5.29 is substantially lower than the maximum pressure distribution without leakage.

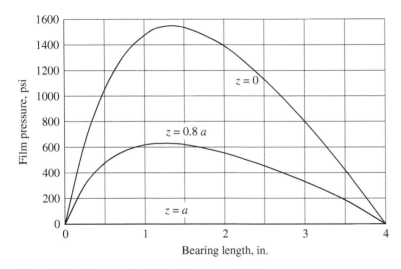

Figure 5.29 Pressure distribution in exponential film.

Hydrodynamic Lubricated Journal Bearings

Testing by Towers [12] showed that two bearing surfaces can be separated by a thin film of liquid under certain operating conditions of load, speed, and viscosity. His test device included a fixed sleeve and a rotating shaft to simulate wheel bearings in locomotives. The sleeve had a hole drilled in the top for injecting different greases. Once grease had been injected, Towers plugged the hole with a wooden plug. During his test, he noticed that the wooden plug kept popping out and oil would leak out of the hole. Following this observation, he measured film pressure distributions around journal bearings and concluded that this pressure is great enough to support bearing loads. Reynolds [13] applied the principles of fluid mechanics to formulate differential equations, which predict fluid film pressure distributions. This pioneering work of both Tabor and Reynolds established the foundation for analytical advances, which has provided a sound basis for the practical engineering design of thick film–lubricated bearings.

Pressure Distribution Around an Idealized Journal Bearing

The mechanism for developing a thin film in journal bearings is the same as for slider bearings. The variation in film thickness is shown in Figure 5.30. In this case, the runner is replaced by the surface of the journal and the shoe is replaced by the outer bearing surface.

The radial clearance between the bearing surface and the journal surface is designated as "*c*." The radius of the journal is r and the radius of the bearing is $r + c$. The distance between the center of the bearing and the center of the journal is "*e*" and is called *eccentricity* of the bearing.

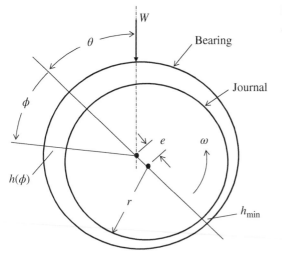

Figure 5.30 Mathematical model of idealized journal bearing.

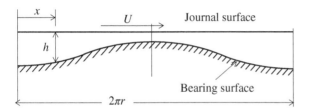

Figure 5.31 Film thickness in journal bearing.

The film between bearing and journal surfaces is laid out in Figure 5.31. The distance x is related to ϕ by

$$x = r$$

The film thickness, h, becomes

$$h(\phi) = c + e \cos \phi$$
$$h(\phi) = c(1 + n \cos \phi) \tag{5.185}$$

where

n	$=$	$\frac{e}{c}$ (eccentricity ratio, also attitude)
c	$-$	radial clearance between journal and bearing (concentric position)
e	$-$	center distance between bearing and journal
ϕ	$-$	independent variable
h	$-$	film thickness at position, i
r	$-$	radius of journal

The derivation of the pressure distribution around the journal starts as before

$$\frac{d}{dx}\left[h^3 \frac{dp}{dx}\right] = 6\mu U \frac{dh}{dx} \tag{5.186}$$

Changing the independent variable to $r\phi$

$$\frac{d}{d\phi}\left[h^3 \frac{dp}{d\phi}\right] = 6\mu U r \frac{dh}{d\phi} \tag{5.187}$$

Integration gives

$$\frac{dp}{d\phi} = 6\mu U r \left[\frac{1}{h^2} - \frac{k}{h^3}\right] \tag{5.188}$$

where k is a constant of integration. Substituting for h gives

$$\frac{dp}{d\phi} = \frac{6\mu r U}{c^2} \left[\frac{1}{(1 + n \cos \phi)^2} - \frac{k}{c(1 + n \cos \phi)^3}\right] \tag{5.189}$$

Reynolds [13] derived this equation in 1886 and found an approximate solution in the form of a Fourier series. The solution was useful only for small eccentricity ratios. The exact solution was later developed by Sommerfeld [14] and recaptured by Radzimovsky [9]. Following Sommerfeld's work, the expression that defines the pressure surrounding a journal for an idealized bearing is

$$p(\phi) = p(0) + \frac{6\mu r U}{c^2}\left[\frac{n(2 + n\cos\phi)\sin\phi}{(2 + n^2)(1 + n\cos\phi)^2}\right]\tag{5.190}$$

$p(0)$ is film pressure at $\phi = 0°$. This pressure can be determined by using a known pressure at the inlet point, i.e. p_i at ϕ_i. The location of the inlet point is not necessarily at $\phi = 0$.

Example It is desired to find the pressure distribution around the journal.

Diameter of bearing	$d = 1.5$ in.
Length of bearing	$b = 2.5$ in.
Speed of journal	$N = 3000$ rpm
Radial clearance	$c = 0.001$ in.
Expected T_{ave} of the film	$t = 165°$ F
Oil pressure at inlet	$p_i = 45$ psi
Location of inlet hole	$\phi_i = 315°$
Attitude ($n = e/c$, eccentric ratio)	$n = 0.8$
Lubricating oil	SAE 20 ($\mu = 2.15 \times 10^{-6}$ reyn)

Note that $p(0)$ can be determined from Eq. (5.190) by the known pressure, p_i at ϕ_i. By substitution, the pressure at $\phi = 0$ is $p(0) = 556$ psi. Film pressure can now be determined at any ϕ from

$$p(\phi) = 556 + 691\left[\frac{(2 + n\cos\phi)\sin\phi}{(1 + n\cos\phi)^2}\right]\tag{5.191}$$

The locations of the maximum and minimum pressure are determined from

$$\cos\phi = -\frac{3n}{n^2 + 2}\tag{5.192}$$

$$\phi_{max} = 155.5° \quad (p_{max} = 5476 \text{ psi})$$

$$\phi_{min} = 205.4° \quad (p_{min} = -4364 \text{ psi})$$

The attitude angle, θ, is not given in the problem statement. It can be shown that for an ideal journal bearing the attitude angle, θ, is 90° from the applied load, W.

The minus pressure over part of the film as shown in Figure 5.32 is ignored. Only the positive pressure is considered in formulating the load capacity of the bearing film.

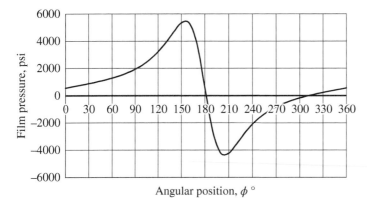

Figure 5.32 Journal bearing pressure distribution.

Load-Carrying Capacity

The load-carrying capacity of journal bearings is based on a minimum required film thickness. The relation between the design variables was established by Sommerfeld [14] and captured by Radzimovsky [9].

$$\left(\frac{r}{c}\right)^2 \frac{\mu N'}{P} = \frac{(2+n^2)\sqrt{1-n^2}}{12\pi^2 n} \tag{5.193}$$

where

n	–	eccentricity ratio (also attitude)
μ	–	viscosity, reyn
N'	–	rotary speed, rev/s
P	–	average journal pressure, psi
r	–	radius of journal, in.
c	–	radial clearance, in.

The left side of this equation is called the Sommerfeld number in recognition of his pioneering work on the lubrication of journal bearings. This function is shown graphically in Figure 5.33.

Example Consider a journal bearing with the following operating conditions:

$\mu = 0.145 \times 10^{-6}$ reyn $= 1$ cp (water)
$r = 1$ in.
$b = 2$ in. (bearing length)
$c = 0.010$ in.
$N = 1000$ rpm
$W = 1$ lb ($P = 0.25$ lb/in.^{2})
$S = \left(\frac{r}{c}\right)^2 \frac{\mu N}{P}$
$S = \left(\frac{1}{0.010}\right)^2 \frac{0.145 \times 10^{-6}(100)}{(0.25)6} = 0.0967 \sim 0.1$

From Eq. (5.193) or Figure 5.33

$n = 0.185$
$\frac{e}{c} = 0.185$

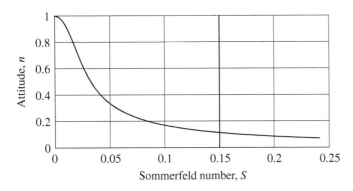

Figure 5.33 Sommerfeld number vs. attitude (eccentricity ratio).

$e = 0.001\ 85$

Minimum film thickness is

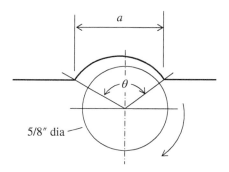

$h_0 = c - e = 0.010 - 0.00185$
$h_0 = 0.0085$

Minimum Film Thickness in Journal Bearings

Experiments conducted Karelitz and Keyon [15], McKee [16], and Stanton [17] have shown that the smallest practical limit for fluid film lubrication is about 0.000 05 in. (1.27 µm). However, for large commercial bearings the minimum film thickness should be kept above 0.0001 in. (2.54 µm). To accommodate possible particles of contamination in the lubricant, elastic, and thermal distortions in the bearing structure, the lower limit might possibly be set at 0.001 in. (25.4 µm).

Karelitz's testing arrangement centered on rotating a shaft against a flat test block. Even with a lubricant, the block showed wear as indicated by a dial gage. At some point in the experiment, wear stopped, indicating a thin film had formed separating the two sliding surfaces.

Measurements on the wear cavity were:

$a = 0.02$–0.06-in.
$\theta = 3.7°$–$10.9°$

Width of the test block was 1 in.

Details of the circular wear profile in the test block were used to calculate the minimum film thickness, which was determined to be approximately 0.00005 in. (1.27 µm). Measurements of the surface asperities showed that the peak-to-valley distance of the asperities was of the same order of magnitude.

The minimum allowable film thickness in journal bearings depends on factors, such as: (i) smoothness or finish of the sliding surfaces, (ii) elastic rigidity of the bearing structure, (iii) thermal distortion, and (iv) size of contamination particles in the lubricant.

Film thickness common in electric motors is $h_{min} = 0.00075$ in. (0.019 mm) at median speeds of 500–1500 rpm. For larger shafts, having running speeds between 1500 and 3600 rpm, the minimum film thickness ranges between 0.003 and 0.005 in. (0.076–0.13 mm). For small automotive and aviation engine bearings, with high finished surfaces, h_{min} ranges between 0.0001 and 0.0002 in. (0.0025–0.005 mm). The smallest allowable film thickness in journal bearings depends on factors, such as: (i) smoothness or finish of the sliding surfaces, (ii) elastic rigidity of the bearing structure, (iii) thermal distortion, and (iv) size of contamination particles in the lubricant.

Friction in an Idealized Journal Bearing

Friction acting on a journal is due to fluid shear stress at the surface of the journal. Shear stress is related to rate of shear as explained earlier. Before discussing Sommerfeld's solution, consider friction torque in concentric journal rotation.

Petroff's Law

Petroff [18] gives a baseline for friction in journal bearings showing that when the journal is concentric within a bearing, friction is due to simple shear in the film. In this case, shear stress on the surface of the bearing is

$$\tau = \mu \frac{du}{dr} \sim \mu \frac{U}{c} = \mu \frac{2\pi r N}{c \, 60} \tag{5.194}$$

where

τ	–	shear, psi
μ	–	viscosity, reyn $\frac{\text{lb·s}}{\text{in.}^2}$ (1 cp = 1.45×10^{-7} reyn)
N	–	rotary speed, rpm
c	–	radial clearance between journal and bearing, in.
r	–	radius of journal, in.
U	–	surface velocity of journal, in./s

Friction torque produced by shear stress on the bearing surface is

$$T = \tau(2\pi r b)r = \frac{4\pi^2 r^3 b \mu N}{c60} = 0.658 \frac{\mu b r^2 N}{m} \tag{5.195}$$

Expressing bearing load in terms of average pressure,

$$P = \frac{W}{2rb} \tag{5.196}$$

Representing the friction force by $F = fW$ and $T = Fr$, then

$$T = rfW = rfP2rb = 2r^2 fbP \tag{5.197}$$

Equating torque expressions gives

$$f = 0.349 \left(\frac{\mu N}{P} \right) \left(\frac{r}{c} \right) \tag{5.198}$$

$$\left(\frac{r}{c} \right) f = 2\pi^2 \left(\frac{\mu N'}{P} \right) \left(\frac{r}{c} \right)^2 \tag{5.199}$$

where

μ	–	viscosity, reyn
N'	–	rotary speed, rev/s
P	–	average journal pressure, psi
r	–	radius of journal, in.
c	–	radial clearance, in.
f	–	coefficient of friction

This expression for the coefficient of friction is known as Petroff's law. It is based on the ideal condition of concentric rotation of a journal inside a bearing.

Sommerfeld's Solution

Sommerfeld developed journal friction for nonconcentric journals also based on fluid shear stress at the journal's surface. His expression for the coefficient of friction is

$$f = \frac{F_j}{W} \tag{5.200}$$

Substituting for the friction force on the journal gives

$$\left(\frac{r}{c}\right)f = \frac{1 + 2n^2}{3n} \tag{5.201}$$

Since the friction term on the left is a function only of attitude (n), it is also a function of the Sommerfeld number. This relation is shown in Figure 5.34.

Example Consider a journal bearing having the following operating conditions.

W = 3000 lb
N = 2000 rpm
r = 2 in. (journal radius)
μ = 4×10^{-6} reyn (viscosity at 100 °F)
c = 0.01 in. (radial clearance)
L = 2 in. (bearing length)

The objective is to determine the minimum operating film thickness and coefficient of friction. Both unknowns depend on the Sommerfeld number.

$$S = \left(\frac{r}{c}\right)^2 \frac{\mu N'}{P} \tag{5.202}$$

$$S = \left(\frac{2}{0.01}\right)^2 \frac{4 \times 10^{-6} 2000}{375(60)} = 0.0142 \tag{5.203}$$

The eccentricity ratio can be determined from Eq. (5.201) or from Figure 5.34, from which $n \sim$ 0.85. Already we know this is a highly loaded bearing and expect a relatively small minimum film thickness and high friction coefficient. Noting that

$$h_{\min} = c - e = 0.01(1 - 0.85) = 0.0015 \text{ in.}$$

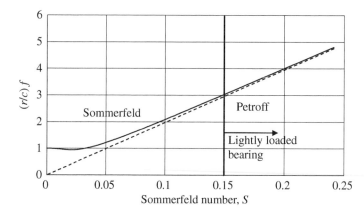

Figure 5.34 Petroff friction and lightly loaded bearings.

The coefficient of friction for this bearing is

$$\frac{r}{c}f = \frac{1 + 2n^2}{3n} = \frac{1 + 2(0.85)}{3(0.85)} = 1.0588$$

from which $f = 0.0053$. Friction torque on the journal is

$$T_j = fWr = 0.0053(3000)2 = 31.8 \text{ in.-lb}$$

Power loss, then, is

$$P_f = \frac{T_j N}{60} = \frac{31.764(2000)}{60} = 1058.8 \text{ in.-lb/s}$$

or 0.16 hp.

The rate at which heat is being generated is

$$Q_f = \frac{1058.8}{12}\frac{1}{778} = 0.1134 \text{ Btu/s}$$

Stribeck Diagram and Boundary Lubrication

While Sommerfeld's work gave much needed insight into the performance of journal bearings, at very high loads (small Sommerfeld numbers) a point is reached where the film thickness is of the same order of magnitude at the surface roughness (peaks and valleys) and the theory of thick film lubrication no longer applies. Under this condition, the peaks between bearing surfaces begin to make intermittent contact.

Stribeck [19] conducted friction tests under conditions of heavily loaded bearings and found there is a certain value of the Sommerfeld number for which boundary lubrication begins. The lowest point in these curves marks the limit of thick film lubrication. Further loads increase friction, indicated by the upward trend of the friction curve. Under this condition, friction and heat increase and usually lead to bearing failure.

Ways to improve the performance of journal bearings focus on ways to move the minimum friction point to the left. Running in bearings at low loads and speeds has been recommended for new machinery. Animal-based oil performs better than mineral oil.

Stribeck recognized this effect and studied ways to shift the minimum friction point farther to the left. Factors that affect the location of the minimum point are run-in time and lubricant chemistry. Recently, Barnhill [20] showed that a significant shift in the minimum friction point can be achieved by use of ionic liquid (IL) additives.

Regions of Friction

Fuller [7] shows four regions of bearing friction (Figure 5.35). Coefficient of friction is plotted against $\frac{ZN}{P}$, where

Z	–	viscosity (cp)
N	–	rotary speed (rpm)
P	–	force over projected area; length (L) × diameter (D) (psi)

This ratio is called the Hersey number. It is not dimensionless as the Sommerfeld number but is a mixture of units for convenience.

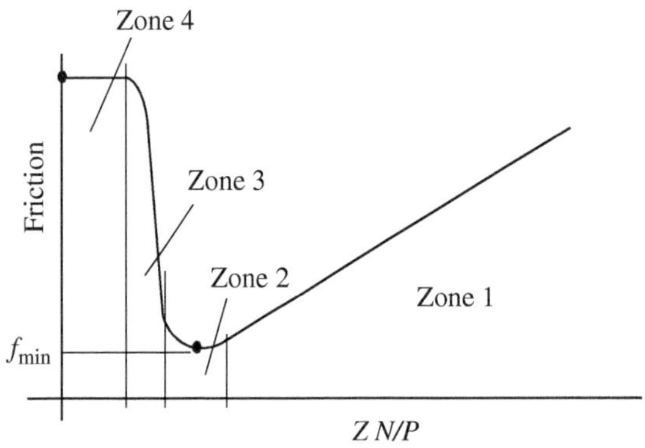

Figure 5.35 Plot of coefficient of friction vs. ZN/P.

1) Thick film region – Bearing surfaces are completely separated by a liquid film. When the lubricant is free of abrasive particles, wear is prevented. The coefficient of friction can be as low as 0.001 or even less. This region begins at a Hersey number within the range of 30–40 [7]. This number could be lower depending on surface finish and run-in time.
2) Thin-film lubrication region – This region represents the lower limit of complete separation by a film. Film thickness varies from around 0.0002 in. down to 0.000 05 in. Bearing rigidity and surface smoothness are important too.
3) Mixed-film region – Loading is so severe that complete surface separation cannot be achieved by the lubricant. Most of the surface experiences rubbing without the benefit of complete separation of the peaks and valleys of the surface roughness. At best, lubrication is achieved by surface separation by a few layers of fluid molecules. The coefficient of friction under this condition may range from 0.02 to 0.08.
4) Boundary-film region – There is no fluid film because of either low viscosity or velocity or because of high bearing load. Film thickness can only be described in terms of a molecule. The coefficient of friction may range between 0.08 and 0.15.

To the right of the minimum point, bearings operate under thick film conditions. Under this condition, a thick film completely separates the bearing surfaces. If the load on the bearing is increased, the operating point moves to the left until thermal equilibrium is again established. Additional increases in load will lower the coefficient of friction until the minimum point has been reached. It is best to operate journal bearings at this low friction point. Further increases in load will cause boundary-film lubrication causing friction and heat to increase.

Two factors can move the minimum friction point to the left. One is surface smoothness, and the other is the "oiliness" of the lubricant. Surface smoothness is sometimes achieved by "running in" bearings, i.e. initially running new machine at lower-than-specified operating speeds. This allows the bearing surfaces to polish themselves.

In some designs, cylindrical roller bearings have been replaced with journal-type bearings. As explained earlier, friction in journal bearings depends on the Hersey number, $\frac{ZN}{P}$.

If the cylindrical bearing is replaced by a journal bearing, local stresses are greatly reduced, but it is unlikely that a lubricant film is developed because of the high load and low relative rotation of the cone on the pin.

Consider, for example, a drill bit journal bearing having the following conditions

Z	–	105 cp (same as SAE W30 motor oil at 100 °F)
N_{rel}	–	200 rpm
P	–	3000 psi

The Hersey number in this case is 7. Since this Hersey number is less than 30, journal surfaces are at best boundary lubricated. Nonetheless sliding friction in the journal bearing is greatly reduced; a reliable dynamic seal is essential, however.

Comparison of Journal Bearing Performance with Roller Bearings

For this comparison, we choose a 2 in. ID (inside diameter) journal rotating at 2000 rpm with an applied load of 1000 lb.

Journal Bearing

Assuming a 2-in. journal with a 1 in. length gives $P = 500$ psi as the average bearing load and $\mu = 16$ $(10)^6$ reyn

$$S = \left(\frac{r}{c}\right)^2 \frac{\mu N'}{P} \tag{5.204}$$

$$S = \left(\frac{1}{0.01}\right)^2 \frac{16 \times 10^{-6} 2000}{500(60)} = 0.0108$$

The eccentricity ratio can be determined from Eq. (5.193) or from Figure 5.33, from which $n = 0.91$. Already we know that this is a highly loaded bearing, and we expect a relatively small minimum film thickness and high friction coefficient. Noting that

$$h_{min} = c - e = 0.01(1 - 0.91) = 0.0009 \text{ in.}$$

The coefficient of friction for this bearing is

$$\frac{r}{c}f = \frac{1 + 2n^2}{3n} = \frac{1 + 2(0.91)^2}{3(0.91)} = \frac{2.86}{2.79} = 0.973$$

From which the coefficient of friction is: $f = 0.0097$.

Journal bearing perform better at high speeds as shown at $N = 4000$ rpm.

Roller Contact Bearing (See Footnote 1)

Selecting a 1.9685 in. ID roller bearing having a dynamic load capacity of 18 300 lb gives (using Eq. (5.1))

$$L = \left(\frac{C}{F}\right)^{\frac{10}{3}} \tag{5.205}$$

$$L = \left(\frac{18\ 300}{2000}\right)^{\frac{10}{3}} = 1601 \times 10^6 \text{ rev} \tag{5.206}$$

The amount of time to reach this level of revolutions at 2000 rpm is

$$h = 1601 \times 10^6 \ \text{rev} \left| \frac{1 \ \text{min}}{2000 \ \text{rev}} \right| \left| \frac{1 \ \text{h}}{60 \ \text{min}} \right| \left| \frac{1 \ \text{d}}{24 \ \text{h}} \right| \left| \frac{1 \ \text{yr}}{365 \ \text{d}} \right| \tag{5.207}$$

$$h = 1.52 \ \text{year}$$

Ball Bearing (See Footnote 1)

Selecting a 1.9685 in. ID ball bearing having a dynamic load capacity of 15 600 lb gives

$$L = \left(\frac{15 \ 600}{2000} \right)^3 = 474.6 \times 10^6 \ \text{rev} \tag{5.208}$$

The amount of time to reach this level of revolutions at 2000 rpm is

$$h = 474.6 \times 10^6 \ \text{rev} \left| \frac{1 \ \text{min}}{2000 \ \text{rev}} \right| \left| \frac{1 \ \text{h}}{60 \ \text{min}} \right| \left| \frac{1 \ \text{d}}{24 \ \text{h}} \right| \left| \frac{1 \ \text{yr}}{365 \ \text{d}} \right| \tag{5.209}$$

$$h = 0.445 \ \text{year}$$

Note

1 http://www.timken.com/catalogs.

References

1 Vogelpohl, G. (1951). *Scientific Lubrication* 3: 9.
2 Peter Jost, H. (1966). *Lubrication (Tribology), A Report on the Present Position and Industry's Needs*, 80 pp. Her Majesty's Stationery Office.
3 Radzimovsky, E.I. and Dareing, D.W. (1964). *Influence of Load Variation Upon Life Duration of Roller-Contact Bearings*. Ukranian Technical-Economical Institute, Scientific Notes.
4 Hughes, B. (2009). Hughes Two–Cone Drill Bit historic mechanical engineering landmark 1909–2009. American Society of Mechanical Engineers (ASME). The Woodlands, Texas.
5 King, G.R. (1959). Effect of fluid environment on rock-bit bearing performance. AAODC Paper.
6 Dareing, D.W. and Radzimovsky, E.I. (1965). Effect of dynamic bit forces on bit bearing life. *Trans. AIME, SPE J.* 5 (4): 272–276.
7 Fuller, D.D. (1984). *Theory and Practice of Lubrication for Engineers*, 2e. Wiley.
8 Timoshenko, S. and Goodier, J.N. (1951). *Theory of Elasticity*, 2e. New York: McGraw-Hill (see page 258).
9 Radzimovsky, E.I. (1959). *Lubrication of Bearings*. New York: Ronald Press.
10 Cameron, A. (1966). *The Principles of Lubrication*. Longmans.
11 Hayes, D.F. (1958). Plane sliders of finite width. *Am. Soc. Lubr. Eng.* 1 (2) (Lubrication Science & Technology): 233–240.
12 Towers, B. (1883). First report on fiction experiments. *Proc. Inst. Mech. Eng.(London)* 34.
13 Reynolds, O. (1886). On the theory of lubrication and its application to mr. beauchamp tower's experiments, including an experimental determination of the viscosity of olive oil. *Phil. Trans. Roy. Soc. London* 177 (Pt. I): 157–234.

14 Sommerfeld, A. (1904). Zur Hydrodynamischen Theorie der Schmiermitelreibung. *Zeitschrift fur Mathematik und Physik* 50.

15 Karelitz, G.B. and Kenyon, J.N. (1937). Oil-film thickness at transition from semifluid to viscous lubrication. *Trans. ASME* 59: 239–246.

16 McKee, S.A. (1928). The effect of running-in on journal bearing performance. *Mech. Eng.* 50: 528–533.

17 Stanton, T.E. (1922). On the characteristics of cylindrical journal lubrication at high values of eccentricity. *Proc. Roy. Soc. Lond. Ser. A* 102.

18 Petroff, N. (1883). Friction in machines and the effect of lubrication (in Russian). *Engineering Journal St. Petersburg*, Nos. 1,2,3 and 4.

19 Stribeck, R. (1902). Die Wesentlichen Eigenschaften der Gleit- und Rollenlager" [the main characteristics of the sliding and roller bearings]. *Zeitschrift des Vereins Deutscher Ingenieure* [Journal of the Association of German Engineers] 36 (B and 46): 1341–1348, 1432–38, and 1463–70.

20 Barnhill, W.C. (2016). Tribological testing and analysis of ionic liquids as candidate anti-wear additives for next-generation engine lubricants. Master thesis. University of Tennessee.

6

Energy Consumption

Machines are designed to perform certain functions to achieve a desired outcome. Energy is consumed in friction and other inefficiencies in the process. The overall efficiency of a machine is the ratio of useful output to input. Industry relies on special machines to accomplish such tasks as machining, moving objects, and pumping fluids.

There are many examples of energy consumption. The oil well drilling rig (Figure 6.1) and process are used here to illustrate various avenues for energy consumption. The "end effect," in this case, is a well bore for reaching and producing oil from a subsurface hydrocarbon reservoir. The path of a well bore can be vertically downward or along a preplanned well path extending downward and laterally to reach a reservoir located at some lateral distance from the drilling site [1].

A considerable amount of mechanical power is required to drill an oil well. Diesel engines and generators are mounted on skid units located some distance away from the central drilling activity. This is arranged for safety and efficiency as electrical power is more transportable than direct mechanical drives. Portable power units are necessary because drilling operations are usually located in remote areas where public utilities are not available. The irony of the drilling operation is while power units supplies some 8000 hp to drilling rigs, only a very small percentage (~50 hp) is used to drill into rock to make a well bore. What, then, happens to the rest of the power and how is it consumed? The answer lies in the subsystems that make up the total rig.

Subsystems of Drilling Rigs

There are five basic subsystems in a drill rig. Each subsystem can be viewed as a separate machine, each having a power source, various means of transmitting power, and an end use.

- Hoisting – The hoisting system includes the derrick, crown block, traveling block, and a power-driven drum. The crown block and traveling block provide a huge mechanical advantage for lifting tubulars (~300 000 lb) in and out of boreholes. This pulley arrangement reduces the force on the fast end (drum) of the cable to around 30 000 lb.
- Rotary drive and drillstring – The drillstring extends from the rig floor to the drill bit. Its two basic tubular components are drill pipe and drill collars. This very long pipe transmits torque from the surface to the drill bit and provides a conduit for drill mud. The rotary table transmits torque to drill pipe by means of a square or hexagonal pipe called the Kelly, which allows the drillstring to advance during the drilling process while maintaining rotary torque. A large amount of energy is lost to friction along the drillstring.

Engineering Practice with Oilfield and Drilling Applications, First Edition. Donald W. Dareing.
© 2022 John Wiley & Sons, Inc. Published 2022 by John Wiley & Sons, Inc.

Figure 6.1 Oil well drilling rig (Northern Alberta, Canada, 1963).

- Hydraulic or circulation – The hydraulic system is central to the rotary drilling method. Surface mud pumps circulate drilling fluid down the drillstring to the drill bit for hole cleaning purposes. Cuttings are carried back to the surface for inspection and analysis. Drill fluid is continually monitored and maintained before being recirculated down hole for bit cleaning. A significant amount of energy is lost to fluid friction over several thousand feet of pipe.
- Well control – The progression of tools available to defend against a well blowout are drilling mud, annular preventer, and hydraulic rams. The annular preventer and hydraulic rams make up the Blowout Preventer (BOP) stack. Flow rate returns are continually monitored for possible formation fluid invasion into the well bore. When this happens, the well is taking a kick from the formation and bad things can happen if this is allowed to continue. Well control measures are taken to control formation fluid invasion.
- Mechanical/electrical power (~8000–10 000 hp) – Diesel is the prime source of power around a drilling rig. These engines power electric generators which supply power to motors around the rig for lifting pipe and tools in and out of the well bore, drive the rotary table, and power mud pumps.

While this level of power is needed to around a drilling rig, ironically, only about 50 hp is used to break rock formations to advance the well bore.

Draw Works in Drilling Rigs

The structural framework of drilling rigs supports the hoisting system. The very top of a drilling rig supports the crown block. This, plus the traveling block, makes up a block and tackle arrangement with a huge mechanical advantage for pulling heavy pipe weight in and out of a well bore. Typical pipe weight is around 300 000 lb. This weight is supported by the hoisting systems during regular drilling and when the drillstring is pulled out of the well bore to replace bottom-hole equipment, such as a drill bit. A typical length of a joint of drill pipe is 30 ft. A drillstring is disconnected in 90 ft sections and leaned inside the derrick during a "tripping" operation. As a 90 ft section is removed, the remaining portion of the drillstring is hung by slips during the disconnect.

The "end use" in this case is hoisting and lowering pipe and equipment in and out of a well bore. This involves several steps, but still energy is required to elevate the drillstring mass. This energy is not retrieved when lowering the drillstring back into the well bore. Much of it is lost in friction between pipe and well bore, and some is lost in controlling the speed of the drillstring mass during reentering steps.

Special pulley arrangement is a block and tackle arrangement (Figure 6.2). This arrangement has a pulley (top) which is free to rotate about a fixed axis. This top pulley is the crown block. The moving pulley is the traveling block. This arrangement of pulleys has a huge mechanical advantage: pull

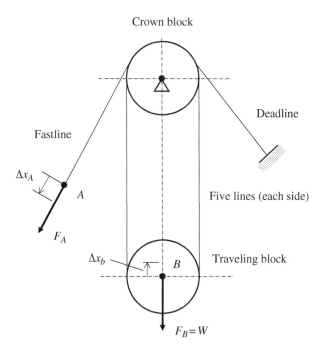

Figure 6.2 Oil rig draw works.

force vs. lifting load. The weight of pipe suspended in a well bore may about 320 000 lb, while the pull force on to a drum is 30 000 lb.

A band brake is used to hold and adjust drillstring position during various activities, such as installing equipment and addition drill pipe. During tripping in and out of the well bore, the band brake is used to set pipe in "slips" inserted within the rotary table. In each case, energy is lost in the band brakes.

Block and Tackle Hoisting Mechanism

There is a huge mechanical advantage in this hoisting subsystem or energy transmission mechanism. Assuming there are 10 lines between crown block and traveling block and applying the principal of virtual work (Figure 6.2)

$$\Delta x_A = 10\Delta x_B \tag{6.1}$$

$$v_A = 10v_B \tag{6.2}$$

Equating work done by both forces

$$F_A\Delta x_A = F_B\Delta x_B \tag{6.3}$$

$$F_A = \frac{1}{10}F_B \tag{6.4}$$

This means that a force of 30 000 lb is required to lift a 300 000 lb load. This load represents the buoyed weight of the drillstring, friction along the drillstring, and possible force to dislodge stuck pipe. The sole purpose of the draw works is to insert or pull pipe in and out of the well bore.

Spring Constant of Draw Works Cables

The axial spring constant of the draw works affects the dynamic behavior of the entire drillstring. It is a boundary constraint in mathematical models. This spring constant is determined on site by establishing a reference point on the Kelly. Raising the drill bit off the bottom transfers the bit force to the draw works.

Measure how much the mark travels as the draw works picks up part of the weight on bit (WOB), say 40 000 lb. Based on the number of strands of the wire cable passing around the crown and traveling block, determine the amount the Kelly moves up. The difference between the theoretical travel and the actual travel is the stretch of the pully system. The spring constant of the pulley system is the ratio of the 40 000 lb and the stretch. The spring constant is typically in the range of 40 000–50 000 lb/in.

Band Brakes Used to Control Rate of Decent

The mechanical arrangement for band brakes similar to the ones used in drilling rigs is shown in Figure 6.3. The relation between T_1 and T_2 is

$$T_2 = T_1e^{\mu\theta}$$

where

θ – angle of contact between belt and pulley
μ – coefficient of friction

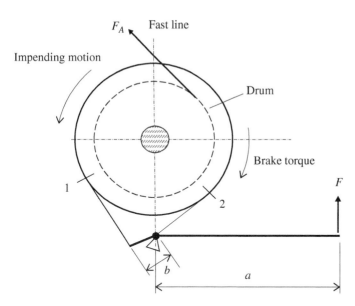

Figure 6.3 Drum brake.

The relation between fast line force, F_A, and T_1 is

$$R(T_2 - T_1) = rF_A$$

By substitution

$$RT_1\left(e^{\mu\theta} - 1\right) = rF_A$$

Applying force, F, adjusts and locks the brake. The tension in the brake band at point 1 is

$$T_1 = \frac{r}{R} \frac{F_A}{\left(e^{\mu\theta} - 1\right)} \tag{6.5}$$

The tension in the band at point 2 is

$$T_2 = \frac{r}{R} \frac{e^{\mu\theta} F_A}{\left(e^{\mu\theta} - 1\right)} \tag{6.6}$$

These equations show that $T_2 \succ T_1$. The force, F, applied to the end of the brake handle is determined by equilibrium of the handle, i.e.

$$aF = bT_1 \tag{6.7}$$

Combining Eqs. (6.5) and (6.7) gives

$$F = \frac{b}{a} \frac{r}{R} \frac{F_A}{\left(e^{\mu\theta} - 1\right)} \tag{6.8}$$

Rotary Drive and Drillstring Subsystem

Kelly and Rotary Table Drive

The rotary table is a means of transmitting torque to the drillstring (Figure 6.4). It is used to support the freely hanging drillstring during tripping out or going into the well bore. During drilling, a special bushing, called the Kelly bushing, is used to transmit torque to the drillstring. The Kelly is a special pipe having a square or hexagonal external shape for the purpose of transmitting torque. The Kelly slides through the rotary table as well bore deepens. Drilling fluid flows through the center of the Kelly. A special sub, called a Kelly saver sub, is attached to the lower end of the Kelly to protect the lower pin-end from multiple connections with every section of drill pipe.

The elements in this subsystem are (i) rotary drive, (ii) Kelly, (iii) drillstring, and (iv) bottom-hole assembly (BHA). The BHA contains a drill bit but may also include other tools such as stabilizers, motors and measurement-while-drilling (MWD) instruments. Again, we see this subsystem has a power source, means of transmitting power, and an "end usage." The end usage in this case is making a hole in subterranean rock formations. Power is transmitted to a drill bit by rotating the entire drillstring. Two type of rotating units may be used, rotary table or a top drive.

Much energy is lost to friction during drillstring rotation and pulling the drilling out of the well bore.

Friction in Directional Wells

Directional wells may extend laterally several thousand feet. The well plan may call for defined paths to be drilled to reach a specific target in a reservoir. Complex well paths can be drilled

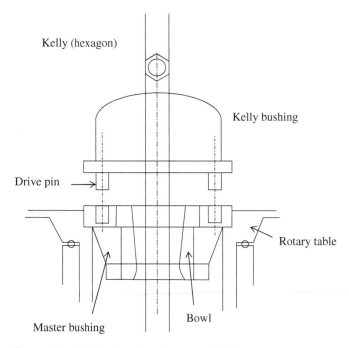

Figure 6.4 Kelly being driven by a rotary table.

Table 6.1 Coefficient of friction.

	Well #1	Well #2	Well #3[a]
Pick-up	0.28	0.31	0.40
Slack-off	0.27	0.31	0.41
Rotating	0.27	0.29	0.41

[a] Well #3 had several severe dog legs in the lower portion of the build zone.

and navigated by rotary steerable tools (RSTs) and MWD monitoring. Friction forces created by complex well paths can create high drillstring torque and pull out forces. These loads can sometimes exceed the capability of the surface equipment and strength of the drillstring. During directional drilling operations, drill pipe is usually limited by rotating torque and not by direct pull. One practice is to allow rotary torque to reach 80% of make-up torque.

Johancsik et al. [2] measured axial pipe force and rotary torque directly below the Kelly. This information, along with a computer model of drillstring friction, was used to determine the coefficient of friction in three different wells (9790, 15 573, and 12 200 ft). The coefficient of frictions for each well is given in Table 6.1.

Through experience, it has been determined that friction forces can be greatly reduced by rotating the pipe and maintaining mud circulation during tripping. Top drives have been designed for this purpose.

Top Drive

Modern drill rigs are now using top drives instead of rotary tables. In this case, power is supplied at the top end of the drillstring by use of an electric motor. The Kelly and rotary table are eliminated; however, the drillstring is still suspended in slips as before during pipe removal and insertion. The swivel and motor are mounted to move along guide rails directing pipe in and out of the well bore (Figures 6.5 and 6.6). Single joints of drill pipe are added as before, but pipe connections are made by robotic mechanisms.

Pull-out or slack-off friction forces can sometimes be excessively high to the point of getting the pipe stuck. The standard practice to minimize pipe sticking especially while tripping is to use top drives to keep drillstrings rotating during tipping. Experience shows that pipe rotation, i.e. shearing contact friction forces in the tangent direction, reduces contact friction forces in the longitudinal direction.

Rotary power is typically transferred to drillstrings through the Kelly by means of the rotary table. Power is delivered to rotary tables by electric motors in the form of torque and rotary speed. Following conventional methods of rotary drilling, a single 30 ft joint of drill pipe is added to a drillstring by

- setting the drillstring in the slips
- breaking out the Kelly saver sub
- attaching the Kelly onto a single 30 ft joint which has been place in the mouse hole
- raising the Kelly assembly and attaching the new single onto the drillstring which is hanging in the slips
- removing the slips
- circulating back to bottom and continue drilling

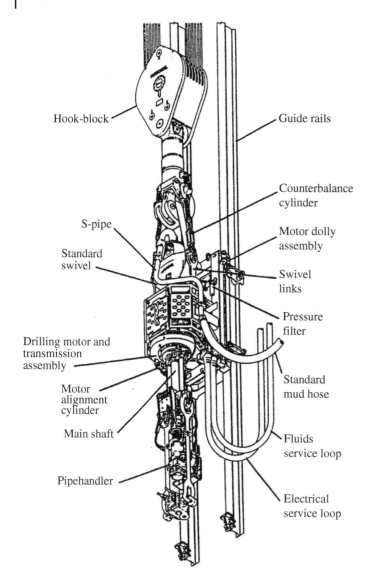

Hook-block

Guide rails

Counterbalance cylinder

S-pipe

Motor dolly assembly

Standard swivel

Swivel links

Pressure filter

Drilling motor and transmission assembly

Standard mud hose

Motor alignment cylinder

Main shaft

Fluids service loop

Pipehandler

Electrical service loop

Figure 6.5 Schematic of top drive. *Source:* Courtesy NOV Inc.

The top drive replaces the Kelly, Kelly bushing, and rotary table. Power is transmitted to drill-strings by an electric motor which travels with the top drive assembly. Triples or 90 ft sections of drill pipe, which have been previously racked in the derrick, can be drilled down without interrupting drilling. The ability to handle 90 ft sections of drill pipe while tripping in and out has distinct advantages over the conventional rotary table approach. Major advantages are

- reduced connection time with both drilling and tripping;
- ability to circulate and rotate out of the hole to reduce drag on the drillstring, reducing the chances of getting stuck;

Figure 6.6 Top drive as mounted on a drilling rig. *Source:* Courtesy of NOV.

- ability to drill through bridges and tight spots without picking up the Kelly (adding pipe);
- helps control tool face orientation in directional drilling by capturing the trapped torque in drill-strings over a 90 ft interval as opposed to a 30 ft interval – reduces downhole motor orientation activity;
- reduces the number of connections required by the rig crew and thus improves rig safety;
- motor drive can be calibrated against tong torque gauges as a means to quantify rotary torque.

Top drives greatly reduce friction in high-angle directional drilling by allowing pipe rotation while pulling pipe out of the well bore. They are essential in the drilling of extended reach wells.

Five basic subsystems are

1) Drilling motor and swivel assembly
2) Guide dolly assembly
3) Pipe handler assembly

4) Counterbalance system

5) Top drive control system

A drilling motor and swivel assembly contain an electric motor, a gear drive, and a main shaft which connects directly onto a standard swivel. This assembly is supported by the swivel and traveling block and mounted on a guide dolly. The guide dolly is constrained to move up and down on vertical guide rails that are rigidly attached to the rig. This assembly also contains an air brake capable of developing 35 000 ft-lb of static braking torque at the output shaft. One motor assembly can develop 30 000 ft-lb of continuous torque at speed up to 175 rpm and can generate intermittent torque as high as 41 500 ft-lb. This assembly can generate torque high enough to make up tool joints (Figure 6.6).

A pipe handler connects and disconnects a stand of pipe from the drilling motor assembly. Mechanical pipe handlers are designed to break out pipe in the derrick at any height with torque capability up to 60 000 ft-lb.

The counterbalance system provides a 6 in. cushioned stroke to prevent damage to tool joint threads while making or breaking connections. Each of these operations can be controlled from a console located at the rig floor.

Drillstrings transmit mechanical rotary power from the surface to drill bits, serve as a conduit for drilling fluid, apply force to drill bits, and affect hole direction. A major portion of drillstrings is made up of drill pipe while the bottom portion (roughly 700 ft) is a heavier pipe called drill collars. The bottom portion (~150 ft) of the drill collar section contains tools (drill bit, positive displacement motors [PDMs], turbine, MWD, and stabilizers) called the BHA. Even though the total length of a drillstring may be several thousand feet long, the BHA affects everything: rate of penetration, footage cost, and well bore direction.

Drillstring Design and Operation

Drillstrings transmit mechanical rotary power from the surface to drill bits, serve as a conduit for drilling fluid, apply force to drill bits, and affect hole direction. A major portion of drillstrings is made up of drill pipe, while the bottom portion (roughly 700 ft) is a heavier pipe called drill collars. The bottom portion (~150 ft) of the drill collar section contains tools (drill bit, PDM, turbine, MWD, and stabilizers) called the BHA. Even though the total length of a drillstring may be several thousand feet long, the BHA affects everything: rate of penetration, footage cost, and well bore direction.

Drillstrings are subjected to many operational loads such as direct pull, torsion, and bending. Each has a static and a dynamic component. They are designed from static load considerations following the state-of-the-art as given in API RP 7G [3]. Since 1960 much has been learned about the dynamic behavior of drillstrings through downhole measurements and analytical studies. Vibrations are known to affect the performance of downhole equipment. Well bore curvature and local dog-legs also affect fatigue life.

Buoyancy

In practice WOB is established by slacking off the desirable bit force from the hook load at the surface. This puts drill collars in compression while drill pipe is in tension. A basic question is how long should the drill collar section be to prevent drill pipe buckling? This is important because of the large difference in structural stiffness between drill pipe and drill collars. From a structural point of view, stiffness attracts bending moment and resulting stresses can be damaging, especially under stress reversals caused by drillstring rotation.

Hook Load

Applying Archimedes' principle to drill pipe hanging freely from the draw works predicts a hook load of

$$H = W - B \tag{6.9}$$

where

H – hook load
W – air weight of total string
B – weight of the drill mud displaced by the string as stated by Archimedes' principle

In this equation, W is a body force and B is a surface force. Equation (6.9) can also be written as

$$H = W\left(1 - \frac{B}{W}\right) = W\left(1 - \frac{\gamma_m}{\gamma_{stl}}\right) \tag{6.10}$$

$$H = WBF \tag{6.11}$$

where

W – air with of pipe
BF – buoyancy factor

Assuming a 12 ppg mud, the buoyancy factor is 0.817 as calculated:

$$\gamma_m = 12\,\text{lb/gal}\left|\frac{7.48\,\text{gal}}{1\,\text{ft}^3}\right| = 89.76\,\text{lb/ft}^3 \tag{6.12}$$

so

$$BF = \left(1 - \frac{89.76}{490}\right) = 0.817$$

The buoyed weight of a drillstring (including tools, etc.) is total air weight multiplied by the buoyancy factor.

The magnitude of buoyancy forces can be quite high. For example, consider 5½ in. (19.2 lb/ft) drill pipe having a cross-section area of 4.9624 in.^{2} If 10 000 ft of 5½ in. drill pipe hangs freely in $\gamma_m = 12$ ppg mud, the hydrostatic force pushing up at the lower open end is determined as follows.

$p = L\gamma_m = 0.052\,(10\,000)\,(12) = 6240$ psi
$F = pA$
$F = 6240\,\text{lb/in.}^2\,(4.9624\,\text{in.}^2)$
$F = 30\,965$ lb

It would appear that this force is great enough to buckle the drill pipe several times. However, this does not happen.

Definition of Neutral Point

Assume the bottom end of an open-ended pipe (Figure 6.7a) is given a virtual sidewise displacement. The static equivalence of the hydrostatic pressure forces on the pipe below section a–a is

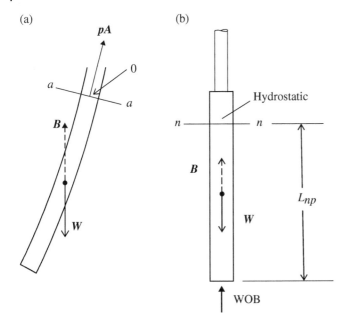

(a)

(b)

Figure 6.7 (a, b) Stability of vertical pipe under hydrostatic loading.

represented by the vectors B and pA. The force B is equal to the weight of fluid displaced by the pipe below section a–a. The vector pA is the product of the local fluid pressure and pipe area. The solid vector, W, represents the air weight of the drill pipe below cross section a–a. W is a body force, while B is a surface force.

Moments about point "0" show there is a restoring moment that always moves the pipe back toward the vertical, assuming the density of the pipe is greater than the density of the fluid. In other words, an open-ended drill pipe will not buckle from hydrostatic pressure alone regardless of well bore depth.

Now assume the drill pipe is attached to the top of a drill collar section (Figure 6.7b) and the compression at the top of the drill collars is equal to the local hydrostatic pressure. According to the discussion above, the drill pipe will not buckle under this condition.

A good definition for the neutral point is the point in the drillstring, where compressive stress is equal to the local hydrostatic pressure. If the hydrostatic pressure point is located within the drill collars, the drill pipe will not buckle. Drill pipe will buckle when the internal force at the lower end reaches a critical level, which is somewhat higher than the hydrostatic force level.

The location of the neutral point depends on bit force (WOB). If bit force is zero, the neutral is located at the drill bit. As bit force is increased, the neutral point moves up the drill collars.

In practice, the neutral point is kept within the drill collars as a safety measure to avoid buckling the drill pipe. The relation between WOB and distance to neutral point is determined by

$$WOB = wL_{np} - w_m L_{np} \tag{6.13}$$

or

$$WOB = wL_{np}\left(1 - \frac{w_m}{w}\right) \tag{6.14}$$

$$WOB = wL_{np}\left(1 - \frac{\gamma_m}{\gamma}\right)$$ (6.15)

where

γ – steel density (490 lb/ft^3 or 65.5 ppg)
γ_m – drilling mud density (ppg)

The distance to the neutral point (or point of hydrostatic compression) from the drill bit is determined by

$$L_{np} = \frac{WOB}{wBF}$$ (6.16)

Note the location of the neutral point does not depend on collar length, only WOB.
Drill collar length is generally computed by

$$L_{np} = 0.85L_c$$ (6.17)

which means that the neutral point is 85% of drill collar length. The extra length represents a safety factor against drill pipe buckling.

$$L_c = \frac{WOB}{0.85wBF}$$ (6.18)

Basic Drillstring: Drill Pipe and Drill Collars

Consider a simple drillstring made up of drill pipe and drill collars. If the drillstring is hanging freely from the derrick, the neutral point is located at the bottom of the drill collars. As WOB is added, the neutral point moves up the string and reaches the top of the drill collar section when

$$WOB = w_1L_1 - A_1L_1\gamma$$ (6.19)

Since $A_1L_1\gamma = w_fL_1$ (where w_f is weight per length of fluid displaced by the collars),

$$WOB = (w_1 - w_f)L_1 = w_1L_1BF$$ (6.20)

Additional bit forces move the neutral point up into the drill pipe. Letting (x) represent distance from top of drill collars to the neutral point, then

$$WOB = w_1L_1 + w_2x - w_{f1}L_1 - w_{f2}x$$ (6.21)
$$WOB = w_1L_1BF + w_2xBF$$ (6.22)
$$WOB = (w_1L_1 + w_2x)BF$$ (6.23)

The neutral point location from the bottom of the drill collars is $L_{np} = L_1 + x$.

Physical Properties of Drill Pipe

Dimensions of three sizes of standard drill pipe are listed in Table 6.2. Each of the listed pipe sizes are made in different weights as dictated by the inside diameter (ID) dimensions. Wall thickness for each can easily be determined as well as the weights (lb/ft) using the density of steel (490 lb/ft) and cross-sectional area. However, pipe weight is expressed in terms of nominal weight which includes the weight of the connectors. For example, 5 in. (19.50 lb/ft) drill pipe has an ID of 4.276 in. and a plain tubular weight of 17.93 lb/ft. Drill pipe is identified by size, weight, grade, and class [3].

Table 6.2 Dimensions of standard drill pipe.

OD (in.)	ID (in.)	Plain weight (lb/ft)	Nominal weight (lb/ft)
4½	3.958	12.24	13.75
	3.825	14.98	16.6
	3.640	18.69	20.00
	3.50	21.36	22.82
5	4.408	14.87	16.25
	4.276	17.93	19.50
	4.000	24.03	25.60
5½	4.892	16.87	19.20
	4.778	19.81	21.90
	4.670	22.54	24.70

- Size – Nominal outside diameter of pipe
- Weight – Weight per foot of drill length, including weight of connections
- Grade – Yield strength of pipe material (E75, X95, G105, S135, numbers refer to yield strength)
- Class – New, premium, used

Load and torque capacity of the four grades can be calculated as shown below. Considering 5 (19.50) and grade E75 drill pipe, the pull capacity is 395 595 lb.

Pull capacity is determined as follows:

$$\sigma_{yld} = \frac{F_{yld}}{Area} \tag{6.24}$$

$$Area = \frac{\pi}{4}\left(5^2 - 4.276^2\right)$$

$$F_{yld} = (5.2760)75\,000$$

$$F_{yld} = 395\,595\,lb$$

Shear strength, τ_{yld}, is a determined from

$$\tau_{yld} = 0.577\sigma_{yld} \tag{6.25}$$

which is based on the von Mises energy of distortion criteria of failure. σ_{yld} is an experimental value, while τ_{yld} is a calculated value.

The maximum allowable torque is determined as follows:

$$\tau_{yld} = \frac{T_{yld}}{S} \tag{6.26}$$

$$S = \frac{J}{c} \quad \text{(torsion section modulus)}$$

$$T_{yld} = (11.415)75\,000(0.577)$$

$$T_{yld} = 493\,984\,\text{in.-lb}$$

$T_{yld} = 41\,165$ ft-lb (checks with API Tables)

Tension and torque limits for the other three grades (X95, G105, S135) can be determined by the above equations as well.

Selecting Drill Pipe Size and Grade

Drill pipe is subjected to various types of loads during any drilling operation. Common loads are direct pull, torque, bending, and internal pressure. Drill pipe selection (size, grade) is typically based on direct pull type loading with a factor of safety to account for uncertainty. Other types of loading, such as dog-leg bending, should be checked for stress reversal type fatigue, especially if the dog legs are severe.

Normally, drillstrings are designed to an allowable load of 90% of yield strength of drill pipe, so the allowable load is

$$H_a = 0.9 H_{yld} \tag{6.27}$$

where

H_{yld} – direct pull causing material yielding
H_a – maximum allowable pull force, i.e. buoyed weight of drillstring plus margin of overpull (MOP)

The corresponding factor of safety (FS) is

$$FS = \frac{H_{yld}}{H_a} = 1.11 \tag{6.28}$$

The level of the safety factor (FS) is subject to modification depending on application. High FS means higher weight and cost, while low FS could lead to unexpected failure. Factors of safety are used to cover uncertainty in the design, such as material properties, applied loads, and accuracy of stress models.

The allowable pull force relates to hook load and MOP by

$$H_a = H + MOP \tag{6.29}$$

where

H – working hook load (based on buoyed weight of drillstring at maximum depth)
MOP – margin of overpull, in case of stuck pipe or excessive friction

The allowable load (H_a) is the basis for selecting drill pipe size and grade for a given class of pipe. Bringing these equations together gives

$$H_a = 0.9 H_{yld} = \left(w_p L_p + w_c L_c \right) BF + MOP \tag{6.30}$$

This equation will be applied to the following two examples.

Select Pipe Grade for a Given Pipe Size

Example As an example, consider the following:

Total depth – 14 000 ft
Mud weight at total depth (TD) – 15 ppg
MOP – 100 000 lb

Drill pipe – 5 (19.50); Premium class

Drill collars – 6¾ in. outside diameter (OD) × 2¹³⁄₁₆ in. ID, 750 ft

The problem is to select the lowest grade of drill pipe that will reach TD and accommodate the specified MOP.

The static hook load (working hook load) at TD is

$$H = \left(w_p L_p + w_c L_c\right) BF \tag{6.31}$$

$$H = [(19.5)13\,250 + (100)750](0.771)$$

$$H = 257\,032\ \text{lb}$$

The selected grade must have the strength to support

$$H_a = H + MOP \tag{6.32}$$

$$H_a = 257\,032 + 100\,000$$

$$H_a = 357\,032\ \text{lb}$$

To account for the FS pipe grade selection must satisfy

$$H_{yld} \geq 1.1(357\,032)$$

$$H_{yld} \geq 396\,702\ \text{lb}$$

Grade X95 (Premium Class) has a yield load limit of 394 612 lb. Grade G105 has a yield load limit of 436 150 lb. Grade G105 would be the proper choice for drill pipe grade because 436 150 is greater than the required 396 702 lb yield strength. In this case, the FS increases from $FS = 1.1$ to 436 150/ 357 032 = 1.22.

Determine Maximum Depth for Given Pipe Size and Grade

Example Now consider a situation requiring maximum depth capability of a given pipe size, class, and grade. Other specifications are

Drill pipe size – 4½ in. (16.6 lb/ft), Grade E75, Premium Class

MOP – 57 000 lb

Drill collar – 6½ in. OD × 2½ in. ID, 700 ft long

Mud weight – 12 ppg

Recall

$$H_a = 0.9 H_{yld}$$

$$H_a = H + MOP$$

$$H = 0.9 H_{yld} - MOP \tag{6.33}$$

The maximum allowable depth is based on the allowable hook load, *H*. From the API RP 7G Standard [3], the pull strength for the specified pipe Grade and Class is

$$H_{yld} = 260\,165\ \text{lb}$$

The working hook load, *H*, is

$$H = 0.9(260\,165) - 75\,000 = 159\,149\ \text{lb}$$

Maximum hole depth is determined from

$$H = \left[w_{dp} L_{dp} + w_c L_c \right] BF \tag{6.34}$$

where

H	=	159 149 lb
BF	=	0.817
w_{dp}	=	16.6 lb/ft
w_c	=	96 lb/ft
L_c	=	700 ft

Substituting these numbers into the above equation gives $L_{dp} = 7687$ ft. The maximum depth capability is

$$TD = L_{dp} + L_c = 8\,387 \text{ ft} \tag{6.35}$$

Roller Cone Rock Bits

There are many different types of roller cone bits and each has unique features and benefits. They can be divided into milled tooth bits and tungsten carbide insert cutters. Bearing performance has been extended over the years through design and seal improvements.

Three components of roller cone bits are cutters, bearings, and bit body. The *cutters* are fixed to cones which rotate on bearings relative to the bit body. Cutter elements are either machined directly on the roller cones (Figure 6.8a) or are tungsten carbide inserts which are pressed into the cone surfaces (Figure 6.8b). Cutters on milled tooth bits are longer, and the cones are offset to create dragging action making these bits aggressive in soft formations. Insert bits are designed for hard formations and break up the rock by crushing action rather than chipping and gouging action.

(a)   (b)

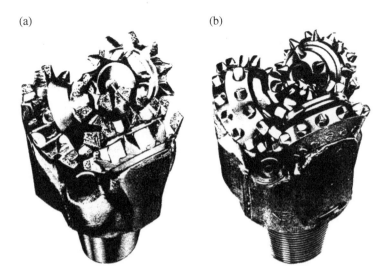

Figure 6.8   (a, b) Milled tooth and insert roller drill bits.

(a)

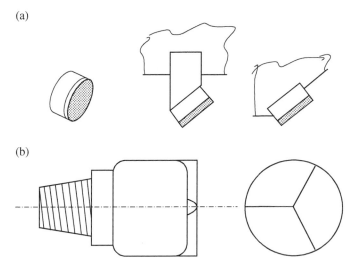

(b)

Figure 6.9 (a) Polycrystalline diamond compact (PDC) cutter. (b) PDC drill bit.

Roller cone rock bits are made with many design features and for a wide range of formation hardness. The selection of any of these bits depends on bit cost, expected rate of penetration, and bit life. Bit performance prediction is based on a database of bit records and is a statistical prediction.

Polycrystalline Diamond Compact (PDC) Drill Bits

Polycrystalline diamond compact (PDC) bits are drag bits containing multiple cutters made of synthetic diamonds. The size of synthetic diamonds is about 175 diamonds per carat. These tiny diamonds are bonded to form a disc shape (about the size of a nickel) and backed by a thick layer of tungsten carbide substrate. The synthetic diamond layer is about 0.025 in. thick. The tungsten carbide backing is about 0.115 in. thick (Figure 6.9a). This cutting structure is then bonded onto a metal stud which is pressed into a steel body, which forms the drill bit. These cutters are also embedded directly into the bit body (Figure 6.9b). When used in tungsten carbide matrix bits, they are bonded onto a short disc which is then bonded into the matrix.

Rock bits of all types fail rock by developing high shear stresses under cutters. PDC cuter develop shear stresses efficiently provided they can be forced into the rock. The application areas for PDC drill bits are soft to medium hard formations. PDC bits were introduced in the early 1970s and became a good companion bit for downhole motors and turbines, which operate at high rotary speeds.

Natural Diamond Drill Bits

Natural diamond drill bits were introduced in the late 1940s. These bits allow exploration of deep reservoirs which usually mean harder and more abrasive formations. Individual diamonds are set in the matrix of the bit so that about one-third of the diamond is exposed and two-third is buried within the matrix (Figure 6.10). The average depth of cut is generally one-third of the exposure of the stone.

Industrial diamonds are handset in a machined graphite mold. A special mixture of matrix granules is placed between the mold and blank, then, heated in a furnace until the matric material

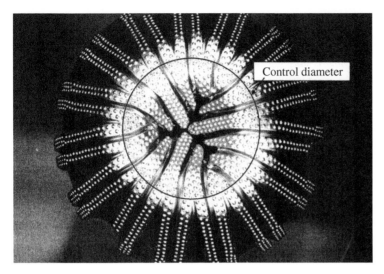

Figure 6.10 Leading face of a natural diamond drill bit. *Source:* Used by permission from Baker Hughes.

melts, capturing the diamonds onto the blank. A shank is then threaded onto the blank and welded to form the diamond drill bit.

Diamond drill bits are used in hard formations typically found in deep wells (15 000 ft and deeper).

Hydraulics of Rotary Drilling

The hydraulic system or circulating system is central to the rotary drilling method. Drilling fluid is necessary to remove rock chips from underneath drill bits and carry them back up to the surface for examination and disposal. Efficiency of cuttings removal depends on the amount of hydraulic horsepower (HHP) in terms of fluid pressure and flow rate delivered to the drill bit nozzles in drill bits and convert this energy into kinetic energy through specially selected nozzles. The effect of nozzle size on pressure drop across drill bits is shown in Table 6.3.

Bit nozzles transform available energy into kinetic energy which is discharged under the bit for cleaning and cuttings removal (Figure 6.11). Flow rate selection and nozzle sizes are important for making the best use of available HHP at the lower end of a drillstring. Only a portion of the HHP supplied by mud pumps at the surface (Figure 6.12) gets to the drill bit. About one-third is lost due to fluid friction as the drill muds moves down the drillstring.

Optimized Hydraulic Horsepower

Another source of lost energy is friction in fluid flow. This is emphasized in friction losses in pumping drill fluids in an oil well.

When drilling fluid is pumped through a drilling system, energy is lost due to fluid friction. This friction loss is taken from the HHP and can be viewed as parasitic. Losses occur in surface equipment, inside, and out of the drillstring. Energy loss across drill bit nozzles is not considered parasitic. Parasitic losses are substantial and must be considered in designing the overall circulating system. A major portion of the total HHP input by the mud pumps is lost to friction. The remaining

Table 6.3 Pressure drop across bit nozzles (12 ppg drilling fluid density).

Nozzle size TFA (sq in.) Q (gpm)	11, 11, 11 0.2784 Δp (psi)	11, 11, 12 0.2961 Δp (psi)	11, 12, 12 0.3137 Δp (psi)	12, 12, 12 0.3313 Δp (psi)
300	1274	1126	1003	900
310	1360	1203	1071	961
320	1449	1281	1142	1024
330	1541	1363	1214	1088
340	1636	1447	1289	1155
350	1734	1533	1366	1224
360	1834	1622	1445	1295
370	1938	1713	1526	1368
380	2044	1807	1610	1443
390	2153	1903	1696	1520
400	2265	2002	1784	1599
410	2379	2103	1874	1680
420	2497	2207	1967	1763
430	2617	2314	2061	1848
440	2740	2423	2158	1935
450	2866	2534	2258	2024

TFA means total flow area
Nozzle sizes are expressed in 32 of an inch (for example 11/32 in.)

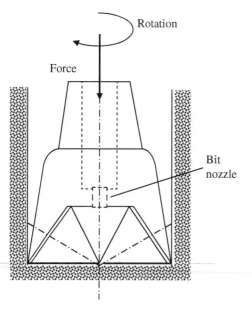

Figure 6.11 Fluid jets for bottom hole cleaning.

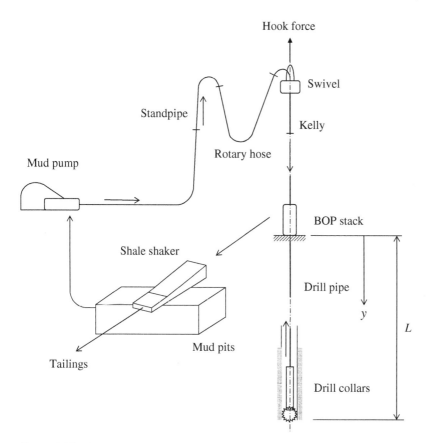

Figure 6.12 Hydraulic system.

part is used for bottom hole cleaning and power for downhole motors or turbines. Parasitic losses occur in

a) Surface equipment
b) Inside drill pipe
c) Inside drill collars
d) Annular space around drill collars
e) Annular space around drill pipe

Chapter 10 gives the analytical tools to predict pressure losses throughout the circulating system. Bit cleaning, however, depends upon HHP that is created underneath the drill bit ($HHP_{bit} = \Delta p_{bit}Q$). This section develops the rationale for maximizing HHP for best bottom hole cleaning.

The difference between the power supplied by the pump and the power consumed by friction is what is left over at the bottom of the drillstring. Similar calculations at other flow rates define a curve showing how the available bottom hole HHP varies with flow rate.

System pressure losses are related to flow rate, Q, by

$$p_s = CQ^{1.8} \tag{6.36}$$

Mathematically, the available HHP reaches a maximum at a particular flow rate [4]. The condition for maximum available hydraulic power is predicted as follows:

$$HHP_a = \frac{1}{1714}\left(P_pQ - CQ^{2.8}\right) \tag{6.37}$$

$$\frac{d(HHP_a)}{dQ} = P_p - 2.8CQ^{1.8} = 0 \tag{6.38}$$

By substitution of Eq. (6.36)

$$P_p = 2.8p_s \tag{6.39}$$

which means that HHP_a is maximum when

$$P_s = \frac{1}{2.8}P_p \tag{6.40}$$

or when system pressure losses are about one-third of the maximum allowable pump pressure. The flow rate at which this condition occurs depends on the hydraulics of the circulating system. Substituting Eqs. (6.36) into (6.40) gives

$$\frac{P_p}{2.8} = CQ_{opt}^{1.8} \tag{6.41}$$

$$Q_{opt} = \left[\frac{P_p}{2.8C}\right]^{\frac{1}{1.8}} \tag{6.42}$$

Using $C = 0.0142$ and assuming

$$P_p = 3000\,\text{psi (maximum pump pressure)}$$

Then

$$Q_{opt} = \left[\frac{3000}{2.8(0.0142)}\right]^{\frac{1}{1.8}} = 513\,\text{gpm}$$

$$P_s = \frac{1}{2.8}\,3000 = 1071\,\text{psi}$$

$$HHP_a = \frac{(3000 - 1071)513}{1714} = 577\,\text{hp (maximum)}$$

These results are shown in Figure 6.13.

Field Application
A practical application of this information is explained below. Assume you go to the rig and find the following drilling data:

- Standpipe pressure is 2500 psi
- Mud density is 12 ppg
- Bit nozzle sizes are 11, 11, 12 (total flow area [TFA] = 0.2961 in.²)
- Flow rate 360 gpm
- Maximum allowable pump pressure, however, is 2800 psi
- A bit change is imminent

The question is – What would be your recommendation for changes in these drilling parameters, if any? The first step is to determine the value of "C" for the hydraulic system at current depth.

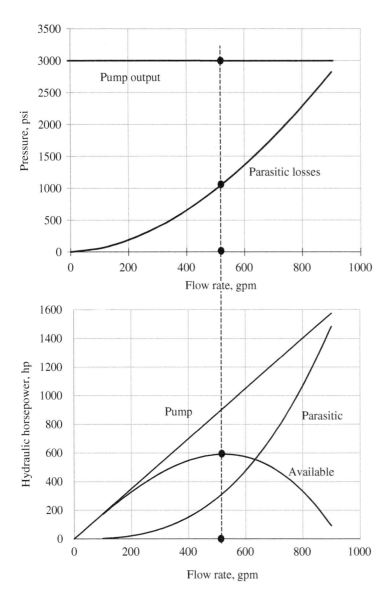

Figure 6.13 System hydraulics.

Using Table 6.3, we see that the pressure drop across the drill bit is 1622 psi. Knowing the standpipe pressure, then the parasitic losses are $p_s = 2500 - 1622 = 878$ psi. Using Eq. (6.36)

$$C = \frac{p_s}{Q^{1.8}} = \frac{878}{360^{1.8}} = 0.022$$

This number defines the total losses throughout the circulating system. Turn now to the drilling parameters that will optimize bit cleaning.

With a maximum allowable pump pressure of 2800 psi, then system losses at maximum hydraulics is

$$p_s = \frac{2800}{2.8} = 1000 \text{ psi}$$

So bit pressure drop is $2800 - 1000 = 1800$ psi. Optimum flow rate is

$$Q_{opt} = \left[\frac{2800}{2.8(0.022)}\right]^{\frac{1}{1.8}} = 387 \text{ pgm}$$

Using Table 6.3 and to select nozzle sizes would suggest staying with the same nozzle size but increase flow rate to 387 gpm. Standpipe pressure should read 2800 psi.

Controlling Formation Fluids

Well control refers to drilling techniques for keeping formation fluids from erupting at the surface [1]. There are three levels of well-control: (i) drilling mud weight, (ii) annular preventer, and (iii) mechanical rams within a blowout preventer stack (BOP). Maintaining mud pressure slightly higher than formation pressures is non-disruptive and is the first line of defense against a blowout. The annular preventer is used to shut in a well and to remove a kick. Mechanical rams (pipe, shear, blind) are used as a last resort.

It is apparent that the consequences of a well blow out are not good. There is loss of the well, equipment, time, and possibly loss of life and personnel injury. There is also damage to the environment and perceived losses in reputation and good will. Since the early days of rotary drilling much has been learned about the early detection and control of formation pressures while drilling.

Hydrostatic Drilling Mud Pressure

Mud weight reduces rate of penetration exponentially. The practical selection of mud weight depends on formation pressure, fracture gradient of the formation and drilling rate. In normal pressure areas, mud weight can be kept constant down to total depth (Figure 6.14). However, formation pressure is not normal and mud weight may have to be increased. Mud weight is limited by formation strength at the end of the last casing. This limit is established by conducting a leak-off test after a casing has been cemented in place.

Hydrostatic pressure is the result of gravity and is always in the drilling mud whether the pumps are running. Total pressure is the algebraic sum of hydrostatic and dynamic pressures. This means that both hydrostatic and dynamic pressures can be analyzed separately and then added together to determine pressure at any point in the circulating system.

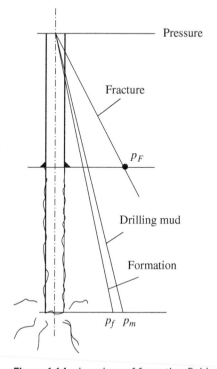

Figure 6.14   Invasions of formation fluids.

Annular Blowout Preventer

The annular preventer is located at the top of the BOP stack (Figure 6.15). A key component of an annular

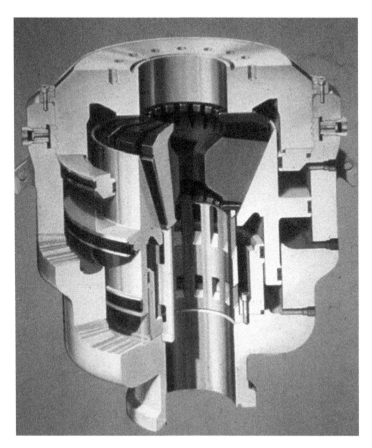

Figure 6.15 Annular preventer. *Source:* Used by permission from Baker Hughes.

preventer is an elastomer (rubber) ring which deforms around drill pipe sealing off the annular space. The elastomer ring is activated mechanically. The pressure limit of annular preventers is around 1500 psi.

Because drilling mud is not heavy enough to equalize formation pressure down hole, back pressure is generated in the standpipe when the well is shut in. In this case, standpipe pressure increases until bottom hole pressure is equal to formation pressure. Bottom hole formation pressure is determined from shut-in pressure by

$$p_f = p_{si} + 0.052\gamma_m L \tag{6.43}$$

where

p_f – formation breakdown pressure, psi
p_{si} – shut-in pressure (standpipe pressure), psi
γ_m – drilling mud weight, ppg
L – hole depth, ft

The constant, 0.052 compensates for mixed units.

The magnitude of the well kick is equal to the shut-in standpipe pressure, p_{si}. The new mud weight required to balance the formation pressure is

$$\gamma_{new} = \gamma_m + \Delta\gamma_m \tag{6.44}$$

where

$$\Delta\gamma_m = 19.23\frac{p_{si}}{L} \tag{6.45}$$

Assuming

- shut-in standpipe pressure is 300 psi (magnitude of kick)
- mud weight inside drill pipe is 12 ppg
- hole depth is 8000 ft

then formation pressure is

$$p_f = 0.052(12)8000 + 300 = 5292\,\text{psi}$$

The mud weight required to balance this formation pressure is

$$5292 = 0.052(\gamma_m + \Delta\gamma_m)8000\,\text{psi}$$

$$\gamma_m + \Delta\gamma_m = 12.72\,\text{ppg}$$

The well took a 0.72 ppg kick and the 12 ppg mud must be brought up to 12.72 ppg to establish a balanced pressure condition. Final mud weight would be 13.22 ppg to have a 0.5 ppg overbalance while drilling.

Hydraulic Rams

In cases where the magnitude of the well kick is very high (10 000 psi), the well may be shut in with ram type closing devices (Figures 6.16 and 6.17).

Types of Rams:

- Pipe ram – Closes around pipe to trap annulus pressure
- Blind ram – Closes well when pipe is not at this location
- Shear ram – Severs pipe or casing and seals well bore

Figure 6.16 Double ram unit. *Source:* Courtesy of NOV.

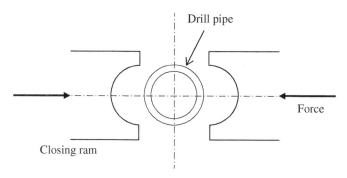

Figure 6.17 Schematic of pipe ram.

Figure 6.18 Blowout preventer (BOP) stack. *Source:* Courtesy of NOV.

Ram units are stacked as shown in Figure 6.18 to provide back-up units and accommodate different pipe sizes. The total unit is known as the BOP stack. Annular preventers are attached to the top of BOP stacks. The total height of BOP stacks may be 20–30 ft, which accounts for drilling rig floors being some 30–40 ft above ground level.

The end use of these devices is to control formation fluids which could migrate into a well bore while drilling and create a blowout at the surface.

Casing Design

A casing program is planned from the bottom up, i.e. the smallest or inside casing depends on size of production tubing or rate of production expected from the reservoir. The size and number of casing tubulars depend on formation pressure and fracture strength of the layered formations [1].

Cementing casing is an important and costly part of setting casing in a well bore. Cement around casing provides support and prevents drilling mud and gas from moving upward from outside of the casing. It is a vital aspect of well control.

Basic casing strings typical of most oil and gas wells are (Figure 6.19)

- *Conductor* is a piece of large pipe (~30 in. OD) installed at the surface to keep the upper part of the hole from caving in. The installation of conductor pipe is referred to as "spudding" in a well. It can be installed prior to setting up the drill rig on land wells and may or may not be cemented in place.
  - Holds back unconsolidated surface formations
  - Contains flow line for mud returns
  - Contains diverter above flow line when shallow gas is probable
  - Up to 30″ in diameter
  - Set between 300 and 1000 ft

- *Surface casing* is set to protect fresh water sands. It is required by law. It contains a well head flange that supports the BOP stack and later, production "Christmas Tree" equipment. Also, surface casing must support the other casing strings. The annular space must be cemented back to the surface to prevent possible gas returns outside of the BOP stack:
  - Holds back unconsolidated shallow formations
  - Protects shallow zones from drilling mud and deeper formation contamination
  - Serves as a base for the BOP stack
  - Casing seat must hold deeper formation pressure
  - Supports the suspended weight of other casing strings
  - Up to $13\frac{3}{8}$ in. diameter
  - Set at depths from 1000 to 5000 ft (~2000 ft is typical)
  - Cemented back to the surface

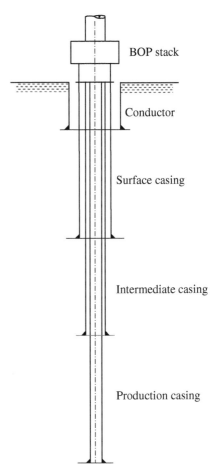

Figure 6.19 Types of casing strings.

- *Intermediate casing* (or liner) is set for the same reasons:
  - Controls open-hole conditions that may prevent a well from being drilled safely to the reservoir
  - Prevents lost circulation
  - Isolates troublesome shale, shallow gas, or water flows
  - Reduces amount of uncased hole, especially in deep wells

- *Production casing* is set through production zone (except in open-hole conditions). It is cemented from the casing shoe back to the surface:

  - Designed to hold maximum shut-in pressure.

The objective in casing design is to select the lowest grade and weight of casing, to withstand specified burst and collapse pressure loads, in order to minimize overall casing cost.

Collapse Pressure Loading (Production Casing)

Collapse pressure is the difference between the outside and inside pressures according to

$$\Delta p = [p_0(0) + \gamma_o x] - [p_i(0) - \gamma_i x] \tag{6.46}$$

It is common practice to assume that the most severe collapse pressure loading condition occurs when the casing is completely evacuated, and external pressure is the result of hydrostatic mud pressure. External pressure for this case increases linearly with depth and is based on the weight of drilling mud when casing is run. This condition is usually assumed in casing design because it gives a conservative design against collapse. The collapse load equation then becomes

$$p(x) = 0.052\gamma_o x \tag{6.47}$$

where

p – pressure, psi
γ_o – mud density (γ_m), ppg (when casing was run-in)
x – vertical distance, ft (downward from surface)

The collapse pressure loading line is illustrated in Figure 6.20.

Burst Pressure Loading (Production Casing)

Burst loading is based on reservoir pressure (p_f) with gas trapped inside the casing from bottom to top. Assuming the formation at the casing shoe can support bottom-hole pressure, then the *internal pressure* distribution is defined by

$$p_i(x) = p_f - 0.052\gamma_g(L - x) \tag{6.48}$$

where γ_g is density of gas.

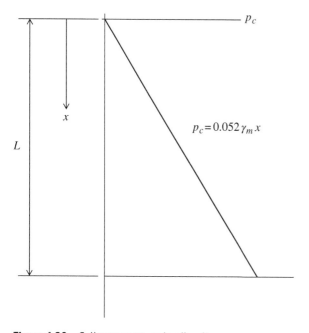

Figure 6.20 Collapse pressure loading line.

Here, p_f is formation or reservoir pressure. At the surface, the shut-in pressure is formation pressure minus the effect of the gas pressure.

In terms of pressure gradients, Eq. (6.48) becomes

$$p_i(x) = p_f - 0.052\gamma_g(L - x) \tag{6.49}$$

The pressure gradient term has units of psi/ft.

The *external pressure* is usually assumed to follow drilling mud hydrostatic pressure.

$$p_o = 0.052\gamma_m x \tag{6.50}$$

The resulting burst pressure, p_b, then becomes the difference between the inside pressure and the outside pressure in the casing:

$$p_b = p_i - p_o \tag{6.51}$$
$$p_b = p_f - 0.052\gamma_g(L - x) - 0.052\gamma_m x \tag{6.52}$$

For the sake of simplicity, the gas pressure gradient is sometimes ignored (a conservative assumption) leaving

$$p_b = p_f - 0.052\gamma_m x \tag{6.53}$$

Taking formation pressure as bottom hole mud pressure

$$p_f = 0.052\gamma_m L \tag{6.54}$$

Then the burst loading line becomes (Figure 6.21)

$$p_b = 0.052\gamma_m(L - x) \tag{6.55}$$

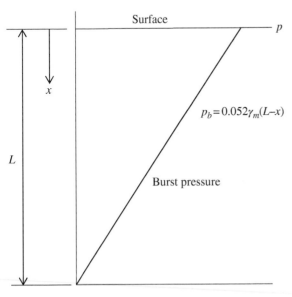

Figure 6.21 Burst pressure loading line.

API Collapse Pressure Guidelines

The API standard on casing design [5] defines three regions of casing properties leading to collapse.

1) Plastic collapse pressure – The equation defining the minimum collapse pressure in this case is

$$p_P = f_{ymn} \left[\frac{A_c}{D/t} - B_c \right] - C_c \tag{6.56}$$

p_P – pressure for plastic collapse
f_{ymn} – specified minimum yield strength
$A_c B_c C_c$ – empirical constants depending on casing grade and D/t ratio

2) Transition collapse pressure – The equation defining the minimum collapse pressure in this case is

$$p_T = f_{ymn} \left[\frac{F_c}{D/t} - G_c \right] \tag{6.57}$$

p_T – pressure for transition collapse
$F_c G_c$ – empirical constants depending on casing grade and D/t ratio

3) Elastic collapse pressure – The equation defining elastic collapse pressure is

$$p_E = 46.95 \times 10^6 \left\{ \frac{1}{[D/t(D/t-1)^2]} \right\} \tag{6.58}$$

Plastic Yielding and Collapse with Tension

Tension also effects plastic yielding and collapse in casing. Tension and compression combine to create a biaxial state of stress (Figure 6.22), which is evaluated for yielding using the von Mises criteria of failure:

$$\sigma' \geq \sigma_{yld} \tag{6.59}$$

where σ_{yld} is material yield strength as determined from uniaxial testing and

$$(\sigma')^2 = \sigma_a^2 - \sigma_a \sigma_\theta + \sigma_\theta^2 \tag{6.60}$$

According to Eq. (6.59), yielding occurs when von Mises stress, $\sigma' \geq \sigma_{yld}$.

$$\sigma_a^2 - \sigma_a \sigma_\theta + \sigma_\theta^2 \geq (\sigma_{yld})^2 \tag{6.61}$$

In dimensionless units, yielding occurs when

$$\left(\frac{\sigma_a}{\sigma_{yld}} \right)^2 - \frac{\sigma_a}{\sigma_{yld}} \frac{\sigma_\theta}{\sigma_{yld}} + \left(\frac{\sigma_\theta}{\sigma_{yld}} \right)^2 \geq 1 \tag{6.62}$$

For example, if

$$\sigma_\theta = -40\,000 \text{ psi}$$

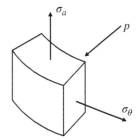

Figure 6.22 Biaxial state of stress.

$$\sigma_a = 20\,000 \text{ psi}$$

$$\sigma_{yld} = 50\,000 \text{ psi}$$

then

$$\frac{\sigma_\theta}{\sigma_{yld}} = -0.8 \ (80\%)$$

$$\frac{\sigma_a}{\sigma_{yld}} = 0.4 \ (40\%)$$

By substituting these numbers into Eq. (6.62), the left side is equal to 1.12, therefore this biaxial stress state will cause plastic yielding.

Strength properties of casing sizes are conveniently display in API tables. Collapse and burst strength capabilities are based on API equations outlined earlier. Casing size is listed in terms of their outside diameter. Weight changes reflect changes in inside diameter. Grade or material strength may vary for each casing size and weight. Grade of material effects collapse, burst, and tension limits. A complete set of casing performance data can be found in Ref. [5].

Summary of Pressure Loading (Production Casing)

Collapse pressure loading is illustrated in Figure 6.22 showing the pressure distribution increasing directly with depth according to

$$p_c = 0.052\gamma_m x \text{ psi} \tag{6.63}$$

The maximum collapse pressure loading is at the bottom end.

Burst pressure load is the difference in internal pressure and the external pressure. An extreme case would exist if formation gas pressure is exerted internally from bottom to top and internal pressure being partially resisted by hydrostatic mud pressure outside the casing. The resulting burst pressure for this assumption is

$$p_b = p_f - 0.052\gamma_m x \tag{6.64}$$

$$p_b = 0.052\gamma_m (L - x) \tag{6.65}$$

The maximum burst pressure occurs at the well head.

The collapse and burst pressure loads are illustrated in Figure 6.23 by the "*X*" loading diagram.

A casing string designed strictly for collapse would be strongest at the bottom and weakest at the top. While this unusual casing string would not collapse, it would be susceptible to burst type failure in the event the well is shut in during a blowout. A casing string designed for burst alone would most likely collapse if the casing were evacuated.

A proper casing design is one which can resist both collapse pressure load and burst pressure load. In addition, it is tapered and has the strongest section near the top and bottom with the weakest casing section near the middle.

The selection of a factor of safety may vary with the level of experience in a given area and the confidence level for each casing load prediction. Small factors of safety not only give lighter and therefore cheaper casing strings but also carry a higher risk element.

Effect of Tension on Casing Collapse

Tension along the axis of casing also affects the *collapse* resisting capability of casing. The collapse pressures given in the casing performance tables are based on a uniaxial state of stress assuming casing is subjected to external pressure only. This pressure creates circumferential or hoop stresses.

Figure 6.23 Typical loading for casing design.

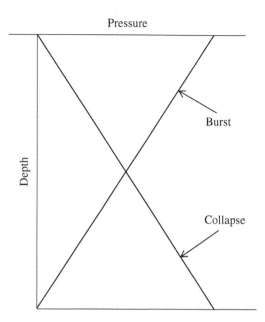

Casing in general can be loaded with a direct pull creating longitudinal stresses simultaneously with circumferential stress producing a biaxial state of stress. According to the von Mises criteria of failure, local yielding in casing will develop when the maximum energy of distortion created by biaxial stress state is equal to the maximum energy of distortion at yielding under a uniaxial state of stress. Equating these two energies gives

$$\sigma_{yld}^2 = \sigma_T^2 - \sigma_T\sigma_C + \sigma_C^2 \tag{6.66}$$

or

$$\left(\frac{\sigma_T}{\sigma_{yld}}\right)^2 - \left(\frac{\sigma_T}{\sigma_{yld}}\right)\left(\frac{\sigma_C}{\sigma_{yld}}\right) + \left(\frac{\sigma_C}{\sigma_{yld}}\right)^2 = 1 \tag{6.67}$$

where

σ_T – tension stress, psi
σ_C – collapse stress, psi
σ_{yld} – yield strength, psi

Equation (6.67) can also be written as

$$\left(\frac{T}{T_{yld}}\right)^2 - \left(\frac{T}{T_{yld}}\right)\left(\frac{p_C}{p_{c,yld}}\right) + \left(\frac{p_C}{p_{c,yld}}\right)^2 = 1 \tag{6.68}$$

This equation establishes the relation between collapse pressure and tension which produce yielding. The plot of this equation (Figure 6.24) shows how much the collapse pressure has to be derated for a given amount of direct pull. The portion of this diagram that is applicable to casing design is the portion in the fourth quadrant. These coordinates can be tabulated for accuracy and convenience.

For example, if the direct pull on 9⅝, 40 lb/ft, N80 casing at a given location is 110 000 lb, then

$$\frac{T}{T_{yld}} = \frac{110\,000}{916\,000} = 0.12$$

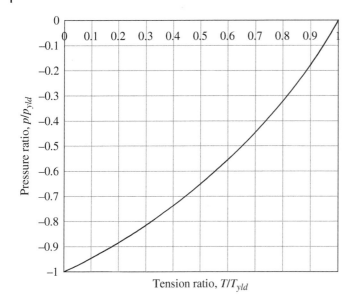

Figure 6.24 Yielding condition for biaxial state of stress.

According to Figure 6.24, the casing's resistance to collapse pressure has to be derated by a factor of 0.93. At this location, collapse strength is derated from 3090 to 2874 psi.

Tension Forces in Casing

The neutral point here is defined as the point in the casing where axial force or stress is zero. This point is different from the neutral point in drill collars, which is based on buckling. Tension calculations are usually based on casings being completely surrounded by drilling mud (inside and out). In reality, this is not true, but the calculation is conservative.

Based on this assumption, the internal force distribution along casing is defined by (see Figure 6.25).

$$F(x) = \gamma A(L - x) - L\gamma_m A \tag{6.69}$$

where x is measure from the top and $w = \gamma A$

$$F = w(L - x) - w_m L \tag{6.70}$$

This equation is used to define the magnitude of the true axial force in casing. Accordingly, at $x = 0$

$$F = W(BF) \tag{6.71}$$

and at $x = L$

$$F = -w_m L \tag{6.72}$$

Figure 6.25 Internal force in casing.

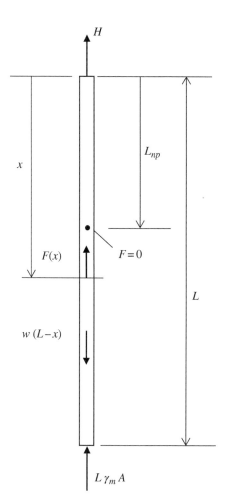

The location of the point where $F = 0$ $(\sigma_a = 0)$

$$0 = \gamma AL - \gamma AL_{np} - \gamma_m AL \tag{6.73}$$

$$0 = (w - w_m)L - wL_{np}$$

$$L_{np} = \left(1 - \frac{w_m}{w}\right)L \tag{6.74}$$

$$L_{np} = (BF)L \tag{6.75}$$

If casing is empty and plugged at the bottom, the force distribution would be completely different from what is predicted by Eq. (6.70). Equation (6.75) defines the location of the neutral point, the point of zero stress. The internal tension above the neutral point is determined from

$$F(x) = w\left(L_{np} - x\right) \tag{6.76}$$

Casing collapse pressure strength has to be derated to account for the effect of tension. The strength ellipse in Figure 6.24 relates the pressure ratio, $\dfrac{p}{p_{yld}}$, to the tension ratio, $\dfrac{T}{T_{yld}}$. The collapse strength is derated from p_{yld} to p which becomes the allowable collapse pressure, p_a before FS is applied.

Design of 9⅝ in. Production Casing

Example As an example, consider the design of a production casing having the following specifications:

- Depth of well – 9500 ft
- Casing size – 9⅝ in.
- Mud weight – 12 ppg

The objective of the design is to select a distribution of casing having lowest weight and grade. The design starts at the bottom by considering collapse loading (Figure 6.26). The maximum collapse pressure load is shown as 5928 psi ($p = 0.052(12)9500 = 5928$ psi).

Design Without Factors of Safety

For the sake of simplicity, FS are not considered at this point. They will be included later. From Table 6.4, the lowest weight and grade combination having a collapse strength equal to or greater than 5928 psi is 47 lb/ft (S-95), which has a collapse strength of $p_c = 7100$ psi (burst strength is $p_b = 8150$ psi).

Each casing section will be tabulated as the design moves upward.

Section #1	9500 − x	47 lb/ft (S95)

At what point can casing weight and grade be reduced? Casing 43.5 lb/ft (S95) has a collapse strength of $p_c = 5600$ psi (burst strength of $p_b = 7510$ psi).

The location at which the applied collapse pressure is 5600 psi is determined from the pressure load function.

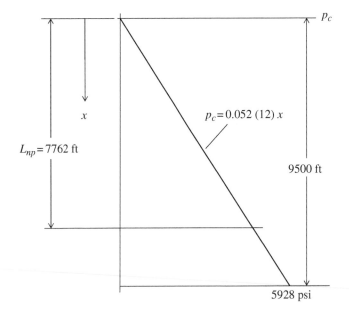

Figure 6.26   Collapse loading (9⅝ casing, 12 ppg mud).

Table 6.4 Strength capability of 9⅝ in. casing (non API grades).

Nominal weight (lb/ft)	Grade	Collapse pressure (psi)	Burst pressure (psi)	Body tension (lb × 1000)	Coupling tension (lb × 1000)
36.00	H40	1740	2540	410	
	K53	2020	3520	564	489
	S80	2980	3520	564	605
40.00	K55	2570	3950	630	561
	S80	4230	3950	630	694
	C75	2980	5390	859	694
	N80	3090	5750	916	737
	SS95	4230	5750	916	837
	S95	4230	6820	1088	858
	C95	3330	6820	1088	847
43.50	C75	3750	5930	942	776
	N80	3810	6330	1005	825
	SS95	5600	6330	1005	936
	S95	5600	7510	1193	959
	C95	4130	7510	1293	948
	P110	4430	8700	1381	1106
47.00	C75	4630	6440	1018	852
	N80	4750	6870	1086	905
	SS95	7100	6870	1086	1027
	S95	7100	8150	1289	1053
	C95	5080	8150	1289	1040
	P110	5310	9440	1493	1213

Source: Condensed from Lone Star Steel casing tables, Ref. [6].

$$p_c(x) = 0.052(12)x = 5600 \text{ psi}$$

$$x = \frac{5600}{0.052(12)} = 8974 \text{ ft}$$

This locates the beginning of Section #2.

Section #1	9500 − 8974	47 (S95)
Section #2	8974 − x	43.5 (S95)

The next lowest weight is 40 (S95) having collapse strength of 4230 psi (burst strength is 6820 psi). The beginning of the third section then is

$$x = \frac{4230}{0.052(12)} = 6778$$

The status of the casing design is now

Section #1	9500 − 8974	47 (S95)
Section #2	8974 − 6778	43.5 (S95)
Section #3	6778 − x	40 (S95)

The next lightest casing weight is 36 (S80) having a collapse strength of 2980 psi. This casing weight and grade is selected for Section #4, which starts at

$$x = \frac{2980}{0.052(12)} = 4776 \text{ ft}$$

The status of the casing design is now

Section #1	9500 − 8974	47 (S95)
Section #2	8974 − 6778	43.5 (S95)
Section #3	6778 − 4776	40 (S95)
Section #4	4776 − x	36 (S80)

Each of these sections is shown in Figure 6.27. This process could continue along the collapse load line, but at this point, burst loading becomes the important consideration. The two dots in Section #4 indicate that 36 (S80) can support collapse loading from 4776 ft and upward. At the same time, this casing can withstand the burst loading up to the upper dot.

The fourth casing weight and grade has a burst strength of $p_b = 3520$ psi. The length of this section is controlled by the burst loading line. The location where the burst load is equal to 3520 psi is determined as (using Eq. (6.65))

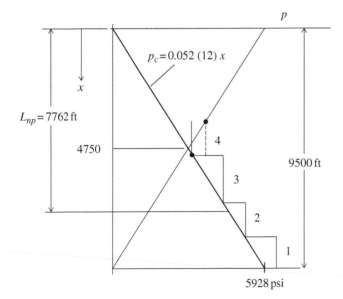

Figure 6.27 Collapse and burst loading.

$$x = L - \frac{p_b}{0.052(12)} = 9500 - \frac{3520}{0.052(12)} = 3859$$

which locates the upper dot. From this point upward, burst pressure controls the design. The casing string has four different casing weights (Figure 6.29):

Section #1	9500 – 8974 ft	47 (S95)	
Section #2	8974 – 6778 ft	43.5 (S95)	
Section #3	6778 – 4776 ft	40 (S95)	
Section #4	4776 – 3859 ft	36 (S80)	(burst controls)

Continuing upward on the *burst pressure* line (Figure 6.28), casing weight of 40 lb/ft gives several grade options. We choose grade S95 because its burst strength is 6820 psi which is greater than the maximum burst load (5928 psi) at the surface. Adding Section #5 completes the casing design.

	Interval (ft)	Weight (grade) (lb/ft)	Weight (lb)	
Section #1	9500 – 8974 ft	47 (S95)	24 722	
Section #2	8974 – 6778 ft	43.5 (S95)	95 526	
Section #3	6778 – 4776 ft	40 (S95)	80 080	
Section #4	4776 – 3859 ft	36 (S80)	33 012	(burst controls)
Section #5	3859 – 0	40 (S95)	154 360	(burst controls)
		Total	387 700 lb	

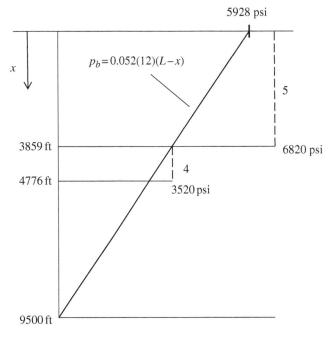

Figure 6.28 Burst load line and casing sections.

Directional Drilling

Directional well paths depend on shape of reservoir and location of a drilling and production platform. As many as 40 directional wells may be drilled from one platform. A platform may be located offshore or in environmentally sensitive land locations. With multiple wells from one platform, well paths of each must be documented with respect to the others. The point of entry into the reservoir is also important.

There are three basic types of well paths (see Figure 6.29). Well paths may be more complicated depending on shape of the reservoir. Broken and faulted reservoirs may require entry from the side or even from below.

Downhole Drilling Motors

Downhole motors were introduced in the late 1960s as a means of deflecting a direction well away from vertical without using whipstocks. A bent sub was placed on top of the motor to bring about lateral drilling (Figure 6.30a). Tool face is within the bend. It was oriented in a desired plane of drilling to bring about a directional change. This arrangement began to replace whipstocks. Tool face, hole direction, and inclination were measured with a downhole sensor lowered on a wire line. Measurements were taken intermittently until the desired hole angle was achieved. PDM emerged as a preferred directional tool during this time.

The advantage of using PDMs in directional drilling was huge. It replaced the time-consuming, multiple steps of setting a whipstock and surveying to establish hole direction. The bent sub and PDM did the same thing faster. At first, surveys were made by placing a survey tool in a side pocket and using a hard wire to bring hole direction and inclination information back to the surface. Later survey instruments were inserted and retrieved from within the drill collars to track progress. Once the angle had been established, a stabilized BHA was used to maintain hole direction to the target.

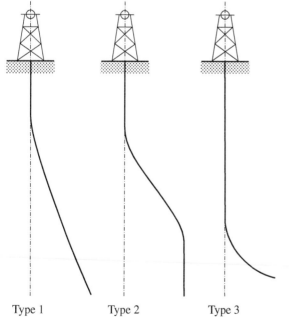

Figure 6.29 Three basic types of directional wells.

Type 1   Type 2   Type 3

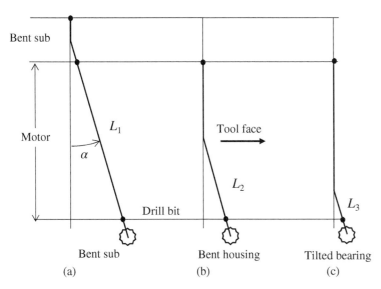

Figure 6.30 (a–c) Effect of bend location on rate of build.

When the bit drilled off course, a bent sub and PDM were used again to bring the hole back on course.

As a rule, a bent sub and motor turn a well path about 75% of the bent sub angle over 30 ft of drilling. This means that a 2° bent sub creates a rate of build of about 1½° per 30 ft or a dog leg severity of 5° per 100 ft. The maximum allowable bent sub angle is limited by the mount of offset at the bit. If this offset is large, it is difficult to get the BHA into the well bore. As a result, the maximum bent sub angle is limited to 2½.

Bit torque is transmitted across motors, and this torque creates a rotational reaction of the tool face, which depends on hole depth, friction, bit torque, and drillstring size. This reactive torque is a major consideration in setting and maintaining tool face orientation during a correction run.

Eventually, the bent sub was replaced by a bend within the motor housing (Figure 6.30b). The bent housing configuration brought about a more responsive direction change. This configuration was subsequently replaced by a tilted bearing assembly (Figure 6.30c) which located the tool face even closer to the drill bit. This configuration produced even a greater rate of directional change, which can be expressed by $\frac{\Delta\theta}{\Delta s} = \frac{\alpha}{L}$. By reducing the distance, L, the rate of directional change is increased. By moving the bend angle close to the drill bit, the angle, α could be reduced and still achieve the same rate of build.

The realigned bearing or tilted bearing assembly gave another advantage, too. The motor housing itself could be used to drill ahead as well as for making hole corrections. By rotating the motor housing, the drill bit would be continually disoriented, and drilling could proceed with motor power. Rotation of the motor housing at low speeds simply disoriented the drill bit. When hole corrections were needed, the tool face in the housing was set, and the correction was made by motor power. This, along with MWD, eliminated costly tripping time.

Rotary Steerable Tools

A vision for many years has been a technology that would allow directional drilling to proceed along a specified path without stopping to make directional changes. Essentially, the drill bit would

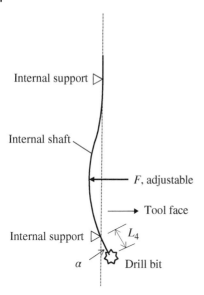

Figure 6.31 Schematic of rotary steerable tool.

advance through layers of various formations much like airplanes flying and navigated along a specified flight path, with the ability to alter the course in the event of strong crosswinds or bad weather. In each case three elements are essential:

- Power to move forward
- Navigation capabilities to define position and well path
- Control of the well path

Technology is now in place to direct drill bits in three-dimensional space. Tool face monitoring and control is central. The ability to monitor bit location and path through MWD and then bring about directional change is equally important. Each of these tasks is now achieved by a RST system. The only limitation appears to equipment reliability.

The schematic in Figure 6.31 illustrates an internal shaft which is deflected by an internal force, F. The direction and magnitude of the force established tool face orientation and rate of change in hole angle. Note that the parameters, α and L_4 have the same role as the ones in Figure 6.33; however, the magnitude and orientation of α is adjusted while drilling. Many mechanical designs for this purpose have been patented. The author proposes another method to orient tool face and magnitude of angle α, which affects dog-leg severity.

Stabilized Bottom-Hole Assemblies

Downhole motors are commonly used to deflect a well at a kick-off point toward a target. The deflection takes place in the plane of the target line. A bent sub/motor assembly is used until hole angle reaches about 6–8°; a stabilized building assembly is not effective below these angles. In hard rock, a bent sub/motor assembly may have to establish even higher angles before a building assembly is place in the hole. Building assemblies are desirable as early as possible because tool face orientation is not a consideration. However, bent sub/motor assemblies have been used to build up to 45° from the kick-off point.

Bent sub/motor assemblies are used to

- Kick off
- Establish rate of build
- Make hole corrections

After hole angle is achieved, drilling proceeds with a stabilized assembly: building, holding, or dropping. These assemblies differ primarily in the number and location of stabilizers.

Stabilizers are special tools that centralize drill collars at specific locations to establish a desired side load to drill bits. Stabilizer blades are roughly 2 ft long and allow fluid to flow upward in the annulus with minor flow restriction. Stabilized assemblies can build, hold, or drop hole angle, depending on stabilizer number and spacing.

Prediction of bit side forces produced by various stabilizer assemblies is a guide for predicting directional performance. Ultimately, stabilizer assemblies have to be tried, tested, and usually modified in a given area until they perform as needed. A good data base of BHA performance in a given area is useful in this regard.

Building assemblies – Building assemblies develop an upward side force on drill bits. A near-bit stabilizer acts as a fulcrum point for the collar weight to pry against (Figure 6.32a). The mechanical advantage of this assembly is greatest when the first stabilizer is close to the drill bit. A second stabilizer is positioned some distance to establish a large free span distance to increase the strength of building assemblies. Positioning of the second stabilizer depends on drill collar size. Stiffer or larger drill collars will allow a longer free span distance.

Holding assemblies – Stabilizers are positioned within drill collars so that relatively small side forces (usually less than 200 lb) are developed at the bit. When drill bits having long shanks

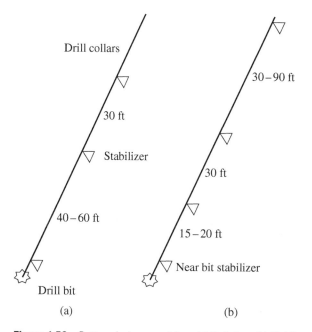

Figure 6.32 Bottom hole assemblies. (a) Building. (b) Holding.

Figure 6.33 Diesel-driven generator.

or large gauge protection areas are used, it is desirable to minimize the tilt angle with stabilizer placement to alleviate bit binding. A typical holding assembly is illustrated in Figure 6.32b. Calculations show that hole angle directly affects the magnitude of the side force for each of these assemblies. Side forces increase with hole angle; examples are given in the Appendix.

Dropping assemblies – Unlike building and holding assemblies, dropping assemblies have no near-bit stabilizers. The drill bit stands alone and is pushed downward by the weight of drill collars suspended between the bit and the first stabilizer which is roughly 60 ft from the bit. The first stabilizer in this case acts much like a hinge allowing the suspended collars to swing downward like a pendulum.

Power Units at the Rig Site

The power source for drilling rigs is a skid unit containing a diesel engine and an electric generator (Figure 6.33). The output power can be as high as 10 000 hp (7 457 000 W or 7.457 MW). These power units are necessary because drilling rigs are usually located in remote area where public utilities are not available.

Electric power is safer and easier to transmit to various power needs around the rig.

As a power unit, the "end usage" of the power source is electric power. This power is consumed by (i) hoisting, (ii) rotary system, (iii) drilling mud pumps, and (iv) well control equipment.

References

1 Dareing, D.W. (2019). *Oilwell Drilling Engineering*. ASME Press.

2 Johancsik, C.A., Friesen, D.B., and Dawson, R. (1984). Torque and drag in directional wells – prediction and measurement. *J. Pet. Technol.* **SPE 11380**: 987–992.

3 API Standard (1987). *Recommended Practice for Drill Stem Design and Operating Limits (API RP 7G)*, 12e. American Petroleum Institute.

4 Kendall, H.A. and Goins, W.C. (1960). Design and operation of jet-bit programs for maximum horsepower, maximum impact force and maximum jet velocity. *Trans. AIME* **219** (01).

5 API Committees. (2008). Technical report on equations and calculations for casing, tubing, and line pipe used as casing or tubing; and performance properties tables for casing and tubing. *ANSI/API Technical Report 5C3*. 1 ed.

6 Lone Star Steel Company (1984). *Casing and Tubing Technical Data*. Dallas, TX.

Part III

Analytical Tools of Design

Engineering Design is about predicting future performance – through science and mathematics. Analytical tools are available for designing structures and machines to a high level of reliability. Insightful application of these tools is essential for safe and reliable products.

Designs from antiquity were based on experience, tradition, geometry, esthetics, and empirical formulas and probably a lot of testing. The material of construction was stone, quarried by hand, moved to the construction site, and lifted in place. Each step required strenuous manual labor using, levers, ramps, and pulleys. Even though Newtonian physics and thermodynamic principles emerged during the 1700s, it was not until the mid-1800s that science-based engineering design was accepted. Also, during this time steel became a structural material.

Engineering science is now commonly used to develop machines and structures to satisfy design specifications. Life expectancy and performance can be predicted with a high degree of accuracy using analytical models and numerically based computer software.

Part III brings together the mechanics of many analytical tools commonly used in design. References are given for further investigation into these topics.

Engineering Practice with Oilfield and Drilling Applications, First Edition. Donald W. Dareing.
© 2022 John Wiley & Sons, Inc. Published 2022 by John Wiley & Sons, Inc.

7

Dynamics of Particles and Rigid Bodies

Dynamics is a science that deals with bodies in motion. The subject stems from Newton's laws of motion, which were first published in 1687. An important aspect of Newton's mathematics is the concept of rate of change, such as speed and acceleration. This led to the principles of calculus. The subject of dynamics is viewed per the following breakdown:

Statics – Study of bodies at rest or in a state of static force equilibrium, having zero acceleration.
Kinematics – Study of motion of objects without regard to the forces that cause the motion.
Kinetics – Study of motion of objects caused by an imbalance of applied forces.

All three, stem from Newton's laws of motion, which apply to both discrete particles and rigid bodies.

Statics – Bodies in Equilibrium

Statics refers to bodies at rest. Equilibrium refers to a system of forces that allows this to happen. If a body is at rest, then the forces applied to the body are in equilibrium. Statics is a special case of the second law of Newtonian physics, i.e. linear and angular accelerations of a body are both zero.
When a body is in equilibrium the force system must balance.

$$\sum \overline{F} = 0 \tag{7.1}$$

$$\sum \overline{M}_0 = 0 \quad \text{(point 0 is arbitrary)} \tag{7.2}$$

There are different types of force systems: concurrent forces, biaxial forces, coplanar forces, and general three-dimensional (3-D) force systems. It is helpful to recognize the type of force system in setting up the equations. Freebody diagrams are most helpful.
In any engineering situation, an analysis based on elementary mechanics is a good starting point. This provides a means of scoping out the problem by defining magnitudes of forces, stresses, flow rates, etc. A more refined analysis with computer software can follow, if necessary. This approach is useful in evaluating the technical feasibility of design concepts and configuring design possibilities. What are the magnitudes of external and internal forces? How do the forces flow through a design configuration and its subsystems? Understanding load magnitudes (internal and external) is essential in design innovation.

Engineering Practice with Oilfield and Drilling Applications, First Edition. Donald W. Dareing.
© 2022 John Wiley & Sons, Inc. Published 2022 by John Wiley & Sons, Inc.

Force Systems

A *concurrent force* is one where all forces intersect at a point. The force vectors can be either two or three dimensional. In either case, the sum of all the force vectors must be zero for equilibrium:

$$\sum \overline{F}_n = 0 \tag{7.3}$$

Three unknown force components can be determined from this vector equation. The three forces acting on the pin A in the simple structure (Figure 7.1b) form a concurrent force system. Summing forces in the x direction gives

$$-T\cos\theta + Q = 0$$

Summing forces in the y direction gives

$$T\sin\theta - F = 0$$

Solving these equations together with $\theta = 30°$ and $F = 1000$ lb, gives $T = 2000$ lb and $Q = 8660$ lb.

Members that are pinned at each end (AB, Figure 7.1a) are examples of *biaxial (or collinear) force* systems. The end connections do not transmit moment, therefore, applied forces at each end are collinear. This condition is useful in determining the direction of forces as illustrated above.

A *coplanar force* system is one in which side forces and/or moments are applied to a structural element. The force vectors are applied in one plane and moments vectors are perpendicular to the plane. In which case

$$\sum \overline{F}_n = 0 \tag{7.4}$$

$$\sum M_a = 0 \tag{7.5}$$

where

n	–	number of force vectors
a	–	fixed point

Consider a freebody diagram left of section a–a in Figure 7.2. It is desired to find the magnitude of internal shear, V, and internal moment, M. Applying Eqs. (7.4) and (7.5)

$$F_A - V - wx = 0$$
$$V = F_A - wx$$

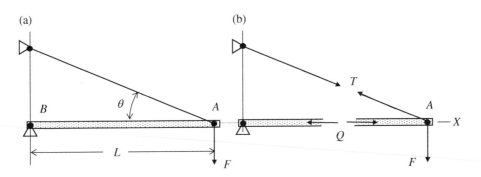

(a)   (b)

Figure 7.1 Concurrent force system.

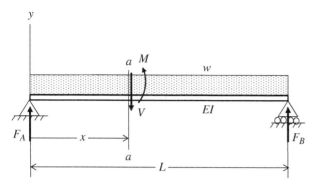

Figure 7.2 Internal force balance.

and

$$M + wx\frac{x}{2} - F_A x = 0$$

$$M = F_A x - \frac{1}{2}wx^2$$

The reactions at points A and B are found by applying Eqs. (7.4) and (7.5) to the total beam.

In general, force vectors and moment vectors can be three dimensional. In which case, equilibrium is established when

$$\sum \overline{F}_n = 0 \tag{7.6}$$

By substitution into Eq. (7.2)

$$\sum \overline{M}_a = 0 \tag{7.7}$$

Consider the bent rod of Figure 7.3. The problem is to determine the force and moment reactions at point O. The six unknowns are O_x, O_y, O_z, M_x, M_y, M_z. Equation (7.6) gives three scalar equations whose solution establishes the three reaction forces:

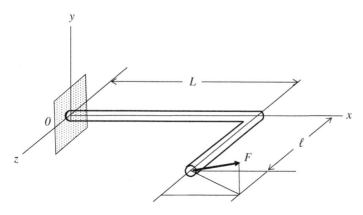

Figure 7.3 Three-dimensional force balance.

$$O_x + F \cos \alpha = 0$$
$$O_y + F \cos \beta = 0$$
$$O_z + F \cos \gamma = 0$$

where $\cos \alpha$, $\cos \beta$, and $\cos \gamma$ are directional cosines of force F.

Equation (7.7) gives

$$\overline{M}_0 + \overline{r} \times \overline{F} = 0$$

where

$$\overline{r} = Li + 0j + lk$$
$$\overline{F} = F(i \cos \alpha + j \cos \beta + k \cos \gamma)$$

$$\overline{M}_0 = F \begin{vmatrix} i & j & k \\ L & 0 & l \\ \cos \alpha & \cos \beta & \cos \gamma \end{vmatrix}$$

The solution to the three scalar equations gives the moment reactions at point 0.

Freebody Diagrams

Freebody diagrams are isolated portions of a structure used to determine external or internal forces and moments. Internal reactive forces (and moments) are needed to determine stresses, buckling, or other responses that may affect the structures performance, including possible failure.

Control volumes are imaginary boundaries defined to study fluid flow parameters such as flow rate, pressure drop, hydraulic horsepower, and friction.

Force Analysis of Trusses

A truss is a structure made up of a network of triangular sections. The overall stiffness of the structure is generated by the combination of each triangular section. The first of all metal trusses was designed and built in 1840 by Squire Whipple [1]. Compression members were cast iron. The chords and diagonals (tension members) were made of wrought iron. Prior truss bridges were made of wood.

Whipple's metal truss bridges, built across several locations along the Erie Canal, were the first scientifically designed truss bridges. His force analysis assumes that the connection points are frictionless pins, bending moments at the joints are ignored. While this is not exactly true in rigid connections, the assumption yields reasonable forces in truss members. This assumption greatly simplifies the force analysis by allowing forces in each beam to be treated as collinear, i.e. internal forces are in-line with the axis of the element. Each member is assumed to be axially loaded with zero bending.

Consider the trusses shown in Figure 7.4. Force of magnitude in each element is needed to conduct a stress and buckling analysis the structure. Whipple gives two approaches: (i) method of joints and (ii) method of sections.

The first step is to view the total structure as a freebody with unknown forces, R_A and R_B. The magnitudes of these two forces are determined from statics:

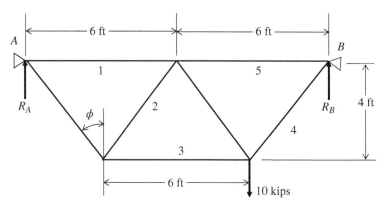

Figure 7.4 Truss example.

$$\sum M_A = 0$$
$$-12R_B + 9(10) = 0 \tag{7.8}$$
$$R_B = 7.5 \text{ kip}$$

$$\sum F_y = 0$$
$$R_A + R_B = 10 \tag{7.9}$$
$$R_A = 2.5 \text{ kip}$$

The next step is to determine the magnitude of the forces in each of the seven (7) members. This information is needed to size each member for bridge performance and safety. Dimensions for this example are selected so that $\sin \phi = 3/5$ and $\cos \phi = 4/5$ for convenience of calculation.

Method of Joints

This method of determining forces in each structural element is based on freebody diagrams of each joint. Forces at each joint form a concurrent force system with each force coming together at each joint as shown at joint B in Figure 7.5.

$$\sum F_y = 0$$
$$R_B - F_4 \cos \phi = 0 \tag{7.10}$$
$$F_4 = 9.375 \text{ kip (tension)}$$

$$\sum F_x = 0$$
$$F_5 - F_4 \sin \phi = 0 \tag{7.11}$$
$$F_5 = 5.625 \text{ kip (compression)}$$

Forces in each element are determined from freebody diagrams of each joint, remembering that the force system at each joint is concurrent. Whipple sometimes used cables for members in tension.

Method of Sections

The method of sections creates freebody diagrams by cutting across the structure with an imaginary line as shown in the left portion of Figure 7.5. Forces in each member are determined using all three of the equations of statics:

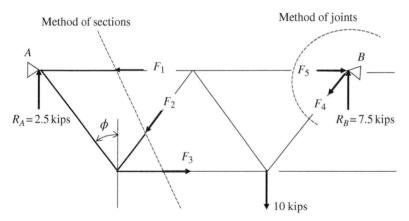

Figure 7.5 Application of freebody diagrams.

$$\sum F_y = 0$$
$$R_A - F_2 \cos\phi = 0$$
$$2.5 - F_2\frac{4}{5} = 0 \tag{7.12}$$
$$F_2 = 3.125 \text{ kip (compression)}$$

$$\sum M_A = 0$$
$$4F_3 - F_2(3\cos\phi + 4\sin\phi) = 0$$
$$F_3 = 1.2F_2 \tag{7.13}$$
$$F_3 = 3.75 \text{ kip (tension)}$$

$$\sum F_x = 0$$
$$-F_1 - F_2\sin\phi + F_3 = 0 \tag{7.14}$$
$$F_1 = 1.875 \text{ kip (compression)}$$

Since each force prediction is plus (+), the assumed directions of F_1, F_2, F_3 are correct.

This method is most useful for finding internal loads in members some distance away from support points.

Kinematics of Particles

Linear Motion

Linear motion refers to motion of a particle along a straight line. Position, velocity, and acceleration are related through calculus. If $s = s(t)$ is known, then

$$v(t) = \frac{ds}{dt} \tag{7.15}$$

$$a(t) = \frac{dv}{dt} \tag{7.16}$$

Combining these two equations gives

$$a = \frac{dv}{ds}\frac{ds}{dt} = \frac{dv}{ds}v$$

$$a\,ds = v\,dv \tag{7.17}$$

This expression is useful when $a(t)$ is known.
Assuming acceleration, a, is *constant* gives

$$a(s_2 - s_1) = \frac{1}{2}(v_2^2 - v_1^2) \tag{7.18}$$

If an object is thrown upward with an initial velocity of v_1, the height of travel is

$$-g(s_2 - 0) = \frac{1}{2}(0 - v_1^2)$$

$$s_2 = \frac{v_1^2}{2g} \tag{7.19}$$

If $a(t)$ is *not constant* and a known function of time, it is best to use $dv = a(t)dt$.

$$v_2 = v_1 + \int_1^2 a(t)dt \tag{7.20}$$

$$s_2 = s_1 + \int_1^2 v(t)dt \tag{7.21}$$

These two equations are useful in tracking velocity and position with time.

Rectangular Coordinates

A particle traveling in two-dimensional space is defined by tracking both $x(t)$ and $y(t)$ components. Projectiles are good examples. Ignoring air friction, the path of a projective is determined as follows (Figure 7.6).

x Component of the Path

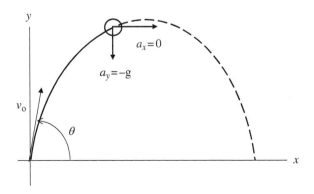

Figure 7.6 Projectile path.

$$a_x = 0$$

$$v_x = C_1$$

$$x(t) = C_1 t + C_2 \tag{7.22}$$

y Component of the Path

$$a_y = -g$$

$$v_y = -gt + C_3$$

$$y(t) = -g\frac{t^2}{2} + C_3 t + C_4 \tag{7.23}$$

Initial conditions are

$$x(0) = 0 \qquad C_2 = 0$$
$$v_x = v_0 \cos\theta \quad C_1 = v_0 \sin\theta$$
$$y(0) = 0 \qquad C_4 = 0$$
$$v_y = v_0 \sin\theta \quad C_3 = v_0 \cos\theta$$

Applying initial conditions gives

$$x(t) = (v_0 \cos\theta)t \tag{7.24}$$

$$y(t) = -g\frac{t^2}{2} + (v_0 \sin\theta)t \tag{7.25}$$

Eliminating the time variable gives the path in terms of $y = y(x)$

$$y(x) = x \tan\theta - \frac{g}{2}\frac{x^2}{(v_0 \cos\theta)^2}. \tag{7.26}$$

The projectile touches ground at a horizontal distance of

$$L = \frac{2\sin\theta\cos\theta v_0^2}{g}$$

$$L = \frac{v_0^2}{g}\sin 2\theta$$

from the launch point. Assuming $\theta = 80°$ and $v_0 = 100$ fps, $L = 1062$ ft.

Polar Coordinates

Some problems can best be defined in terms of independent variables r and θ as shown in Figure 7.7. In the left case, the point moves in a circle with constant radius R. Angular position $\theta(t)$ may vary with time. In this case, the position of a point on the circle is defined by $s(t) = R\,\theta(t)$.
Its tangent velocity is equal to

$$v = R\frac{d\theta}{dt} = R\omega \tag{7.27}$$

Its vector is tangent to the circle. Acceleration of the point has two components, a normal component and a tangent component. At any point on the circle

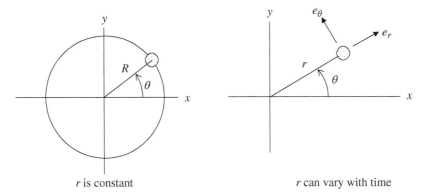

Figure 7.7 Polar coordinates.

$$\bar{a} = -R\omega^2 \bar{e}_r + R\alpha\bar{e}_\theta \tag{7.28}$$

The position of the particle may be determined by arc travel, $s(t)$, in which case:

$$s(t) = R\theta$$
$$\dot{s}(t) = R\omega$$
$$\ddot{s}(t) = R\alpha$$

If $\theta(t)$ is not continuous, then position, velocity, and acceleration can be determined like the straight line motion formulas discussed earlier.

In general, both the angular and radial positions can change with time (Figure 7.7, right drawing). In this case

$$\bar{r} = r\bar{e}_r \tag{7.29}$$

$$\bar{v} = r\frac{d\bar{e}_r}{dt} + \dot{r}\bar{e}_r \tag{7.30}$$

as shown in Figure 7.8.

Since

$$\frac{d\bar{e}_r}{dt} = \dot{\theta}\bar{e}_\theta \quad \text{and} \quad \frac{d\bar{e}_\theta}{dt} = -\dot{\theta}\bar{e}_r$$

$$\bar{v} = r\dot{\theta}\bar{e}_\theta + \dot{r}\bar{e}_r \quad \text{or} \quad \bar{v} = v_\theta\bar{e}_\theta + v_r\bar{e}_r \tag{7.31}$$

By differentiation

$$\bar{a} = r\dot{\theta}\left(-\dot{\theta}\bar{e}_r\right) + \bar{e}_\theta\left(r\ddot{\theta} + \dot{\theta}\dot{r}\right) + \dot{r}\dot{\theta}\bar{e}_\theta + \bar{e}_r\ddot{r}$$

$$\bar{a} = \left(\ddot{r} - r\dot{\theta}^2\right)\bar{e}_r + \left(r\ddot{\theta} + 2\dot{r}\dot{\theta}\right)\bar{e}_\theta \tag{7.32}$$

Consider the mechanism shown in Figure 7.9 where the collar is constrained to follow along a path defined by

$$r = Ae^{a\theta} \tag{7.33}$$

By differentiation:

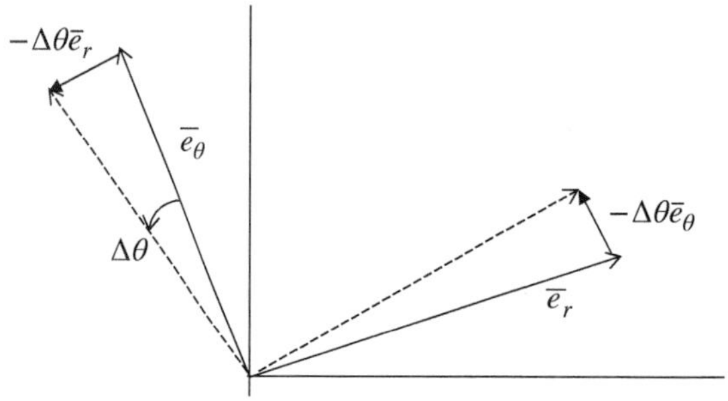

Figure 7.8 Differential change in unit vectors.

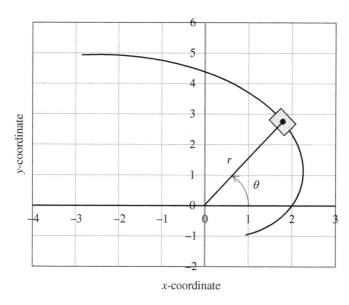

Figure 7.9 Exponential cam drive.

$$\dot{r} = A\, a\, e^{a\theta}\, \dot{\theta} \tag{7.34}$$

$$\ddot{r} = Aa\left(e^{a\theta}\, \ddot{\theta} + \dot{\theta}ae^{a\theta}\right) \tag{7.35}$$

Assuming r and θ are related by

$$r = 2e^{0.1\theta}$$

and the arm rotates with constant angular velocity $\dot{\theta} = 2$ rad/s. The problem is to determine the velocity vector and acceleration vector of the collar for $\theta = 60°$ or 1.047 rad.

For this case,

$A = 2$ in.

$a = 0.1$

$r = 2.22$ in.

$$v_r = \dot{r} = Aae^{a\theta} \; \dot{\theta} = 0.4442 \text{ in./s}$$
$$\ddot{r} = Aa(\dot{\theta}ae^{a\theta}) = Aa^2\dot{\theta}e^{0.1\theta} = 0.0444 \text{ rad/s}^2$$
$$v_\theta = r\dot{\theta} = 2.22(2) = 4.44 \text{ in./s}^2$$

Velocity Vector

Applying Eq. (7.41) gives

$$\bar{v} = v_\theta \bar{e}_\theta + v_r \bar{e}_r$$
$$\bar{v} = 4.44\bar{e}_\theta + 0.444\bar{e}_r$$

Acceleration Vector

Applying Eq. (7.42) gives

$$\bar{a} = \left(\ddot{r} - r\dot{\theta}^2\right)\bar{e}_r + \left(2\dot{r}\dot{\theta}\right)\bar{e}_\theta$$
$$\bar{a} = \left(0.0444 - 2.22(2)^2\right)\bar{e}_r + \left(2(0.444)2\right)\bar{e}_\theta$$
$$\bar{a} = -8.835\bar{e}_r + 1.776\bar{e}_\theta$$

Curvilinear Coordinates

In this case, a particle moves along a prescribed path, $y(x)$ (Figure 7.10). Its location along this path is defined by the arc distance, $s(t)$. Tangent velocity is determined by $\dfrac{ds}{dt}$. Tangent acceleration is determined by $\dfrac{d^2s}{dt^2}$. The total acceleration has a normal component, too, as explained below:

$$\bar{v} = v\bar{e}_t \tag{7.36}$$
$$\bar{a} = \frac{d\bar{v}}{dt} = v\frac{d\bar{e}_T}{dt} + \frac{dv}{dt}\bar{e}_T \tag{7.37}$$

but

$$\Delta\bar{e}_T = \frac{\Delta s}{\rho}\bar{e}_N \tag{7.38}$$

$$\frac{d\bar{e}_T}{dt} = \frac{v}{\rho}\bar{e}_N \tag{7.39}$$

so

Figure 7.10 Curvilinear motion.

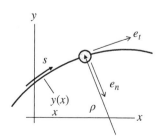

$$\bar{a} = \frac{v^2}{\rho}\bar{e}_N + a\bar{e}_T \tag{7.40}$$

where

$$\frac{1}{\rho} = \frac{\dfrac{d^2y}{dx^2}}{\left[1 + \left(\dfrac{dy}{dx}\right)^2\right]^{\frac{3}{2}}} \quad \text{(curvature)} \tag{7.41}$$

Example   Consider the acceleration components of a point moving along a path defined by

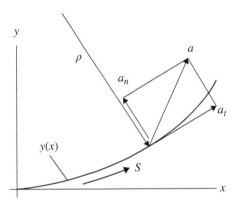

$$y = 3x^2 - 2x$$

at coordinates $(x = 2, y = 8)$. Assuming the velocity and acceleration of a moving point at this coordinate are $\dot{s} = 30$ fps and $\ddot{s} = 20$ fps^2, its acceleration components are

$$a_N = \frac{v^2}{\rho} = 0.0059(30^2) = 5.32 \text{ fps}^2$$

where

$$\frac{1}{\rho} = \frac{6}{[1 + 10^2]^{\frac{3}{2}}} = 0.0059$$

The tangent component of acceleration is given as

$$a_t = 20 \text{ fps}^2$$

The direction of the normal acceleration component is always toward the center of curvature of the path.

Figure 7.11 gives another two-dimensional space path, which could represent the terrain on a hill or roller coaster. The path is represented by one half of a sin wave:

$$y(x) = H \sin \frac{\pi x}{L}$$

Figure 7.11 Sinusoidal ramp.

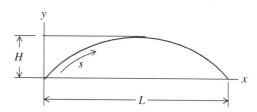

Assuming a constant velocity of 60 mph (88 ft/s), the acceleration at the top of the hill ($x = L/2$) would be

$$a_n = \frac{v^2}{R}$$

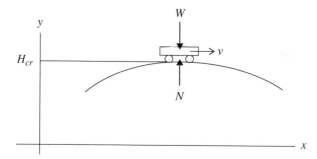

The radius of curvature of the road at this point is

$$\left.\frac{dy}{dx}\right|_{x=\frac{L}{2}} = H\frac{\pi}{L}\cos\frac{\pi}{2} = \frac{\pi H}{L}(0) = 0$$

$$\left.\frac{d^2y}{dx^2}\right|_{x=\frac{L}{2}} = -H\left(\frac{\pi}{L}\right)^2\sin\frac{\pi}{2} = -H\left(\frac{\pi}{L}\right)^2$$

Therefore, from Eq. (7.52)

$$\frac{1}{R} = -H\left(\frac{\pi}{L}\right)^2$$

The minus sign means that the $y(x)$ at $L/2$ is cupping down away from the $+y$ axis. The acceleration of a vehicle at the top of the hill is (note $a_t = 0$)

$$a = a_n = \frac{v^2}{R} = Hv^2\left(\frac{\pi}{L}\right)^2 \text{ fps}^2$$

Now consider the contact force between the tires and road of a 2000 lb vehicle going over the peak point at 60 mph.

Applying the second law,

$$W - N = ma$$

Solving for N gives

$$N = W - \frac{W}{g}\left[Hv^2\left(\frac{\pi}{L}\right)^2\right]$$

An interesting situation occurs when $N = 0$ or the wheels start to leave the road. That condition is reached when

$$\left[Hv^2\left(\frac{\pi}{L}\right)^2\right] = g$$

Assuming a wavelength $L = 1000$ ft and vehicle velocity is 60 mph (88 fps), the critical rise (H_{cr}) in the hill is

$$H_{cr}(88)^2\left(\frac{\pi}{1000}\right)^2 = 32.2 \text{ ft/s}^2$$

$$H_{cr} = 421.3 \text{ ft}$$

Under this condition, passengers would also feel no force with the seat and a sense of zero gravity.

Navigating in Geospace

The early 1960s marked the beginning of major research efforts to improve drilling technology. A visionary goal was to navigate drill bits along a specified path and intersect a given target at any subterranean location. Also, during this time, major efforts were under way to explore outer space requiring space capsules to be placed in specified orbits or intersect specified space targets. Requirements placed on directional drilling and space travel are ironically similar. Both require engines of thrust, navigation, and the ability to make course changes.

Even though there are similarities, navigating through geospace offers the challenge of unpredictable forces, while space travel navigates through a well-defined variable gravity field and environmental drag. Both technologies have matured so that trajectories and targets are achievable with high degrees of accuracy and reliability.

The key to drill bit navigation is the ability to monitor drill bit parameters, while drilling several thousand feet into the earth, allowing well path monitoring and control. Accurately navigating drill bit in geospace is vital in developing complex oil and gas reservoir structures and for capping dangerous fluid eruptions. Multiple directional production wells are often drilled from one offshore platform.

Tracking Progress Along a Well Path

The path into the earth's crust that drill bits make is calculated incrementally. Three parameters are needed to determine incremental steps of the path, Δx, Δy, Δz. They are (i) well bore inclination, which is determined by a plumb line, (ii) azimuth of the tangent, which is determined by a compass, and (iii) incremental distance drilled, which is determined by drill pipe movement down the well bore.

The tangent to the well path at any location is defined by

$$\bar{a} = \sin\alpha\sin\varepsilon\bar{i} + \sin\alpha\cos\varepsilon\bar{j} - \cos\alpha\bar{k} \tag{7.42}$$

This unit vector is referenced from the right hand x, y, z axes as shown in Figure 7.12.

Three methods have been developed for determining well path shape while drilling. They are

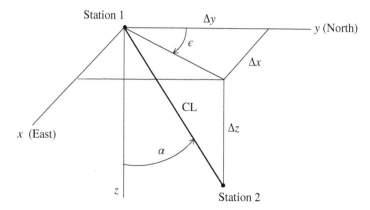

Figure 7.12 Incremental steps between two survey stations.

1) Average angle method
2) Radius of curvature method
3) Minimum curvature method

Only the minimum curvature method is discussed here.

Minimum Curvature Method

This method is also called the circular arc method because the well path between stations 1 and 2 is assumed to be an arc of a circle. The arc, in general, lies in a space plane (Figure 7.13) which is defined by unit vectors tangent to the well path at survey stations 1 and 2.

The two unit vectors are defined in terms of hole angle (α) and direction (ε):

$$\bar{a}_1 = \sin\alpha_1 \sin\varepsilon_1 i + \sin\alpha_1 \cos\varepsilon_1 j + \cos\alpha_1 k \tag{7.43}$$

$$\bar{a}_2 = \sin\alpha_2 \sin\varepsilon_2 i + \sin\alpha_2 \cos\varepsilon_2 j + \cos\alpha_2 k \tag{7.44}$$

Arc distance between stations 1 and 2 is the course length, CL, and can be related to radius of curvature of the arc by

$$\cos\beta = \bar{a}_1 \cdot \bar{a}_2 \quad \text{(vector dot product)} \tag{7.45}$$

Since

$$R\beta = CL = s \tag{7.46}$$

$$R = \frac{CL}{\beta°} \frac{180}{\pi} \tag{7.47}$$

The latitude, departure, and vertical distances between stations 1 and 2 are the components of vector, $\bar{B}$. This vector is

$$\bar{B} = \bar{A}_1 + \bar{A}_2 \tag{7.48}$$

where

$$\bar{A}_1 = R\tan\frac{\beta}{2}\bar{a}_1 \tag{7.49}$$

$$\bar{A}_2 = R\tan\frac{\beta}{2}\bar{a}_2 \tag{7.50}$$

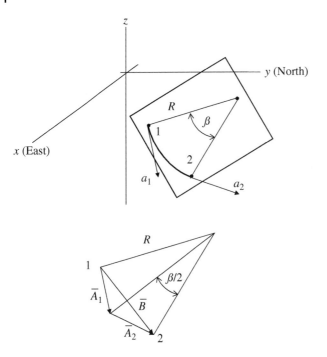

Figure 7.13 Variables used in minimum curvature method.

Example As an example, consider the survey data as follows:

$$\alpha_1 = 40° \quad \alpha_2 = 38°$$
$$\varepsilon_1 = 25° \quad \varepsilon_2 = 29°$$

It is desired to determine incremental displacements of the well bore between location 1 and 2 given the following information. Assume a drilling interval of 93 ft:

$$\bar{a}_1 = \sin 40 \sin 25i + \sin 40 \cos 25j - \cos 40k$$
$$\bar{a}_1 = \sin 39 \sin 29i + \sin 38 \cos 29j - \cos 38k$$

$$\bar{a}_1 = 0.2717i + 0.5826j - 0.766k$$
$$\bar{a}_2 = 0.2985i + 0.5385j - 0.788k$$

$$\cos \beta = 0.9984$$
$$\beta = 3.2394°$$
$$R = \frac{93}{3.2394} \frac{180}{\pi} = 1645 \text{ ft}$$

$$\bar{A}_1 = 1645(0.0283)\bar{a}_1 = 46.55a_1$$
$$\bar{A}_1 = 46.55(0.2717i + 0.5826j - 0.766k)$$

$$\bar{A}_2 = 1645(0.0283)\bar{a}_2 = 46.55a_2$$
$$\bar{A}_2 = 46.55(0.2985i + 0.5385j - 0.788k)$$
$$\bar{B} = 46.55(0.5702i + 1.121j - 1.554k)$$
$$\bar{B} = 26.54\hat{i} + 52.18\hat{j} - 72.34\hat{k}$$

The components of vector $\overline{B}$ define the incremental steps:

$$\Delta x = 26.54 \text{ ft}$$

$$\Delta y = 52.18 \text{ ft}$$

$$\Delta z = -72.34 \text{ ft}$$

The severity of the curvature of this section of the well bore is

$$\text{Deg/ft} = \frac{3.239°}{93 \text{ ft}} = 0.0348°/\text{ft} = 3.48°/100 \text{ ft}$$

This is considered a high curvature.

Dogleg Severity

Dogleg severity (DLS) is a term used to quantify the severity of well path curvature, either intentional or unintentional. DLS affects bending in drilling tools and casing as well as drill string friction, applied weight-on-bit (WOB), and fatigue. The rate of change of hole angle with respect to distance drilled is (see Figure 7.14):

$$s = \beta R \tag{7.51}$$

$$s = \beta° \frac{\pi}{180} R \tag{7.52}$$

$$\frac{\beta°}{s} = \frac{180}{\pi R} \quad (°/\text{ft}) \tag{7.53}$$

DLS is defined as the change in hole angle over a given drilled length. It is expressed in degrees/100 ft of drilled hole:

$$DLS = \frac{\beta°}{s} 100 \tag{7.54}$$

By substitution

$$DLS = \frac{18\,000}{\pi R} \quad (°/100 \text{ ft}) \tag{7.55}$$

DLS defines the curvature of the well bore and affects bending of all tubulars passing through the dogleg.

Vectors which define DLS are, in general, 3-D vectors. Once the angle, β, has been established between two survey stations:

$$R = \frac{CL}{\beta(\text{rad})} \tag{7.56}$$

and

$$\beta(\text{rad}) = \frac{(CL)DLS}{100} \left| \frac{\pi}{180} \right| \tag{7.57}$$

$$R = \frac{18\,000}{\pi} \frac{1}{DLS} \text{ ft} \tag{7.58}$$

For example

Figure 7.14 Dogleg severity.

$$R = 1500 \text{ ft} \quad (DLS = 3.8°/100 \text{ ft})$$
$$R = 150 \text{ ft} \quad (DLS = 3.8°/100 \text{ ft})$$

The vector method provides a mathematical basis for calculating DLS in 3D space. The DLS based on the minimum curvature method (previous example) is

$$DLS = \frac{18\ 000}{\pi} \frac{1}{1645} = 3.48°/100 \text{ ft}$$

Projecting Ahead

The change in well bore inclination or direction required to turn a well toward a target cannot be achieved instantaneously. These changes have to be achieved over an interval of drilling while staying within a specified maximum dogleg severity. The actual angle change will be greater than the estimated angle because of the footage drilled continually shortens the distance to the target. It is therefore important to make corrections as early as possible.

Each survey station, in a way, is a new kick-off point. If the last survey station has a unit target vector, $\bar{a}$, then it is essential to be able to turn the hole in the plane of drilling toward the target and stay within maximum dogleg severity limits. The situation depicted in Figure 7.15 shows that the target can still be reached with a correction run having a radius of curvature, R. However, the maximum DLS must not the exceeded.

If drilling continues along direction, $\bar{a}$, a greater dogleg will be required to turn the hole toward the target. However, if drilling continues along direction $\bar{a}$ and the hole reaches point m, the hole cannot be turned toward the target. Point "m" is a critical point and prior to reaching this critical point, the well would have to be plugged back and sidetracked.

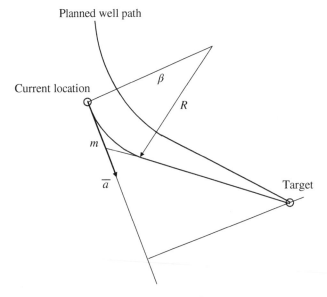

Figure 7.15 Needed correction to reach target.

Kinematics of Rigid Bodies

Kinematics analysis of rigid bodies gives acceleration of the center of gravity and angular acceleration. These two accelerations directly affect magnitude and direction of forces at connection joints.
  Rigid bodies can experience three types of motion:

1) Translation – Any line within the body remains parallel.
2) Rotation – Every point in the body describes a circle.
3) General – Combination of translation and rotation.

  The problem in each case is to determine the velocity and acceleration of any point in a rigid body compatible with physical constraints. Knowing the velocity and acceleration of one point in a rigid body along with angular velocity and angular accelerations is enough information to determine the velocity and acceleration of any other point in the body.

Rigid Body Translation and Rotation

An example of all three types of motions in one system is the slider crank mechanism. Another example is illustrated in the mechanism shown in Figure 7.16. A spring-mass contains a disc which can be rigidly attached to the bar or it can be pinned at the bar attachment. In either case, the equation of motion is

$$I_0\ddot{\theta} + a^2 k\theta = 0 \tag{7.59}$$

  The natural circular frequency can be pulled directly from this equation, but question pertains to the mass moment of inertia with respect to the pivot point, O for the two cases.

Case #1 – If the disc is *pinned* to the bar, the disc *translates* as a rigid body and its mass can be viewed as a discrete mass of M. In this case,

$$I_0 = a^2(m + M) \tag{7.60}$$

Case #2 – If the disc is *fixed* to the bar, the disc *rotates* about fixed point, O, and the moment of inertia is

$$I_O = a^2 m + \left[a^2 M + \frac{r^2}{2} M \right] \tag{7.61}$$

  There can be a significant difference between the two natural frequencies.

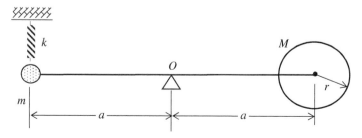

Figure 7.16 Mass moment of inertia with disc.

Case #3 – If the pivot support is replaced with a spring, the system vibrates at two degrees of freedom system with general plane motion.

General Plane Motion

Another example of general plane motion is Figure 7.17. The velocity of points A and B is known and has horizontal directions. Accelerations of both points are assumed to be zero. The problem here is to determine the angular velocity, ω, of the wheel. The relative velocity equation given earlier is one approach. The instantaneous center method is used below.

The instantaneous center (IC) is a point at which the velocity of zero. Linear velocities at any point in the body can be determined by multiplying radial distance by angular velocity. Angular velocity can be determined by dividing linear velocity by radial distance.

The governing equations in this example are

$$v_A = (d + r_A)\omega \tag{7.62}$$

$$v_B = (r_B - d)\omega \tag{7.63}$$

Example For example, assume

$$r_B = 6 \text{ in.}$$
$$r_A = 4 \text{ in.}$$
$$V_A = 5 \text{ fps}$$
$$V_B = 2 \text{ fps}$$

Then

$$\frac{V_A}{V_B} = \frac{d + r_A}{r_B - d} \tag{7.64}$$

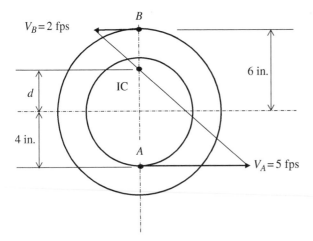

Figure 7.17 Use of instantaneous centers.

$$\frac{5}{2} = \frac{d+4}{6-d}$$

$$5(6-d) = 2(d+4)$$

$$d = 3.14 \text{ in.}$$

$$\omega = \frac{5 \text{ fps}}{7.14 \text{ in.}} \left| \frac{12 \text{ in.}}{1 \text{ ft}} \right| = 8.4 \text{ rad/s} \quad (CCW) \tag{7.65}$$

Dynamics of Particles

Dynamics refers to the motion of particles and rigid bodies when an imbalance of forces is applied to them. The engineering problem is one of finding the resulting motion caused by an imbalance of forces. The subject of dynamics is usually separated into kinetics and kinematics. Kinetics deals with motion response caused by a force imbalance and is based on Newtonian physics, $F = ma$. Kinematics is a study of motions without regard to forces. Kinetics is covered in this section.

Units of Measure

Two units of measure used in mechanics are Systems International (SI) and English System. Both are used in nearly every industry around the world. There is a move toward the SI system internationally because its units are subdivided into tenths, thousands, micro, etc. Within each system, there are base units and derived units. The derived unit in each case is based on Newton's second law. Base units for both systems are given in Table 7.1.

Force (Newton) in the SI system is the derived unit, the rest are base units. Mass (slugs) in the English system is the derived unit, and the rest are base units. Once the base units are arbitrarily set, corresponding derived units must satisfy the second law, $F = ma$. For example

$$F(\text{N}) = M(\text{kg})a(\text{m/s}^2) \quad (\text{SI})$$

Therefore, force (N) has units of kg(m)/s². Similarly, mass is the derived unit in the English system:

$$F(\text{lb}) = m(\text{slugs})a(\text{ft/s}^2)$$

Therefore, mass (slugs) has units of lb(s²)/ft.

The acceleration at sea level due to gravity in each system as determined from Newton's gravitation law is

$$g = 32.2 \text{ ft/s}^2 \quad (\text{English})$$

Table 7.1 Comparison of units of measure.

	Force	Mass	Length	Time
SI	N[a]	kg	meter	t, s
English	lb	slug[a]	feet	t, s

[a] Refers to derived units, the others are base units.

$$g = 9.81 \, \text{m/s}^2 \quad (\text{SI})$$

International Bureau of Weights and Measures in Sevres, France defines the standard kilogram by the mass of platinum volume having a diameter of 39 mm and a length of 39 mm:

1 lb – weight of 0.4536 kg at sea level

1 lb – 4.4482 N

Application of Newton's Second Law

The simplest application of Newton's law is a particle in free fall:

$$W = Mg \tag{7.66}$$

Mass of an object is constant in any gravity field. Its weight, however, depends on local gravity. When mass and g (at a given location in space) are known, then weight can be determined from this equation. On the other hand, when weight is known near the surface of the earth, its mass can also be determined from Eq. (7.66).

In any design, dynamic forces should be considered along with static forces. Total loads and stresses are affected by both. Consider the two masses (Figure 7.18) connected by a cable and pulley. Here, we need to know the total force in the cable in order to determine its required size and strength.

Static Analysis

A static analysis establishes whether block A slides (break friction) or not. It will also provide a reference load in the cable to compare with the predicted dynamic force.

Example   Consider block A

$$T = f + 100 \sin 30$$

$$N = 100 \cos 30 = 86.6 \, \text{lb}$$

The friction force is

$$f = N\mu = 86.6(0.2) = 17.32 \, \text{lb} \quad (\text{assuming } \mu = 0.2)$$

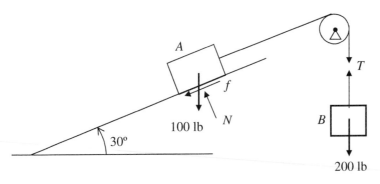

Figure 7.18   Internal dynamic loads.

Static tension in the cable required to break friction is

$$T = 17.32 + 50 = 67.32\,\text{lb}$$

Under static conditions, a 200 lb tension in the cable is enough to break friction and cause both masses to move and accelerate.

Dynamic Analysis

Freebody of mass A Freebody of mass B

$$M_A a = T - f - 100\sin 30 \quad M_B a = 200 - T \tag{7.67}$$

$$\frac{100}{g} a = T - f - 50 \qquad\qquad \frac{200}{g} a = 200 - T \tag{7.68}$$

Also

$$N = 100\cos 30$$

$$N = 86.6\,\text{lb}$$

$$f = \mu N = 0.2(86.6) = 17.32$$

Adding both equations:

$$\frac{300}{32.2} a = 150 - 17.32$$

$$a = 132.68\frac{32.2}{300} = 14.24\,\text{ft/s}^2$$

Tension force in the cable is

$$T = 200 - \frac{200}{32.2} 14.24 = 111.55\,\text{lb}$$

The dynamic tension (111.55 lb) is much greater than the static tension (67.32 lb) required to break friction.

Work and Kinetic Energy

Energy is a useful tool in engineering analysis. Energy is not a vector, so we wind-up with one energy equation accounting for all mechanical work and changes in kinetic energy over a given travel distance. The basic formula stems from Newton's second law. Starting with

$$F = m\frac{d^2x}{dt^2} = m\frac{d}{dt}\left(\frac{dx}{dt}\right)\frac{dx}{dx} \tag{7.69}$$

$$\int_1^2 F\,dx = m\int_1^2 v\,dv \tag{7.70}$$

The left integral represents the work done by force F between location 1 and location 2. The right integral represents the change in kinetic energy between the two points:

$$\text{Work} = \frac{1}{2}m\left[V_2^2 - V_1^2\right] \tag{7.71}$$

While the force does work on a mass to change its kinetic energy, the reverse is also true. The kinetic energy of a mass can do work on something:

$$\text{Work} = KE_2 - KE_1 \tag{7.72}$$

Work can be performed by gravity, applied force, friction, and springs.

The arrangement of Figure 7.19 shows a mass being pulled downward by gravity toward a spring of constant, k. The problem here is to determine the maximum compression (δ_{max}) of the spring when the mass moves into it and is stopped. The work of friction must be considered, as well as gravity and the elastic compression of the spring. The kinetic energies at locations 1 and 2 are zero so according to Eq. (7.72), total work on the system during the travel between locations 1 and 2 is zero:

$$\text{Work} = W(l + \delta_{max})\sin\theta - f(l + \delta_{max}) - \frac{1}{2}k\delta_{max}^2 = 0 \tag{7.73}$$

$$\frac{1}{2}k\delta_{max}^2 - \delta_{max}(W\sin\theta - f) - (W\sin\theta - f)l = 0 \tag{7.74}$$

Example Assuming

$W = 100 \, \text{lb}$

$k = 500 \, \text{lb/in.}$

$l = 15 \, \text{in.}$

$\theta = 45°$

$\mu = 0.1$

The friction force, f is

$$f = \mu N = 0.1(W\cos\theta) = 0.1(100\cos 45) = 7.07 \, \text{lb}$$

The quadratic equation then becomes

$$250\delta_{max}^2 - 63.64\delta_{max} - 954.6 = 0$$

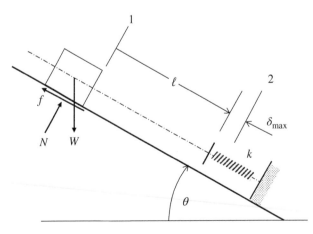

Figure 7.19 Work and kinetic energy.

The solution is

$$\delta_{\max} = \frac{63.64 \pm \sqrt{63.64^2 - 4(250)(-954)}}{2(250)} = 2.08 \text{ in. (minus solution is trivial)}$$

Ignoring the middle term gives $\delta_{\max} = 1.95$ in.

Potential Energy

Potential energy is also a useful concept. Potential energy may be stored in mechanical springs or related to elevation of mass. Loss in potential represents work done by the potential energy source. Stated mathematically

$$\text{Work (by the potential energy source)} = PE_1 - PE_2 \tag{7.75}$$

For example, assuming the elevation of a mass (Figure 7.19) is measured by distance, y, from a datum line. The work done by gravity is

$$W_{gravity} = PE_1 - PE_2 = W(y_1 - y_2) \tag{7.76}$$

where $y_1 \succ y_2$. By substitution into Eq. (7.72):

$$PE_1 - PE_2 = KE_2 - KE_1 \tag{7.77}$$

giving

$$PE_1 + KE_1 = PE_2 + KE_2 \tag{7.78}$$

which shows that total energy (potential plus kinetic) is conserved.

Next consider simple harmonic motion of a spring-mass system (right drawing, Figure 7.20). Assuming at position 1, the spring is not stretched, then the potential energy at position 1 is

$$PE_1 = Wy_1$$

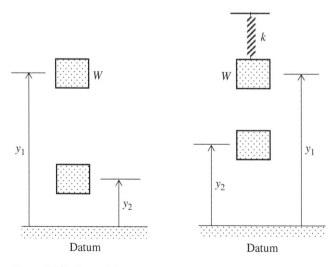

Datum   Datum

Figure 7.20 Potential energy.

while the potential energy at position 2 is

$$PE_2 = Wy_2 + \frac{1}{2}k(y_1 - y_2)^2$$

Still, Eq. (7.78) applies, allowing the velocity of the mass at position 2 to be determined.

The advantage of the conservation of energy is that we track energy, a scalar quantity, and not force vectors. One energy equation accommodates only one unknown, however. For example, assume we wish to fine the velocity of the mass at any position y_2 starting with zero velocity at y_1. From Eq. (7.78)

$$Wy_1 + 0 = Wy_2 + \frac{1}{2}k(y_1 - y_2)^2 + \frac{1}{2}mV_2^2 \tag{7.79}$$

$$W(y_1 - y_2) - \frac{1}{2}k(y_1 - y_2)^2 = \frac{1}{2}\frac{W}{g}V_2^2 \tag{7.80}$$

$$V_2^2 = 2g(y_1 - y_2) - \frac{k}{m}(y_1 - y_2)^2 \tag{7.81}$$

No vectors are involved.

Example Now consider the application of a safety belt tethered to a steel structure and attached to a harness worn by a steel worker. Assume the belt is nylon 5 ft long and has a spring rate of 50 lb/in. We examine the impact force applied to a 200 lb worker resulting from an accidental fall.

The change in potential from location 1 to the end and extension of the safely belt (location 2) is 60 in. plus δ, the stretch of the belt. Since the KE and both location is zero then the potential energy at both locations must be the same:

$$PE_1 = PE_2$$

where

$$PE_1 = Wy_1 \tag{7.82}$$

$$PE_2 = Wy_2 + \frac{1}{2}k\delta^2 \tag{7.83}$$

Equating the two equations and noting that $(y_1 - y_2) = 60 + \delta$ ft gives

$$200(60 + \delta) = \frac{1}{2}50\delta^2$$

$$400(60 + \delta) = 50\delta^2$$

$$\delta^2 - 8\delta - 480 = 0 \tag{7.84}$$

$$\delta = \frac{-b \pm \sqrt{b^2 - 4ac}}{2a} \tag{7.85}$$

$$\delta = 26.27 \text{ in.}$$

Force applied to the worker by the safety belt is

$$F = k\delta = 50(26.27) = 1314 \text{ lb} \quad \text{(a very large and perhaps harmful force)}$$

Drill Bit Nozzle Selection

One of the most useful applications of conservation of energy is in fluid flow. Bernoulli's equation, which is based on the conservation of energy, accounts for energy going in and out of a control volume. Consider the fluid flow of drilling fluid flow through drill bit nozzles. The practical drilling problem is to determine nozzles (usually three nozzles) that would convert available hydraulic horsepower into maximum kinetic energy for cleaning cuttings from under a drill bit. The total cross-sectional area of flow is called total flow area or TFA.

Bit pressure drop depends on TFA of all nozzles in a drill bit. Applying Bernoulli's energy equation between sections 1 and 2 (see Figure 7.21) gives

$$\frac{p_1}{\gamma} + \frac{V_1^2}{2g} = \frac{p_2}{\gamma} + \frac{V_2^2}{2g} \tag{7.86}$$

Since V_1 is small by comparison with V_2

$$p_1 - p_2 = \Delta p = \frac{\gamma}{2g} V^2 \quad (V_2 \text{ is replaced by } V) \tag{7.87}$$

In oil field units

$$\Delta p = \frac{7.48\gamma}{2(32.2)(144)} V^2 \tag{7.88}$$

$$\Delta p = \frac{\gamma V^2}{1240} \tag{7.89}$$

where

Δp – psi
V – fps (note the unit change)
γ – ppg

Applying a nozzle coefficient of 0.95 [2] gives

$$\Delta p = \frac{\gamma V^2}{1120} \tag{7.90}$$

The continuity flow equation is

$$V = \frac{0.32Q}{A_n} \text{ fps} \tag{7.91}$$

Figure 7.21 Fluid flow through bit nozzles.

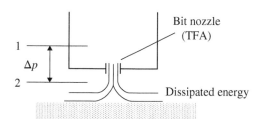

By substitution

$$\Delta p = \frac{\gamma Q^2}{10\,938 A_n{}^2} \tag{7.92}$$

where

Δp – psi (scales directly to mud weight)
Q – gpm
γ – ppg (pressure scales directly with γ)
A_n – total nozzle area (TFA), in.^{2}

which predicts pressure drop (Δp_{bit}) across bit nozzles.

Consider, for example, the pressure drop across three (3), $^{11}\!/_{32}$ nozzles ($TFA = 0.2784$ in.^{2}) is $\Delta p_{bit} = 1887$ psi assuming mud weight of $\gamma = 10$ ppg and a flow rate of 400 gpm. It is desirable to use up available horsepower for best bottom hole cleaning. This is accomplished by selecting the best flow rate and then solving for total nozzle flow area.

Dividing bit hydraulic horsepower $\left(HHP = \dfrac{\Delta p Q}{1714} \right)$ by the cross-sectional area of the hole gives hydraulic horsepower per square inch of hole or HSI. An HSI of 5–7 is considered a high level of hydraulic horsepower across roller cone drill bits.

Impulse–Momentum

The principle of impulse–momentum is derived directly from Newton's second law. Assuming linear motion

$$F = m\frac{d}{dt}\frac{dx}{dt} = m\frac{dv}{dt} \tag{7.93}$$

$$\int_1^2 F\,dt = \int_1^2 m\,dv \tag{7.94}$$

After integration

$$\int_1^2 F\,dt = m(V_2 - V_1)$$

$$Im = mV_2 - mV_1 \tag{7.95}$$

$$Im = H_2 - H_1 \tag{7.96}$$

The letter "H" represents linear momentum.

The impulse–momentum method is useful when the applied force is known as a function of time. A good example is the take-off and landing of aircraft. In this case, the thrust capability of jet engines is a better parameter than a horsepower rating. Impulse and momentum are vector quantities and are related as follows (Figure 7.22).

Figure 7.22 Linear impulse and momentum.

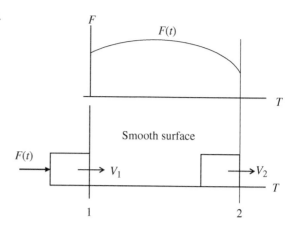

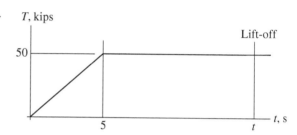

Figure 7.23 Thrust from gas turbine.

For example, assume a lift-off requirement of an airplane is 150 mph. Weight of airplane = 100 kip. Maximum engine thrust capability is 50 000 lb. Assume total engine thrust at take-off varies according to Figure 7.23.

Impulse in this example is the area under the triangle and rectangle. Since 60 mph = 88 fps

$$\frac{1}{2}5(50) + (t-5)50 = \frac{100}{g}\left[\left(150\frac{88}{60} - 0\right)\right]$$

$$t = 16.17 \text{ seconds}$$

The airplane would achieve the 150-mph lift-off speed in 16.17 seconds.

Remember, impulse and momentum are vector quantities, so in planar motion, both may have x and y components. Jet and rocket engines are rated in terms of thrust, which is easy to convert into changes in momentum and velocity.

Impulse–Momentum Applied to a System of Particles

Linear momentum is useful for determining the total force required to change direction of fluid flow by a diverter, such as a turbine blade. The general discussion is made by considering flow within a control volume (Figure 7.24), which shows a fluid plug ($A_1\Delta s_1$) moving into the control volume past section 1.

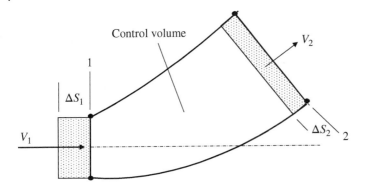

Figure 7.24 Resultant force required to deflect fluid.

The same amount of fluid ($A_2 \Delta s_2$) exits the control volume past section 2. The sum of total change in momentum of all fluid particles internal to the control volume is unchanged except for volume of fluid entering and leaving the control volume:

$$\text{Fluid momentum in} = \overline{V}_1 \Delta s_1 A_1 \rho$$

$$\text{Fluid momentum out} = \overline{V}_2 \Delta s_2 A_2 \rho$$

Applying impulse–momentum principles to the system of fluid particles gives

$$\sum (\overline{F}_i + \overline{f}_i) \Delta t = \overline{V}_2 \Delta s_2 A_2 \rho_2 - \overline{V}_1 \Delta s_1 A_1 \rho_1 \qquad (7.97)$$

Internal forces, $\overline{f}_i$ between elemental masses cancel because they exist in equal and opposite pairs leaving only the sum of the external forces. Also, to satisfy continuity of flow

$$\frac{\Delta s_2}{\Delta t} A_2 \rho = \frac{\Delta s_1}{\Delta t} A_1 \rho = Q_m$$

Therefore,

$$\sum \overline{F}_i = \overline{F} = Q_m (\overline{V}_2 - \overline{V}_1)$$

This force represents the total force applied to the control volume to bring about the change in momentum between sections 2 and 1. According to Newton's third law, there is an equal reaction (R) on the engine frame and airplane:

$$\overline{R} = Q_m (\overline{V}_2 - \overline{V}_1) \qquad (7.98)$$

where Q_m is mass flow rate through the control-volume. $\overline{V}_1$ and $\overline{V}_2$ are absolute fluid velocities (vectors) into and out of the control volume.

If the control volume moves with a given velocity, the reaction equation remains the same keeping in mind Q_m is mass flow rate through the control volume, while $\overline{V}_1$ and $\overline{V}_2$ are absolute velocities of the fluid entering and leaving the control volume. A major application of this theory is the prediction of torque and power generated by fluid across turbine blades.

Mechanics of Hydraulic Turbines

Downhole turbines were introduced into drilling programs before PDMs. They generate power at a much higher rotary speed. Turbine speed is approximately double that of PDMs. Assuming an output speed of 800 rpm and bit torque of 1500 ft-lb

$$P_{bit} = \frac{2\pi(1500)800}{33\,000} = 228.4\,\text{hp}$$

a factor of 8 increase over normal rotary speeds of 100 rpm. However, maximum power of drilling turbines is developed at one rotational speed, and this optimum speed is sometimes difficult to monitor and control.

Downhole drilling turbines develop torque fundamentally differently from positive displacement motors (PDMs). Torque is developed by fluid momentum, while torque in a PDM is developed by direct application of fluid pressure. Output speed of a turbine depends directly on output torque, while output speeds of PDMs depend on flow rate and is independent of torque. Maximum power is developed at a particular rotor speed and herein lies a monitor/control issue. Monitoring and operating turbines at their optimum power speed is necessary for best use of drilling turbines. Since turbine torque is the result of fluid momentum, torque output depends on both drilling mud density and flow rate.

Each turbine stage has a fixed set of blades (stator) and a rotating set of blades (rotor). Stator blades direct drilling fluid onto rotor blades where output torque is generated. Output torque, output power, and overall pressure drop of a turbine is a direct multiple of the number of stages in the turbine.

When the rotor blade is stationary, fluid momentum changes the most, and the torque is the greatest. When the rotor blade is moving, fluid momentum is not changed as much so torque drops off. The faster the rotor turns, the lower the torque generated by the rotor blades. At a high speed of rotor rotation, fluid leaving the stator blades cannot catch the rotor blade, so its momentum does not change. This speed is the runaway speed, and at this speed, no torque is developed on the rotor blades.

In practice, turbine rotational speed is not preset; output speed responds to applied torque or bit weight. Torque is applied, and rotational speed automatically adjusts to the torque. Both stall torque and runaway speed change with mud weight and flow rate. Turbines are rated at maximum power. Whether maximum power is developed by downhole turbines depends on the ability to monitor and control bit torque.

The mechanics of turbine power is explained by considering momentum changes across the control volume, *abcd* (Figure 7.25). The rate of change of fluid momentum across *cd* and *ab* requires a force vector, $\overline{R}$.

$$\overline{R} = Q_m\left(\overline{V}_2 - \overline{V}_1\right) \tag{7.99}$$

where

$\overline{V}_1$ – absolute velocity of fluid leaving stator blade boundary *ab*
$\overline{V}_2$ – absolute velocity of fluid leaving rotor blade boundary *cd*
Q_m – mass flow rate through control volume
$\overline{R}$ – resultant force on fluid in control volume
U – tangent velocity of rotor blade (center of blade)

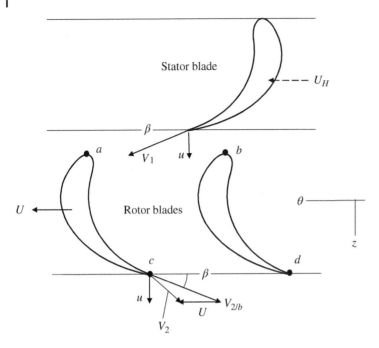

Figure 7.25 Rotor and stator turbine blades.

The component of this vector in the tangent direction is

$$R_\theta = Q_m(U - 2V \cos \beta)$$ (7.100)

or

$$R_\theta = Q_m\left(U - \frac{2u}{\tan \beta}\right)$$ (7.101)

Since this is the force acting on the fluid in the control volume, the fluid force applied to the blade is

$$F_\theta = Q_m\left(\frac{2u}{\tan \beta} - U\right)$$ (7.102)

This force produces output torque. Total torque generated by one turbine stage is

$$T = rQ_m\left(\frac{2u}{\tan \beta} - U\right)$$ (7.103)

where Q_m is now the total mass flow rate through the turbine and U is the tangent speed of the rotor and r is the average radial distance. Also

$$u = \frac{(1-\varepsilon)Q}{A\lambda} = \frac{1-\varepsilon}{\lambda}\frac{Q}{A}$$ (7.104)

where

---

u – axial velocity of fluid entering (and leaving) control volume
ε – coefficient of fluid loss around outside of rotor
λ – contraction coefficient due to flow area occupied by blade thickness
A – cross-sectional area of space occupied by rotor (or stator) blade

---

In consideration of these factors, the torque generated by one turbine stage is

$$T = r\rho Q \left[\frac{2(1-\varepsilon)Q}{A\lambda \tan \beta} - U \right] \tag{7.105}$$

where Q_m has been replaced by ρQ. U relates to rotational speed by

$$U = r\omega = r\frac{2\pi N}{60}$$

The actual rotational speed, N, depends on applied torque. In practice, rotational speed of the drill bit depends on the torque developed by the drill bit (or WOB).

From Eq. (7.105), maximum torque occurs when $U = 0$ or the rotor blade is stalled:

$$T_{\max} = r\frac{\gamma}{g} \frac{2(1-\varepsilon)Q^2}{A\lambda \tan \beta} \quad (\text{stall torque, } T_{st}) \tag{7.106}$$

On the other hand, output torque is zero when

$$U = \frac{2(1-\varepsilon)Q}{A\lambda \tan \beta} \tag{7.107}$$

This velocity is the runaway speed of the turbine. In terms of angular velocity of the turbine,

$$N_{rs} = \frac{60}{2\pi r} \frac{2(1-\varepsilon)Q}{A\lambda \tan \beta} \quad (\text{runaway speed}) \tag{7.108}$$

When the blade tip angle (β) decreases from 30° to 20°, maximum torque and runaway speed increase by a factor of 1.59.

Output mechanical horsepower is formulated in terms of rotor/stator design parameters, drilling fluid density and flow rate.

Starting with

$$T = r\frac{\gamma}{g}Q\left[\frac{2(1-\varepsilon)Q}{A\lambda \tan \beta} - U \right] \tag{7.109}$$

and

$$P_{out} = T\omega = T\frac{U}{r} \tag{7.110}$$

Power output is defined by

$$P_{out} = \frac{\gamma}{g}Q\left[\frac{2(1-\varepsilon)Q}{A\lambda \tan \beta} - U \right]U \tag{7.111}$$

Maximum power occurs at

$$\frac{d(P_{out})}{dU} = \frac{2(1-\varepsilon)Q}{A\lambda \tan\beta} - 2U = 0 \qquad (7.112)$$

or

$$U = \frac{(1-\varepsilon)Q}{A\lambda \tan\beta} \quad (1/2\,\text{runaway speed}) \qquad (7.113)$$

which gives

$$P_{max} = \frac{\gamma}{g}\left[\frac{(1-\varepsilon)}{A\lambda \tan\beta}\right]^2 Q^3 \;(\text{mechanical output}) \qquad (7.114)$$

Note that maximum power is achieved at ½ runaway speed.

Pressure drop across one turbine stage is determined by applying Bernoulli's equation to the control volume shown in Figure 7.26:

$$\frac{p_1}{\gamma} = E_{out} + \frac{p_2}{\gamma} \;\left(\text{Bernoulli, }\frac{\text{ft-lb}}{\text{lb}}\right)$$

$$P = \frac{\Delta p}{\gamma}\gamma Q \;(\text{conversion to power})$$

At maximum power output,

$$P_{max} = \Delta p Q \qquad (7.115)$$

and

$$\Delta p = \frac{\gamma}{g}\left[\frac{(1-\varepsilon)}{A\lambda \tan\beta}\right]^2 Q^2 \qquad (7.116)$$

Total pressure drop across multirotor turbines creates a downward force that must be supported by thrust bearings.

The above equations predict the performance for drilling turbines. Actual performance is established experimentally as explained earlier. Mechanical output is typically less than hydraulic power input to turbines. Efficiency of the power transformation is defined by

$$\eta = \frac{HP_{out}}{HHP_{in}} \qquad (7.117)$$

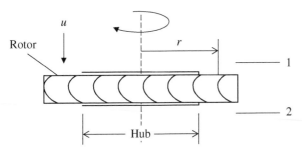

Figure 7.26 Control volume of flow through a turbine rotor.

Performance Relationships

If performance information is known for one set of operating parameters, the performance at other operating parameters can be calculated using

$$N_2 = \frac{Q_2}{Q_1} N_1 \tag{7.118}$$

$$T_2 = \left(\frac{Q_2}{Q_1}\right)^2 \left(\frac{\gamma_2}{\gamma_1}\right) T_1 \tag{7.119}$$

$$\Delta p_2 = \left(\frac{Q_2}{Q_1}\right)^2 \left(\frac{\gamma_2}{\gamma_1}\right) \Delta p_1 \tag{7.120}$$

$$HP_2 = \left(\frac{Q_2}{Q_1}\right)^3 \left(\frac{\gamma_2}{\gamma_1}\right) HP_1 \tag{7.121}$$

These formulas are based on the preceding derivation.

In many cases, performance data is given for one mud weight or a given flow rate. The above equations are useful to predict performance at different mud weights or flow rates. For example, consider the data in Table 7.2. Assume that speed ($N_1 = 1010$ rpm), torque ($T_1 = 1430$ ft lb), pressure drop ($\Delta p_1 = 1450$ psi), and output power ($HP_1 = 275$ hp) are known at a flow rate of 475 gpm. Furthermore, it is desired to determine these parameters at a flow rate of 400 gpm with density staying the same at 12 ppg. The conversion is as follows:

$$N_2 = \frac{400}{475} 958 = 807 \text{ rpm}$$

$$T_2 = \left(\frac{400}{475}\right)^2 2390 = 1695 \text{ ft-lb}$$

$$\Delta p_2 = \left(\frac{400}{475}\right)^2 2241 = 1589 \text{ psi}$$

$$HP_2 = \left(\frac{400}{475}\right)^3 436 = 260 \text{ hp} \quad \text{(exponent)}$$

These results compare exactly with the numbers in Table 7.2. Converting the data to a different mud weight is accomplished using the above equations.

Table 7.2 Performance date for 6¾ in. turbine (12 ppg).

Pump rate (gpm)	Bit speed (rpm)	Power output (hp)	Torque (ft-lb)	Pressure drop (psi)	Thrust force (1000 lb)
250	504	64	662	621	10
300	605	110	953	894	14
350	706	175	1297	1217	19
400	807	260	1695	1589	25
450	906	371	2145	2011	31
500	1009	509	2648	2483	39

Maximum Output of Drilling Turbines

The torque and rotary speed are related graphically as a straight line (Figure 7.27). Both stall torque and runaway speed change with mud weight and flow rate. Mechanical power delivered by turbines can be derived from the torque-speed line as follows:

$$T = T_{st} - \frac{T_{st}}{N_{rs}} N \qquad (7.122)$$

where

T_{st} – stall torque
N_{rs} – runaway speed

The mechanical output power is

$$P = \frac{TN}{5252} \text{ hp} \qquad (7.123)$$

When Eq. (7.149) is substituted in to this equation, the result is a parabola that reaches a maximum value at ½ stall torque and ½ runaway speed (Figure 7.27):

$$P = \frac{T_{st}}{5252} N \left(1 - \frac{N}{N_{rs}}\right) \text{ hp} \qquad (7.124)$$

This is a parabolic function that reaches a maximum value at one-half stall torque and one-half runaway speed. Because turbine speed depends on applied torque, mechanical power delivered by turbines depends on applied torque. When applied torque changes, output speed changes, and power is delivered at particular combinations of torque and speed. A means of monitoring turbine speed is necessary to make maximum use of turbine power under drilling operations.

Consider the performance data given in Table 7.2 for a 6¾ in turbine operating in 12 ppg mud. Assuming a flow rate of 400 gpm:

- Bit speed is 807 rpm
- Output torque is 1695 ft lb
- Maximum mechanical power is 260 hp
- Pressure drop across the turbine is 1589 psi

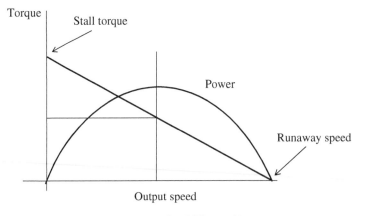

Figure 7.27 Performance curves for drilling turbines.

According to the equations given above:

- Runaway speed is 1614 rpm
- Stall torque is 3390 ft lb

The amount of hydraulic horsepower consumed under the above conditions is

$$HHP = \frac{\Delta pQ}{1714} = \frac{1589(400)}{1714} = 371 \text{ hp}$$

$$HP = \frac{TN}{5252} = \frac{1695(807)}{5252} = 260 \text{ hp}$$

Maximum efficiency of the turbine then is

$$\eta = \frac{HP}{HHP} = \frac{260}{371} = 0.7$$

or 70%. The efficiency curve is also parabolic because HHP is constant and HP is parabolic. Efficiency is maximum when mechanical power is maximum.

Dynamics of Rigid Bodies

Planar motion of rigid bodies is considered here. A rigid body can be viewed as a composite of many particles connected into one body. In this case, the equations of motion of rigid bodies are determined from the sum of the effects of all particles. Applying the second law to each particle and then summing all forces both internal and external gives

$$\sum \bar{F} = m\bar{a}_G \tag{7.125}$$

where $\sum \bar{F}$ is the vector sum of all external forces applied to the rigid body and $\bar{a}_G$ is the acceleration of the center of gravity of the rigid body. The sum of all internal forces is zero since they occur in equal and opposite pairs. This single vector equation produces two scalar equations for planar motion.

A second equation is developed from consideration of rate of change of angular momentum and moments measured about an axis (point P) perpendicular to the plane of motion (Figure 7.28).

Consider the axis passing through point P, attached to the moving body. By summing all moments applied to the body about P and equating them to all angular inertia forces gives

$$\sum M_P = I_P\alpha - \bar{y}m(a_P)_x + \bar{x}m(a_P)_y \tag{7.126}$$

This general expression reduces to

$$\sum M_P = I_P\alpha \tag{7.127}$$

when any one of the following three conditions are met.

$\bar{a}_P = 0$
$d = 0 \quad (\text{for } \bar{a}_P)$
$\bar{a}_P$ passes through the center of gravity, G

Each are useful in setting up equations of motion for certain cases.

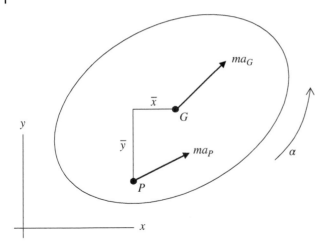

Figure 7.28 Rigid body in planar motion.

Rigid Bodies in Plane Motion

Equation (7.126) can be transformed into another very useful form, by kinematic relationships:

$$\sum M_P = I_G \alpha - \bar{y} m (a_G)_x + \bar{x} m (a_G)_y$$

In this case, unlike I_P, I_G is a property of the body. This means that point P can be located either within the body or outside the body provided moments of both the external forces and inertia forces are taken with respect to the same point P (Figure 7.29). In this case, the moment equation can be viewed as equating a force diagram to a kinetic diagram

$$\sum M_P = \sum \mathcal{M}_P \tag{7.128}$$

where

$\sum M_P$ is the summation of all moments with respect to P (left diagram)
$\sum \mathcal{M}_P$ is the sum of all kinetic effects (right diagram)
The location of point P is arbitrary and is factored into both sides of the equation as shown in Figure 7.29.

This useful concept is based on the force and kinetic equivalence. This equality can easily be visualized by equating a free body force diagram to a free body kinetic diagram (Figure 7-29).

Two reference points are illustrated in Figure 7.30: (i) Point P is fixed on the plane of rolling and (ii) point P is on the roller in contact with the rolling plane. The advantage of the first case is that the friction force on the roller has no moment with respect to point P so

$$2rF = I_G \alpha + r(r\alpha)m$$

$$2rF = \frac{1}{2} r^2 m\alpha + r^2 m\alpha$$

$$F = \frac{3}{4} rm\alpha$$

$$\alpha = \frac{4F}{3rm}$$

The friction force is also removed by use of the second approach:

$$2rF = \frac{3}{2} r^2 m\alpha$$

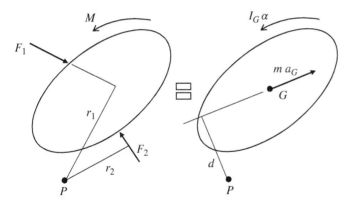

Figure 7.29 Force – kinetic equivalence.

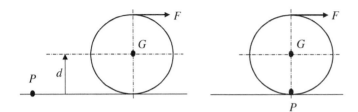

Figure 7.30 Reference point "*P*" alternatives.

$$\alpha = \frac{4F}{3rm}$$

These diagrams graphically illustrate the above equations showing that

$$\sum M_P = \sum \mathcal{M}_P \tag{7.129}$$

Giving

$$\sum M_P = I_{cg}\alpha + dma_{cg} \tag{7.130}$$

and

$$\sum \overline{F} = m\overline{a}_{cg} \tag{7.131}$$

These two equations of dynamics will now be applied to rigid bodies in plane motions under the following conditions:

- Plane translation
- Pure rotation
- General motion

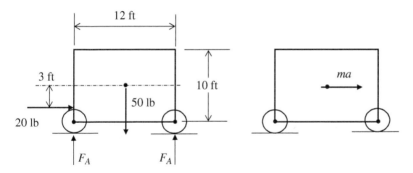

Figure 7.31 Rigid body translation.

Translation of Rigid Bodies

Bodies in pure translation have no rotational component; any line in the body stays parallel to its initial orientation. The force and kinetic diagrams are useful in determining reactions throughout (Figure 7.31):

$$\sum F_y = 0; \quad F_A + F_B = 50 \quad \text{(y component of Eq. (7.125))}$$

$$\sum M_G = 0; \quad 6F_B + 3(20) - 6F_A = 0$$

Combining gives, $F_A = 30$ lb and $F_B = 20$ lb. The x component of Eq. (7.125) gives

$$\sum F_x = ma$$
$$20 = \frac{50}{32.2} a$$
$$a = 12.88 \text{ ft/s}^2$$

Rotation About a Fixed Point

In this case, each point in the rigid body travels along a circular path around a center of rotation. A special case of rigid body dynamics is a bar pinned at one end with freedom to rotate or oscillate (Figure 7.32). Each point in the rigid body travels along a circular path about the pinned end. Motion could be induced by $\bar{a}_P = 0$ an applied torque by the force of gravity. Reaction forces at the support and motion of the body and angular motion, θ, are to be determined.

The three unknowns in this type of problem are force reaction components at the support and angular acceleration, α.

The equation of motion is (noting that $\bar{a}_P = 0$)

$$\sum M_0 = I_0 \ddot{\theta} \tag{7.132}$$

$$-\left(WL + w\frac{L}{2} \right) \sin \theta = I_0 \ddot{\theta} \tag{7.133}$$

which defines the angular acceleration of the body. A useful form of the equation is

$$\ddot{\theta} + \frac{1}{I_0}\left(WL + w\frac{L}{2}\right)\sin\theta = 0 \tag{7.134}$$

The solution of this differential equation defines the angular motion, θ, as a function of time, t. In its present form, the equation is nonlinear. However, for small angular displacements, it is written as

$$\ddot{\theta} + \frac{L}{I_0}\left(W + \frac{w}{2}\right)\theta = 0 \tag{7.135}$$

The natural circular frequency is

$$\omega^2 = \frac{L}{I_0}\left(W + \frac{w}{2}\right)$$

An interesting aspect of this pendulum is how the disc is mounted on the bar. If the disc is fixed to the bar and rotates with the bar, then

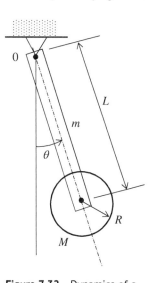

Figure 7.32 Dynamics of a two component pendulum.

$$I_0 = \frac{1}{3}mL^2 + \left(L^2M + \frac{1}{2}R^2M\right) \tag{7.136}$$

However, if the disc is pinned to the bar by a bearing, then

$$I_0 = \frac{1}{3}mL^2 + L^2M \tag{7.137}$$

and the disc translates as a rigid body.

Center of Gravity of Connecting Rod

A simple method of obtaining this information is as follows. The center of gravity can be determined by supporting the connecting rod (Figure 7.33) and measuring the forces, R_A and R_B. The location of the center of gravity is

$$aW = R_B L \tag{7.138}$$

$$a = \frac{R_B}{W}L \tag{7.139}$$

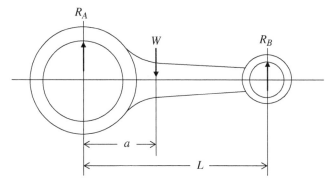

Figure 7.33 Model for determining center of gravity.

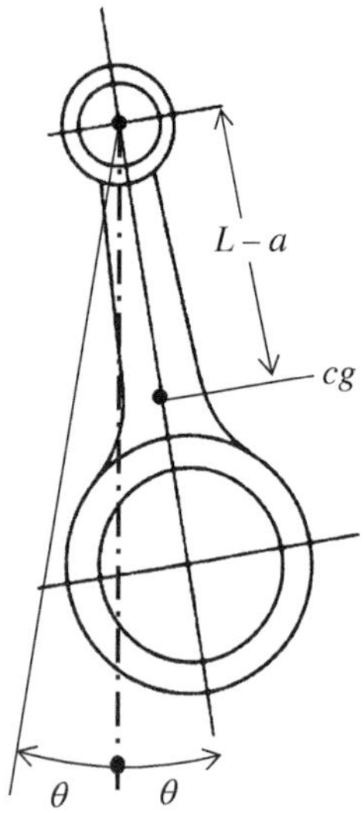

Figure 7.34 Oscillation of a connecting rod.

Mass Moment of Inertia of Connecting Rod

It is difficult to analytically calculate the mass moment of inertia of connection rods because of its complex geometry. This information is typically determined experimentally. One method is based on the period of oscillation of a simple pendulum as shown in Figure 7.34. The period of a rigid body pendulum is determined from the equation of motion:

$$I_0 \frac{d^2\theta}{dt^2} - (L-a)W\theta = 0 \tag{7.140}$$

$$\omega^2 \frac{(L-a)W}{I_0} \frac{1}{s^2} \tag{7.141}$$

Since ω is rad/s and there are 2π rad per cycle

$$f = \frac{\omega}{2\pi} \text{ cps} \tag{7.142}$$

The period of each oscillation is

$$T = \frac{1}{f} \text{ seconds} \tag{7.143}$$

Example Consider

$$W = 6 \text{ lb}$$

$$L - a = 6 - 2 = 4 \text{ in.}$$

$$T = 1.5 \text{ seconds (determined experimentally)}$$

Back calculations give I_0 from the period of oscillation:

$$\omega_n = \frac{2\pi}{T} = 4.19 \text{ rad/s}$$

$$I_0 = \frac{(L-a)W}{\omega^2}$$

$$I_0 = \frac{4(6)}{17.556} = 1.3671 \text{ in.-lb-s}^2$$

The mass moment of inertia with respect to the center of gravity is determined from the transfer formula:

$$I_{cg} = I_0 - (L-a)^2 m \tag{7.144}$$

$$I_{cg} = 1.3671 - 4^2 \frac{6}{386} = 1.1183 \text{ in.-lb-s}^2$$

General Motion of Rigid Bodies

In general, rigid bodies may move in planar motion with a combination of translation and rotation. Movement from one position to another can be viewed as separate motions of translation and then rotation.

Example A disc translates and rotates simultaneously as shown in Figure 7.35.

The spool has a mass of 8 kg and a radius of gyration of $k_g = 0.35$ m with respect to its center. Determine the spools angular acceleration and the force in the cord if a force of 100 N is applied as shown.

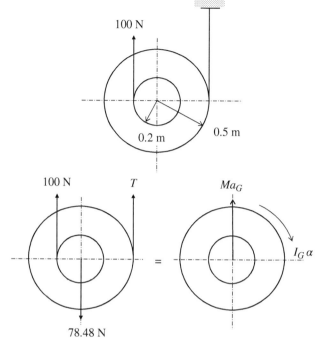

Figure 7.35 General motion of a rigid body.

Equations of motion are

$$\sum T_G = I_G \alpha$$

$$0.2(100) - 0.5T = I_G \alpha \tag{7.145}$$

and

$$\sum F = Ma_G$$

$$100 + T - 78.48 = Ma_G \tag{7.146}$$

Note that

$$I_G = Mk_G^2 = 8\,\text{kg}(0.35\,\text{m})^2 = 0.98\,\text{kg-m}^2$$

The relation between a_G and α is

$$a_G = 0.5\alpha$$

Combining Eq. (7.145)

$$20 - 0.5T = 0.98\alpha$$

And Eq. (7.146)

$$21.52 + T = 8(0.5)\alpha$$

gives

$$\alpha = 10.32 \text{ rad/s}^2$$

$$a_G = 5.16 \text{ m/s}^2$$

$$T = 19.77 \text{ N}$$

The *static* tension, T, and force, F (left cable), are $T = 22.42$ N and $F = 56.06$ N. The increase in applied force from 56.06 to 100 N produced the upward acceleration and angular acceleration causing tension T to drop from 22.42 to 19.77 N.

Example In another example, consider a disc that is attached to a rail accelerating horizontally at 3 fps². Determine the reactive force components at support A and the angular acceleration of the disc in its current position.

Here a disc is pinned to a runner having an acceleration of 3 fps². The disc weighs 15 lb having a mass moment of inertia with respect to the center of gravity, G, of

$$I_G = \frac{1}{2}mr^2$$

It is desired to find the five unknowns, A_x, A_y, a_{Gy}, a_{Gx}, α.
The freebody diagram and kinetic diagrams, shown in Figure 7.36, yield

$$\sum F_y = ma_{Gy} \tag{7.147}$$

$$A_y - W = ma_{Gy}$$

$$\sum F_x = ma_{Gx} \tag{7.148}$$

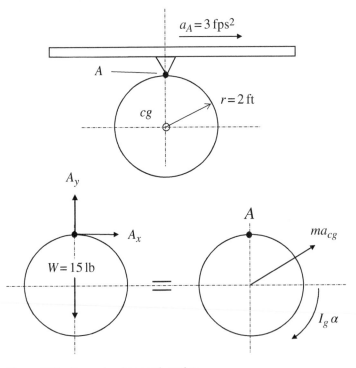

Figure 7.36 Example of general motion.

$$A_x = ma_{Gx}$$

$$\sum M_A = I_G\alpha - rma_{Gx} \tag{7.149}$$

$$0 = I_G\alpha - rma_{Gx}$$

$$\alpha = \frac{rm}{I_G}a_{Gx}$$

Kinematic relationships

$$\bar{a}_G = \bar{a}_A + \bar{a}_{G/A} \tag{7.150}$$

$$a_{Gx}i + a_{Gy}j = 3i + (a_{GA})_n j - (a_{G/A})_t i \tag{7.151}$$

$$a_{Gx} = 3 - r\alpha$$

$$a_{Gy} = 0$$

Dynamic Forces Between Rotor and Stator

PDMs are commonly used in bottom-hole assemblies to provide greater power to drill bits and to help navigate drill bits in directional drilling operations. Internal components are shown in Figure 7.37.

These motors (and pumps) contain a rotor and a stator each possessing special gear-shaped geo-metries which mess together in such a way to form cavities, which progress in the longitudinal direction. These motors have multiple cavities or stages. The number of lobes establishes the rela-tion between volume flow rate and rotor speed. The stator always has one more lobe than the rotor. In general, the high lobes perform with lower flow rate, the higher the torque. Moineau pumps are often used in petrochemical production facilities.

Side loads are generated by the rotor against the elastomer molding of the stator as the rotor moves with planetary motion within the stator. These side loads can cause (i) premature failure of the elastomer lining at high rotor speeds and (ii) excite lateral modes of vibration over the drill

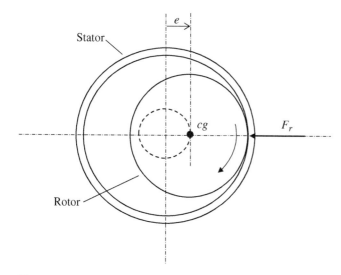

Figure 7.37 Pitch circles of rotor and stator.

Table 7.3 Planetary relationships.

	Arm, e	Rotor, r	Stator, s
With e	1	1	1
Relative to e	0	$-\dfrac{r_s}{r_r}$	-1
Total	1	$1-\dfrac{r_s}{r_r}$	0

collar section, depending on force frequency and magnitude. Figure 7.37 shows the pitch circles of both rotor and stator. The force, F_r, represents a radial force interactive force between rotor and stator liner. It is the reactive force on the stator that can excite lateral modes of vibration. It passes through the center of gravity of the rotor according to the earlier discussion of rigid bodies.

A common lobe arrangement is the 1 : 2 motor/pump. In this case, the diameter of the pitch circle of the rotor is one-half that of the stator. As the rotor travels with planetary motion within the stator, the center of the rotor progresses along a circular path. In general, the relation between arm rotation of (e) and rotation of the rotor is

$$\frac{\omega_r}{\omega_e} = 1 - \frac{r_s}{r_r}$$

as shown by Table 7.3.

The darker circle represents the cross section of the rotor lobe. The rotor moves with general type motion, i.e. simultaneous rotation and translation. At any one location, the acceleration of the center of gravity of the rotor is

$$a_G = \omega^2 e \tag{7.152}$$

where

e = pitch radius of the rotor
ω = angular velocity of the rotor

Note that any cross section of the 1 : 2 rotor travels within a transverse plane per the following equations:

$$x = r(\cos\theta + \cos\phi)$$
$$y = -r(-\sin\theta + \sin\phi)$$

These equations show that the path of center of the rotor lobe is a straight line; however, the path of the center of the pitch circle is a circle of radius r.

Consider the following parameters:

$$\frac{n_s}{n_r} = 2 \text{ (lobe ratio)}$$

$$N = 200 \text{ rpm}$$

$$\omega = \frac{2\pi N}{60} \text{ rad/s}$$

$$W_r = 200 \text{ lb}$$

$$r_r = \text{pitch radius of rotor } (2 \text{ in.})$$

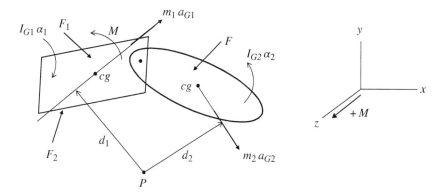

Figure 7.38 Interconnecting bodies.

From the kinematics of the rotor/stator,

$$\omega = \left(1 - \frac{n_s}{n_r}\right)\omega_e \tag{7.153}$$

meaning that vector, e, rotates counterclockwise with angular velocity of ω. The acceleration of the rotor center is $a_{cg} = \omega^2 e = \omega^2 r_r$. The force, $F_r = Ma_{cg}$ or

$$F_r = \frac{200}{32.2}\left(\frac{2\pi 200}{60}\right)^2 \frac{2}{12} = 454 \text{ lb}$$

The force vector rotates with $\omega_e = 20.94$ rad/s or a frequency of 3.33 cps. The force magnitude and frequency are both important in determining possible resonance of lateral modes of vibration.

Interconnecting Bodies

In many mechanical situations, rigid bodies are interconnected. Each rigid body can be analyzed separately. However, the analysis is greatly simplified by applying the equations of motion to the assembly (Figure 7.38). The connecting forces are not involved in the equations because they are interactive and cancel out. The equations of motion are applied as follows:

$$\sum M_P = \sum I_{Gi}\alpha_i + \sum d_i m_i a_{Gi} \tag{7.154a}$$

$$\sum F_{ix} = \sum m_i(a_{Gi})_x \tag{7.154b}$$

$$\sum F_{iy} = \sum m_i(a_{Gi})_y \tag{7.154c}$$

where the subscript "i" refers to each body.

The advantage in this approach is that internal forces are not involved in the calculation and can greatly simplify a dynamic analysis. The last three scalar equations, however, allow for only three unknowns. If there are more than three, then the system then part of the system may have to be isolated.

Gear Train Start-Up Torque

As an example, consider a simple gear set driven by a motor as shown in Figure 7.39. Assume that each gear has relatively large angular moment of inertia which has to be considered during the

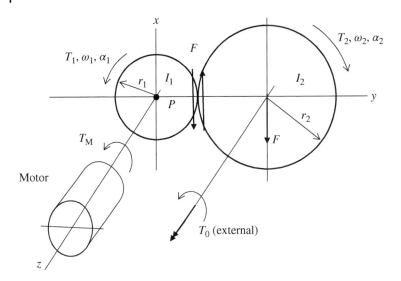

Figure 7.39 Start-up torque.

start-up of the gear drive. These inertias will affect the startup time required to reach the desired transmitted torque. The angular acceleration during the start-up period will now be discussed by applying the above equations.

From Eq. (7.154a) and including the motor armature as part of I_1,

$$T_M + T_o - (r_1 + r_2)F = I_1\alpha_1 - I_2\alpha_2 \tag{7.155}$$

Moments are taken with respect to point P.
From kinematics,

$$r_1\theta_1 = r_2\theta_2 \tag{7.156}$$

$$r_1\alpha_1 = r_2\alpha_2 \tag{7.157}$$

Accounting for directions,

$$\alpha_2 = \frac{r_1}{r_2}\alpha_1 \tag{7.158}$$

Bringing these equations together gives

$$T_M + T_o - (r_1 + r_2)F = \left[I_1 - \frac{r_1}{r_2}I_2\right]\alpha_1 \tag{7.159}$$

From a freebody diagram of gear 1,

$$T_M - r_1F = I_1\alpha_1$$

$$F = \frac{T_M - I_1\alpha_1}{r_1}$$

The progression of substitution and simplification follows:

$$T_M + T_o - \left(1 + \frac{r_1}{r_2}\right)(T_M - I_1\alpha_1) = \left(I_1 - \frac{r_1}{r_2}I_2\right)\alpha_1$$

$$T_o + I_1\alpha_1 - \frac{r_2}{r_1}T_M + \frac{r_1}{r_2}I_1\alpha_1 = \left(I_1 - \frac{r_1}{r_2}I_2\right)\alpha_1$$

$$T_M = \frac{r_1}{r_2}T_0 + \left[I_1 + I_2\left(\frac{r_1}{r_2}\right)^2\right]\alpha_1 \tag{7.160}$$

If inertia is not considered, then

$$T_M = \frac{r_1}{r_2}T_0 \tag{7.161}$$

Motor limitations may require that gear inertia be overcome first before applying the external load. This would also allow the large gears to serve as flywheels. Eq. (7.160) could be used to determine time to reach rated speed of the motor.

Kinetic Energy of Rigid Bodies

Considering rigid bodies are made up of many particles, then Eq. (7.72) also applies except total kinetic energy is equal to

$$KE = \frac{1}{2}Mv_g^2 + \frac{1}{2}I_g\omega^2 \tag{7.162}$$

Work may be produced by either torque or forces or both.

$$W = F\Delta x = T\Delta\theta \tag{7.163}$$

Friction plays an important role in maintaining pure rolling of a disc. However, friction in this case does *no work* provided there is no sliding. Recall

$$dW = F\,dx = F\frac{dx}{dt}dt = FV\,dt \tag{7.164}$$

Forces at points of zero velocity do no work.

Example   Consider the disc and spring arrangement in Figure 7.40. The potential energies in this problem are gravity and the elastic energy of the spring:

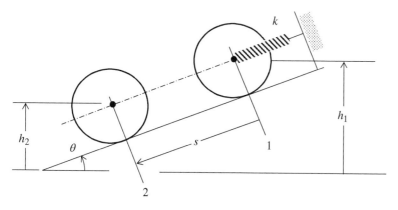

Figure 7.40   Principal of work and kinetic energy.

$$PE_1 = h_1 W + \tfrac{1}{2} k(\delta_{st})^2$$

$$PE_2 = h_2 W + \tfrac{1}{2} k (s + \delta_{st})^2$$

The kinetic energies are

$$KE_1 = 0 \quad \text{(disc starts at rest)}$$

$$KE_2 = \frac{1}{2} M V_2^2 + \frac{1}{2} I_G \omega_2^2 = 0 \quad \text{(maximum compression of spring)}$$

Since rolling friction does zero work

$$PE_1 + KE_1 = PE_2 + KE_2 \tag{7.165}$$

In this case,

$$PE_1 = PE_2$$

$$h_1 W + \tfrac{1}{2} k(\delta_{st})^2 = h_2 W + \tfrac{1}{2} k(s + \delta_{st})^2$$

$$W(s \sin \theta) = \tfrac{1}{2} k(s + \delta_{st})^2 - \tfrac{1}{2} k(\delta_{st})^2$$

This equation is quadratic. If, however, $\delta_{st} = 0$, then

$$W \sin \theta = \frac{1}{2} ks$$

and

$$s = \frac{2W \sin \theta}{k} \tag{7.166}$$

In this case, the friction force at the contact point does zero work.

The Catapult

Catapults of antiquity stored elastic energy in flexible beams, such as in composite bows made up of animal sinew (tension), wood, and bone (compression). This energy is released suddenly to propel a warhead. In modern times, catapults are used on aircraft carriers to assist in launching aircraft.

When the composite bow reached its limits, new elastic systems were needed to increase range and size of projectiles. Eventually, the bow was replaced with *torsion* springs. Each torsion spring was a bundle of sinew strands with a wooden arm thrust through its center. Each sinew strand was thread size. The torsion spring was lighter and more powerful (energy storage – 20 times greater than steel per pound).

When the trigger was pulled of, the energy stored in the torsion springs was used to accelerate the projectile, but a significant portion was also used to accelerate the wooded throwing arm. The forward motion of the arms was eventually arrested by the sling snapping taut. Since the tightening of the sling also contributed to the forward motion of the projectile, nearly 100% of the elastic energy was transmitted to kinetic energy of the projectile. A *U*-joint at the support point gave the mechanism more flexibility in aiming the projectile. Catapults could throw 50 lb weights. These improvements accelerated the arms race at that time.

Impulse–Momentum of Rigid Bodies

The equations of motion for rigid bodies can be extended to yield angular impulse and angular momentum expressions (Figure 7.41). Starting with

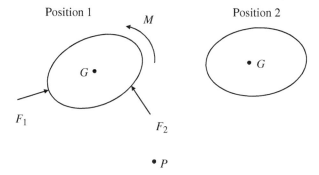

Figure 7.41   Impulse–momentum of rigid bodies.

$$\sum \overline{F} = m\bar{a}_G \tag{7.167}$$

$$\sum \overline{F}\, dt = m\left(\bar{v}_{cg,2} - \bar{v}_{cg,1}\right) \tag{7.168}$$

$$\sum \overline{F}\, dt = \overline{H}_2 - \overline{H}_1 \tag{7.169}$$

and

$$\sum M_P = I_G \alpha + dma_G \tag{7.170}$$

$$\sum M_P = I_G \frac{d\omega}{dt} + dm\frac{dv_G}{dt} \tag{7.171}$$

$$\sum \int M_P\, dt = I_G(\omega_2 - \omega_1) + m(d_2 v_{G2} - d_1 v_{G1}) \tag{7.172}$$

Linear Impulse and Momentum

Equation (7.169) provides two algebraic equations:

$$\sum \int F_x\, dt = H_{x2} - H_{x1} \tag{7.173}$$

$$\sum \int F_y\, dt = H_{y2} - H_{y1} \tag{7.174}$$

where H is linear momentum.

Angular Impulse and Momentum

The angular impulse–momentum equation provides a third equation:

$$\sum \int M_P\, dt = G_{P2} - G_{P1} \tag{7.175}$$

which is rewritten as

$$G_{P2} = \sum \int M_P\, dt + G_{P1} \tag{7.176}$$

where G is angular momentum reference from an arbitrary point P.

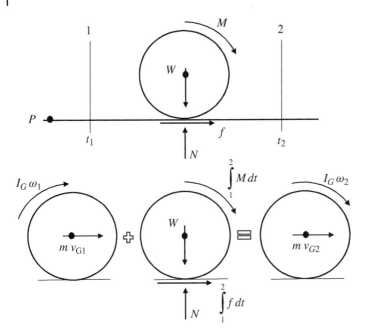

Figure 7.42 Impulse–momentum (two equations).

The application of these expressions is illustrated with the situation in Figure 7.42. Here a disc rolls on a horizontal flat surface with an applied moment, M applied to it. At time, t, the disc has angular velocity, ω_1. The problem is to determine, ω_2 at time t_2. We have three momentum equations to use.

Linear impulse–momentum (two equations):

$$mv_{G1} + \int f \, dt = mv_{G2} \tag{7.177}$$

and

$$0 + \int (N - W)dt = 0 \tag{7.178}$$

The second equation tells us that $N = W$. The first equation gives

$$\int_1^2 f \, dt = mv_2$$

We cannot say at this point that slippage is impending.

Angular impulse–momentum (one equation):

$$I_0\omega_1 + \int_1^2 (M - rf)dt = I_0\omega_2 \tag{7.179}$$

$$Mt_2 - r \int_1^2 f \, dt = I_0 \omega_2$$

$$Mt_2 - rmv_2 = I_0 \omega_2$$

but $v_2 = r\omega_2$, so

$$Mt_2 = \left(I_0 + r^2 m\right)\omega_2 \tag{7.180}$$

and

$$\omega_2 = \frac{Mt_2}{I_G + r^2 m} \tag{7.181}$$

Example The disc shown in Figure 7.43 accelerates to the right due to force, F. The magnitude of this force increases with time according to $F = t + 10$ N.

The problem is to determine the angular velocity of the disc after five seconds. Assume the disc starts from rest. The disc weighs 981 N and has a mass moment of inertia about its center of $I_G = 12.25$ kg m^2.

The angular impulse–momentum equation is

$$G_{P2} = \sum \int M_P \, dt + G_{P1} \tag{7.182}$$

By substitution,

$$I_P \omega_1 + \int_0^5 1.15 F(t) dt = I_P \omega_2 \tag{7.183}$$

$$I_P \omega_1 + \int_0^5 1.15(t + 10) dt = I_P \omega_2 \tag{7.184}$$

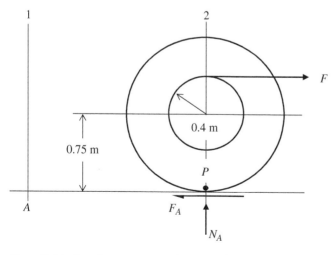

Figure 7.43 Impulse–momentum (one equation).

With $\omega_1 = 0$,

$$1.15\left(\frac{1}{2}t^2 + 10t\right)\bigg|_0^5 = I_P\omega_2$$

$$1.15(12.5 + 50) = \left(12.25 + 0.75^2 100\right)\omega_2$$

$$71.87 = 68.5\omega_2$$

$$\omega_2 = 1.05\,\text{rad/s}$$

Angular Impulse Caused by Stabilizers and PDC Drill Bits

Tooth breakage on PDC drill bits is possible due to large impulsive forces caused by momentum changes.

Consider a disc rotating at a constant angular velocity about its geometric center (Figure 7.44). Over a very short time interval, $\Delta t = t_2 - t_1$, the edge of the disc is caught by a fixed pin as shown. At time t_2, the disc suddenly rotates about its instantaneous center, point "a."

Over this short time interval (Δt), angular momentum with respect to point "a" is conserved, there is no angular impulse. However, linear momentum with respect to the center of gravity is not conserved. Based on the conservation of angular momentum:

$$G_2 = G_1 \tag{7.185}$$

from which

$$\left(I_G + Mr^2\right)\omega_2 = I_G\omega_1 \tag{7.186}$$

and

$$\omega_2 = \frac{I_G}{I_G + Mr^2}\omega_1 = \frac{1}{3}\omega_1 \tag{7.187}$$

assuming a solid disc. The velocity of the center of gravity is now $v_{02} = r\omega_2$. Considering the component of linear impulse to the right:

$$Im_x = H_{x2} - H_{x1} \tag{7.188}$$

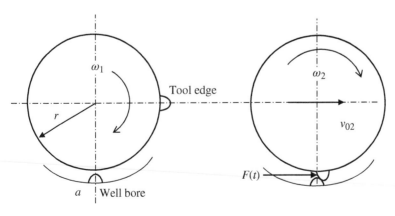

Tool edge

$F(t)$

a Well bore

Figure 7.44   Cutter impulsive force.

giving

$$\int_1^2 F(t)dt = v_{o2}M - 0 = r\omega_2 M \tag{7.189}$$

This change in linear momentum takes place over a very short period of time. The magnitude of the maximum impulsive force can be approximated by

$$\int_1^2 F(t)dt = F_{ave}\Delta t \tag{7.190}$$

Dividing the impulse over time Δt gives the average impact force:

$$F_{ave} = \frac{r\omega_2 M}{\Delta t} \tag{7.191}$$

Example Consider a rigid disc weighing 40 000 lb, having a 7-in. diameter and rotating at a speed of 100 rpm. These numbers were chosen to simulate drill collars. Adjusting the units for substitution into Eq. (7.191)

$$r = \frac{3.5}{12} = 0.2917 \text{ ft}$$

$$M = \frac{40\,000}{32.2} = 1242 \text{ slugs}$$

$$\omega_2 = \frac{1}{3}\frac{2\pi 100}{60} = 3.49 \text{ rad/s}$$

Putting these numbers into Eq. (7.191) gives

$$F_{ave} = \frac{0.2917(3.49)1242}{0.05} = 79\,496 \text{ lbs}$$

Here, Δt is arbitrarily chosen to be 0.05 second. A smaller Δt increases the predicted average force. The peak impact force on the pin would be much higher. Cutter damage and failure is possible from this level of force.

Accounting for Torsional Flexibility in Drill Collars

Now consider the torsion flexibility of drill collars of length, L (Figure 7.45). Assume the drill collars rotate clockwise as in drilling. Once a cutter contacts an obstruction, creating impact force, $F(t)$, a shear wave starts to travel to the right with acoustic velocity $c = 10\,600$ ft/s $\left(c = \sqrt{\frac{G}{\rho}} = 10\,600 \text{ fps}\right)$.

Over a short span of time, the leading edge of the shear wave has traveled a distance, $x = ct$. The shape of the impulse force, $F(t)$, will be sustained within the collars as illustrated. The space interval Δx is related to the time interval, Δt by $\Delta x = c\Delta t$. As the shear wave travels to the right, linear momentum is changed according to

$$\int_0^{\Delta t} F(t)dt = r\omega_2 m\Delta x \quad (m \text{ is mass per unit length}) \tag{7.192}$$

In terms of the average impulse force,

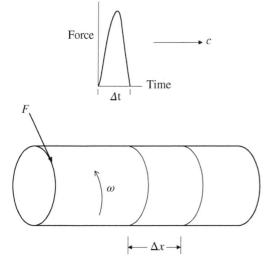

Figure 7.45   Torsion shear wave set up by cutter impact.

$$F_{ave}\Delta t = r\omega_2 mc\Delta t \tag{7.193}$$

$$F_{ave} = r\omega_2 mc \tag{7.194}$$

Assuming the drill collars weigh 114 lb/ft ($m = 3.54$ slugs/ft) and applying the previous numbers to Eq. (7.194) gives

$$F_{ave} = 0.291(3.49)(3.54)(10\ 600) = 37\ 980\ \text{lb}$$

This number is substantially less than for the rigid disc because not all of the linear momentum of the collars is changed simultaneously.

Interconnecting Bodies

Interconnecting bodies can also be analyzed as one unit using the equations of motion for interconnecting bodies explained earlier. In terms of impulse–momentum:

$$\sum\int M_A\, dt = \sum(G_A)_2 - \sum(G_A)_1 \tag{7.195}$$

$$\sum I_{Gi}\omega_2 + \sum d_i m_i V_2 = \sum\int M_A dt + \sum I_{Gi}\omega_1 + \sum d_i m_i V_1 \tag{7.196}$$

Example   Consider the arrangement shown in Figure 7.46. Relevant numbers are given in the figure.

For this example, Eq. (7.196) reduces to

$$(I_A\omega_2 + r m_B V_{B2}) = M_A\Delta t + (I_A\omega_1 + r m_B V_{B1})$$

The moment, $M_A = 0.2(58.86) = 11.77$ N · m

Using the numbers in the figure and noting that $V_B = 0.2\omega$ the equation yields

$$\omega_2 = 65.17\ \text{rad/s}$$

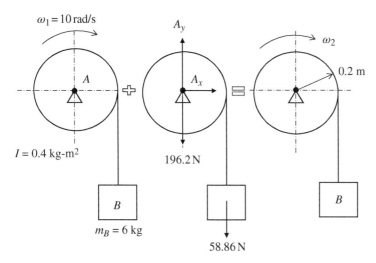

$\omega_1 = 10 \, \text{rad/s}$

A_y

ω_2

A

A_x

0.2 m

$I = 0.4 \, \text{kg-m}^2$

196.2 N

B

$m_B = 6 \, \text{kg}$

58.86 N

B

Figure 7.46 Interconnecting system.

and

$$V_{B2} = 13 \, \text{m/s}$$

when $\Delta t = 3$ seconds

This approach gives a direct solution without dividing the total system into two separate freebody diagrams.

Conservation of Angular Momentum

Gear pairs can also be analyzed as interconnecting bodies using angular impulse-momentum principles.

$$\sum \int M_0 \, dt = \sum (G_0)_2 - \sum (G_0)_1 \tag{7.197}$$

If angular impulse is zero (left side), then angular momentum is conserved.

$$\sum (G_0)_2 = \sum (G_0)_1 \tag{7.198}$$

Example Consider the conservation of angular momentum during the activation a flywheel (Figure 7.47). The gear train, clutch, and flywheel are considered as a group of interconnecting bodies.

A simple gear train transmits power from an electric motor through a gear pair to location A. Initial power from the motor is 3 hp delivered at 1000 rpm. Determine the torque and speed delivered to point A. If the clutch is engaged to active a flywheel, determine the output speed of the motor immediately after the clutch is engaged assuming the two torques are unchanged. How much time is required to bring the system up to its operating speed? Assume mass polar moments of inertia are

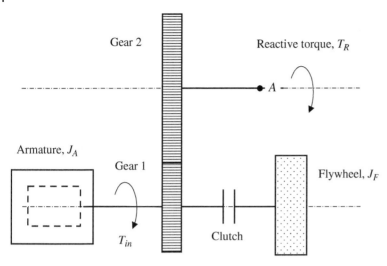

Figure 7.47 Angular momentum changes during flywheel engagement.

$$J_A = 0.1 \text{ lb-s}^2/\text{in. (armature)}$$

$$J_1 = 0.05 \text{ lb-s}^2/\text{in. (gear 1)}$$

$$J_2 = 0.2 \text{ lb-s}^2/\text{in. (gear 2)}$$

$$J_F = 0.4 \text{ lb-s}^2/\text{in. (flywheel)}$$

$$R_1 = 2 \text{ in. (pitch radius of gear 1)}$$

$$R_2 = 4 \text{ in. (pitch radius of gear 2)}$$

Determine:

a) Torque and speed at point A.
b) Rotational speeds after clutch has been engaged.
c) Time required to bring the system back up to its original speed

Part a

The torque delivered by the motor is

$$HP = \frac{T_M N}{5252} \quad T_M = \frac{3(5252)}{1000} = 15.756 \text{ ft-lb} = 189.1 \text{ in.-lb} \tag{7.199}$$

The reactive torque is $T_R = 2T_{in} = 31.512$ ft-lb at $N_2 = 500$ rpm.

Part b

This part of the problem will be addressed using the equation of motion for the gear pair shown in Figure 7.39. The dynamic behavior of the gear pair is described by:

$$T_M = \frac{r_1}{r_2} T_0 + \left[I_1 + I_2 \left(\frac{r_1}{r_2} \right)^2 \right] \frac{d\omega}{dt}$$

In terms of impulse-momentum

$$\left(T_M - \frac{r_1}{r_2} T_o\right) \Delta t = \left[J_A + J_1 + J_F + \left(\frac{r_1}{r_2}\right)^2 J_2\right] \Omega_0 - \left[J_A + J_1 + \left(\frac{r_1}{r_2}\right)^2 J_2\right] \omega$$

The change in motion caused by the activation of the flywheel takes place instantaneously such that impulse is zero. This means that angular momentum is conserved and

$$\left[J_A + J_1 + J_F + \left(\frac{r_1}{r_2}\right)^2 J_2\right] \Omega_0 = \left[J_A + J_1 + \left(\frac{r_1}{r_2}\right)^2 J_2\right] \omega \tag{7.200}$$

where Ω_0 is new motor speed at time zero.

$$\Omega_0 = \frac{(J_A + J_1 + 0.25 J_2)}{(J_A + J_1 + J_F + 0.25 J_2)} \omega_1 \tag{7.201}$$

By substituting the numbers

$$\Omega_0 = \frac{(0.1 + 0.05 + 0.25*0.2)}{(0.1 + 0.05 + 0.4 + 0.25*0.2)} \omega_1$$

$$\Omega_0 = \frac{\omega_1}{3} \quad \text{or} \quad 0.333(1000) = 333 \text{ rpm}$$

The flywheel has the effect of lower the motor speed from 1000 to 333 rpm immediately after the clutch is engaged.

Part c

At time zero (immediately after clutch engagement), assuming motor power is constant (30 hp), motor torque is

$$T(0) = \frac{HP(5252)}{N} = \frac{3(5252)}{333} = 52.52 \text{ ft-lb (input torque at } t = 0) \tag{7.202}$$

Please note motor performance curve will define the true torque–speed relationship. We now apply Eq. (7.160) to determine time to reach original speed, ω_1 with the flywheel engaged.

$$T_M = \frac{r_1}{r_2} T_0 + \left[J_1 + J_2 \left(\frac{r_1}{r_2}\right)^2\right] \frac{d\Omega}{dt} \tag{7.203}$$

Assuming the motor power remains the same, i.e. 30 hp then by

$$HP = \frac{T\Omega}{5252} = 3 \text{ hp} \tag{7.204}$$

Torque and speed are related by

$$T\Omega = 15\,756 \quad (T - \text{ft-lb}, \quad \Omega - \text{rpm}) \tag{7.205}$$

$$T\Omega = 19\,800 \quad (T - \text{in-lb}, \quad \Omega - \text{rad/s})$$

By substitution

$$T(t) - \frac{1}{2} T_R = (J_A + J_1 + J_F + 0.25 J_2) \frac{d\Omega}{dt} \tag{7.206}$$

which reduces to

$$\frac{19\,800}{\Omega} - \frac{1}{2}T_R = (J_A + J_1 + J_F + 0.25J_2)\frac{d\Omega}{dt} \tag{7.207}$$

This equation is nonlinear and best solved numerically. However, if we assume speed recovery is made without the reactive torque the equation simplifies to a form that is easily solved analytically:

$$\Omega\frac{d\Omega}{dt} = \frac{19\,800}{(J_A + J_1 + J_F + 0.25J_2)} = \frac{19\,800}{0.6} = 33 \times 10^3 \tag{7.208}$$

$$\frac{1}{2}\frac{d\Omega^2}{dt} = 33 \times 10^3$$

$$\Omega^2 = 33 \times 10^3(2)t + C$$

When $t = 0$, $\Omega = \Omega_0$, so $C = \Omega_0^2$. Solving for t when $\Omega = \omega_1$ gives

$$t = \frac{\omega_1^2 - \Omega_0^2}{66 \times 10^3} = \frac{(1 - 0.0123)}{66 \times 10^3}\omega_1^2 \tag{7.209}$$

$$\omega_1 = \frac{2\pi}{60}1000 = 104.72 \text{ rad/s}$$

By substitution, $t = 0.164$ second. The recovery time is quite short for this set of numbers. However, for larger gears, time of recovery would be greater.

References

1 Whipple, S. (1883). *An Elementary Treatise on Bridge Design*, 4e. New York (USA): D. Van Nostrand.
2 Schuh, F. (1977). *Drilling Equations*. Petroleum Engineering Publishing Company.
3 Garrett, W.R. and Rollins, H.M. (1926). Steering wheel for rock bits, *Presented at the 9th Annual Petroleum Mechanical Engineering Conference*, Los Angeles, Calif.

8

Mechanics of Materials

Romans arches were used to form bridges, aqueducts, and buildings, including the Coliseum. Arches accommodate uniform loads quite well, but concentrated load, especially at the center point, cause instability and possible collapse [1]. Instability was solved by building wall-type extension, called spandrels, outside of arches. The Romans also discovered how to make concrete to solidify the spandrels. Horizontal support at the base of arches was also essential. Stone arches were used throughout Europe until the early 1700s.

Iron became the new structural material replacing masonry and wood by the early 1700s. The British iron maker, Abraham Darby (1678–1717), began to produce better iron by using coke, a derivative of coal, instead of charcoal, a derivative of wood. His iron business (1709) was in Coalbrookdale, England. In those days, there was a much greater supply of coal than timber, plus Darby produced a higher quality of iron.

In 1777 his grandson Darby III was contracted to build a bridge spanning 100 ft across the Severn River. The bridge, which was completed in 1779, was made completely of cast iron. Each member was casts into desired shapes, thus eliminating costly hand work required to chisel and carve stone. The cast iron shape was an arch which closely resembled the geometry of Roman arches, putting the cast iron components in compression. Historically, this bridge marked the end of arch bridges made of stone and timber.

Two factors began to change the design of engineering structures: (i) stronger steel and (ii) scientific tools of analysis. Scientific reason began to replace the empirical approach. The first science-based engineering structure is the Eads Bridge built over the Mississippi River at St. Louis. This bridge was designed and constructed by James Eads between 1867 and 1874. It was the world's longest arch bridge and was made of steel. The bridge is still in use today.

Engineering design deals with loads, stresses, deflections, and failure of structural members. The challenge in design is to configure member sizes and shapes that can withstand external loads without failing. As a rule, cost relates to weight, so the lighter the structure, the lower the cost. Weight is reduced through engineering analysis. Cost increases with member complexity and number of components. Design is all about making equipment cost-effective, and this is done through achieving functionality through simplicity.

Equipment can contain simple components under direct tension/compression, shear, or twist. In general beams, relative long members that carry transverse loads, can develop high stresses, and require special analyses. In simple terms, it is easier to break a stick by bending than to pull it apart. This section summarizes important aspects of stress analysis used in engineering design.

Engineering Practice with Oilfield and Drilling Applications, First Edition. Donald W. Dareing.
© 2022 John Wiley & Sons, Inc. Published 2022 by John Wiley & Sons, Inc.

Stress Transformation

When a rod or strap is pulled (Figure 8.1), the average normal stress across the section is simply $\sigma = \frac{E}{A}$ as determined during a standard tension test. Considering the forces on an inclined plane, we see that a normal force (N) and a shear force (V) are required for equilibrium:

$$\sum F_n = 0 \quad \sum F_t = 0$$
$$N - P\cos\theta = 0 \quad V + P\sin\theta = 0$$
$$N = P\cos\theta \quad V = -P\sin\theta$$

Corresponding stresses (normal and shear) on the inclined plane are

$$\sigma_n = \frac{N}{A_n} = \frac{P}{2A}(1 + \cos 2\theta) \quad \text{(normal stress)} \tag{8.1}$$

$$\tau_n = \frac{V}{A_n} = -\frac{P}{2A}\sin 2\theta \quad \text{(shear stress)} \tag{8.2}$$

The sign convention of each is set by the direction of the n,t axis with θ measured counterclockwise from the x-axis. The stress values in Figure 8.2 have been unitized, i.e.

$$\frac{\sigma_n}{P/A} = \frac{1}{2}(1 + \cos 2\theta) \quad \text{(normal stress)}$$

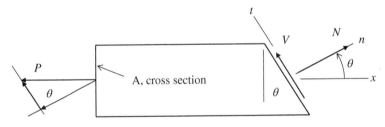

Figure 8.1 Stresses across a strap.

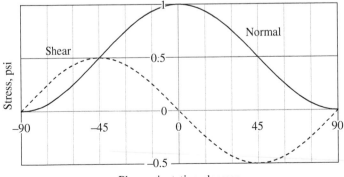

Figure 8.2 Effect of plane orientation.

$$\frac{\tau_n}{P/A} = -\frac{1}{2} \sin 2\theta \quad \text{(shear stress)}$$

The equations show how normal and shear stresses on any plane vary with θ. The maximum normal stress occurs on the transverse plane ($\theta = 0$). The maximum shear stress occurs on planes inclined at 45°. The stress condition in Figure 8.1 can also be viewed from a small element as shown in Figure 8.3, where *forces* are in equilibrium – not stress.

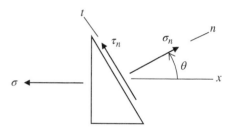

Figure 8.3 Stress element.

Theory of Stress

In general, local stress is defined by three stress components, σ_x, σ_y, τ_{xy} which are applied to this small element. This condition (Figure 8.4) defines stress at a point even though surfaces are required to yield forces, which must be in equilibrium. Stresses are shown in their "plus" direction.

From moments about point "a," $\tau_{xy} = \tau_{yx}$, showing that shear stresses on perpendicular planes have the same magnitude. These stress components may be established analytically or experimentally.

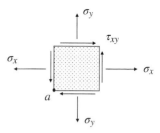

Figure 8.4 Stress components at a point ($\sigma_z = 0$).

Normal and Shear Stress Transformations

Normal and shear stresses on inclined planes vary as before. Forces associated with these stresses are in equilibrium. Equilibrium equations expressed in terms of the stress components are given below:

$$\sum F_n = 0 \tag{8.3}$$

$$\sigma_n = \sigma_x \cos^2\theta + \sigma_y \sin^2\theta + 2\tau_{xy} \sin\theta \cos\theta$$

In terms of double angle:

$$\sigma_n = \frac{\sigma_x + \sigma_y}{2} + \frac{\sigma_x - \sigma_y}{2} \cos 2\theta + \tau_{xy} \sin 2\theta \tag{8.4}$$

$$\sum F_t = 0 \tag{8.5}$$

$$\tau_{nt} = -\frac{\sigma_x - \sigma_y}{2} \sin 2\theta + \tau_{xy} \cos 2\theta \tag{8.6}$$

The sign convention for each stress is plus (+) as shown in Figure 8.5.

To illustrate the use of Eqs. (8.4) and (8.6), consider the stress condition shown in Figure 8.6.

The stress components are

$\sigma_x = 80$ MPa
$\sigma_y = -100$ MPa

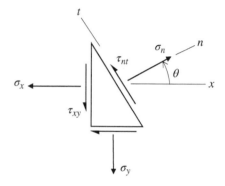

Figure 8.5 Stress on an inclined surface.

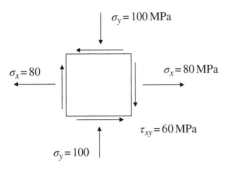

Figure 8.6 Plane stress at a point (MPa).

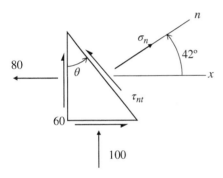

Figure 8.7 Orientation of a biaxial state of stress.

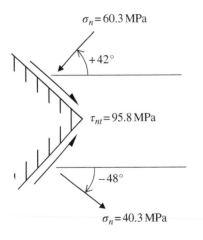

Figure 8.8 Stress components on an inclined plane at 42°.

$\tau_{xy} = -60$ MPa

We wish to determine normal and shear stresses on a plane whose normal axis is 42° counterclockwise from the x-axis (Figure 8.7).

Substituting these stress components into Eqs. (8.4) and (8.6), gives

$$\sigma_n = -60.26 \text{ MPa}$$

$$\tau_{nt} = -95.78 \text{ MPa}$$

These stresses are shown on perpendicular surfaces in Figure 8.8.

The normal stress, σ_n, and shear stress, τ_{nt}, both vary with angle, θ. The variations of both (Figure 8.9) show orientations where normal stresses are maximum and minimum. Likewise, shear stress reaches maximum values at certain inclinations, θ. Normal and shear stress components for the 42° orientation are identified by the heavy dots.

Maximum Normal and Maximum Shear Stresses

Maximum normal and maximum shear stresses and their orientation are of interest in design. Material failure may depend on both, depending on criteria of failure. The location and magnitude of maximum stresses are determined by calculus. Starting with the normal stress,

$$\sigma_n = \frac{\sigma_x + \sigma_y}{2} + \frac{\sigma_x - \sigma_y}{2} \cos 2\theta + \tau_{xy} \sin 2\theta \tag{8.7}$$

$$\frac{d\sigma_n}{d\theta} = 0 \tag{8.8}$$

gives

$$\tan 2\theta_p = \frac{2\tau_{xy}}{\sigma_x - \sigma_y} \tag{8.9}$$

By substitution

$$\sigma_{p1,p2} = \frac{\sigma_x + \sigma_y}{2} \pm \sqrt{\left(\frac{\sigma_x - \sigma_y}{2}\right)^2 + \tau_{xy}^2} \tag{8.10}$$

Similarly, to determine maximum shear and starting with

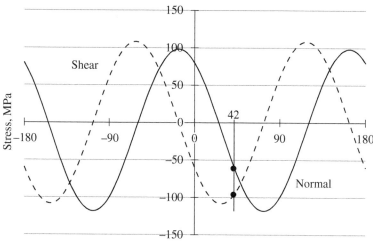

Figure 8.9 Stress components vs. surface orientation.

$$\tau_{nt} = -\frac{\sigma_x - \sigma_y}{2}\sin 2\theta + \tau_{xy}\cos 2\theta \qquad (8.11)$$

$$\frac{d\tau_{nt}}{d\theta} = 0 \qquad (8.12)$$

$$\tan 2\theta_\tau = -\frac{\sigma_x - \sigma_y}{2\tau_{xy}} \qquad (8.13)$$

$$\tau_{max} = \pm\sqrt{\left(\frac{\sigma_x - \sigma_y}{2}\right)^2 + \tau_{xy}^2} \qquad (8.14)$$

Observations:

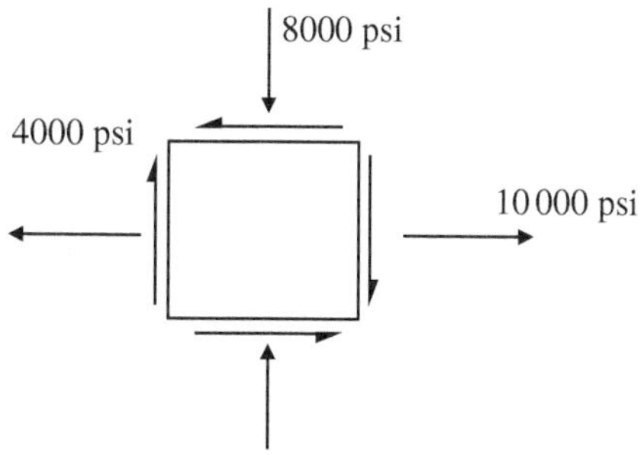

Figure 8.10 Example stress components.

1) Shear stresses are zero on planes of max and min normal stress (principal planes)
2) Principal planes are 45° from planes of max shear stress
3) The two principal planes are 90° apart; perpendicular to each other

Example Consider the stress condition given in Figure 8.10.

$$\sigma_x = 10\,000 \text{ psi}$$

$$\sigma_y = -8000 \text{ psi}$$

$$\tau_{xy} = -4000 \text{ psi}$$

Substituting these numbers into Eq. (8.10) gives, $\sigma_{p1,\,p2} = -8849$ and $+10\,850$ psi (Figure 8.11).

$$\tan 2\theta_p = \frac{2\tau_{xy}}{\sigma_x - \sigma_y}$$

$$\tan 2\theta_p = \frac{2(-4000)}{10\,000 - (-8000)} = -0.444$$

Two values of θ_p satisfy this equation

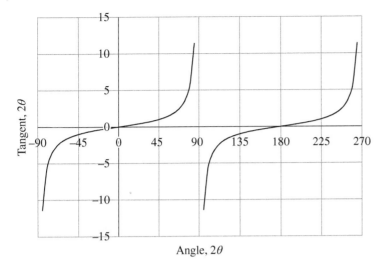

Figure 8.11 Tangent function.

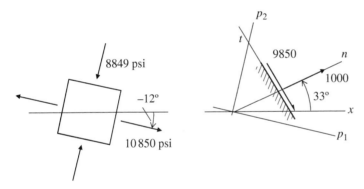

Figure 8.12 Maximum normal and shear stresses.

$$\theta_p = -12°$$

and

$$\theta_p = 78°$$

To determine which principal stress applies to which plane, substitute, say $\theta_p = -12°$ into

$$\sigma_n = \sigma_x \cos^2\theta + \sigma_y \sin^2\theta + 2\tau_{xy} \sin\theta \cos\theta$$

gives $\sigma_p = 10\,850$ psi. Both principal stresses are shown on the element in Figure 8.12. Now consider maximum shear stress:

$$\tan 2\theta_\tau = -\frac{\sigma_x - \sigma_y}{2\tau_{xy}}$$

$$\tan 2\theta_\tau = -\frac{10\,000 - (-8000)}{2(-4000)} = 2.25$$

$$2\theta_{max} = 66°$$

$$\theta_{max} = 33°$$

$$\tau_{max} = -9850 \text{ psi}$$

$$\sigma_n = 1000 \text{ psi}$$

These stresses are also shown in Figure 8.12.

Mohr's Stress Circle

In 1882, Otto Mohr, a German engineer, recognized that transformation of stresses at a point can be represented by a special circle [2]. Mohr's circle gives a graphical representation of the transformation equations and is easy to remember. Normal stresses are located on the horizontal axis (Figure 8.13). Shear stresses are located on the vertical axis. Stress components at a point are located on the circle as follows. σ_x and σ_y are marked on the horizontal axis. The average of these stresses is the center of the circle. Shear stress is measured from these two normal stress points. The coordinates, σ_x, τ_{xy} determine the radius of the circle. In this figure, $\sigma_x \succ \sigma_y$. When τ_{xy} is plus (moment is CCW), it is plotted down in the Mohr diagram:

$$\sigma_x \succ \sigma_y$$

τ_{xy} is plus and moment is CCW. It is plotted down in the Mohr diagram

Both principal stresses can be formulated from the diagram:

$$\sigma_n = \frac{\sigma_x + \sigma_y}{2} + \frac{\sigma_x - \sigma_y}{2}\cos 2\theta + \tau_{xy}\sin 2\theta \tag{8.15}$$

$$\sigma_{p1,p2} = \frac{\sigma_x + \sigma_y}{2} \pm \sqrt{\left(\frac{\sigma_x - \sigma_y}{2}\right)^2 + \tau_{xy}^2} \tag{8.16}$$

$$\tan 2\theta_p = \frac{2\tau_{xy}}{\sigma_x - \sigma_y} \tag{8.17}$$

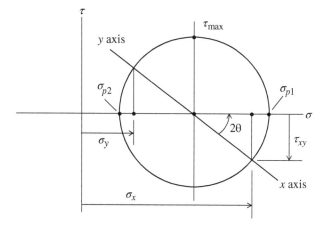

Figure 8.13 Mohr's stress circle.

Maximum shear stress is formulated from the diagram:

$$\tau_{nt} = -\frac{\sigma_x - \sigma_y}{2}\sin 2\theta + \tau_{xy}\cos 2\theta \tag{8.18}$$

$$\tau_{max} = \pm\sqrt{\left(\frac{\sigma_x - \sigma_y}{2}\right)^2 + \tau_{xy}^2} \tag{8.19}$$

$$\tan\theta_\tau = -\frac{\sigma_x - \sigma_y}{2\tau_{xy}} \tag{8.20}$$

Example The use of Mohr's circle is illustrated with the following example. Assume the components of stress at a point are $\sigma_x = +8$ ksi, $\sigma_y = -6$ ksi, and $\tau_{xy} = +4$ ksi (Figure 8.14). The objective is to determine the principal normal stresses, maximum shear stress, σ_{p1}, σ_{p2}, τ_{max}, and the orientation of the elements on which they act.

The center of the circle is

$$\frac{\sigma_x + \sigma_y}{2} = \frac{8 - 6}{2} = 1$$

The radius of the circle is

$$R = \left[\left(\frac{\sigma_x - \sigma_y}{2} + \tau_{xy}^2\right)\right]^{\frac{1}{2}} = \left[\frac{8 + 6}{2} + 4^2\right] = 8.06$$

The x axis is drawn between (1, 0) and (8, −4).

$$\tan 2\theta_p = \frac{4}{7} \tag{8.21}$$

$$2\theta_p = 29.74°$$

$$\theta_p = 14.87°$$

From the circle, the principal normal stresses are

$$\sigma_{p1} = 9.06 \text{ ksi}$$

$$\sigma_{p2} = -7.06 \text{ ksi}$$

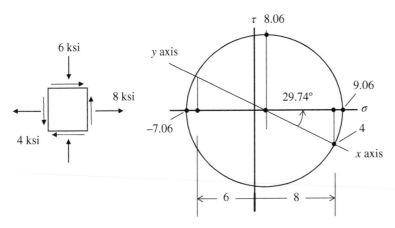

Figure 8.14 Solution by Mohr's circle (ksi stress units).

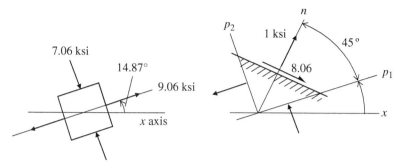

Figure 8.15 Plane of maximum shear stress.

Also, $\tau_{max} = -8.06$ ksi and acts on a plane oriented $45°$ from the principal axis. Principal planes and planes of maximum shear are shown in Figure 8.15.

Theory of Strain

The theory of strain is developed independently from the theory of stress. Both stress and strain are related by Hooke's law, which will be cover later.

There are two types of strain, normal and shear, and they both relate to the two types of stress, normal and shear as previously discussed.

Normal strain is simple elongation (or compression) as shown in Figure 8.16. By definition

$$\varepsilon = \frac{\delta}{L} \quad \text{(dimensionless)} \tag{8.22}$$

In general

$$\varepsilon = \frac{du}{dx} \tag{8.23}$$

(a)   (b)

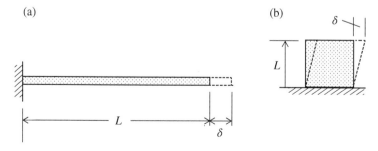

Figure 8.16 Two types of strain. (a) Normal strain. (b) Shear strain.

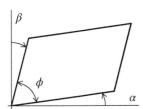

Because internal tension/compression may not be uniform.
The magnitude of normal strain is roughly

$$\varepsilon = \frac{\sigma_{yld}}{E} = \frac{50 \times 10^3}{30 \times 10^6} = 1.67 \times 10^{-3}$$

Normal strain is commonly expressed in micro units,

$$\varepsilon = 1670\,\mu\,\text{strain}$$

$$\varepsilon = 1670 \times 10^{-6}$$

A normal strain of 2000 μ strain is a fairly large strain and a good reference number.
Shear strain is a measure of the amount of distortion as opposed to stretching:

$$\gamma = \frac{\delta}{L} = \tan\alpha \approx \alpha \tag{8.24}$$

Figure 8.16 shows shear strain is a measure of distortion indicated by angle, α. A more general condition of shear distortion is shown in the side drawing. In this case

$$\gamma = \alpha + \beta \tag{8.25}$$

Both angles add to define the total distortion. Shear strain is also expressed by angle, ϕ.

$$\gamma = \frac{\pi}{2} - \phi \tag{8.26}$$

The sign convention for shear strain is

$$\phi < \frac{\pi}{2}\quad \text{shear strain is plus}$$

$$\phi \succ \frac{\pi}{2}\quad \text{shear strain is negative}$$

Strain Transformation

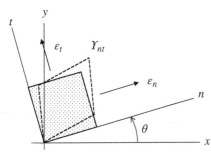

Strain at a point includes both normal and shear strains simultaneously. We will now transform these strains on an element oriented by θ from the x-axis (Figure 8.17).

Normal strains, referenced from the n,t reference frame, can be expressed in terms of ε_x, ε_y, γ_{xy} by

$$\varepsilon_n = \varepsilon_x \cos^2\theta + \varepsilon_y \sin^2\theta + \gamma_{xy}\sin\theta\cos\theta \tag{8.27}$$

In terms of double angle

Figure 8.17 Strain transformation.

$$\varepsilon_n = \frac{\varepsilon_x + \varepsilon_y}{2} + \frac{\varepsilon_x - \varepsilon_y}{2} \cos 2\theta + \frac{\gamma_{xy}}{2} \sin 2\theta \tag{8.28}$$

Maximum normal strains are determined by use of calculus.

$$\varepsilon_{p1}, \varepsilon_{p2} = \frac{\varepsilon_x + \varepsilon_y}{2} \pm \sqrt{\left(\frac{\varepsilon_x - \varepsilon_y}{2}\right)^2 + \left(\frac{\gamma_{xy}}{2}\right)^2} \tag{8.29}$$

$$\tan 2\theta_p = \frac{\gamma_{xy}}{\varepsilon_x - \varepsilon_y}$$

Shear strains, reference from the n,t coordinate reference frame, can be expressed in terms of ε_x, ε_y, γ_{xy} by

$$\frac{\gamma_{nt}}{2} = -\frac{\varepsilon_x - \varepsilon_y}{2} \sin 2\theta + \frac{\gamma_{xy}}{2} \cos 2\theta \tag{8.30}$$

Maximum strain at a point is determined from calculus.

$$\frac{\gamma_{max}}{2} = \pm \sqrt{\left(\frac{\varepsilon_x - \varepsilon_y}{2}\right)^2 + \left(\frac{\gamma_{xy}}{2}\right)^2} \tag{8.31}$$

$$\tan 2\theta_{\gamma\,max} = -\frac{\varepsilon_x - \varepsilon_y}{\gamma_{xy}} \tag{8.32}$$

Note the mathematical similarity between strain and stress transformations.

$$\sigma_n = \sigma_x \cos^2\theta + \sigma_y \sin^2\theta + 2\tau_{xy} \sin\theta \cos\theta \tag{8.33}$$

$$\varepsilon_n = \varepsilon_x \cos^2\theta + \varepsilon_y \sin^2\theta + 2\left(\frac{\gamma_{xy}}{2}\right) \sin\theta \cos\theta \tag{8.34}$$

$$\tau_{nt} = -\frac{\sigma_x - \sigma_y}{2} \sin 2\theta + \tau_{xy} \cos 2\theta \tag{8.35}$$

$$\frac{\gamma_{nt}}{2} = -\frac{\varepsilon_x - \varepsilon_y}{2} \sin 2\theta + \frac{\gamma_{xy}}{2} \cos 2\theta \tag{8.36}$$

The transformation equations for stress and strain are mathematically the same except shear strain is divided by 2. This means, analytical expressions for maximum normal and shear stresses also apply to strain. Mohr's circle also applies to shear strain except the shear strain axis is divided by 2. The sign convention for both stress and stain are compared in Figure 8.18.

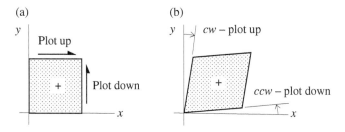

Figure 8.18 Sign convention for shear strain. (a) Stress. (b) Strain.

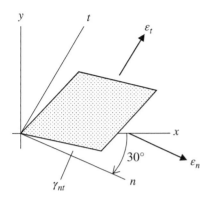

Figure 8.19 Strain transformation.

Example Consider

$$\varepsilon_x = 800\mu$$

$$\varepsilon_y = -1000\mu$$

$$\gamma_{xy} = -600\mu$$

What are the normal strains and shear strain on an element oriented $-30°$ from the x-axis?

By substitution into Eqs. (8.34) and (8.36)

$$\varepsilon_n = 800 \cos^2(-30) - 1000 \sin^2(-30)$$
$$- 600 \sin(-30) \cos(-30)$$

$$\varepsilon_n = 610\mu$$

$$\varepsilon_t = 800 \cos^2(60) - 1000 \sin^2(60) - 600 \sin(60) \cos(60)$$

$$\varepsilon_t = -810\mu$$

$$\gamma_{nt} = -(800 + 1000) \sin(-60) + [(-600) \cos(-60)]$$

$$\gamma_{nt} = 1259\mu$$

These strains are shown in Figure 8.19.

Mohr's Strain Circle

The use of Mohr's strain circle to determine principal strains and maximum shear strains is illustrated below. Local strain components are

$$\varepsilon_x = 1200\mu$$

$$\varepsilon_y = -600\mu$$

$$\gamma_{xy} = 900\mu$$

Mohr's circle for this example is shown in Figure 8.20.

Again, planes of maximum stress and strain are the same. No shear stress and no shear strain (Figure 8.21).

$$\tan 2\theta = \frac{450}{900} = 0.5$$

$$2\theta = 26.6 \quad (\theta = 13.3°)$$

Principal Axes of Stress and Strain

From the theories of stress and strain, we know that there is no shear stress on the principal planes; therefore, there is no shear strain as shear stress and shear strain are related through

$$\tau = G\gamma \tag{8.37}$$

This means that the principal planes of normal stress are the same as the principal planes of normal strain (Figure 8.22).

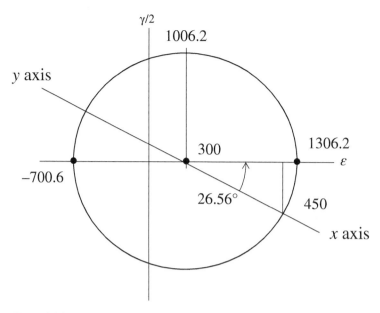

Figure 8.20 Mohr's strain circle.

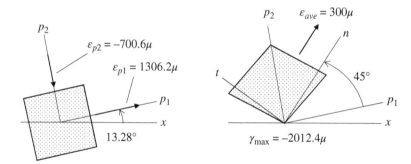

Figure 8.21 Planes of maximum normal and shear strains.

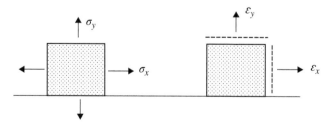

Figure 8.22 Comparison of normal stress and strain.

Generalized Hooke's Law

The theories of stress and strain were developed independent of each other. Both variables are related through Hooke's law. Laboratory tests show that when a load is applied to a test rod, the rod stretches elastically. The relation between applied stress and resulting strain in a uniaxial state of stress is expressed by

$$\varepsilon = \frac{\sigma}{E} \tag{8.38}$$

The strain has the same direction as the stress.

Tests show that stress in one direction not only produces strain in the same direction but also a shrinking normal strain in transverse directions. The amount of shrinking is related to the strain in the direction of the load by

$$\nu = -\frac{\varepsilon_T}{\varepsilon_A} = -\frac{\text{transverse}}{\text{axial}} \tag{8.39}$$

This ratio is called Poisson's ratio and is a property of the material. It is determined from a simple uniaxial state of stress. Equations (8.38) and (8.39) are basic to the generalized Hooke's law for a triaxial state of stress.

$$\varepsilon_x = \frac{1}{E}\left[\sigma_x - \nu\left(\sigma_y + \sigma_z\right)\right] \tag{8.40}$$

$$\varepsilon_y = \frac{1}{E}\left[\sigma_y - \nu\left(\sigma_x + \sigma_z\right)\right] \tag{8.41}$$

$$\varepsilon_z = \frac{1}{E}\left[\sigma_z - \nu\left(\sigma_y + \sigma_x\right)\right] \tag{8.42}$$

Be aware that in general, the ratio of transverse strains is *not* equal to Poisson's ratio. The rules for finding Poisson's ratio must follow the procedure described above.

Hooke's law includes the effects of direct stress plus Poisson's effect due to stresses in the other two transverse directions. Two common engineering situations will now be addressed: (i) plane stress and (ii) plane strain.

Theory of Plain Stress

Under plane stress, $\sigma_z = 0$; however $\varepsilon_z \neq 0$ according to Figure 8.25a. An example is zero stress on a free surface. Stresses lie in the plane of the surface, but stress normal to the surface is zero. Applying Hooke's law gives

$$\varepsilon_x = \frac{1}{E}\left[\sigma_x - \nu\sigma_y\right] \tag{8.43}$$

$$\varepsilon_y = \frac{1}{E}\left[\sigma_y - \nu\sigma_x\right] \tag{8.44}$$

$$\varepsilon_z = -\frac{\nu}{E}\left[\sigma_x + \sigma_y\right] \tag{8.45}$$

By substitution gives

$$\varepsilon_z = -\frac{\nu}{1-\nu}\left(\varepsilon_x + \varepsilon_y\right) \tag{8.46}$$

Note that

$$\varepsilon_z \neq \nu\left(\varepsilon_x + \varepsilon_y\right) \tag{8.47}$$

Rearranging Eqs. (8.43) and (8.44) gives expressions for stress in terms of strain:

$$\sigma_x = \frac{E}{1 - \nu^2}\left[\varepsilon_x + \nu\varepsilon_y\right] \tag{8.48}$$

$$\sigma_y = \frac{E}{1 - \nu^2}\left[\varepsilon_y + \nu\varepsilon_x\right] \tag{8.49}$$

Shear stress is not involved in either equation but relates to shear strain by a separate equation:

$$\tau_{xy} = G\gamma_{xy} \tag{8.50}$$

The modulus of elasticity (E) is determined experimentally from a simple tension test. The modulus of rigidity (G), however, is determined mathematically by

$$G = \frac{E}{2(1 + \nu)} \tag{8.51}$$

From this example, using $E = 30 \times 10^6$ and $\nu = 0.25$, with

$$G = \frac{E}{2(1 + \nu)} = \frac{30}{2(1 + 0.25)} 10^6 = 12 \times 10^6$$

Orientation of Principal Stress and Strain

Consider the state of stress shown in Figure 8.23. The center diagram shows principal stresses. The right diagram shows principal strain. The planes on orientation are the same, indicating zero shear stress and zero shear strain.

- Maximum normal stress is 8.71 ksi
- Maximum normal strain is 330μ
- Principal directions for both stress and strain are the same
- Maximum shear stress is 6.71 ksi
- Maximum shear strain is 559μ

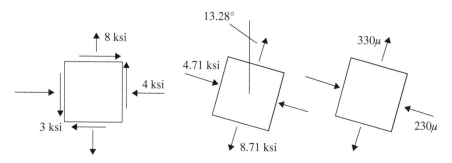

Figure 8.23 Plane stress and corresponding strain.

Example Strain components at a point in a structure are given as

$$\varepsilon_x = 2000\mu$$
$$\varepsilon_y = 1000\mu$$
$$\gamma_{xy} = 1000\mu$$

Using Mohr's strain circle (Figure 8.24) to determine the principal stains and principal stresses assuming plane stress conditions.

The radius, R, of the circle is

$$R = \left[500^2 + 500^2\right]^{\frac{1}{2}} = 707.1$$

Principal strains are

$$\varepsilon_1 = 1500 + 707 = 2207\mu$$
$$\varepsilon_2 = 1500 - 707.1 = 793\mu$$

Shear strain is zero on these planes.
Principal stresses for *plane stress* are determined from

$$\sigma_{p1} = \frac{E}{1 - \nu^2}\left(\varepsilon_{p1} + \nu\varepsilon_{p2}\right)$$

$$\sigma_{p1} = \frac{30 \times 10^6}{1 - 0.28^2}(2207 + 0.28 \times 793)10^{-6}$$

$$\sigma_{p1} = 79\,000 \text{ psi}$$

$$\sigma_{p2} = \frac{30 \times 10^6}{1 - 0.28^2}(793 + 0.28 \times 2207)10^{-6}$$

$$\sigma_{p2} = 45\,900 \text{ psi}$$

Shear stress is zero on these planes.

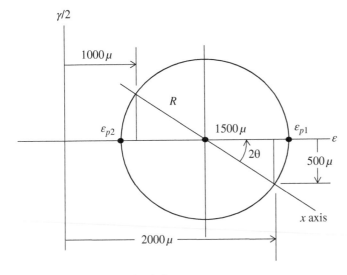

Figure 8.24 Mohr's strain circle.

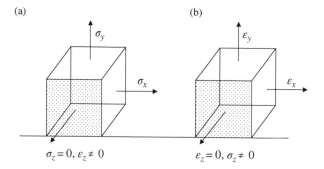

Figure 8.25 Elements of plain stress and plane strain. (a) Plane stress. (b) Plane strain.

Theory of Plain Strain

Plain strain exists when an element is not allowed to move in the z-direction ($\varepsilon_z = 0$), creating stress in the z-direction (Figure 8.25b). When $\varepsilon_z = 0$, generalized Hooke's law gives

$$\sigma_x = \frac{E}{(1+v)(1-2v)}\left[(1-v)\varepsilon_x + v\varepsilon_y\right] \tag{8.52}$$

$$\sigma_y = \frac{E}{(1+v)(1-2v)}\left[(1-v)\varepsilon_y + v\varepsilon_x\right] \tag{8.53}$$

$$\sigma_z = \frac{E}{(1+v)(1-2v)}\left[v\left(\varepsilon_x + \varepsilon_y\right)\right] \tag{8.54}$$

Still

$$\tau_{xy} = G\gamma_{xy} \tag{8.55}$$

Pressure Vessel Strain Measurements

Analytical Predictions of Stress and Strain

One of the laboratory experiments at the University of Tennessee involved attaching strain rosettes to a pressure vessel to gain experience in the use of strain gauges. The experiment allowed strain measurements to be checked against theoretically calculated strains (and stresses) as predicted by thin-wall vessel theory. The pressure vessel had a mean diameter of 5.585 in. and a wall thickness of 0.04 in. (~20-gauge sheet metal having thickness of 0.0375 in.). The sheet metal was rolled into a cylinder and welded longitudinally. Spherical caps, of the same thickness, were welded to the cylinder. The radius to thickness ratio was

$$\frac{r_{mean}}{t} = 69.81 \succ 10$$

so the pressure vessel was well within thin-wall theory range. The cylinder and end caps were formed of steel having yield strength of 60 000 psi. Thin wall stress equations are

$$\sigma_\theta = \frac{r}{t}p \quad \text{(cylinder)} \tag{8.56}$$

$$\sigma_a = \frac{r}{2t}p \quad \text{(cylinder)} \tag{8.57}$$

$$\sigma_s = \frac{r}{2t}p \quad \text{(sphere)} \tag{8.58}$$

Before pressuring the tank, it is important to know the pressure limit or the pressure which would cause material yielding. Using the von Mises' criteria of failure

$$\sigma' \geq \sigma_{yld} \quad \text{(yielding occurs)} \tag{8.59}$$

$$(\sigma')^2 = \sigma_A^2 - \sigma_A\sigma_B + \sigma_B^2 \tag{8.60}$$

$$(\sigma')^2 = \left(\frac{r}{t}p\right)^2\left[1^2 - \frac{1}{2} + \left(\frac{1}{2}\right)^2\right]$$

$$60 = \frac{r}{t}p(0.866)$$

$$p = 992.5 \text{ psi}$$

Cylinder – Yielding occurs when $p = 992.5$ psi.
Sphere – Yielding occurs when $p = 1{,}719$ psi.

From this, we see that a pressure limit of 400 psi is well within the yield pressure limit. It is also a good idea to get a feel for expected levels of strain before starting to gather strain data. Recall that in a standard uniaxial stress–strain test the strain corresponding to a yield stress of 60 000 psi is

$$\varepsilon = \frac{60 \times 10^3}{30 \times 10^6} = 0.002 \text{ in./in. or } 2000\mu\text{(micro strain)}$$

Therefore, we would expect strain data in the range of 1000μ or less.

Assume the vessel is subjected to 400 psi. From the thin wall equations:

$$\sigma_\theta = \sigma_{p1} = 27\,924 \text{ psi}$$

$$\sigma_a = \sigma_{p2} = 13\,962 \text{ psi}$$

Assuming $E = 30 \times 10^6$ and $\nu = 0.29$, the expected principal strain levels are

$$\varepsilon_{p1} = \frac{1}{E}\left[\sigma_{p1} - \nu\sigma_{p2}\right] = 795.8\mu$$

$$\varepsilon_{p2} = \frac{1}{E}\left[\sigma_{p2} - \nu\sigma_{p1}\right] = 195.5\mu$$

These points are located on the Mohr's strain circle shown in Figure 8.26. The diagram also projects the expected strains to be measured by the strain rosette.

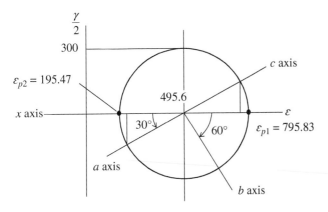

Figure 8.26 Mohr's circle.

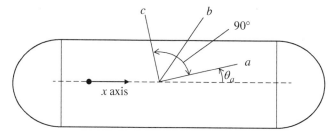

Figure 8.27   Strain rosette orientation.

Consider the strains measured by the strain rosette oriented by 15° as shown in Figure 8.27. The three gauges are oriented as follows:

$\theta_a = 15°$

$\theta_b = 60°$

$\theta_c = 105°$

The expected normal strain values in each of the three individual gauges with vessel pressure of 400 psi are

$$\varepsilon_a = 495.6 - 300 \cos 30 = 235.8\mu$$

$$\varepsilon_b = 495.6 + 300 \cos 60 = 645.6\mu$$

$$\varepsilon_c = 495.6 + 300 \cos 30 = 755.4\mu$$

which are determined directly from Mohr's strain circle.

Strain in the Spherical Cap

Since stress (and strain) in the spherical caps is the same in all directions, Mohr's strain circle for at any point in the end caps is simple a point. This also means that strain measurements by a three-gauge rosette are the same in each of the three gauges. Also, there is no shear stress or strain in the spherical cap.

According to Eq. (8.43), the strain in each rosette gauge should be

$$\varepsilon = \frac{1}{E}[1 - v]\frac{r}{2t}p \tag{8.61}$$

For the example above, assuming $v = 0.29$,

$$\varepsilon = \frac{1}{30}(0.71)\frac{69.81}{2}400 = 330\mu$$

Conversion of Strain Measurements to Principal Strains and Stresses

Strain gauges are used to experimentally determine local strain at a point. A single gauge may be used in a uniaxial loaded member. Only two gauges, 90° apart, are required to determine strain at a point if aligned with known principal strain/stress directions. However, strain at a point usually requires a strain rosette containing three gauges because principal strain directions are unknown.

The basic construction of a strain gauge is shown in Figure 8.28. The "Grid" is composed of a metallic wire with a known electrical resistance. Manufactured strain gauges have a known relation

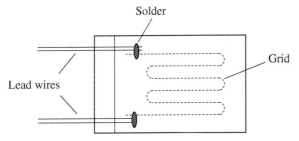

Figure 8.28 Strain gauge schematic.

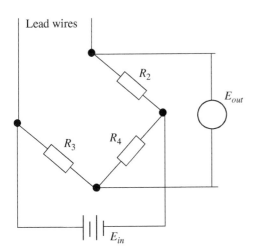

Figure 8.29 Strain gauge used as the fourth resistance in a Wheatstone bridge.

between their change in elongation and associated change in electrical resistance. This relation is known as the "gauge factor," or GF. The "lead wires" may be sent directly to an Ohmmeter to measure resistance; however, the lead wires are more commonly used in a Wheatstone bridge configuration. The "backing" typically contains an adhesive for locking the gauge to the material of interest. Thus, as the material is strained, the wires in the grid are strained by the same amount.

Wheatstone bridge circuits are commonly used in conjunction with strain gauges because they allow small changes in resistance to be measured accurately. This is important because elongation-induced strains in the gauge result in very small resistance changes as compared to the gauge's initial resistance. The Wheatstone bridge is shown in Figure 8.29. Resistance R_1 (not shown) in the bridge is the strain gauge. The output voltage, E_0, may be measured to determine the change in resistance of the strain gauge. When the output voltage equals 0, the bridge is said to be *balanced*, meaning the resistances in opposite legs are equal, and there is no measurable strain [3].

Strain rosettes may come in different gauge arrangements (Figure 8.30). They differ by the orientation of the gauges.

Rosette (a) $\theta_a = 0°$ $\theta_b = 45°$ $\theta_c = 90°$
Rosette (b) $\theta_a = 0°$ $\theta_b = 90°$ (used when principal directions known)
Rosette (c) $\theta_a = 0°$ $\theta_b = 60°$ $\theta_c = 120°$

Two things to note about strain gauges: (i) they do not measure shear strain and (ii) they do not measure stress. They only measure normal strain in one direction. But knowing three normal

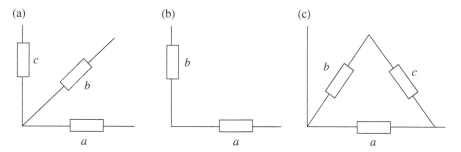

Figure 8.30 Strain rosette arrangements. (a) Rosette *a*. (b) Rosette *b*. (c) Rosette *c*.

strains in three arbitrary directions allows us to *calculate* shear strain and principal stresses at a point.

If normal strains are measured in three arbitrary directions, they can be used to find strains (ε_x, ε_y, γ_{xy}) associated with any *xy* coordinate system. From the theory of strain discussed earlier

$$\varepsilon_a = \varepsilon_x \cos^2\theta_a + \varepsilon_y \sin^2\theta_a + \gamma_{xy} \sin\theta_a \cos\theta_a \tag{8.62}$$

$$\varepsilon_b = \varepsilon_x \cos^2\theta_b + \varepsilon_y \sin^2\theta_b + \gamma_{xy} \sin\theta_b \cos\theta_b \tag{8.63}$$

$$\varepsilon_c = \varepsilon_x \cos^2\theta_c + \varepsilon_y \sin^2\theta_c + \gamma_{xy} \sin\theta_c \cos\theta_c \tag{8.64}$$

Theta, θ, is the angular orientation of each strain gauge.

Strain measurements on a free surface are useful for defining strain at a given location. Once the state of strain is determined, stresses are calculated from Hooke's law.

In matrix form

$$\begin{pmatrix} \varepsilon_a \\ \varepsilon_b \\ \varepsilon_c \end{pmatrix} = \begin{bmatrix} A \end{bmatrix} \begin{pmatrix} \varepsilon_x \\ \varepsilon_y \\ \gamma_{xy} \end{pmatrix} \tag{8.65}$$

Strain referenced from the *xy* frame are determined from inverting Eq. (8.65).

$$\begin{pmatrix} \varepsilon_x \\ \varepsilon_y \\ \gamma_{xy} \end{pmatrix} = \begin{bmatrix} A \end{bmatrix}^{-1} \begin{pmatrix} \varepsilon_a \\ \varepsilon_b \\ \varepsilon_c \end{pmatrix} \tag{8.66}$$

Once ε_x, ε_y, γ_{xy} have been determined, ε_{p1}, ε_{p2}, γ_{max} can be determined analytically or by using Mohr's strain circle as discussed above. Principal stresses are determined from Hooke's law, according to the theory of plane stress as previously discussed.

Example Consider a biaxial state of plane stress ($\sigma_z = 0$) is shown in Figure 8.31, along with a strain rosette. Determine the strains (ε_a, ε_b, ε_c) that would be indicated in the strain rosette.

Given

$$E = 100 \text{ GPa}$$

$$\nu = 0.28$$

$$G = \frac{E}{2(1+\nu)} = 39.06 \text{ GPa}$$

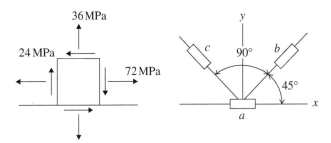

Figure 8.31 Strain rosette.

Approach: Determine $(\varepsilon_x, \varepsilon_y, \gamma_{xy})$, then determine $(\varepsilon_a, \varepsilon_b, \varepsilon_c)$.

$$\varepsilon_x = \frac{1}{E}\left(\sigma_x - \nu\sigma_y\right)$$

$$\varepsilon_x = \frac{1}{100 \times 10^3}[72 - 0.28(36)] = 619.2\mu$$

$$\varepsilon_a = \varepsilon_x = +619.2\,\mu\ \text{strain}$$

$$\varepsilon_y = \frac{1}{E}\left(\sigma_y - \nu\sigma_x\right) = 158.4\mu\ \text{strain}$$

$$\varepsilon_x = \frac{1}{100 \times 10^3}[36 - 0.28(72)] = 158.4\mu$$

$$\gamma_{xy} = \frac{\tau_{xy}}{G} = \frac{-24}{(39.06)10^3} = -614.4\mu\ \text{strain}$$

Using

$$\varepsilon_n = \varepsilon_x \cos^2\theta + \varepsilon_y \sin^2\theta + \gamma_{xy} \sin\theta \cos\theta$$

$$\varepsilon_b = +81.6\mu\ \text{strain}$$

$$\varepsilon_c = +696\mu\ \text{strain}$$

Beam Deflections

Shear and bending moment diagrams were discussed earlier along with shear and bending stresses. This section revisits beam behavior by looking at beam displacements. The earlier discussion showed that

$$\frac{EI}{\rho} = M \tag{8.67}$$

Recall from calculus, local curvature of any function, $y(x)$ is determined from

$$\frac{1}{\rho} = \frac{\frac{d^2y}{dx^2}}{\left[1 + \left(\frac{dy}{dx}\right)^2\right]^{\frac{3}{2}}} \tag{8.68}$$

In applying this expression, we assume beam deflections are small so, $\frac{dy}{dx} \approx 0$. Applying this assumption to Eq. (8.68) leaves

$$\frac{1}{\rho} \cong \frac{d^2y}{dx^2} \tag{8.69}$$

In some cases, such as pipelines suspended in the ocean, this assumption is invalid and the total expression for curvature must be used.

By substitution

$$EI\frac{d^2y}{dx^2} = M \tag{8.70}$$

This equation is linear and amendable to mathematical solutions.

Cantilever Beam with Concentrated Force

Consider a cantilever beam with concentrated load at the end (Figure 8.32).

$$M(x) = -(L-x)F \tag{8.71}$$

$$\frac{d^2y}{dx^2} = -\frac{F}{EI}(L-x) \tag{8.72}$$

By direct integration

$$\frac{dy}{dx} = \frac{F}{EI}\frac{1}{2}(L-x)^2 + C_1$$

$$y(x) = -\frac{F}{6EI}(L-x)^2 + C_1x + C_2$$

Applying boundary conditions: $\left.\frac{dy}{dx}\right|_{x=0} = 0$ and $y(0) = 0$.

$$C_1 = -\frac{FL^2}{2EI}$$

$$C_2 = \frac{FL^3}{6EI}$$

Combining everything

$$y(x) = -\frac{F}{6EI}\left[(L-x)^3 + 3L^2x - L^3\right]$$

Figure 8.32 Deflection of cantilever beam.

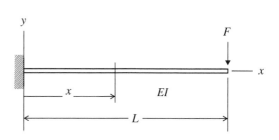

which reduces to

$$y(x) = -\frac{Fx^2}{6EI}(3L - x) \tag{8.73}$$

The maximum displacement and slope occur at $x = L$.

$$y_{max} = -\frac{FL^3}{3EI}$$

The slope along the beam is

$$\theta(x) = -\frac{F}{6EI}\left[-x^2 + 2x(3L - x)\right] \tag{8.74}$$

with maximum slope at $x = L$.

$$\theta_{max} = -\frac{FL^2}{2EI}$$

Cantilevered Beam with Uniform Load

If the applied load is distributed as shown in Figure 8.33, then

$$EI\frac{d^2y}{dx^2} = -w(L - x)\frac{1}{2}(L - x) = -\frac{w}{2}(L - x)^2 \tag{8.75}$$

The solution to this equation is determined by direct integration.

$$EI\frac{dy}{dx} = \frac{w}{2}\frac{(L - x)^3}{3} + C_1$$

$$EIy(x) = -\frac{w}{2}\frac{(L - x)^4}{12} + C_1 x + C_2$$

Boundary conditions for this beam are

$$EI\frac{dy}{dx}\bigg|_{x=0} = 0 \quad \text{yielding} \quad C_1 = -\frac{w}{2}\frac{L^3}{3}$$

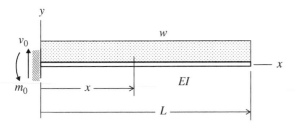

Figure 8.33 Deflection of a uniformly loaded beam.

$$y(0) = 0 \quad \text{yielding} \quad C_2 = +\frac{w}{2}\frac{L^4}{12}$$

After applying the boundary conditions, the final solution becomes

$$y(x) = -\frac{1}{24}\frac{w}{EI}\left[(L-x)^4 - L^4 + 4L^3 x\right] \tag{8.76}$$

Expanding the first term by the binomial series and collecting terms gives

$$y(x) = -\frac{wx^2}{24EI}\left[x^2 + 6L^2 - 4Lx\right]$$

The maximum deflection and slope occur at $x = L$.

$$y_{max} = -\frac{1}{8}\frac{wL^4}{EI} \quad \text{and} \quad \theta_{max} = -\frac{wL^3}{6EI}$$

Simply Supported Beam with Distributed Load

The differential equation of bending can also be casts into another form as follows:
Since $\frac{dV}{dx} = w$ and $\frac{dM}{dx} = V$,

$$EI\frac{d^3y}{dx^3} = \frac{dM}{dx} = V \quad (\text{using Eq.(8.70)}) \tag{8.77}$$

giving

$$EI\frac{d^4y}{dx^4} = \frac{dV}{dx} = w \tag{8.78}$$

Applying this equation to the beam loading shown in Figure 8.34 gives

$$EI\frac{d^4y}{dx^4} = -w \tag{8.79}$$

By direct integration

$$EI\frac{d^3y}{dx^3} = -wx + C_1$$

Figure 8.34 Simply supported beam with uniform load.

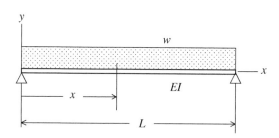

$$EI\frac{d^2y}{dx^2} = -\frac{wx^2}{2} + C_1 x + C_2$$

$$EI\frac{dy}{dx} = -\frac{wx^3}{2} + \frac{1}{2}x^2 C_1 + C_2 x + C_3$$

$$EIy(x) = -w\frac{x^4}{24} + C_1\frac{x^3}{6} + C_2\frac{x^2}{2} + C_3 x + C_4 \qquad (8.80)$$

Boundary conditions for this beam are

$$y(0) = 0$$

$$\left.\frac{d^2y}{dx^2}\right|_{x=0} = 0$$

$$y(L) = 0$$

$$\left.\frac{d^2y}{dx^2}\right|_{x=L} = 0$$

The first two boundary condition yield, C_4 and C_2 both equal zero. The second two boundary conditions produce

$$0 = -w\frac{L^4}{24} + C_1\frac{L^3}{6} + C_3 L$$

and

$$0 = -w\frac{L^2}{2} + C_1 L$$

Solving both equations simultaneously gives

$$C_1 = \frac{wL}{2} \quad \text{and} \quad C_3 = -\frac{wL^3}{24}$$

Collecting terms gives

$$EIy(x) = -\frac{w}{24}x^4 + \frac{wL}{12}x^3 - \frac{wL^3}{24}x \qquad (8.81)$$

$$y(x) = -\frac{wx}{24\,EI}\left(L^3 - 2\,Lx^2 + x^3\right)$$

The maximum displacement and slope are $x = L/2$.

$$y_{max} = -\frac{5wL^4}{384EI} \quad \text{occurs at } x = L/2 \qquad (8.82)$$

$$\theta_{max} = \frac{wL^3}{24EI} \quad \text{occurs at } x = 0 \text{ and } x = L \qquad (8.83)$$

Statically Indeterminate Beams

When a static analysis defines all forces, the procedure for determining shear and moment diagrams and forces are determined as explained above. In some cases, beam forces cannot be determined by a static analysis alone and may require the use of deflection equations.

Figure 8.35 Statically indeterminate beam.

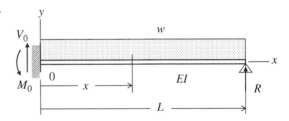

Consider the beam shown in Figure 8.35, with a fixed boundary at the left end and a simple support at the right end. Three unknowns are M_0, V_0, R. These external loads are needed before shear and bending moment diagrams (as well as shear and bending stresses) can be established. Three equations are needed. Two come from a force analysis.

$$\sum F_y = 0 \quad V_0 + R - wL = 0 \tag{8.84}$$

$$\sum M_0 = 0 \quad LR - w\frac{L^2}{2} + M_0 = 0 \tag{8.85}$$

The third comes from the deflection equation:

$$EI\frac{d^2y}{dx^2} = (L-x)R - \frac{w}{2}(L-x)^2 \tag{8.86}$$

By direct integration:

$$EI\frac{dy}{dx} = -\frac{1}{2}(L-x)^2R + \frac{w}{6}(L-x)^3 + C_1$$

$$EIy(x) = \frac{R}{6}(L-x)^3 - \frac{w}{24}(L-x)^4 + C_1x + C_2 \tag{8.87}$$

Appling boundary conditions:

$$y(L) = 0 \text{ gives } C_2 = -C_1 L$$

$$y(0) = 0 \text{ gives } C_2 = -\frac{RL^3}{6} + \frac{wL^4}{24}$$

With these constants, Eq. (8.87) defines beam displacements in terms of the reaction force, R. The force, R, is determined from a third boundary condition,

$$\left.\frac{dy}{dx}\right|_{x=0} = 0$$

Giving

$$R = \frac{3}{8}wL \tag{8.88}$$

Substituting R into Eqs. (8.84) and (8.85) gives

$$V_0 = \frac{5}{8}wL \quad \text{and} \quad M_0 = \frac{1}{8}wL^2 \tag{8.89}$$

Multispanned Beam Columns

The effect of stabilizer location on magnitude and direction of drill bit side force can be calculated by treating the stabilized section as a multispanned beam column. One method is outlined in Ref. [4]. The goal is to produce the needed side force on drill bits to maintain or affect a desired well bore direction. This example is based on the following input data:

Hole size – 8½ in.
Drill collar size – 6¾ × 2 · 13/16 in.
Drilling mud weight – 12 ppg
Inclination of the well bore – 30°
WOB – 40 000 lb

    Two stabilizers used, one at 20 ft the other at 60 ft from the drill bit. Two other support points are at the drill bit and the contact point at the tangent point at the upper end, which is to be determined.

    The output gives an accurate prediction of drill collar deflection from the drill bit to point of contact with the wall. Collar deflections between drill bit and the first stabilizer are shown in Figures 8.36 and 8.37.

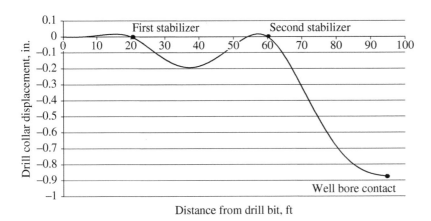

Figure 8.36   Deflections over a drill collar span.

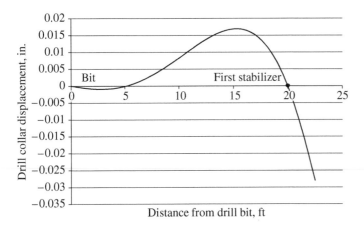

Figure 8.37   Drill collar deflections between drill bit and first stabilizer.

Table 8.1 Transverse side forces.

Location	Transverse force (lb)
Drill bit	181
First stabilizer	1088
Second stabilizer	1481
Wall contact	471[a]

[a] Shear is actually minus at the upper tangent location.

Predicted side forces developed at drill bit, the stabilizers and wall contact are listed in Table 8.1. Since the side force at the drill bit is only 181 lb while weight on bit (WOB) is 40 000 lb, the stabilizer arrangement is considered a holding assembly.

Large Angle Bending in Terms of Polar Coordinates

In some cases, bending is deflected by circular guides, such as pipe in a curved bore hole. It is convenient in this case to set-up equations of bending in polar coordinates. This avoids the complexity of nonlinearities associated with rectangular coordinates.

The differential element of length $ds = R\, d\theta$ is used to establish the necessary bending equations in terms of polar coordinates (Figure 8.38).

Summing forces in the "n" direction gives

$$\frac{dV}{ds} = q + \frac{T}{R} \tag{8.90}$$

Summing forces in the "t" direction gives

$$\frac{dT}{ds} = -\frac{V}{R} \tag{8.91}$$

Summing moments about the center of curvature of the well bore gives

$$\frac{dM}{ds} - \frac{d(Tw)}{ds} + R\frac{dT}{ds} = 0 \tag{8.92}$$

By substitution

$$\frac{dM}{ds} - V - T\frac{dw}{ds} = 0 \tag{8.93}$$

$$\frac{d^2M}{ds^2} = \frac{dV}{ds} + \frac{d}{ds}\left[T\frac{dw}{ds}\right]$$

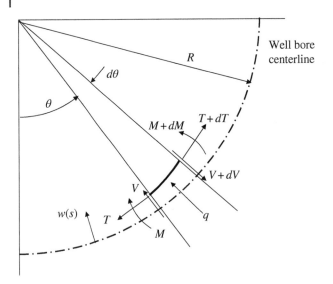

Figure 8.38 Equations of bending in terms of polar coordinates.

$$\frac{d^2M}{ds^2} = q + \frac{T}{R} + \frac{d}{ds}\left[T\frac{dw}{ds}\right]$$

The curvature of the displaced beam is expressed in terms of beam deflection (see Timoshenko and Woinowsky-Krieger [5], pp. 503 and 504)

$$\frac{1}{\rho} = \frac{1}{R} + \frac{w}{R^2} + \frac{d^2w}{ds^2} \tag{8.94}$$

Assuming Euler bending

$$M = EI\frac{1}{\rho} \tag{8.95}$$

and combining the above equations gives

$$q + \frac{T}{R} + \frac{d}{ds}\left[T\frac{dw}{ds}\right] = \frac{EI}{R^2}\frac{d^2w}{ds^2} + EI\frac{d^4w}{ds^4} \tag{8.96}$$

$$EI\frac{d^4w}{ds^4} + \frac{EI}{R^2}\frac{d^2w}{ds^2} - \frac{d}{ds}\left[T\frac{dw}{ds}\right] = q + \frac{T}{R} \tag{8.97}$$

In general, T can vary with s; however, in many cases, T can be assumed constant over stepwise pipe intervals. In which case, Eq. (8.97) simplifies to

$$\frac{d^4w}{ds^4} - \lambda^2\frac{d^2w}{ds^2} = \zeta \tag{8.98}$$

where

$$\lambda^2 = \frac{T}{EI} - \frac{1}{R^2}$$

$$\zeta = \frac{Rq + T}{REI}$$

In this discussion q, the distributed load is assumed. The solution to the differential equation is

$$w(s) = -\frac{1}{2}\chi s^2 + As + B + C\cosh\lambda s + D\sinh\lambda s \tag{8.99}$$

Letting

$$\chi = \frac{T}{RT - \dfrac{EI}{R}}$$

Once $w(s)$ has been determined, shear and bending moment can be established. Shear force along the pipe

$$V = \frac{dM}{ds} - T\frac{dw}{ds} \tag{8.100}$$

Bending moment along the pipe is

$$M = EI\left(\frac{1}{R} + \frac{w}{R^2} + \frac{d^2w}{ds^2}\right) \tag{8.101}$$

In this case, λ^2 is assumed to be plus. In some cases, this coefficient could be minus making the bending equation:

$$\frac{d^4w}{ds^4} + \lambda^2\frac{d^2w}{ds^2} = \frac{q}{EI} \tag{8.102}$$

For example, T could be a compressive force. The second term now has a $(+)$ sign and this leads to a complimentary solution with sines and cosines.

Bending Stresses in Drill Pipe Between Tool Joints

Bending caused by local dog legs can be significant. Consider the schematic (Figure 8.39) showing tool joints lying flat against the inside of a curved well bore. A drill pipe section of length, L, is assumed to be under uniform tension, T. This condition creates high localized bending at each tool joint as explained below. In this case, pipe deflection, which yields bending moments and stress are best determined by use of polar coordinates.

Consider a joint of drill pipe bent within a dog leg. The magnitude of bending stresses depends both on local hole curvature and tension in the drill pipe. When curved pipe is pulled, the middle portion tends to straighten out and shift more curvature toward tool joints. Bending stresses are greatest near tool joints. Total stress is the sum of stress caused by direct pull and bending stress. The analysis starts with

$$\frac{d^4w}{ds^4} - \lambda^2\frac{d^2w}{ds^2} = \zeta \tag{8.103}$$

where

$$\lambda^2 = \frac{T}{EI} - \frac{1}{R^2}$$
$$\zeta = \frac{Rq + T}{REI}$$

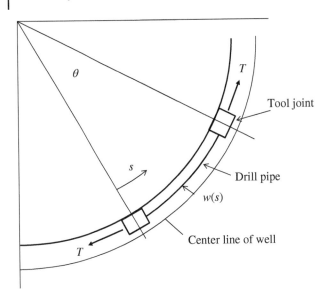

Figure 8.39 Bending of a pipe section between tool joints.

R – radius of centerline

Furthermore, let

$$\chi = \frac{Rq + T}{RT - \frac{EI}{R}}$$

In applying Eq. (8.103), we assume side-loading, q, is negligible. The solution to Eq. (8.103) is

$$w(s) = -\frac{1}{2}\chi s^2 + As + B + C\cosh \lambda s + D\sinh \lambda s \tag{8.104}$$

Letting

$$\chi = \frac{T}{RT - \frac{EI}{R}}$$

$$\lambda^2 = \frac{T}{EI} - \frac{1}{R^2}$$

and applying the following boundary conditions:

$$w(0) = w_0 \quad w(\ell) = w_0$$

$$\left.\frac{dw}{ds}\right|_{s=0} = 0 \quad \left.\frac{dw}{ds}\right|_{s=\ell} = 0$$

establish the four constants A, B, C, and D.

1) $B + C = 0$
2) $A\ell + B + C\cosh \lambda\ell + D\sinh \lambda\ell = \frac{1}{2}\chi\ell^2$
3) $A + D\lambda = 0$

4) $A + C\lambda \sinh \lambda l + D\lambda \cosh \lambda l = \chi l$

Combining (1)–(3) gives

$$Al\left(1 - \frac{1}{\lambda l} \sinh \lambda l\right) - C(1 - \cosh \lambda l) = \frac{1}{2}\chi l^2 \tag{8.105}$$

Combining (3) and (4) gives

$$A(1 - \cosh \lambda l) + C\lambda \sinh \lambda l = \lambda l$$

or

$$Al(1 - \cosh \lambda l) + C\lambda l \sinh \lambda l = \lambda l^2 \tag{8.106}$$

Equations (8.105) and (8.106) are two algebraic equation with A and C as unknowns:

$$\begin{bmatrix} a_{11} & a_{12} \\ a_{21} & a_{22} \end{bmatrix} \begin{Bmatrix} A \\ C \end{Bmatrix} = \begin{Bmatrix} b_1 \\ b_2 \end{Bmatrix}$$

where

$a_{11} = l\left(1 - \frac{1}{\lambda l}\right) \sinh \lambda l$
$a_{12} = \cosh \lambda l - 1$
$a_{21} = l(1 - \cosh \lambda l)$
$a_{22} = \lambda l \sinh \lambda l$

Using Cramer's rule with $l = 30$ ft, $\lambda = 0.0153$ and $\lambda l = 5.502$

$A = 0.015$

$C = 0.9895$

$$D = -\frac{1}{\lambda}A = -0.9815$$

$B = -C = 0.9895$

Displacement along the pipe.

$$w(s) = -\frac{1}{2}\chi s^2 + As + B + C\cosh \lambda s + D\sinh \lambda s \quad \left(\text{with } \chi = 8.334 \times 10^{-5}\right)$$

From

$$M = EI\left(\frac{1}{R} + \frac{w}{R^2} + \frac{d^2 w}{ds^2}\right)$$

Bending moment at $s = 0$ is

$$M_{max} = EI\left(\frac{1}{R} + \frac{w_0}{R^2} - \chi\right) + C\lambda^2 EI \tag{8.107}$$

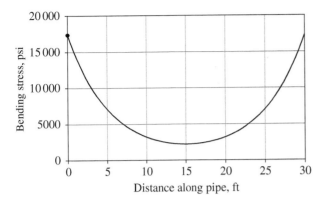

Figure 8.40 Bending stress along a section of 5 in. (16.25 lb/ft) drill pipe (R = 1000 ft, T = 100 000 lb).

Example   Parameters for this example are

$\lambda = 0.0153$
$\chi = 8.334 \times 10^{-5}$
$C = 0.9895$
$R = 1000 \text{ ft} = 12\,000 \text{ in.}$
$I = 13 \text{ in.}^4$
$E = 30 \times 10^6 \text{ psi}$
$C = 2.5 \text{ in.}$

By substitution and dropping w_o/R^2

$$\sigma_{max} = \frac{M_{max}c}{I}$$

$$\sigma_{max} = Ec\left(\frac{1}{R} - \chi + C\lambda^2\right)$$

$$\sigma_{max} = 30 \times 10^6 (2.5)\left(\frac{1}{12\,000} - 8.334 \times 10^{-5} + 0.9895(0.0153)^2\right)$$

$$\sigma_{max} = 17\,372 \text{ psi}$$

These equations were programed to show how bending stress varies over a 30 ft pipe joint when the radius of hole curvature is $R = 1000$ ft. Tension is assumed to be 100 000 lb. The highest bending occurs at the tool joints 17 372 psi (see Figure 8.40).

Application to Pipe Bending in Curved Well Bores

A major challenge is analysis of bending and friction on drill strings in doglegs. Several parameters factor into the complexity of this problem: location of points of contact, magnitude contact forces and resulting friction, shape of the dogleg, tension, and flexural stiffness of pipe to mention a few.

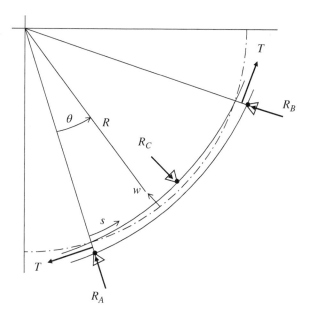

Figure 8.41 Model of pipe with three points of contact.

Consider a pipe making contact at three points as shown in Figure 8.41. This problem is similar to a straight beam supported at three points. It is statically indeterminate and requires the solution to beam deflection as well as statics. The same is true of the model in the figure. In general, the center contact point will not be in the center between points A and B requiring solutions to sections on both sides of point C with matching boundary conditions at C.

For the sake of simplicity, we assume point C is midway between A and B, to take advantage of symmetry. The problem in this case becomes a cantilever beam problem, using polar coordinates.

The bending equations for the CB section are given as follows:

$$w(s) = -\frac{1}{2}\chi s^2 + As + B + C\cosh \lambda s + D\sinh \lambda s \tag{8.108}$$

The boundary conditions that apply are

$$w(0) = \delta \quad w(\ell) = -\delta$$

$$\left.\frac{dw}{ds}\right|_{s=0} = 0 \quad \left.\frac{1}{\rho}\right|_{s=\ell} = 0, \text{which leads to} \left.\frac{d^2w}{ds^2}\right|_{s=\ell} + \left.\frac{w}{R}\right|_{s=\ell} = -\frac{1}{R}$$

where ℓ is arc length between points C and B. Distance, δ, is inward $(+)$ radial displacement of point C and outward $(-)$ radial displacement of point B.

The boundary conditions quantify the constants, A, B, C, D. The moment at point C is determined by Eq. (8.101). Forces V and R_B are determined by statics. Reaction R_C is equal to $2\,V$.

Multispanned Beam in Terms or Polar Coordinates

Another approach for solving the multispanned beam problem is explained by Huang et al. [6]. Results are given for strait beams, but the method applies equally well to polar coordinates.

A stepwise approach was used by Dareing and Ahlers [7] and Rocheleau and Dareing [8] predict drill pipe deflection between tool joints as well as contact forces and friction. The model in both papers assume contact is made by each tool joint within a curved portion of the well bore. The solutions give insight as to how bending stiffness enters the friction calculations. The basic equations for (i) pulling out of the well bore and (ii) going into the well bore are explained below.

In both papers, pipe is analysis by considering deflections of segments between tool joints. It is also assumed that tool joints contact the well bore (Figure 8.42).

Pulling Out of the Well Bore
The differential equation of bending for each pipe section is

$$\frac{d^4w}{ds^4} - \lambda^2 \frac{d^2w}{ds^2} = \zeta \tag{8.109}$$

where

$$\lambda^2 = \frac{T}{EI} - \frac{1}{R^2}$$

$$\zeta = \frac{Rq + T}{REI}$$

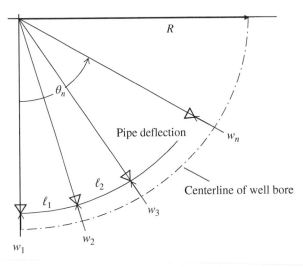

Figure 8.42 Multiple spanned pipe in a dogleg.

Table 8.2 Pull out force over a 90° turn; R = 200 ft (4½ in. 16.5 lb/ft drill pipe).

		Top tension	
Back pull (lb)	Free-free (lb)	Fixed-fixed (lb)	No stiffness (lb)
2 000	6 162	5 544	4 234
4 000	9 421	8 773	7 438
6 000	12 794	12 361	10 642
8 000	16 164	15 849	13 846
10 000	19 614	19 337	17 050
12 000	23 117	22 826	20 254

The general solution to Eq. (8.109) is

$$w(s) = -\frac{1}{2}\chi s^2 + As + B + C\cosh \lambda s + D \sinh \lambda s$$

Following the above procedure given in Ref. [8], pullout forces corresponding to six (6) different pull back forces were calculated, and the results given in Table 8.2.

Columns and Compression Members

Column Buckling Under Uniform Compression

Column is a structure member which carries compressive loads. It can be a member in a truss or an independent vertical support. Leonhard Euler (1707–1783), a Swiss mathematician, was the first to study the buckling of columns. He derived the critical buckling equation,

$$P = \frac{C}{4}\left(\frac{\pi}{\ell}\right)^2 \tag{8.110}$$

stating

> Therefore, unless the load P to be borne be greater than $\frac{C}{4}\left(\frac{\pi}{\ell}\right)^2$, there will be absolutely no fear of bending; on the other hand, it the weight P be greater, the column will be unable to resist bending. Now when the elasticity of the column and likewise its thickness remain the same, the weight P which it can carry without danger will be inversely proportional to the square of the height of the column; and a column twice as high will be able to bear only one-fourth of the load [9].

The constant, C, represents the elastic property and cross-section dimensions, which translate into

$$P_{cr} = \left(\frac{\pi}{L}\right)^2 EI \tag{8.111}$$

Euler called attention to the difference between compression failure and failure by buckling.

Buckling is a condition, under which a critical force, Q_{cr}, is reached causing the column to become unstable, i.e. the column takes a different configuration. Columns are usually assumed to have uniform compression throughout. End supports, or boundary conditions may vary as shown in Figure 8.43.

Consider a column simply supported as shown in Figure 8.44. The differential equation of bending from the freebody diagram of the lower end is

$$EI\frac{d^2y}{dx^2} + Qy = 0 \tag{8.112}$$

$$\frac{d^2y}{dx^2} + \beta^2 y = 0 \tag{8.113}$$

where

$$\beta^2 = \frac{Q}{EI}$$

Here, Q, is a compressive load. The solution to Eq. (8.112) is

$$y = A\sin\beta x + B\cos\beta x \tag{8.114}$$

For simply supported type boundary conditions, $y(0) = 0$ giving $B = 0$.

$$y = A\sin\beta x \tag{8.115}$$

Applying $y(L) = 0$ gives

$$0 = A\sin\beta L \tag{8.116}$$

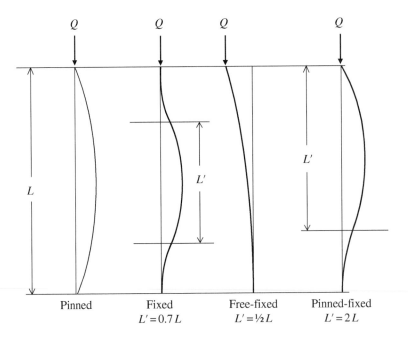

| | Pinned | Fixed
$L' = 0.7L$ | Free-fixed
$L' = \frac{1}{2}L$ | Pinned-fixed
$L' = 2L$ |

Figure 8.43 Columns with different end constraints.

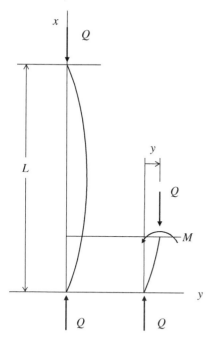

Figure 8.44 Simply supported columns.

For a nontrivial solution

$$\beta L = n\pi, \quad n = 1, 2, 3 \tag{8.117}$$

For the first buckling mode, $n = 1$ and

$$\frac{Q}{EI} = \left(\frac{\pi}{L}\right)^2 \tag{8.118}$$

giving

$$Q_{cr} = \left(\frac{\pi}{L}\right)^2 EI \tag{8.119}$$

In terms of compressive stress, Eq. (8.119) becomes

$$\sigma_{cr} = \frac{\pi^2}{(L/r)^2} E \tag{8.120}$$

where L/r is the slenderness ratio of the column cross section and r is the radius of gyration of the smallest moment of inertia.

Figure 8.45 shows how critical buckling stress increases with reduction in slenderness ratio. However, for short compressive members, material yielding will occur prior to buckling, as indicated by the horizontal line (assuming $\sigma_{yld} = 60$ ksi; $L/r = 70.2$). This condition establishes a limit to Euler's buckling equation.

Testing of columns show failure follows along a curved line between compression failure and Euler buckling (Figure 8.46). The transition between these two failure modes is defined by three slenderness ratio ranges:

- Compression block range
- Intermediate range
- Slender range

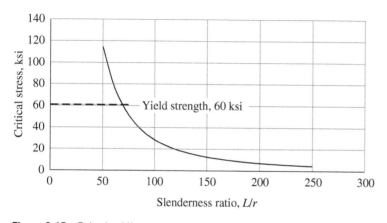

Figure 8.45 Euler buckling curve.

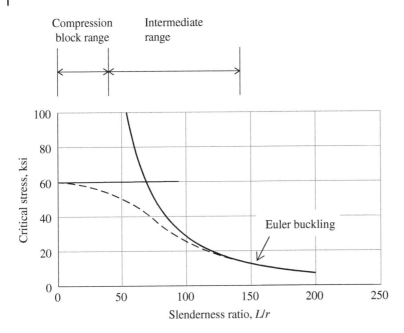

Figure 8.46 Failure theory ranges.

As a guideline, the compression block theory covers $L/r < 40$. Euler buckling theory applies to $L/r > 140$. Failure over the intermediate range is defined by empirical formulas, depending on material type (steel, aluminum, wood). These empirical formulas are defined in column codes [10–12].

The above equations apply to simple supports. The buckling equations also apply for other boundary conditions if L is replaced by L' (see Figure 8.43).

Example Consider a section of heavy pipe ($6'' \times 2''$, 85 lb/ft) supported vertically between pinned connections separated by 60 ft. The minimum yield strength for the pipe is $\sigma_{yld} = 110$ ksi.

$$I = \frac{\pi}{4}\left(R_0^4 - R_i^4\right) = 62.83 \text{ in.}^4$$

$$A = \pi\left(R_0^2 - R_i^2\right) = 25.13 \text{ in.}^2$$

$$r = 1.58 \text{ in.} \quad \text{(radius of gyration)}$$

$$L/r = 455.69 \quad \text{(slenderness ratio)}$$

Using Eq. (9.120)

$$\sigma_{cr} = \left(\frac{\pi}{455.69}\right)^2 30 \times 10^6 = 1426 \text{ psi}$$

This stress-load level is well within the yield limits of the collar material so Euler buckling theory would apply. The corresponding collar force is $F_{cr} = 1426(25.13) = 35\,832$ lb. This force level is within typical drill bit force levels, so buckling is possible.

Next consider a larger pipe size ($8'' \times 2''$, 160 lb/ft; $\sigma_{yld} = 100$ ksi).

$$I = 200.27 \text{ in.}^4$$

$$A = 47.12 \text{ in.}^2$$

$$r = 2.06 \text{ in.} \quad \text{(radius of gyration)}$$

$$L/r = 349.5 \quad \text{(slenderness ratio)}$$

Using Eq. (9.146)

$$\sigma_{cr} = \frac{\pi^2}{(349.5)^2} 30 \times 10^6 = 2424 \text{ psi}$$

This stress-load level is well within the yield limits of the collar material so Euler buckling theory would apply. The corresponding collar force is $F_{cr} = 2424(25.13) = 60\ 915$ lb.

Columns of Variable Cross Section

Columns of antiquity indicate that designers had an empirical understanding of the effect of cross-section properties of column stability. Stone columns of the Greco–Roman era show columns having a more robust cross section over the center portion.

This section develops the critical force for columns having a variable cross section. The cross-sectional inertia is assumed to vary linearly from each end to a maximum value in the center of twice the end values:

$$I = I_0\left(1 + \frac{2x}{L}\right) \tag{8.121}$$

Following the Rayleigh method of approximation, we assume a mode shape of

$$y(x) = y_0 \sin \frac{\pi x}{L} \tag{8.122}$$

By the principle of minimum potential energy

$$V = U + '\Omega \tag{8.123}$$

where

$$U = \frac{1}{2}\int_0^L EI(y'')^2 dx \tag{8.124}$$

$$\Omega = -P\frac{1}{2}\int_0^L (y')^2 dx \tag{8.125}$$

For stability, $\frac{d^2V}{dy_0^2} \geq 0$

$$P_{crit} = \frac{\int_0^L EI(y'')^2 dx}{\int_0^L (y')^2 dx} \tag{8.126}$$

By substitution of the assumed deflection function, $y(x)$ and the expression for the moment of inertia $I(x)$, is [13]

$$P_{crit} = \frac{1.70\ \pi^2\ EI_0}{L^2} \tag{8.127}$$

See Timoshenko and Gere [14] for a comprehensive discussion of elastic stability.

Tubular Buckling Due to Internal Pressure

An interesting and important aspect of column buckling is found in the mechanics of long pipe, such as drill pipe, marine risers, and production tubing, used in the oil industry. The bending of these pipes is greatly affected by fluid pressure inside and out. This phenomenon is discussed in the following sections.

Consider the buckling of pipe with internal pressure acting alone as illustrated in Figure 8.47. The surface pressure force is shown in the left drawing. The right element represents the statically equivalent force system which is obtained by adding and subtracting pressure forces over the inside area across an imaginary plane. The net effect of the pressure within an enclosed volume is zero, leaving the end force vectors as shown. The equivalent force system is easily incorporated into the bending equation.

This is a special case of Euler's equation. The solution, using Eq. (8.119) and noting that $Q = p_i A_i$ is

$$(p_i)_{cr} A_i = \left(\frac{\pi}{L}\right)^2 EI \tag{8.128}$$

where

p_i – pressure inside pipe, psi
A_i – inside area of pipe, in.^{2}

The pipe is assumed to be rigidly and simply supported and not capped. The pipe in this case cannot be a compression block, but burst pressure limits may have to be considered.

Consider a section pipe ($4'' \times 3.5''$) 100 ft long simply supported and axially constrained at each end. Applying Eq. (8.128) gives

$$p_{cr} = \left(\frac{\pi}{L}\right)^2 \frac{EI}{A_i} \tag{8.129}$$

$$p_{cr} = \left(\frac{\pi}{1200}\right)^2 \frac{30 \times 10^6 (5.20)}{9.621} = 111.13 \text{ psi}$$

This pressure is equivalent to a buckling force of

$$F_{cr} = (111.3)9.621 = 1069 \text{ lb}$$

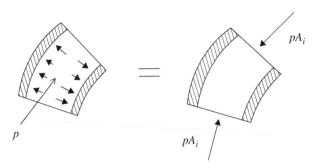

Figure 8.47 Statically equivalent constant pressure effects.

Effective Tension in Pipe

Pressure applied to the outside surface acts to stabilize pipe. Figure 8.48 shows the statically equivalent force system (right drawing) accounts for external pressure, internal pressure, and internal tension in the walls of the pipe.

The net effect of both internal and external pressures combined with an axial force is an effective tension defined by

$$T_{eff} = T - (p_i A_i - p_o A_o) \tag{8.130}$$

where

T is the actual internal force (plus is tension and a minus T is compressive force)
$A_i = \frac{\pi}{4} d_i^2$
$A_0 = \frac{\pi}{4} d_0^2$

This term is called *effective tension* because it occupies the space in the differential equation of bending reserved for tension. It varies with vertical location in regard to drill pipe.

A common application of effective tension is in the bending vertical pipe in a hydrostatic situation. This term shows that axial tension and external pressure tend to stabilize while internal pressure encourages buckling.

In this case, the inside and outside hydrostatic pressures are equal, but vary with depth. The pressure terms can be replaced by

$$p_0 A_0 = z \gamma_m A_0$$
$$p_i A_i = z \gamma_m A_i$$

giving

$$p_0 A_0 - p_i A_i = z \gamma_m A_c \tag{8.131}$$

where A_c is the cross-sectional area of the pipe.

Noting that the volume displaced by a unit length of pipe is

$$\Delta L (A_c \gamma_m) = \Delta L w_m \tag{8.132}$$

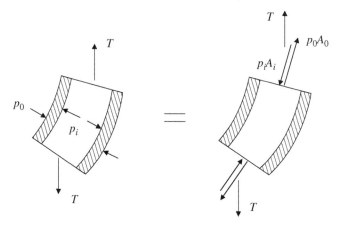

Figure 8.48 Statically equivalent internal/external pressure system.

The pressure can now be written as

$$p_0 A_0 - p_i A_i = z w_m \tag{8.133}$$

where w_m is the weight of drill fluid displaced by a unit length of pipe. Effective tension, in the equation of drill pipe where drilling mud is the same inside and out becomes

$$T_{eff} = T + w_m z \tag{8.134}$$

where z is measure from the top downward. If the pipe is hanging freely, then

$$\left(T_{eff}\right)_{top} = H \quad \text{(hook load)}$$

At the lower end, with the pipe open,

$$\left(T_{eff}\right)_{bottom} = T + L w_m = -L w_m + L w_m = 0 \tag{8.135}$$

If the pipe is not open at the lower end but is subjected to an internal force of F_B (tension), then effective tension at location L is

$$T_{eff} = F_B + L w_m \tag{8.136}$$

Effective compression is also a useful parameter. It is defined as

$$Q_{eff} = Q + (p_i A_i - p_0 A_0)$$

where plus numbers mean compression. For the case of drilling mud inside and outside the pipe

$$Q_{eff} = Q - (L - x) w_m$$

where x is measured from the bottom upward. Q is the actual internal compression in the pipe.

Buckling of Drill Collars

Example Consider the buckling of drill collars between a drill bit and the first stabilizer as depicted in Figure 8.49. Dimensions of the drill collars are

OD = 6 ½ in.

ID = 2 in.

$w = 102 \, \text{lb/ft}$

$L = 90 \, \text{ft}$

We wish to determine the WOB that will buckle the 90 ft section under the following two conditions:

a) Air

b) 11 ppg drilling mud ($BF = 0.832$)

Boundary conditions are assumed to be pinned at both drill bit and stabilizer locations. The actual boundary condition at the stabilizer is somewhere between fixed and pinned so the solution will be an approximation. Also, axial compression in the collars between these two locations varies linearly. We simplify this by using the average compression between the two locations [15].

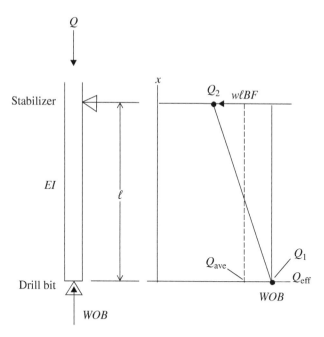

Figure 8.49 Buckling of drill collars between drill bit and stabilizer.

The effective compression at any location, x, in the drill collars is

$$Q_{eff} = (Q - wx) - wx - (L - x)w_m$$

But the actual force at the drill bit is

$$Q = WOB + Lw_m$$

By substitution, the effective compression at any location in the drill collars is

$$Q_{eff} = WOB + xBFw$$

Buckling will occur when

$$Q_{cr} = \left(\frac{\pi}{L}\right)^2 EI$$

$$I = \frac{\pi}{4}\left(3.25^4 - 1^4\right) = 86.84 \text{ in.}^4$$

For the first case (a), buckling in air

$$\left(Q_{eff}\right)_{ave} = \left(WOB - \frac{1}{2}wL\right)$$

$$\left(WOB - \frac{1}{2}wL\right)_{cr} = \left(\frac{\pi}{L}\right)^2 EI$$

$$(WOB)_{cr} = \left(\frac{\pi}{L}\right)^2 EI + \frac{1}{2}wL$$

$$(WOB)_{cr} = \left(\frac{\pi}{90 \times 12}\right)^2 30 \times 10^6 (86.84) + \frac{1}{2}102(90) = 26\,634 \text{ lb}$$

For the second case (b), buckling in drilling fluid

$$(WOB)_{cr} = \left(\frac{\pi}{L}\right)^2 EI + \frac{1}{2}wL(BF)$$

$$(WOB)_{cr} = 22\,044 + 3818$$

$$(WOB)_{cr} = 25\,862\,\text{lb}$$

By comparison, buckling of a slick assembly (no stabilizers) based on Lubinski's solution [16] is

$$(WOB)_{cr} = 1.94(wBF)^{\frac{2}{3}}\left(\frac{EI}{144}\right)^{\frac{1}{3}}$$

$$(WOB)_{cr} = 1.94(102 \times 0.832)^{\frac{2}{3}}\left(\frac{30 \times 10^6(86.84)}{144}\right)^{\frac{1}{3}} = 9835\,\text{lb}$$

which is much lower than the stabilized assembly.

Combined Effects of Axial Force and Internal/External Pressure

Using the previous notation for effective compressive force when the axial force is compressive,

$$Q_{eff} = Q + (p_iA_i - p_oA_o) \quad \text{(effective compression)} \tag{8.137}$$

where $+Q$ is compression. By substitution

$$EI\frac{d^2y}{dx^2} + Q_{eff}y = 0 \tag{8.138}$$

$$\frac{d^2y}{dx^2} + \beta^2y = 0 \tag{8.139}$$

where

$$\beta^2 = \frac{Q_{eff}}{EI}$$

and following the earlier solution, buckling occurs when

$$[Q + (p_iA_i - p_oA_o)]_{cr} \geq \left(\frac{\pi}{L}\right)^2 EI \tag{8.140}$$

where the left term is effective compression. In this case, all loads are uniform with x. Note that external pressure stabilizes the beam while internal pressure destabilizes the beam.

Buckling of Drill Pipe

Lateral buckling consideration starts with the differential equation of bending, which contains the effective tension term:

$$T_{eff} = F_B + wx + (L - x)w_m \tag{8.141}$$

For this case, we assume that the bottom force, F_B, is composed of two terms

$$F_B = -F - w_mL \quad \text{(note both terms are compressive)} \tag{8.142}$$

where

$w_m L$ — fluid force caused by local hydrostatic pressure; recall $w_m L = \gamma_m L A_c$

F — internal force beyond local hydrostatic pressure (stress)

The reason for this is how the lower end of drill pipe is axially loaded at its point of connection with drill collars. If the neutral point is located at the top of the drill collars, then the internal stress in the drill pipe at this conjuncture is hydrostatic or $w_m L$. Length L is the vertical distance from the rig floor to the top of the drill collars. Under hydrostatic open-ended loading, drill pipe will not buckle. The analysis seeks to determine the additional force, F, beyond hydrostatic that buckles drill pipe connected to the drill collars.

The effective tension term becomes

$$T_{eff} = -F + (w - w_m)x \tag{8.143}$$

The mathematical model for this analysis is shown in Figure 8.50.

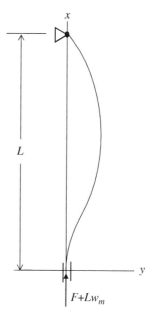

Figure 8.50 Mathematical model of drill pipe.

$$EI\frac{d^4y}{dx^4} - \frac{d}{dx}\left\{[(w - w_m)x - F]\frac{dy}{dx}\right\} = 0 \tag{8.144}$$

Boundary condition imposed on the solution are

$$y(0) = 0$$

$$\frac{dy}{dx}(0) = 0$$

$$y(L) = 0$$

$$\frac{d^2y}{dx^2}(L) = 0$$

In dimensionless form, Eq. (8.144) is written as

$$\frac{d^4y}{d\zeta^4} - \alpha\zeta\frac{d^2y}{d\zeta^2} + \beta\frac{d^2y}{d\zeta^2} - \alpha\frac{dy}{d\zeta} = 0 \tag{8.145}$$

where

$$\alpha = \frac{(w - w_m)L^3}{EI}$$

$$\beta = \frac{FL^2}{EI}$$

Following the derivation in Ref. [17] critical values of β, which contain buckling force, F, are given in Table 8.3 for the first four modes of buckling. The critical buckling force for the first mode defines the on-set of buckling.

Values of α are limited to 4000 because of the high degree of numerical precision required to execute the calculations. The relation between α and β_{cr} for higher values of α can be determined as follows.

Table 8.3 Buckling parameters.

α	β_{cr1}	β_{cr2}	β_{cr3}	β_{cr4}
0	20.19	59.68	118.90	197.86
1	20.84	60.21	119.42	198.37
2	21.50	60.73	119.94	198.88
5	23.44	62.31	121.50	200.40
10	26.66	64.94	124.12	202.95
20	32.99	70.22	129.36	208.07
50	51.06	86.18	145.31	223.51
100	77.92	113.25	172.67	249.60
200	121.61	166.59	230.50	303.21
500	218.06	297.87	403.85	474.41
1000	339.59	465.36	634.66	740.10
2000	529.83	729.55	997.20	1162.37
4000	827.87	1146.48	1569.43	1830.50

Table 8.4 Relation between ϕ values and α.

α	ϕ_1	ϕ_2	ϕ_3	ϕ_4
1	20.89	60.21	119.42	198.37
2	13.54	38.26	75.56	125.82
5	8.02	21.31	41.55	68.54
10	5.74	13.99	26.74	43.73
20	4.48	9.53	17.56	28.24
50	3.76	6.35	10.71	16.47
100	3.62	5.26	8.02	11.59
200	3.56	4.87	6.74	8.87
500	3.46	4.73	6.41	7.53
1000	3.40	4.65	6.35	7.40
2000	3.34	4.60	6.28	7.32
4000	3.29	4.55	6.23	7.26

When the values of β_{cr} vs. α (Table 8.4) are plotted on log–log scale, they become straight and parallel with a slope of about $\tfrac{2}{3}$. A straight line on log–log graph can be expressed as

$$\ln \beta_{cr} = \frac{2}{3} \ln \alpha + \ln \phi \tag{8.146}$$

which simplifies to

$$\beta_{cr} = \phi \alpha^{\frac{2}{3}} \tag{8.147}$$

giving

$$\phi_{cr} = \frac{\beta_{cr}}{\alpha^{\frac{2}{3}}} = \frac{F_{cr}}{(EI)^{\frac{1}{3}}(w - w_m)^{\frac{2}{3}}} \tag{8.148}$$

Pipe length, L, does not show up in this equation. The buckling force becomes independent of pipe length as would be expected for long pipe. The same conclusion applies to other boundary conditions at $x = 0$. According to Table 8.4, ϕ_1 is 3.29 for the first mode; ϕ_n for higher buckling modes are also given in this table.

The critical buckling force equation for long pipe can now be predicted by

$$F_{cr} = \phi \left(\frac{EI}{144}\right)^{\frac{1}{3}} (w - w_m)^{\frac{2}{3}} \tag{8.149}$$

where F_{cr} is a force over and above local hydrostatic compression at $x = 0$. Also the section modulus is assumed to be in units of lb-in.^{2} The critical buckling forces for the first four modes can be determined from

$$F_{cr1} = 3.29 \left(\frac{EI}{144}\right)^{\frac{1}{3}} (w - w_m)^{\frac{2}{3}} \tag{8.150a}$$

$$F_{cr2} = 4.55 \left(\frac{EI}{144}\right)^{\frac{1}{3}} (w - w_m)^{\frac{2}{3}} \tag{8.150b}$$

$$F_{cr3} = 6.23 \left(\frac{EI}{144}\right)^{\frac{1}{3}} (w - w_m)^{\frac{2}{3}} \tag{8.150c}$$

$$F_{cr4} = 7.26 \left(\frac{EI}{144}\right)^{\frac{1}{3}} (w - w_m)^{\frac{2}{3}} \tag{8.150d}$$

Example Consider 5½ in. (19.2 lb/ft) drill pipe along with the following parameters:

$$E = 30(10)^6 \text{ psi}$$

$$I = 16.8045 \text{ in.}^4$$

$$A = 4.9624 \text{ in.}^2$$

$$\gamma_m = 10 \text{ ppg}$$

$$\gamma_{stl} = 490 \text{ lb/ft}^3$$

$$L = 1000 \text{ ft}$$

Applying these numbers to

$$\alpha = \frac{(w - w_m)L^3}{EI} = \frac{BFwL^3}{EI} \tag{8.151}$$

gives

$$\alpha = \frac{0.847(19.2)1000^3}{30(10)^6 16.8045} \left| \frac{144 \text{ in.}^2}{1 \text{ ft}^2} \right| = 4645.2$$

Since α is greater than 4000, critical buckling parameter β_{cr} is determined by Eq. (8.147), where ϕ_1, ϕ_2, ϕ_3, ϕ_4 are 3.29, 4.55, 6.23, and 7.26, respectively. The critical buckling parameter for the first mode is

$$(\beta_{cr})_1 = 3.29(4645.2)^{\frac{2}{3}} = 916$$

Applying

$$\beta_{cr} = \frac{F_{cr}L^2}{EI}$$

gives

$$(F_{cr})_1 = 916 \frac{30(10)^6 16.8045}{1000^2} \left| \frac{1\text{ ft}^2}{144\text{ in.}^2} \right| = 3207\text{ lb} \tag{8.152}$$

The application of the higher modes of drill pipe buckling can only be viewed as approximate since in reality the pipe makes contact with the well bore, but even this could be helpful in horizontal drilling where the vertical portion of drill pipe is sometimes put in compression.

Mode shapes and bending moment distribution are given in Ref. [17].

Bending Equation for Marine Risers

The mechanics of buckling of marine risers is similar to that of drill pipe except for the effective tension term, since in general fluid densities of drilling mud and sea water are different. These fluid forces have a significant effect on marine riser bending. The differential equation of bending containing these forces will now be discussed.

Unique Features of the Differential Equation of Bending

The differential equation of bending for long vertical pipe is developed from the differential element shown in Figure 8.51. The solid vectors represent the actual applied forces and body force of gravity, while the dashed vectors represent the statically equivalent hydrostatic forces both inside and outside the riser pipe. The independent variable, x, is measured from the lower end upward.

The parameters in the drawing are defined below:

F_x – actual internal force in pipe, lb
F_B – internal force at location at the bottom end, lb
V – shear force, lb
M – bending moment, lb-ft
θ – slope of deflection at x location, rad
$P = A_o p_o - A_i p_i = (A_o \gamma_o - A_i \gamma_i)(L - x)$ lb
$b_o = A_o \gamma_o$ lb/ft
$b_i = A_i \gamma_i$ lb/ft
γ_o – weight density of external fluid, lb/ft^3
γ_i – weight density of internal fluid, lb/ft^3
γ – weight density of pipe material
w – air weight of pipe per length(γA_c)
q – side load per unit length, lb/ft

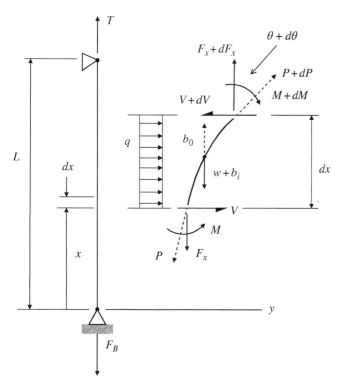

Figure 8.51 Freebody diagram of a differential marine riser section.

Summation of forces in the x-direction gives

$$\frac{dF_x}{dx} = w \tag{8.153}$$

$$F_x = wx + F_B \tag{8.154}$$

Summation of forces in the y-direction gives

$$\frac{dV}{dx} = \frac{d}{dx}\left[(A_o\gamma_o - A_i\gamma_i)(L-x)\frac{dy}{dx}\right] + q \tag{8.155}$$

where q represents transverse loads applied over length, dx. The effect of an addition uniform constant pressure, p, throughout the pipe can be included by adding $-pA_i\frac{d^2y}{dx^2}$ to the right side of Eq. (8.155). This term is important in analyzing bending of production tubing. Summation of moments gives

$$\frac{dM}{dx} - F_x\frac{dy}{dx} - V = 0 \tag{8.156}$$

Combining Eqs. (8.154)–(8.156) gives

$$\frac{d^2M}{dx^2} - \frac{d}{dx}\left[F_x\frac{dy}{dx}\right] - \frac{dV}{dx} = 0 \tag{8.157}$$

Applying Euler's equation, $M = EI\frac{d^2y}{dx^2}$, the differential equation of bending for riser pipes with variable internal tension (or compression) and exposed to inside and outside hydrostatic pressure is [4]

$$EI\frac{d^4y}{dx^4} - \frac{d}{dx}\left\{[(F_B + wx) + (L-x)(A_o\gamma_o - A_i\gamma_i)]\frac{dy}{dx}\right\} = q \tag{8.158}$$

The variable coefficient in the second term is special and unique to this equation. This term includes fluid pressure inside and out as well as the weight of the pipe. Each of these parameters vary linearly with vertical distance along the pipe. This term creates a challenge in solving the differential equation and is called "effective tension" a unique feature of the mathematical solution to this bending problem.

Effective Tension

The effective tension term is clearly shown within the brackets of Eq. (8.158). When uniform internal pressure is included, the effective tension becomes

$$T_{eff} = [(F_B + wx) + (L-x)(A_o\gamma_0 - A_i\gamma_i) - A_ip] \tag{8.159}$$

Effective tension is defined for mathematical convenience. It occupies the space in the differential equation set aside for the actual tension, $F_B + wx$. Effective tension alone accounts for the effects of hydrostatic pressure inside and outside riser pipes as well as uniform pressure throughout the pipe from top to bottom. If, for example, a vertical pipe is suspended in air, then

$$T_{eff} = F_B + wx \tag{8.160}$$

This is the true internal tension.

If a vertical pipe is suspended in a given fluid with density γ_m (both inside and outside) effective tension becomes

$$T_{eff} = F_B + wx + (L-x)w_m \tag{8.161}$$

where $w_m = A_c\gamma_m$, having units of force per length, which represents the weight of the fluid (drilling mud) displaced per unit length of pipe.

When the variable terms are combined

$$T_{eff} = F_B + w_mL + x(w - w_m) \tag{8.162}$$

Buckling of Marine Risers

The difference between the buckling of marine risers in the ocean and drill pipe buckling is that fluid outside risers is sea water while fluid inside risers is drilling fluids. Drilling fluids are significantly greater that sea water. The differential equation of marine riser bending accounting for these effects on effective tension is [4]

$$EI\frac{d^4y}{dx^4} - \frac{d}{dx}\left\{[(F_B + wx) + (L-x)(A_o\gamma_o - A_i\gamma_i)]\frac{dy}{dx}\right\} = 0 \tag{8.163}$$

The dimensionless form of this equation is

$$\frac{d^4y}{d\zeta^4} - \alpha\zeta\frac{d^2y}{d\zeta^2} - \beta\frac{d^2y}{d\zeta^2} - \alpha\frac{dy}{d\zeta} = 0 \tag{8.164}$$

where

$$\zeta = \frac{x}{L} \quad \text{and}$$

$$\alpha = \frac{(w - A_0\gamma_0 + A_i\gamma_i)L^3}{EI}$$

$$\beta = \frac{(F_B + LA_0\gamma_0 - LA_i\gamma_i)L^2}{EI}$$

The results of the drill pipe buckling analysis, discussed earlier, is now applied to marine risers. In this case, the expressions for α and β account for the difference between fluid densities inside and outside risers. The sign on β is adjusted to account for the sign difference in the third term in Eqs. (8.145) and (8.164).

$$\beta = -\frac{(F_B + LA_0\gamma_0 - LA_i\gamma_i)L^2}{EI}$$

where F_B is the actual internal force in the riser at the mud line connection ($x = 0$).
The relation between the critical value of β_{cr} and α, according to Eq. (8.147) becomes

$$-\left(\frac{(F_B + LA_0\gamma_0 - LA_i\gamma_i)L^2}{EI}\right)_{cr} = \phi\left(\frac{(w - A_0\gamma_0 + A_i\gamma_i)L^3}{EI}\right)^{\frac{2}{3}} \tag{8.165}$$

Solving for F_B gives the critical internal force at $x = 0$ causing risers to buckle.

$$F_{B,cr} = -\left(\frac{EI}{144}\right)^{\frac{1}{3}}\phi(w - A_0\gamma_0 + A_i\gamma_i)^{\frac{2}{3}} - (LA_0\gamma_0 - LA_i\gamma_i) \tag{8.166}$$

Note that in this formula F_B is the actual force at the lower end of the riser and can be plus (tension) or minus (compression).
As a check we apply this formula to the drill pipe buckling problem where
$F_B = -(F + w_mL)$ and $\gamma_i = \gamma_0 = \gamma_m$
Recall that $Lw_m = L\gamma_mA_c$ or $w_m = \gamma_mA_c$. By substitution

$$-(F + w_mL) = -\phi\left(\frac{EI}{144}\right)^{\frac{1}{3}}(w - w_m)^{\frac{2}{3}} - LA_c\gamma_m \tag{8-167}$$

$$F_{cr} = \phi\left(\frac{EI}{144}\right)^{\frac{1}{3}}(w - w_m)^{\frac{2}{3}}$$

In this case (drill pipe model) the force F, is the compressive force above local hydrostatic at the top of the drill collars.

For values of α greater than 4000, the equation below can be used.

$$\beta_{cr} = \phi_1 \alpha^{\frac{2}{3}} = 3.29\alpha^{\frac{2}{3}} \quad \text{(1st Mode)} \tag{8-168}$$

Accounting for units

$$\alpha = \frac{(w - A_0\gamma_0 + A_i\gamma_i)L^3}{EI} \left|\frac{144in^2}{ft^2}\right|$$

$$\beta_{cr} = -\frac{(F_{B,cr} + LA_0\gamma_0 - LA_i\gamma_i)L^2}{EI} \left|\frac{144in^2}{1ft^2}\right|$$

For values of α less than 4000, ϕ should be taken from Table 8.5.

Table 8.5 Relation between ϕ_1 values and α (first mode).

α	ϕ_1
1	20.89
2	13.54
5	8.02
10	5.74
20	4.48
50	3.76
100	3.62
200	3.56
500	3.46
1000	3.40
2000	3.34
4000	3.29

Example Critical forces, F_B, were calculated using equation (8.166) and plotted in Fig 8-52 for the operational data listed below.

Riser OD   - 22 in.
Riser ID   - 20.75 in.
A_0 - 2.64 ft^2
A_1 - 2.35 ft^2
w - 142.81 lb/ft
I - 2398 in^4
E - 29 × 10^6 psi
γ_0 - 64 lb/ft^3 (sea water)
γ_i - variable, ppg (7.48 gal/ft^3)
L - variable, ft

This figure gives riser tension (compression) at the bottom connection, at impending buckling. It is curious that each line passes through a common point.

If $A_0\gamma_0 = A_i\gamma_i$, then Eq. (8.166) becomes

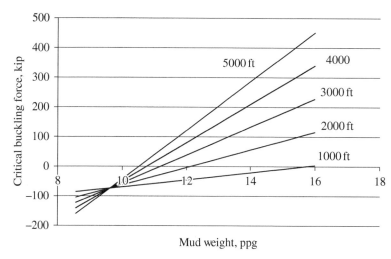

Figure 8.52 Critical buckling force, $(F_B)_{cr}$, for marine risers (first mode).

$$F_{B,cr} = -\left(\frac{EI}{144}\right)^{\frac{1}{3}} 3.29(w)^{\frac{2}{3}} \tag{8.169}$$

In this case, the critical buckling force is independent of L. For the example case, this point corresponds to

$$\gamma_i = \frac{2.6398}{2.3483} 64 = 70.897 \, \text{lb/ft}^3 \text{ or } 9.48 \, \text{ppg}$$

The critical buckling force in this special case is

$$F_{B,cr} = -\left(\frac{29 \times 10^6 (2398)}{144}\right)^{\frac{1}{3}} 3.29(142.8)^{\frac{2}{3}} = -70.5 \, \text{kip}$$

A separate calculation showed that the critical buckling force for each case in air is $F_{B,cr}$ is 70.52 kip (compression). This number is determined from Eq. (8.166) with $\gamma_i = \gamma_o = 0$.

Length, L, is not a factor assuming $\alpha \succ 4000$. Note that for lengths of 2000 ft and 1000 ft, ϕ is slightly higher that 3.29 and in the range of 3.56 so these lines are slightly affected.

Numbers that are minus, represent the amount of compressive force at the bottom end, that initiate buckling. Plus numbers define the magnitude of tension at the lower end that correspond to the onset of buckling. This data shows the effects of higher mud weight on buckling especially in long riser pipes. Top tension in each case is determined from $T_{top} = F_B + wL$.

In anticipation of threatening weather, such as hurricanes, risers are sometimes disconnected from mud line equipment creating a hydrostatic condition inside and outside the riser pipe $\gamma_i = \gamma_o$. This operational step eliminates the possibility of axial buckling.

Tapered Flex Joints

It is fundamental that structural stiffness attracts bending moment. This makes the top and bottom connections critical areas in the design and operation of risers. Ball joints eliminate bending, but the angle across ball joints could create problems for drill strings and production tubulars and

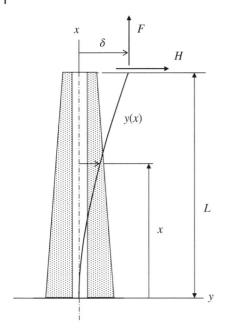

Figure 8.53 Tapered flex joint (add delta).

equipment. Most of the riser is relatively flexible. Bending (static and dynamic) over much of the riser is controlled by top tension.

Tapered flex-joints provide a more balanced transition between the flexible riser and fixed mudline equipment. Taper flex joints are typically truncated cones with constant inside diameter, but the outside diameter varies linearly from top to bottom (see Figure 8.53). The method of analysis developed by Dareing [4] allows tapered stress joints to be analyzed by solving only one differential equation, which provides a closed form solution to bending deflections and stresses.

Equation of Bending

Consider a tapered cantilever beam subjected to a transverse force, H and a pull full, V at the free end as shown in Figure 8.53. The differential equation of bending for the beam is

$$EI(x)\frac{d^2y}{dx^2} - Fy = (L-x)H - F\delta \qquad (8.170)$$

Gravity and hydrostatic forces are dropped because their effects are small by comparison with forces applied at the top (V and H). The solution to this equation is straightforward if there is no pull force, but this force is central to riser operations.

The solution to Eq. (8.170) is also complicated by the variable coefficient, $I(x)$, which accounts for the variation of the cross-sectional moment of inertia. By approximating the straight-tapered stress joint with a tubular having a parabolic moment of inertia distribution, the bending equation can be cast into the Euler type differential equation that is amenable to a closed form solution.

Parabolic Approximation to Moment of Inertia

If the distribution of the moment of inertia along the column is assumed to be of the form

$$I(x) = I_p\left(\frac{h-x}{a}\right)^2 \qquad (8.171)$$

where

$$h = \frac{L}{1 - \sqrt{\frac{I_p}{I_{SJ}}}}$$

and

$$a = h\sqrt{\frac{I_p}{I_{SJ}}}$$

where

I_p – moment of inertia of riser pipe
I_{SJ} – moment of inertia of bottom end of tapered stress joint

the differential equation of bending becomes

$$\frac{EI_p}{a^2}(h-x)^2 \frac{d^2y}{dx^2} - Fy = (L-x)H - F\delta \tag{8.172}$$

Because $I(x)$ is assumed to vary parabolic, the associated stress joint will be referred to as a "parabolic stress joint." Table 8.6 compares the parabolic moment of inertia with the moment of inertia of a straight taper over a 30 ft length.

Table 8.6 Comparison of wall thickness and moment of inertia over a 30 ft flex joint (23.5 in. OD base, 22 in. OD at top and having a 20.5 in. ID throughout).

Joint height (ft)	Wall thickness (in.)		Moment of inertia (in.^{4})	
	Parabolic	Straight	Parabolic	Straight
0	1.500	1.500	6301	6301
1	1.473	1.475	6164	6174
2	1.446	1.450	6027	6048
3	1.419	1.425	5892	5923
4	1.392	1.400	5759	5798
5	1.365	1.375	5627	5674
6	1.339	1.350	5497	5551
7	1.313	1.325	5369	5429
8	1.286	1.300	5241	5308
9	1.260	1.275	5116	5187
10	1.234	1.250	4992	5067
11	1.208	1.225	4869	4948
12	1.183	1.200	4748	4830
13	1.157	1.175	4629	4712
14	1.132	1.150	4511	4596
15	1.106	1.125	4394	4480
16	1.081	1.100	4279	4365
17	1.056	1.075	4166	4250
18	1.032	1.050	4054	4136
19	1.007	1.025	3943	4023
20	0.983	1.000	3835	3911
21	0.959	0.975	3727	3800
22	0.935	0.950	3621	3689
23	0.911	0.925	3517	3579

(Continued)

Table 8.6 (Continued)

	Wall thickness (in.)		Moment of inertia (in.^{4})	
Joint height (ft)	Parabolic	Straight	Parabolic	Straight
24	0.887	0.900	3414	3470
25	0.864	0.875	3313	3361
26	0.841	0.850	3213	3254
27	0.818	0.825	3115	3147
28	0.795	0.800	3018	3040
29	0.772	0.775	2923	2935
30	0.750	0.750	2830	2830

Solution to Differential Equation

The solution to Eq. (8.172) has particular and complimentary components:

$$y(x) = y_P(x) + y_C(x) \tag{8.173}$$

The particular solution is

$$y_P(x) = -(L-x)\frac{H}{F} + \delta \tag{8.174}$$

To determine the complementary solution, it is necessary to perform certain substitutions. By definition

$$\gamma^2 = \frac{a^2 F}{EI_P} \tag{8.175}$$

The differential equation giving the complementary solution is

$$(h-x)^2 \frac{d^2 y}{dx^2} + \gamma^2 y = 0 \tag{8.176}$$

This is a linear differential equation containing a variable coefficient. This equation is of the Euler type.

The earlier approximation for $I(x)$, using a parabolic function, was made to take advantage of the classic method of solving the Euler type differential equation.

The solution starts by eliminating the variable coefficient as follows:

$$(h-x) = e^z \tag{8.177}$$

Differentiating, gives

$$-dx = e^z dz$$

and

$$\frac{dz}{dx} = -\frac{1}{e^z} \tag{8.178}$$

Note that

$$\frac{dy}{dx} = \frac{dy}{dz}\frac{dz}{dx} = -\frac{1}{e^z}\frac{dy}{dz} \tag{8.179}$$

$$\frac{d^2y}{dx^2} = -\frac{1}{e^{2z}}\left(\frac{d^2y}{dz^2} - \frac{dy}{dz}\right) \tag{8.180}$$

and

$$(h-x)^2 = e^{2z} \tag{8.181}$$

Combining Eqs. (8.179) and (8.180) with Eq. (8.176) gives

$$\frac{d^2y}{dz^2} - \frac{dy}{dz} - \gamma^2 y = 0 \tag{8.182}$$

The variable coefficient is gone, and the solution to Eq. (8.182) can be obtained by assuming $y = Ce^{Dz}$. It follows that

$$\left(D^2 - D - \gamma^2\right)Ce^{Dz} = 0 \tag{8.183}$$

which leads to

$$D_{1,2} = \frac{1}{2}\left(1 \pm \sqrt{1 + 4\gamma^2}\right) \tag{8.184}$$

The complementary solution is therefore

$$y_C(x) = C_1 e^{D_1 z} + C_2 e^{D_2 z} = C_1(h-x)^{D_1} + C_2(h-x)^{D_2} \tag{8.185}$$

The total solution to Eq. (4.75) is

$$y(x) = C_1(h-x)^{D_1} + C_2(h-x)^{D_2} - (L-x)\frac{H}{F} + \delta \tag{8.186}$$

Equation (8.186) defines the deflection in the tapered flex joint. This closed form solution simplifies the simultaneous solution of stress joints coupled with marine risers as well as with offshore pipe lines and flow lines.

Example Consider the tapered flex joint defined in Table 8.6. The pull force and horizontal force at the top are assumed to be $F_B = 75\,000$ lb and $H = 50\,000$ lb. Parameters that enter into the deflection and bending stress calculations are

$$h = \frac{L}{1 - \sqrt{\frac{I_p}{I_{SJ}}}} = 90.94$$

$$a = h\sqrt{\frac{I_p}{I_{SJ}}} = 60.94$$

$$\gamma^2 = \frac{a^2 F}{EI_p} = \frac{60.94^2(75000)}{29(10)^6(2830)}\left|\frac{144\ \text{in.}^2}{1\text{ft}^2}\right|$$

$$\gamma = 0.699$$

Applying boundary conditions

$$y(0) = 0 \tag{8.187a}$$

$$\frac{dy}{dx}(0) = 0 \tag{8.187b}$$

$$y(L) = \delta \tag{8.187c}$$

gives

$$h^{D_1}C_1 + h^{D_2}C_2 + \delta = L\frac{H}{F_B} \tag{8.188}$$

$$D_1 h^{D_1-1}C_1 D_2 h^{D_2-1}C_2 = \frac{H}{F_B} \tag{8.189}$$

$$(h-L)^{D_1}C_1 + (h-L)^{D_2}C_2 = 0 \tag{8.190}$$

The solution to these three algebraic equations gives: $C_1 = 0.0856$, $C_2 = -100.079$ and $\delta = 0.417$ ft. The stress distributions for his special case are given in Table 8.7. These numbers apply to an assumed stand-alone tapered flex joint with assumed vertical and horizontal forces at the top. These equations will be used to interface tapered flex joints with uniform riser pipes.

Table 8.7 Deflection and stress in tapered flex joint (corresponds to data in Table 8.6; $H = 50\,000$ lb, $F_B = 75\,000$ lb).[a]

Deflection y (x) (ft)	Bending moment (ft-lb)	Bending stress (psi)	Axial stress (psi)	Total tension stress (psi)
0.0000	1 468 742	32 865	723	33 588
0.0006	1 418 785	32 331	737	33 068
0.0023	1 368 914	31 778	750	32 528
0.0051	1 319 127	31 203	764	31 967
0.0091	1 269 424	30 607	779	31 386
0.0141	1 219 803	29 988	794	30 782
0.0203	1 170 262	29 344	809	30 153
0.0275	1 120 801	28 674	826	29 500
0.0357	1 071 418	27 977	842	28 820
0.0449	1 022 110	27 251	860	28 111
0.0551	972 877	26 495	878	27 373
0.0663	923 717	25 705	897	26 602
0.0785	874 627	24 881	917	25 798
0.0915	825 605	24 019	937	24 957
0.1054	776 650	23 118	959	24 077
0.1202	727 759	22 175	981	23 156
0.1358	678 930	21 187	1 005	22 191
0.1522	630 160	20 150	1 029	21 179
0.1694	581 446	19 061	1 055	20 116

Table 8.7 (Continued)

Deflection y (x) (ft)	Bending moment (ft-lb)	Bending stress (psi)	Axial stress (psi)	Total tension stress (psi)
0.1873	532 786	17 917	1 082	18 999
0.2058	484 175	16 712	1 110	17 822
0.2249	435 612	15 442	1 140	16 582
0.2447	387 092	14 103	1 172	15 274
0.2649	338 612	12 687	1 205	13 892
0.2857	290 168	11 189	1 240	12 429
0.3069	241 756	9 602	1 276	10 878
0.3284	193 370	7 917	1 316	9 232
0.3502	145 007	6 125	1 357	7 482
0.3723	96 662	4 216	1 401	5 617
0.3945	48 328	2 179	1 448	3 627
0.	0	0	1 498	1 498

[a] Straight taper geometry used in stress calculations.

Application to Marine Risers

The following example (Figure 8.54) illustrates the effect of flex joints by comparison with a plain riser pipe. Calculations [4] of the two deflections are based on the following input data:

Ocean depth is 1000 ft
Riser dimensions are 22 in. OD by 20.5 in. ID
Pull force at the bottom of riser is 150 000 lb
Top lateral rig offset is 50 ft
Mud wt (12 ppg) and sea water wt (64 lb/ft^3)
Cross section inertia of riser is 2830 in.^{4}
Cross section inertia at bottom of flex joint is 6301 in.^{4}
Top dimension of flex joint matches riser dimensions

Only the lower 65 ft is shown in the figure.

Torsional Buckling of Long Vertical Pipe

The governing differential equations of bending stem from Huang and Dareing [18]. Figure 8.55 illustrates the model and shows the coordinates used in the analysis. There are two differential equations of bending caused by torsion buckling. One equation relates to bending in the xz plane. The other relates to bending in the yz plane.

$$EI\frac{d^4x}{dz^4} - T\frac{d^3y}{dz^3} - \frac{d}{dz}\left[T_{eff}\frac{dx}{dz}\right] = 0 \qquad (8.191a)$$

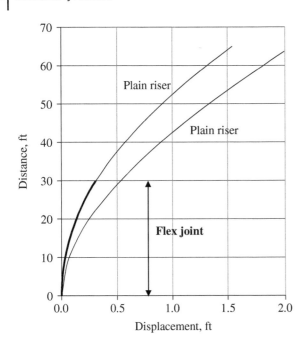

Figure 8.54 Comparison of riser deflections.

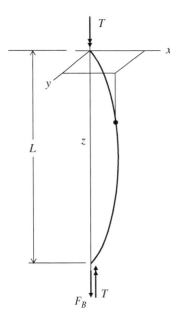

Figure 8.55 Torsion stability model.

$$EI\frac{d^4y}{dz^4} + T\frac{d^3x}{dz^3} - \frac{d}{dz}\left[T_{\text{eff}}\frac{dy}{dz}\right] = 0 \qquad (8.191b)$$

The symbol, T, without the subscript, represents applied torque, which is assumed constant over the pipe length. Effective tension, T_{eff}, however, is assumed to vary along the pipe.

Force, F_B represents the actual internal force in the pipe at $z = L$. Effective tension varies along the pipe and is defined in terms of the independent variable, z.

$$T_{\text{eff}} = [F_B + w(L-z)] + zw_m \qquad (8.192)$$

The first term is the actual tension in the pipe at location z, while the last term represents the contribution of hydrostatic fluid pressure at location z. Assuming fluid density is the same both inside and outside the pipe, the force F_B at the lower exposed end of the pipe is $F_B = -Lw_m$ giving

$$T_{\text{eff}} = (L-z)(w - w_m) \qquad (8.193)$$

Going back to the more general case and rearranging Eq. (8.192)

$$T_{\text{eff}} = (F_B + w_m L) + (w - w_m)(L - z) \qquad (8.194)$$

Boundary Conditions

Boundary conditions for pinned conditions at both top and bottom are

$$x(0) = y(0) = 0 \qquad (8.195a)$$

$$x(L) = y(L) = 0 \tag{8.195b}$$

$$EI\frac{d^2x}{dz^2}(0) = T\frac{dy}{dz}(0) \tag{8.195c}$$

$$EI\frac{d^2y}{dz^2}(0) = -T\frac{dx}{dz}(0) \tag{8.195d}$$

$$EI\frac{d^2x}{dz^2}(L) = T\frac{dy}{dz}(L) \tag{8.195e}$$

$$EI\frac{d^2y}{dz^2}(L) = -T\frac{dx}{dz}(L) \tag{8.195f}$$

Equations (8.191a) and (8.191b) are now combined using the complex variable

$$u = x(z) + iy(z) \tag{8.196}$$

Multiplying Eq. (8.191b) by the imaginary number "*i*" and adding the resulting equation to Eq. (8.191a) gives

$$EI\frac{d^4u}{dz^4} - iT\frac{d^3u}{dz^3} - \frac{d}{dz}\left\{[(F_B + w_mL) + (w - w_m)(L - z)]\frac{du}{dz}\right\} = 0 \tag{8.197}$$

Now replace z with the dimensionless number:

$$\varsigma = \frac{L - z}{L} \tag{8.198}$$

Noting that $dz = -Ld\zeta$ and dividing through by EI gives

$$\frac{d^4u}{d\zeta^4} - i\Theta\frac{d^3u}{d\zeta^3} - \frac{d}{d\zeta}\left[(\beta + \alpha\zeta)\frac{du}{d\zeta}\right] = 0 \tag{8.199}$$

which expands into

$$\frac{d^4u}{d\zeta^4} - i\Theta\frac{d^3u}{d\zeta^3} - \beta\frac{d^2u}{d\zeta^2} - \alpha\zeta\frac{d^2u}{d\zeta^2} - \alpha\frac{du}{d\zeta} = 0 \tag{8.200}$$

where

$$\alpha = \frac{(w - w_m)L^3}{EI} = \frac{wBFL^3}{EI}, \quad BF = 1 - \frac{w_m}{w}$$

$$\beta = \frac{(F_B + w_mL)L^2}{EI}$$

$$\Theta = \frac{TL}{EI}$$

The force, F_B, is the *actual* force at the lower end of the pipe. It is replaced by

$$F_B = F - w_mL$$

Recall

$$\gamma_m\Delta LA_c = w_m\Delta L$$

$$\gamma_mA_c = w_m \quad \text{(weight per unit length of pipe)}$$

So

$$w_m L = \gamma_m A_c L$$

which is the hydrostatic force acting on the lower end of an open pipe. The dimensionless β is now expressed as

$$\beta = \frac{(F_B + w_m L)L^2}{EI} = \frac{FL^2}{EI}$$

Here F can be tension $(+)$ or compression $(-)$ above hydrostatic compression at the bottom of pipe. When drill pipe is attached to the top of the drill collars, this implies the neutral point is located at the top of the collars.

Both Top and Bottom Ends Pinned

The boundary conditions transform into

$$u(0) = 0 \tag{8.201a}$$

$$u(1) = 0 \tag{8.201b}$$

$$\frac{d^2 u}{d\zeta^2}(0) = i\Theta \frac{du}{d\zeta}(0) \tag{8.201c}$$

$$\frac{d^2 u}{d\zeta^2}(1) = i\Theta \frac{du}{d\zeta}(1) \tag{8.201d}$$

Assuming the solution to the differential equation of the form

$$u(\varsigma) = a_0 + a_1 \varsigma + a_2 \varsigma^2 + \cdots + a_n \varsigma^n + \cdots \tag{8.202}$$

where the a's are complex constants for a given set of boundary conditions, and ζ is the independent real variable. $u(\zeta)$ is a complex function as defined earlier.

Substituting Eq. (8.202) into Eq. (8.200), gives the recurrence equation:

$$a_n = i\frac{\Theta}{n} a_{n-1} + \frac{\beta}{n(n-1)} a_{n-2} + \frac{(n-3)\alpha}{n(n-1)(n-2)} a_{n-3}, \quad n \geq 4 \tag{8.203}$$

The complex constants are dependent on α and β, which are real numbers. Using Eq. (8.202) and its derivatives, the boundary conditions yield

$$a_0 = 0 \tag{8.204a}$$

$$\sum_{n=1}^{\infty} a_n = 0 \tag{8.204b}$$

$$a_2 = i\frac{\Theta}{2} a_1 \tag{8.204c}$$

$$\sum_{n=1}^{\infty} n(n-1)a_n = i\Theta \sum_{n=1}^{\infty} n a_n \tag{8.204d}$$

These conditions show that $a_0 = 0$ and a_2 depends on a_1. From these conditions and by successive application of the recurrence equation (Eq. (8.203)), the coefficient a_n can be expressed as a linear combination of a_1 and a_3 or

$$a_n = (F_n + iJ_n)a_1 + (H_n + iL_n)a_3 \tag{8.205}$$

where F_n, H_n, J_n, L_n are real numbers. It is apparent that for $n = 0, 1, 2, 3$ and using boundary conditions Eqs. (8.204a) and (8.204c):

$$F_0 = 0 \quad J_0 = 0 \quad H_0 = 0 \quad L_0 = 0$$
$$F_1 = 1 \quad J_1 = 0 \quad H_1 = 0 \quad L_1 = 0$$
$$F_2 = 0 \quad J_2 = \Theta/2 \quad H_2 = 0 \quad L_2 = 0$$
$$F_3 = 0 \quad J_3 = 0 \quad H_3 = 1 \quad L_3 = 0$$

The remaining values for F_n, H_n, J_n, L_n ($n \geq 4$) are determined from the recurrence equations derived from Eqs. (8.203) and (8.205). The starting values are used to obtain higher-order values of F_n, J_n, H_n, L_n by repeated application of Eqs. (8.206a)–(8.206d), which are each functions of Θ, α, and β.

$$F_n = -\frac{\Theta}{n}J_{n-1} + \frac{\beta}{n(n-1)}F_{n-2} + \frac{(n-3)\alpha}{n(n-1)(n-2)}F_{n-3} \tag{8.206a}$$

$$J_n = +\frac{\Theta}{n}F_{n-1} + \frac{\beta}{n(n-1)}J_{n-2} + \frac{(n-3)\alpha}{n(n-1)(n-2)}J_{n-3} \tag{8.206b}$$

$$H_n = -\frac{\Theta}{n}L_{n-1} + \frac{\beta}{n(n-1)}H_{n-2} + \frac{(n-3)\alpha}{n(n-1)(n-2)}H_{n-3} \tag{8.206c}$$

$$L_n = +\frac{\Theta}{n}H_{n-1} + \frac{\beta}{n(n-1)}L_{n-2} + \frac{(n-3)\alpha}{n(n-1)(n-2)}L_{n-3} \tag{8.206d}$$

Next consider the second and fourth conditions. Substituting Eq. (8.205) into the boundary conditions Eqs. (8.204b) and (8.204d), the following two homogeneous equations are obtained:

$$\left(\sum F_n + i\sum J_n\right)a_1 + \left(\sum H_n + i\sum L_n\right)a_3 = 0 \tag{8.207a}$$

and

$$\left\{\left[\sum n(n-1)F_n + \Theta\sum nJ_n\right] + i\left[\sum n(n-1)J_n - \Theta\sum nF_n\right]\right\}a_1$$
$$+ \left\{\left[\sum n(n-1)H_n + \Theta\sum nL_n\right] + i\left[\sum n(n-1)L_n - \Theta\sum nH_n\right]\right\}a_3 = 0 \tag{8.207b}$$

For nontrivial solutions of a_1 and a_3, the determinant of the coefficients in Eqs. (8.207a) and (8.207b) must be zero. The expansion of the determinate produces a real part and an imaginary part. Both parts contain the eigenvalue. The real part of the expanded determinate is

$$\left(\sum F_n\right)\left[\sum n(n-1)H_n + \Theta\sum nL_n\right] - \left(\sum J_n\right)\left[\sum n(n-1)L_n - \Theta\sum nH_n\right]$$
$$- \left(\sum H_n\right)\left[\sum n(n-1)F_n + \Theta\sum nJ_n\right] + \left(\sum L_n\right)\left[\sum n(n-1)J_n - \Theta\sum nF_n\right] = 0 \tag{8.208a}$$

and the imaginary part is

$$\left(\sum J_n\right)\left[\sum n(n-1)H_n + \Theta\sum nL_n\right] + \left(\sum F_n\right)\left[\sum n(n-1)L_n - \Theta\sum nH_n\right]$$
$$- \left(\sum L_n\right)\left[\sum n(n-1)F_n + \Theta\sum nJ_n\right] - \left(\sum H_n\right)\left[\sum n(n-1)J_n - \Theta\sum nF_n\right] = 0 \tag{8.208b}$$

Let the left sides of Eqs. (8.208a) and (8.208b) be denoted by P and Q, respectively, then

$$P(\Theta, \alpha, \beta) = 0 \qquad\qquad (8.209a)$$

$$Q(\Theta, \alpha, \beta) = 0 \qquad\qquad (8.209b)$$

Both are real value functions containing the eigenvalue Θ. Assuming α and β are given only one equation is needed. We choose $P(\Theta, \alpha, \beta) = 0$ from which to determine the eigenvalues. This is done by trial.

Simply Supported at Both Ends with no End Thrust

Eigenvalues generated from Eq. (8.209a) are listed in Table 8.8. Pipe length, L, is reflected through the dimensionless number, α. Critical torque is determined from the dimensionless number, Θ. Note that as α gets smaller Θ approaches 2π as expected.

The first example is designed to compare the results with Greenhill's solution. Both pipes are assumed suspended in air for which $\beta = 0$ and buoyancy factor, $BF = 1$. Input values for these calculations are

OD – 5½ in.

ID – 4.892 in.

w – 19.2 lb/ft

I – 18.27 in.^{4}

E – 30 × 10^6 psi

Numerical results are plotted in Figure 8.56 along with results from Greenhill's equation. Pipe length and critical torque were extracted from Table 8.8.

Critical torques predicted for the drill pipe case (linear tension) is greater than torques predicted Greenhill's first mode. These two curves converge for short pipe as tension approaches zero. The linear tension case does have a second mode, but it is somewhat higher than for the Greenhill

Table 8.8 Eigenvalues for torsional buckling.

α	Θ
1	6.5
2	6.6
10	7.5
20	7.9
40	8.3
50	8.37
100	8.552
200	8.66
500	8.783
1000	8.865
2000	8.935

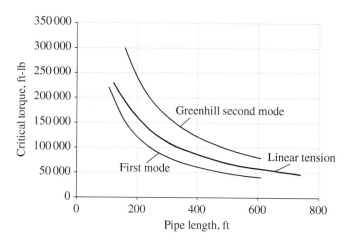

Figure 8.56 Comparison of results with Greenhill's classic solution.

second mode. In each case, there is no force applied to the pipe at the lower end. However, internal tension, which increases linearly in vertical pipe, increases critical torque as expected.

Force Applied to Lower End

The math model allows for a force F_B at the lower end. This force appears in the β term. Two cases were considered: $F = 5000$ lb and $F = -3000$ lb. Buoyancy factor is zero (no fluid) in both cases. Results show moderate amounts of tension does not affect critical torque especially for long pipe. It loses its effect compared to linear increase in tension toward the surface.

However, if pipe is put in compression by a relatively small amount (-3000 lb), critical torque reduces significantly (Figure 8.57).

This is due in part to the instability of the *axial mode*. A separate calculation of drill pipe buckling [17] shows the critical buckling force for the axial mode is

$$F_{cr} = 3.29 \left(\frac{EI}{144} \right)^{\frac{1}{3}} (wBF)^{\frac{2}{3}}$$

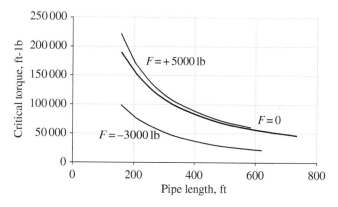

Figure 8.57 Effects of applied force at lower end.

$$F_{cr} = 3683 \text{ lb}$$

This calculation is based on conditions, i.e. $BF = 1$. Assuming a 12 ppg drilling fluid ($BF = 0.817$), the critical buckling force is $F_{cr} = 3219 \text{ lb}$.

Effect of Drilling Fluid on Torsional Buckling

Calculations showed that drilling mud density has negligible effect on torsion buckling. However, it does affect instability caused by axial forces.

Lower Boundary Condition Fixed

The data above was developed for a pinned-pinned boundary condition. We now consider the case where the lower end is fixed and the top end pinned. This more accurately represents the rigidity at the drill pipe – drill collars interface.

In this case, the initial values for F, J, H and L are

$$F_0 = 0 \quad J_0 = 0 \quad H_0 = 0 \quad L_0 = 0$$
$$F_1 = 0 \quad J_1 = 0 \quad H_1 = 0 \quad L_1 = 0$$
$$F_2 = 1 \quad J_2 = 0 \quad H_2 = 0 \quad L_2 = 0$$
$$F_3 = 0 \quad J_3 = 0 \quad H_3 = 0 \quad L_3 = 0$$

A comparison to critical torques for both sets of boundary conditions is given in Figure 8.58. Critical torque for the fixed-pinned case is greater than for the pinned-pinned case because of its rigidity. However, this effect becomes less important for long pipe as indicated.

Operational Significance

Three events take place in drill pipe directly above the drill collars. The effects of each can jointly or independently lead to lateral instability of the drill string, which could affect stick–slip torsional vibrations and overall friction along drill strings:

1) Buckling of axial mode
2) Torsional buckling
3) Drill pipe whirl

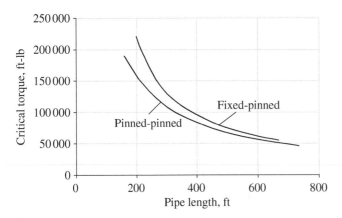

Figure 8.58 Comparison of critical torques for two sets of boundary conditions.

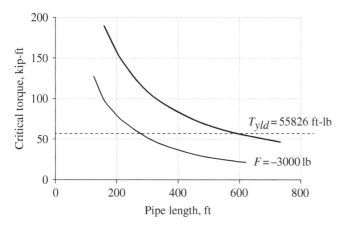

Figure 8.59 Effect of neutral point in drill pipe on critical torque.

Buckling of axial modes are predicted by

$$F_{cr} = \phi \left(\frac{EI}{144} \right)^{\frac{1}{3}} (BFw)^{\frac{2}{3}} \quad \text{(first mode)}$$

For 5½ in. drill pipe (19.2 lb/ft) is predicted by

$$F_{cr} = 3.29 \left(\frac{30 \times 10^6 (18.27)}{144} \right)^{\frac{1}{3}} (19.2)^{\frac{2}{3}} = 3683 \, \text{lb}$$

A 3000 lb axial compressive force is slightly lower than this critical force of 3683 lb. The result is a lower torsional buckling torque as shown.

The mode shape of torsional buckling can be visualized as a coil trapped inside a well bore.

Yield torque	44 074 ft lb (E75)
	55 826 ft lb (X95)
	61 703 ft lb (G105)
	79 332 ft lb (S135)

The yield torque or maximum possible torque that can be applied to this particular drill pipe is 55 286 ft lb (Figure 8.59). Critical torques are within this limit under typical drilling operations. Torsional buckling is also enhanced by axial compression.

Pressure Vessels

Stresses in Thick Wall Cylinders

The theory of hoop and radial stresses in thick wall cylinders is derived from the theory of elasticity and is documented in the literature [19]. These stresses are defined in terms of internal and external pressures by Lames' equations (Figure 8.60).

$$\sigma_\theta = \frac{p_i a^2 - p_o b^2 - a^2 b^2 (p_o - p_i)/r^2}{b^2 - a^2} \tag{8.210}$$

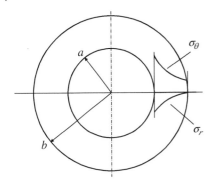

Figure 8.60 Hoop and radial stress distributions ($p_0 = 0$).

$$\sigma_r = \frac{p_i a^2 - p_o b^2 + a^2 b^2 (p_0 - p_i)/r^2}{b^2 - a^2} \qquad (8.211)$$

Stress distributions for the special case of $p_0 = 0$ are shown in Figure 8.60.

The stress state within the cylinder's cross section is biaxial if axial stresses are not included. The maximum stress levels for both hoop and radial are located on the inside surface.

Stresses in Thin-Wall Cylinders

The Lame equations show that for $\frac{D}{t} \geq 20$, the maximum hoop stress is only 5% higher than the average hoop stress across the cylinder wall [20]. The thin wall model gives a simple view of how hoop stresses are developed (see the freebody diagram in Figure 8.61).

Summing forces in the radial direction gives

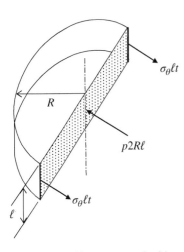

Figure 8.61 Hoop stresses in thin wall pipe.

$$\sigma_\theta = \frac{R}{t} p = \frac{D}{2t} p \qquad (8.212)$$

This simple formula gives reasonable engineering results in many cases.

Many pressure vessels can be analyzed using thin-walled equations. However, axial stresses caused by end loads create an additional stress component in the vessel walls. This axial stress component is

$$\sigma_a = \frac{R}{2t} p \qquad (8.213)$$

or one half of the hoop stress. Both hoop and axial stress components are the principal stresses. If the end cap is spherical, then the stress at each point in the cap is

$$\sigma_\phi = \frac{R}{2t} p \quad (\phi \text{ is arbitrary}) \qquad (8.214)$$

Each point in the spherical cap experiences the same stress level in all directions. A state of hydrostatic stress exists in the spherical cap and Mohr's stress diagram is a point.

Stresses Along a Helical Seam

Pressure vessels are sometimes manufactured by rolling flat sheet metal into a helical shape and then welding the seam as shown in Figure 8.62.

Figure 8.62   Stress along a pipe steam.

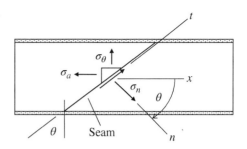

Normal stress to the seam and shear stress in the seam is determined from Eqs. (8.4) and (8.6) by setting $\tau_{xy} = 0$ (x and y are principal axes)

$$\sigma_n = \frac{\sigma_x + \sigma_y}{2} + \frac{\sigma_x - \sigma_y}{2} \cos 2\theta \tag{8.215}$$

$$\tau_{nt} = -\frac{\sigma_x - \sigma_y}{2} \sin 2\theta \tag{8.216}$$

where

$$\sigma_x = \sigma_a = \frac{R}{2t}p \quad \text{(axial stress)}$$

$$\sigma_y = \sigma_\theta = \frac{R}{t}p \quad \text{(hoop stress)}$$

By substitution

$$\sigma_n = \frac{1}{4}\frac{R}{t}p(3 - \cos 2\theta) \tag{8.217}$$

$$\tau_{nt} = \frac{1}{4}\frac{R}{t}p \sin 2\theta \tag{8.218}$$

Following the earlier sign convention, θ in the above figure is minus. For simplicity, we assume the hoop stress $\left(\frac{R}{t}\right)p = 10\,000$ psi and $\theta = 45°$ then by substitution

$$\sigma_n = 7500 \text{ psi} \quad \text{and} \quad \tau_{nt} = -2500 \text{ psi}$$

Interference Fit Between Cylinders

Thin-Wall Cylinders

First consider the effect of an interference fit between two thin wall cylinders. This situation is depicted by the expansion of the outer cylinder, δ_o, and compression of the inner cylinder, δ_i, resulting in hoop stress on both. As a result, pressure is developed between the two contacting surfaces (Figure 8.63).

The radial displacement of thin-walled cylinders due to either external or internal pressure is determined from

$$\delta = \frac{a^2}{hE}p \tag{8.219}$$

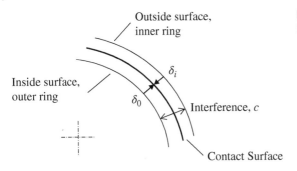

Outside surface, inner ring

Inside surface, outer ring

δ_i

δ_0

Interference, c

Contact Surface

Figure 8.63 Radial displacements of interference surfaces.

where

a – average radius of cylinder
h – wall thickness
δ – radial displacement of cylindrical surface

The basis of this simple equation relates to pressure and stress as follows:

$$\varepsilon_\theta = \frac{2\pi(a + \delta) - 2\pi a}{2\pi a} = \frac{\delta}{a} \tag{8.220}$$

$$\sigma_\theta = E\varepsilon_\theta \quad \text{and} \quad \sigma_\theta = \frac{a}{h}p$$

Combining these equations gives the deflection equation relating δ to pressure, p (Eq. (8.219)). For the interference case, radial deflection of each cylinder is

$$\delta_i = \frac{a_i^2}{h_i E}p \quad \text{(inside cylinder)}$$

$$\delta_0 = \frac{a_0^2}{h_0 E}p \quad \text{(outside cylinder)} \tag{8.221}$$

The given interference between the two cylinders relates to these displacements by

$$c = \delta_i + \delta_0$$

$$c = \left(\frac{a_i^2}{h_i E} + \frac{a_0^2}{h_0 E}\right)p \tag{8.222}$$

Assume dimensions as

$$a_0 = 9.92 \text{ in.} \quad h_0 = 0.125 \text{ in.} \quad c = 0.08 \text{ in.}$$
$$a_i = 10 \text{ in.} \quad h_l = 0.125 \text{ in.}$$

Then the pressure between the two cylinders over the contacting surfaces is

$$0.08 = \left(\frac{10^2}{0.125} + \frac{9.92^2}{0.125}\right)\frac{p}{E} = \frac{1587}{30 \times 10^6}p \tag{8.223}$$

$$p = 1512 \text{ psi}$$

This means that the hoop stress in the cylinder walls is

$$\sigma_\theta = \frac{a}{h}p = \frac{9.92}{0.125}1512 = 120 \text{ ksi} \tag{8.224}$$

Surface Deflection of Thick-Walled Cylinders

A similar problem exists when a thin wall cylindrical encloses a thick wall cylinder with an interference fit. In this case the radial displacement of the outside surface is formulated as follows.

At any radial location, ρ, there is no shear on the principal planes. Strain in the tangent direction is

$$\varepsilon_\theta = \frac{(\sigma_\theta - \nu\sigma_r)}{E} \tag{8.225}$$

This equation applies to any location, ρ, including both inside and outside surfaces of a thick-walled cylinder. The radial displacement at any location, ρ within a thick-walled cylinder is determined from

$$\varepsilon_\theta = \frac{2\pi(\rho + \delta) - 2\pi\rho}{2\pi\rho} = \frac{\delta}{\rho}$$

where

$$\frac{\delta}{\rho} = \frac{(\sigma_\theta - \nu\sigma_r)}{E} \tag{8.226}$$

The stress condition at the outer surface of the left drawing in Figure 8.64 is

$$\sigma_\theta = -p\frac{b^2 + a^2}{b^2 - a^2} \tag{8.227}$$

$$\sigma_r = -p \tag{8.228}$$

By substitution into Eq. (8.226), the radial deflection of the outside surface of the thick wall cylinder is

$$\delta = -\frac{pb}{E}\left(\frac{b^2 + a^2}{b^2 - a^2} - \nu\right) \tag{8.229}$$

The radial displacement is inward.

Next consider the right drawing in Figure 8.64.

$$\sigma_\theta = p\frac{b^2 + a^2}{b^2 - a^2}$$

$$\sigma_r = -p$$

and by substitution the radial deflection of the inside surface is

$$\delta = \frac{pa}{E}\left(\frac{b^2 + a^2}{b^2 - a^2} + \nu\right)$$

The radial displacement is outward

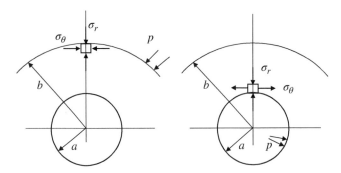

Figure 8.64 Radial displacement in thick wall cylinders.

Thick Wall Cylinder Enclosed by Thin Wall Cylinder

Example   Consider the previous case of two thin wall cylinders except in this case the inside cylinder is thick-walled. As before interference between the two cylinder relates to these displacements by

$$c = \delta_i + \delta_0 \tag{8.230}$$

where

$$\delta_i = -\frac{pb}{E}\left(\frac{b^2 + a^2}{b^2 - a^2} - \nu\right) \quad \text{(inward)} \tag{8.231}$$

and

$$\delta_0 = \frac{a_{ave}^2}{h_0 E}p \tag{8.232}$$

Assume dimensions as

$$a_0 = 10 \text{ in.} \quad h_0 = 0.125 \text{ in.} \quad c = 0.08 \text{ in.}$$
$$a = 5 \text{ in.} \quad b = 10.08 \text{ in.} \quad a_{ave} = 10.0625 \text{ in.}$$

Applying $c = \delta_i + \delta_0$

$$0.08 = \frac{p10.08}{E}\left(\frac{10.08^2 + 5^2}{10.08^2 - 5^2} - 0.25\right) + \frac{p}{E}\frac{10.06^2}{0.125} \tag{8.233}$$

$$0.08 = \frac{p}{E}(14.139 + 810)$$

$$\frac{p}{E} = 0.000\,097$$

$$p = 2913 \text{ psi}$$

compared with $p = 1512$ psi for the two thin wall cylinders with the same interference. The difference is the stiffness of the outer surface (ring) within the thick wall cylinder.

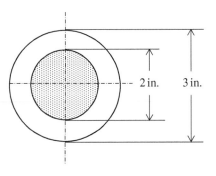

Figure 8.65   Sleeve shrunk on a solid bar.

Thick Wall Cylinder Enclosed by Thick Wall Cylinder

Now consider the case where a sleeve pressed fitted around a solid 2-in. diameter shaft (Figure 8.65). The outside diameter of the sleeve is 3 in. Determine the radial interference that will cause the sleeve to yield. Base your answer in the maximum shear stress criteria of failure. Yield strength of the sleeve material is 60 ksi. Accordingly, $\tau_{yld} = 30$ ksi.

The outside cylinder has a $\frac{D}{t}$ ratio of 1.667 which is much less than 20 must be treated as a thick wall cylinder. The approach is to first determine the relation between the interference and contact pressure between the shaft and sleeve.

The radial deflection of the outside surface of the shaft is

$$\delta_{shaft} = \frac{-pb}{E}\left(\frac{b^2 + a^2}{b^2 - a^2} - \nu\right) = \frac{-p}{E}1(1 - 0.25) = -0.75\frac{p}{E} \quad \text{(inward)}$$

where $a = 0$ and $b = 1$ in. As before, pressure, p, is a plus number and the minus means radial deflection of the shaft surface is inward.

The expression for the radial deflection of the inside surface of the sleeve is obtained by applying Lames' equation to the outer cylinder which now has an inside pressure of p and zero outside pressure:

$$\delta_{sleeve} = \frac{pa}{E}\left(\frac{b^2 + a^2}{b^2 - a^2} + \nu\right) = \frac{p}{E}1\left(\frac{1.5^2 + 1^2}{1.5^2 - 1^1} + 0.25\right) = 2.85\frac{p}{E}$$

The interference is the sum of the two radial deflections.

$$c = \delta_{shaft} + \delta_{sleeve}$$

$$c = 0.75\frac{p}{E} + 2.85\frac{p}{E} = 3.6\frac{p}{E}$$

$$p = 8.33 \times 10^6 c$$

Now we return to the stress components at the inside surface of the sleeve.

$$\sigma_\theta = p\left(\frac{b^2 + a^2}{b^2 - a^2}\right) = 2.6p$$

$$\sigma_r = -p$$

The maximum shear stress is

$$\tau_{max} = \frac{1}{2}(\sigma_\theta - \sigma_r) = 1.8p$$

With a material yield shear strength of 30 ksi

$$30\,000 = 1.8(8.33 \times 10^6)c$$

Giving $c = 0.002$ in. and $p = 16\,660$ psi. The maximum shear stress is 29 988 psi. Corresponding hoop stress is $\sigma_\theta = 2.6(16\,660) = 43\,318$ psi.

This example shows that a small interference between a shaft and thick-walled cylinder creates very high contact pressure and shear stress.

Elastic Buckling of Thin Wall Pipe

One way to view the buckling of thin-walled cylinders is to consider the circumference as a straight beam with end loads. This model is suggested by the shape of the primary mode as depicted in Figure 8.66. Assuming Euler's equation applies,

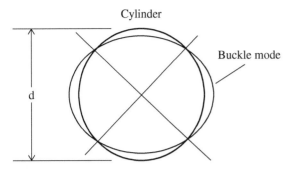

Figure 8.66 Loaded circular ring.

$$F_{cr} = \left(\frac{4\pi}{L}\right)^2 EI = \sigma_\theta A = \left(\frac{d}{2t}p\right)t\ell \quad \text{(four quarter-lobes)} \tag{8.234}$$

$$\left(\frac{4\pi}{\pi d}\right)^2 E\frac{t^3\ell}{12} = \left(\frac{d}{2t}p_{cr}\right)t\ell \tag{8.235}$$

After simplification,

$$p_{cr} = \frac{8}{3}E\frac{1}{(d/t)^3} \tag{8.236}$$

This solution treats the circumference of the cylinder as a straight beam.

Bresse's Formulation

Bresse [21] gives a more realistic model and solution to the buckling of circular rings. His formulation was developed to predict buckling of straps of unit length. Figure 8.67 shows one fourth of the ring of radius, r. The radial deflection, $w(s)$, is expressed in polar coordinates. Curvature of the deflection arc relative to the initial curvature $(1/r)$ is

$$\frac{1}{\rho} = \frac{d^2w}{ds^2} + \frac{w}{r^2} \tag{8.237}$$

Assuming Euler bending

$$EI\left(\frac{d^2w}{ds^2} + \frac{w}{r^2}\right) = -M \tag{8.238}$$

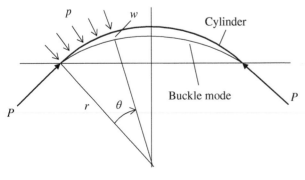

Figure 8.67 Freebody diagram of segment of cylinder.

The bending moment at location s or θ is $M = rwp$. Therefore,

$$EI\left(\frac{d^2w}{ds^2} + \frac{w}{r^2}\right) = -rwp \quad (p \text{ is force per arc length of strap}) \tag{8.239}$$

Collecting terms

$$\frac{d^2w}{ds^2} + \left(\frac{1}{r^2} + \frac{rp}{EI}\right)w = 0 \tag{8.240}$$

$$\frac{d^2w}{d\theta^2} + k^2w = 0 \tag{8.241}$$

where

$$k^2 = \left(1 + \frac{r^3p}{EI}\right)$$

The solution to Eq. (8.240) is

$$w(\theta) = C_1 \sin k\theta + C_2 \cos k\theta \tag{8.242}$$

Applying boundary conditions, $w(0) = 0$ and $w(\pi/2) = 0$ leads to

$$C_2 = 0$$

and

$$0 = C_1 \sin k\frac{\pi}{2}$$

Leading to a nontrivial solution of

$$k\frac{\pi}{2} = \pi$$

Instability or buckling occurs when $k = 2$.

$$1 + \frac{pr^3}{EI} = 4 \tag{8.243}$$

$$P_{cr} = \frac{3EI}{r^3} \quad \text{with } I = \frac{t^3}{12}$$

$$P_{cr} = \frac{E}{4}\left(\frac{t}{R}\right)^3 \quad \text{(viewed as pressure since strap as unit width)}$$

$$P_{cr} = 2E\frac{1}{(D/t)^3} \quad \text{(Bresse's equation)} \tag{8.244}$$

Compared with Eq. (8.236), the Bresse formula predicts a smaller buckling pressure.

Application to Long Cylinders

When applying Bresse's equation to long cylinders E is modified as follows. In the bending of plates and shells, longitudinal (z axis) strain at a point is zero. Consider plate bending as shown in Figure 8.68.

$$\varepsilon_z = \frac{1}{E}(\sigma_z - \nu\sigma_x) \tag{8.245}$$

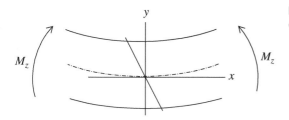

Figure 8.68 Bending of flat plate (z axis out of paper).

In the bending of plates and shells, strain in the z direction is zero. The means

$$\sigma_z = \nu\sigma_x \quad \text{(out of paper)} \tag{8.246}$$

By substitution

$$\varepsilon_x = \frac{1}{E}\left(\sigma_x - \nu^2\sigma_x\right) \tag{8.247}$$

$$\sigma_x = \frac{E}{1-\nu^2}\varepsilon_x$$

This affects the buckling of long tubes and requires that E be replaced by $\frac{E}{1-\nu^2}$ in Euler bending. Therefore, Bresse's formula becomes

$$P_{cr} = \frac{2E}{1-\nu^2}\frac{1}{(d/t)^3} \quad \text{(compare with Eq.(8.244))} \tag{8.248}$$

Consider a long, thin-walled pipe with outside diameter of 8 in. and a wall thickness of 0.25 in.: $\frac{d}{t} = \frac{8}{0.25} = 32$. If the yield strength of pipe material is 75 000 psi, determine the critical buckling pressure. Will the pipe fail by buckling or material yielding? Assume $E = 30 \times 10^6$ and $\nu = 0.25$.

$$P_{cr} = \frac{2 \times 30 \times 10^6}{1-0.25^2}\frac{1}{(32)^3} = 1953 \text{ psi} \tag{8.249}$$

Since

$$p = \frac{2\sigma t}{d}, \quad \sigma_\theta = \frac{d}{2t}p \tag{8.250}$$

Circumferential stress in pipe wall is

$$\sigma_\theta = \frac{d}{2t}P_{cr} = \frac{8}{2(0.25)}1953 = 31\,248 \text{ psi} \tag{8.251}$$

The pipe will buckle before reaching the yield stress level of 75 000 psi.

Thin Shells of Revolution

Thin shells of revolution are generated by rotating the geometry of the meridian curve about the axis of symmetry (Figure 8.69).

A freebody of a surface element gives

$$\frac{\sigma_m}{r_m} + \frac{\sigma_t}{r_t} = \frac{p}{t} \tag{8.252}$$

Figure 8.69 Thin shell of revolution.

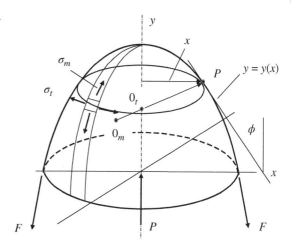

This equation contains two unknowns, σ_m and σ_t. The stress, σ_m, is determined independently from equilibrium of global forces in the y direction.

$$\sum F_y = 0$$

$$p(\pi x^2) - (2\pi xt)\sigma_m \cos \phi = 0$$

Therefore,

$$\sigma_m = \frac{x}{2t}\frac{p}{\cos \phi} \tag{8.253}$$

where

$$\tan \theta = \frac{dy}{dx}\bigg|_r \quad \text{and} \quad \phi = 90 - \theta$$

The radius of curvature, r_m, is determined from the curvature of the meridian. The tangent radius, r_t relates to x by $x = r_t \cos \phi$. Both radii are defined by

$$\frac{1}{r_m} = \frac{\frac{d^2y}{dx^2}}{\left[1 + \left(\frac{dy}{dx}\right)^2\right]^{\frac{3}{2}}} \tag{8.254}$$

$$r_t = \frac{x}{\cos \phi}$$

In the diagram

$$r_m = 0_m P \quad \text{and} \quad r_t = 0_t P$$

Example   Consider a shell of revolution as shown in Figure 8.70. Its profile is defined by

$$y = 8 - 0.5x^2 \qquad \frac{dy}{dx} = -x \qquad \frac{d^2y}{dx^2} = -1$$

Internal pressure is $p = 1500$ psi. We wish to find the stress, σ_m and σ_t at $y = 6$ in. The first step is to find σ_m.

$$\sigma_m = \frac{x}{2t} \frac{p}{\cos\phi} \tag{8.255}$$

The elements in this equation are

$$x = 2\ \text{in.} \quad y = 6\ \text{in.}$$

$$t = 0.125\ \text{in.}$$

$$p = 1500\ \text{in.}$$

$$\left.\frac{dy}{dx}\right|_{x=2} = -2 \quad \theta = 63.4° \quad \phi = 26.6°$$

The *meridian stress* is

$$\sigma_m = \frac{2}{2(0.125)} \frac{1500}{\cos 26.6°} = 13\,420\ \text{psi} \tag{8.256}$$

We now turn to the tangent stress.

$$\frac{\sigma_m}{r_m} + \frac{\sigma_t}{r_t} = \frac{p}{t} \tag{8.257}$$

Elements in this equation are

$$\frac{1}{r_m} = \frac{\frac{d^2y}{dx^2}}{\left[1 + \left(\frac{dy}{dx}\right)^2\right]^{\frac{3}{2}}} = \frac{-1}{\left[1 + (-2)^2\right]^{\frac{3}{2}}} = \frac{-1}{11.18} \tag{8.258}$$

$$r_m = 11.18\ \text{in.}$$

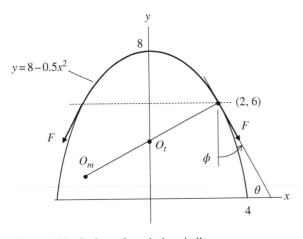

Figure 8.70   Surface of revolution shell.

Also

$$r_t = \frac{x}{\sin \theta} = \frac{2}{\sin 63.4°} = 2.24 \text{ in.} \tag{8.259}$$

By substitution, the *tangent stress* is

$$\sigma_t = \left(\frac{p}{t} - \frac{\sigma_m}{r_m} \right) r_t = \left(\frac{1500}{0.125} - \frac{13\,420}{11.18} \right) 2.24 \text{ psi} \tag{8.260}$$

$$\sigma_t = 21\,815 \text{ psi}$$

As a check on this number, hoop stress in the wall of a cylinder with $R = 2$ in. and $t = 0.125$ in.

$$\sigma_\theta = \frac{R}{t} p = \frac{2}{0.125} 1500 = 24\,000 \text{ psi} \quad \text{(in the same ballpark)}$$

Curved Beams

Winkler [5] developed the theory of curved beams in 1858. Bending stresses in curved beams are based on transverse planes remaining planes after bending as in the case of straight beams, i.e. there is no warping in the plane. With reference to Figure 8.71, normal strain at location y is

$$\varepsilon = \frac{\Delta \theta}{\Delta \phi} \frac{y}{r} = \frac{\Delta \theta}{\Delta \phi} \frac{r_n - r}{r} \tag{8.261}$$

For a given applied moment, M

$$\frac{\Delta \theta}{\Delta \phi} = C \quad \text{(constant)} \tag{8.262}$$

where

$\Delta \phi$ – differential arc length
$\Delta \theta$ – angular rotation of transverse plane (caused by bending moment, M)

And since

$$\sigma = E\varepsilon$$

$$\sigma = E \frac{(r_n - r)}{r} C \tag{8.263}$$

Location of Neutral Axis

As with the straight beam theory, the sum of normal forces across the face of the transverse plane is zero:

$$\int_A \sigma(r) dA = 0 \tag{8.264}$$

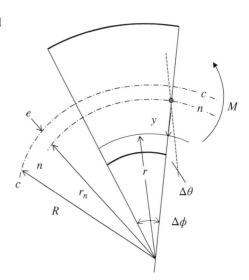

Figure 8.71 Curved beam parameters.

Substitution for σ yields the location of the neutral axis:

$$\int_A \left(\frac{r_n}{r} - 1\right) dA = 0$$

$$r_n = \frac{A}{\int_A \frac{dA}{r}} \qquad (8.265)$$

The neutral axis is not the centroid axis of the cross section as with the straight beam.

Stress Distribution in Cross Section

The stress distribution across the beam depends on the magnitude of the bending moment. Moments of internal stresses are equal to the external moment, M:

$$M = \int_A y(\sigma \, dA) = \int (r_n - r)\sigma \, dA \qquad (8.266)$$

There are several steps leading to the stress distribution equation. The steps include

$$M = EC\int_A \frac{(r_n - r)^2}{r} dA \qquad (8.267)$$

$$M = EC\int_A \left[\frac{r_n^2 - 2r_n r + r^2}{r}\right] dA$$

$$M = EC\left[r_n^2 \int_A \frac{dA}{r} - 2r_n A + \int_A (r_n - y)dA\right] \qquad (8.268)$$

But $R = r_n + e$ and $R - e = r_n$. Further substitution and collecting terms give

$$M = EC(R - r_n)A \qquad (8.269)$$

Replacing C with

$$C = \frac{\sigma r}{E(r_n - r)} \qquad \text{(from Eq.(8.263))}$$

By substitution

$$M = \left[\frac{\sigma r}{(r_n - r)}\right](R - r_n)A \qquad (8.270)$$

$$\sigma = \frac{My}{Ae(r_n - y)} \qquad (8.271)$$

which defines the stress distribution across the beams section (Figure 8.72). Stress distribution and magnitude depend on A, e, and r_n.

Example To illustrate the procedure, consider the curved beam with rectangular cross section shown in Figure 8.73. We assume the following parameters for this example:

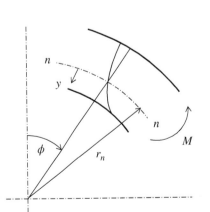

Figure 8.72 Stress distribution in a curved beam.

Figure 8.73 Rectangular cross section.

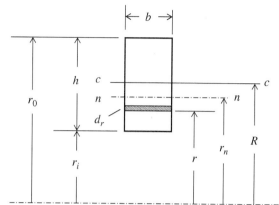

$h = 2$ in.

$r_0 = 3$ in.

$r_i = 1$ in.

$R = 2$ in.

$b = 0.75$ in.

By substitution

$$r_n = \frac{h}{\ln \frac{r_o}{r_i}} = 1.82 \text{ in.}$$

$$e = 2 - 1.82 = 0.18 \quad (e = R - r_n)$$

$$M = 1000 \text{ ft-lb}$$

$$A = 1.5 \text{ in.}^2$$

By substitution

$$\sigma = \frac{My}{Ae(r_n - y)}$$

$$\sigma = \frac{12\,000y}{1.5(0.18)(1.82 - y)}$$

The plot of this equation is shown in Figure 8.74.

Bending stress is a maximum at $y = r_n - r_i = 1.82 - 1 = 0.82$.

$$\sigma_{max} = \frac{12\,000(0.82)}{1.5(0.18)(1)} = 36\,440 \text{ psi}$$

By comparison, the straight beam formula gives a maximum stress of

$$\sigma_{max} = \frac{Mc}{I} \qquad I = \frac{bh^3}{12} = 0.5 \text{ in.}^4$$

$$\sigma_{max} = \frac{1000(12)(1)}{0.5} = 24\,000 \text{ psi}$$

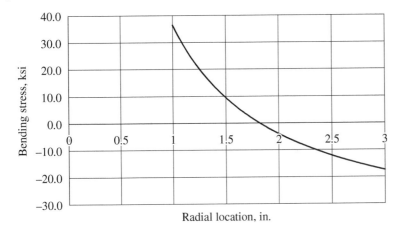

Figure 8.74 Bending stress across section.

The exact solution to the curved beam problem was developed by Golovin [22] based on the theory of elasticity. The exact solution shows there is also a radial stress component not considered by the Winkler solution.

For a curved beam having a *rectangular cross section*, it can be shown that the normal stress component can be represented by

$$\sigma_\theta = m\frac{M}{a^2} \tag{8.272}$$

where

a – inside radius of curved beam
b – outside radius of curved beam
m – numerical factor as shown in Table 8.9

The numbers show that the Winkler model gives very accurate results.

Example The hook shown in Figure 8.75 is to carry a 2000 lb load. Determine the factor of safety against material yielding of 80 000 psi using the von Mises criteria of failure.

$a = 2$ in.   (inside radius)
$c = 3.5$ in.   (outside radius)
$b_1 = 1$ in.   (inside width)
$b_2 = 0.5$ in. (outside width)

Table 8.9 Coefficient "*m*."

		Winkler model [5]		Exact solution [22]	
b/a	Linear stress (straight beam)	Inside	Outside	Inside	Outside
1.3	±66.67	+72.98	−61.27	+73.05	−61.35
2	±6.0	+7.725	−4.863	+7.755	−4.917
3	±1.5	+2.285	−1.095	+2.292	−1.13

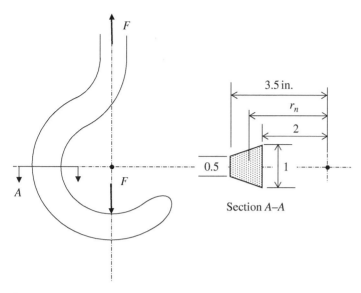

Figure 8.75 Bending stresses in a curved hook.

The elements in the stress equation are listed below:

$$\sigma = \frac{My}{Ae(r_n - y)} \quad \text{(bending stress, } y \text{ distance from neutral axis)}$$

$$A = \frac{b_1 + b_2}{2}(c - a) = 1.125 \text{ in.}^2 \quad \text{(cross sectional area)}$$

$$\int \frac{dA}{r} = \frac{b_1 c - b_2 a}{c - a}\ln\frac{c}{a} - b_1 + b_2 = 0.4327 \text{ in.}$$

$$r_n = \frac{A}{\int \frac{dA}{r}} = 2.6 \text{ in.} \quad \text{(radial distance to neutral axis of bending)}$$

The determination of r_n is often a difficult task for many cross-section shapes. These values can be found for various shapes in tabular form [23, 24]:

$$R = \frac{a(2b_1 + b_2) + c(b_1 + 2b_2)}{3(b_1 + b_2)} = 2.67 \text{ in.} \quad \text{(radial distance to centroid)}$$

$$e = R - r_n = 0.07 \text{ in.}$$

$$y = r_n - a = 0.6 \text{ in.} \quad \text{(location of max bending stress)}$$

Substituting these numbers into the bending stress equation gives the maximum stress at the inside surface of

$$\sigma_{max} = \frac{2000(2.6)0.6}{1.125(0.07)(2.6 - 0.6)} = 19\,810 \text{ psi}$$

Total stress

$$\sigma = \sigma_b + \sigma_a = 19\,810 + \frac{2000}{1.125} = 21\,587 \text{ psi}$$

Factor of safety

$$FS = \frac{80\,000}{21\,587} = 3.71$$

Shear Centers

Beams made up of thin members as shown in Figure 8.76 are susceptible to twisting as well as bending [23]. There is, however, a longitudinal bending axis through which transverse bending loads must pass to avoid twisting of the beam. The intersection of the longitudinal bending axis and transverse plane is called the shear center, indicated by point 0. The shear center is a fixed point in the beams cross section and is determined as follows.

When a transverse force is applied to a beam, it produces shear stresses in each section of the beams cross section. The vector sum or resultant of the internal shear forces (V_1, V_2, V_2) are equal to the total internal shear force generated by the external shear force, V.

$$V = \sum_{1,2,3} V_i = V_R \tag{8.273}$$

V_R is the resultant of all three shear forces. The line of action V_R is determined from the principle of moments, i.e.

$$eV_R = \sum_{1,2,3} d_i V_i = \sum_{1,2,3} T_i \tag{8.274}$$

The reference point is arbitrary, but it is convenient to choose point "*a*" in this case. Distance, *e*, locates the line of action of V_R. It also locates point 0, a point where V_R has zero torque. This point is called the shear center. This means that if the external shear force is applied through the shear center, it will produce zero twist on the beam. If it is applied away from the shear center, V_R and V produce a twisting couple as shown.

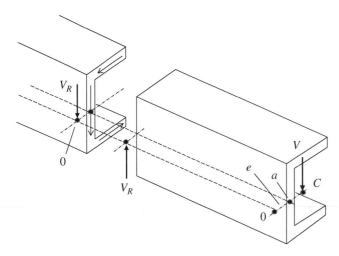

Figure 8.76 Shear center.

Bending axis and shear centers are of special interest in beams made up of thin parts. The location of shear centers is important in the use of angle and channel beams as well as other nonsymmetrical cross-section geometries containing thin members. Beams of this type are often used in flooring, roofing, and even aircraft wings. To avoid twisting these types of beams the line of action of applied load or loads must pass through the shear center. The following explains how to locate shear centers.

The earlier discussion of shear stress showed that shear stress in beams is determined by

$$\tau(y) = \frac{VQ}{Ib} \tag{8.275}$$

Following its derivation, shear stress occurs in vertical and horizontal planes. The maximum shear stress typically occurs at the neutral axis of bending.

An assumption in applying this formula to beams made up of thin members is: shear stresses lie within a plane transverse to thin members regardless of orientation. These stresses are still generated by the gradient of the bending moment.

If any section is horizontal as shown in Figure 8.77, it is viewed as having the ability to contain shear flow within the walls of the thin members. Shear flow in the transverse plane is determined as follows.

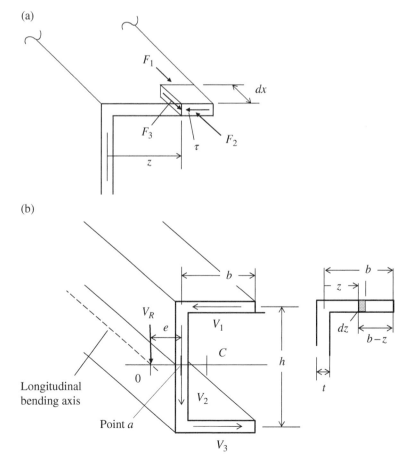

Figure 8.77 (a) Shear flow in horizontal member. (b) Shear flow around channel cross section.

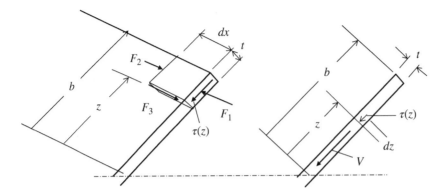

Figure 8.78 Shear flow.

Shear flow for a channel beam made up of three thin members is shown in Figure 8.78:

$$V_1 = \frac{V}{It} \int_0^b \frac{1}{2}(b-z)tht\, dz \tag{8.276}$$

$$V_1 = \frac{Vthb^2}{4I} \tag{8.277}$$

Noting that the moment of the resultant is equal to the sum of the moments of the parts, both respect to point "a" gives

$$eV_R = \frac{h}{2}(V_1 + V_2) = hV_1$$

$$eV_R = h\frac{Vthb^2}{4I} \tag{8.278}$$

Since $V_R = V$

$$e = \frac{\frac{1}{2}b}{1 + \frac{1}{6}(a_w + a_f)} \tag{8.279}$$

where

$a_w = wh$
$a_f = bt$

Angle iron is another common beam with cross section shown in Figure 8.78. Its shear center is determined as follows.

The local shear stress in a cross section at any angle from vertical is determined the same as for a vertical section. The shear flow, V_2 in the thin member is

$$Q = (b-z)\frac{(b-z)}{2} \sin\theta \tag{8.280}$$

$$dV = \tau t\, dz$$

$$V_2 = \frac{V}{It} \int_0^b \left[(b-z)t(b+z)\frac{1}{2} \sin\theta \right] t\, dz \tag{8.281}$$

$$V_2 = \frac{Vt}{I} \frac{\sqrt{2}}{6} b^3 \tag{8.282}$$

$$I = \int (z \sin \theta)^2 t \, dz = t\left(\frac{1}{2}\right)\frac{b^3}{3} \tag{8.283}$$

$$V_2 = \frac{\sqrt{2}}{2} V \tag{8.284}$$

where $\theta = 45°$. When θ is zero, the shear force along the thin leg is zero. In this case, the angle strip is horizontal and at the neutral axis of bending; therefore, normal forces are zero.

The horizontal force components of V_1 and V_2 cancel out. The resultant of the vertical components is

$$V_R = \sqrt{\left(\frac{\sqrt{2}}{2}V\right)^2 + \left(\frac{\sqrt{2}}{2}V\right)^2} \tag{8.285}$$

or

$$V_R = V \tag{8.286}$$

The resultant of the two shear forces in each section passes through point 0 (Figure 8.79). The twisting moment of the resultant with respect to point 0 is zero; therefore, point 0 is the shear center. If the transverse force is applied away from the shear center, say at the centroid C, then the resulting internal shear force, V_R, along with the external shear load, V, will cause the angle iron to twist.

Example Consider a 5-mm plate of steel formed into the semicircular shape as shown in Figure 8.80. The distance, e, to the shear center is determined as follows.

Local shear at any location, θ, is

$$\tau(\theta) = \frac{VQ}{It} \tag{8.287}$$

where

$$Q = \int_0^\theta y \, dA$$

$$y = r \cos \theta$$

$$dA = r \, d\theta(t)$$

$$Q = t \int_0^\theta r^2 \cos \theta \, d\theta = tr^2 \sin \theta$$

so

$$\tau(\theta) = \frac{V}{It} r^2 t \sin \theta \tag{8.288}$$

Moments with respect to point O of the shear flow, $dV_s = \tau(\theta)dA$ is

$$dM = r \, dV_s = r(r \, d\theta)\tau(\theta)t$$

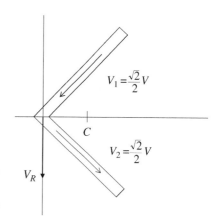

Figure 8.79 Angle iron beam with internal shear.

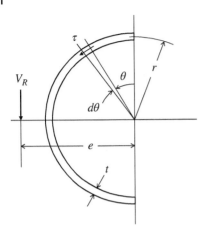

$$dM = \frac{V}{It} t^2 r^4 \sin\theta \, d\theta$$

$$M = 2\frac{Vt}{I} r^4$$

$$I = \frac{\pi}{2} r^3 t$$

After substitution, and integration, the internal moment of all differential shear forces is

$$M = \frac{4}{\pi} rV \tag{8.289}$$

Equating this to the moment of the internal shear forces to the moment of the resultant gives

$$eV_R = M$$

Figure 8.80 Semicircular beam cross sections.

and noting that the magnitude of $V_R = V$, gives

$$e = \frac{4}{\pi} r \tag{8.290}$$

Assuming $r = 250$ mm then $e = 318$ mm.

Unsymmetrical Bending

Previous discussions of bending stresses in beams assumed that applied loads were applied in the planes of principal axis of inertia. Beam cross sections were usually symmetrical, and loads were applied along an axis of symmetry. There are cases in design where loads are not applied in planes of symmetry such as shown in Figure 8.81, producing unsymmetrical bending. Unsymmetrical bending is defined as bending caused by applying loads that do not lie in or are parallel to principal *centroidal* axis of inertia [23]. Principal centroidal axis are designated by the U, V axis. The external load, F, may or may not pass through the centroid of the cross section. If the is thin walled, it is assumed that the external passed through the shear center.

Principal Axis of Inertia

Area moment of inertia is a property of any cross section of a beam. In our discussion of symmetrical bending, moments of inertia were obtained with respect to the neutral or centroid axis of the cross section. This axis is the location of zero bending and bending stresses increase linearly from the neutral axis. This is also the case in unsymmetrical bending, but the neutral axis is not as obvious.

Consider the transformation of the moments of inertia from x,y axis to x',y' axis (refer to Figure 8.82)

The radial distance from point 0 to differential area dA is

$$\bar{r} = x\hat{i} + y\hat{j} = x_1\hat{\alpha} + y_1\hat{\beta} \tag{8.291}$$

Note that

Figure 8.81 Unsymmetrical bending.

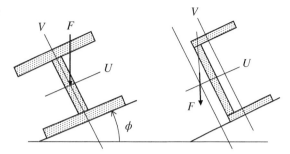

Figure 8.82 Transformation axes for area moments of inertia.

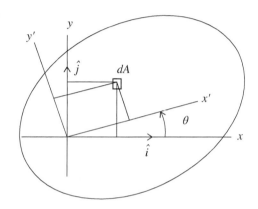

$$\hat{\alpha} = \hat{i}\cos\theta + \hat{j}\sin\theta$$

$$\hat{\beta} = -\hat{i}\sin\theta + \hat{j}\cos\theta$$

$$\hat{i} = \hat{\alpha}\cos\theta - \hat{\beta}\sin\theta$$

$$\hat{j} = \hat{\alpha}\sin\theta + \hat{\beta}\cos\theta$$

Substituting for $\hat{i}$ and $\hat{j}$ and collecting terms

$$x_1 = x\cos\theta + y\sin\theta \tag{8.292}$$

$$y_1 = -x\sin\theta + y\cos\theta \tag{8.293}$$

By definition

$$I_{x1} = \int_A y_1^2 \, dA, \quad I_{y1} = \int_A x_1^2 \, dA, \quad I_{x1y1} = \int_A x_1 y_1 \, dA \tag{8.294}$$

Substituting for y_1 gives

$$I_n = I_x \cos^2\theta + I_y \sin^2\theta + \left(-2I_{xy}\right)\sin\theta\cos\theta \tag{8.295}$$

It is helpful to note that area moments of inertia (I_x, I_y, I_{xy}) are calculated with respect to given coordinates, say x and y. Moments of inertia with respect to $x'y'$ each change with orientation, θ. There is a remarkable similarity of the transformation equations for stress, strain, and area moment of inertia:

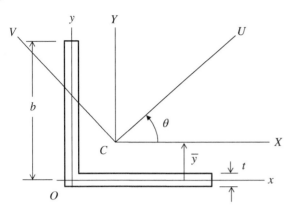

Figure 8.83 Angle iron cross section.

$$\sigma_n = \sigma_x \cos^2\theta + \sigma_y \sin^2\theta + 2\tau_{xy} \sin\theta \cos\theta \text{ given by (8.33)}$$

$$\varepsilon_n = \varepsilon_x \cos^2\theta + \varepsilon_y \sin^2\theta + 2\left(\frac{\gamma_{xy}}{2}\right) \sin\theta \cos\theta \text{ given by (8.34)}$$

$$I_n = I_x \cos^2\theta + I_y \sin^2\theta + \left(-2I_{xy}\right) \sin\theta \cos\theta \text{ given by (8.295)}$$

Notice the similarity between the three equations. Each of these equations has the same form. This means that the earlier discussion of stress (and strain) applies directly to inertia transformations. Mohr's circle also applies and gives a convenient way determine principal axis of inertia. Because of the minus sign attached to the product of inertia in Eq. (8.295), positive values of product of inertia are plotted up and negative values of product of inertia are plotted down.

Consider the cross section of the angle iron shown in Figure 8.83. It is desired to find the area moment of inertia about any axis passing through the center of gravity, point c. The approach is to find the moments of inertia, I_X, I_Y, I_{XY} and then use Mohr's inertia circle to find I_N oriented from the x-axis by angle θ.

Using moment of inertia transfer equations:

$$\bar{x}2bt = \frac{b}{2}bt \quad \bar{x} = \bar{y} = \frac{b}{4} \tag{8.296}$$

$$I_x = \frac{b^3 t}{12} + bt\left(\frac{b}{2}\right)^2 = \frac{b^3 t}{3} \quad I_y = \frac{b^3 t}{3} \quad I_{xy} = 0 \tag{8.297}$$

$$I_x = I_X + \bar{y}^2 A \quad I_X = \frac{b^3 t}{3} - \left(\frac{b}{4}\right)^2 2bt = \frac{5b^3 t}{24} \tag{8.298}$$

$$I_y = I_Y + \bar{x}^2 A \quad I_Y = \frac{5b^3 t}{24} \tag{8.299}$$

$$I_{xy} = I_{XY} + \bar{x}\bar{y}A \quad I_{XY} = -\frac{b}{4}\frac{b}{4}2bt = -\frac{b^3 t}{8} \tag{8.300}$$

Example Assuming $b = 4$ in. and $t = \frac{1}{4}$ in.

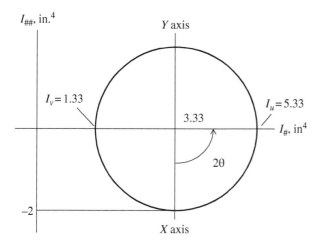

Figure 8.84 Mohr's circle for inertia.

$$I_X = I_Y = 3.33 \text{ in.}^4$$

$$I_{XY} = -2 \text{ in.}^4$$

Using Mohr's circle for area moment of inertia (Figure 8.84),

$$I_u = 5.33 \text{ in.}^4 \ (\theta = 45°)$$

$$I_v = 1.33 \text{ in.}^4$$

Example Consider another cross-sectional area shown in Figure 8.85. Here we wish to determine the principal axis of inertia and principal area moments of inertia with respect to these axes.
Cross product of inertia

$$\left(I_{xy}\right)_1 = 4\left(\frac{1}{4}\right)(-2)(4) = -8 \text{ in.}^4$$

$$\left(I_{xy}\right)_2 = 0$$

$$\left(I_{xy}\right)_3 = 4\left(\frac{1}{4}\right)(2)(-4) = -8$$

Area moment of inertia with respect to the x axis:

$$\left(I_x\right)_1 = 4\left(\frac{1}{4}\right)^3 \frac{1}{12} + 4^2(4)\frac{1}{4} = \frac{1}{(16)12} + 16 \sim 16 \text{ in.}^2$$

$$\left(I_x\right)_2 = \frac{1}{4}\frac{7.5^3}{12} = 8.79 \text{ in.}^4$$

$$\left(I_x\right)_3 = 16 \text{ in.}^4$$

Area moment of inertia with respect to the y axis:

$$\left(I_y\right)_1 = \frac{1}{4}\frac{4^3}{12} + 2^2\frac{1}{4}(4) = \frac{16}{12} + 4 = 5.33 \text{ in.}^4$$

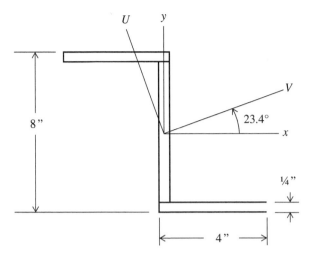

Figure 8.85 Principal Axis of Inertia.

$$\left(I_y\right)_2 \sim 0$$

$$\left(I_y\right)_3 = 5.33 \text{ in.}^4$$

$$I_x = 40.79 \sim 41 \text{ in.}^4$$

$$I_y = 10.67 \sim 11 \text{ in.}^4$$

$$I_{xy} = -16 \text{ in.}^4 \quad \text{(plotted down)}$$

Using Mohr's circle of inertia (Figure 8.86):

$$\tan 2\theta = \frac{16}{15}$$

$$2\theta = 46.8° \quad \theta = 23.4°$$

$$I_V = 26 + 22 = 48 \text{ in.}^4$$

$$I_U = 26 - 22 = 4 \text{ in.}^4$$

Neutral Axis of Bending

When a force, P, is applied at angle ϕ measured from a principal axis of inertia, the neutral axis of bending is determined by

$$\tan \alpha = \frac{I_u}{I_v} \tan \phi \tag{8.301}$$

is measured from a principal axis of inertia
α is measured from the other principal axis of inertia

Force P does not have to be applied through the centroid, C; however, ϕ is still measured from a principal axis of inertia.

Example Consider the symmetrical beam cross section in Figure 8.87.
We wish to determine the orientation of the neutral axis of bending.
The location of the centroid the composite area with respect to the baseline is

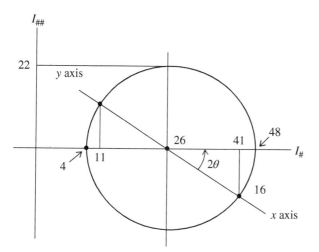

Figure 8.86 Mohr's circle of inertia (numbers have been rounded).

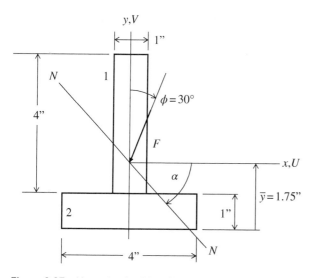

Figure 8.87 Neutral axis of bending.

$$(4+4)\bar{y} = \frac{1}{2}4 + 3(4)$$

$$\bar{y} = \frac{7}{4} = 1.75 \text{ in.}$$

The moment of inertia with respect to the x axis (passing through cg) is

$$(I_x)_1 = \frac{(1)4^3}{12} + (1.25)^2 4 = 11.58 \text{ in.}^4$$

$$(I_x)_2 = \frac{(4)1^3}{12} + (1.25)^2 4 = 6.58 \text{ in.}^4$$

$$I_x = (I_x)_1 + (I_x)_2 = 18.17 \text{ in.}^4$$

$$I_y = \frac{4(1)^3}{12} + \frac{1}{12}(4)^3 = 5.67 \text{ in.}^4$$

So

$$\bar{y} = \frac{7}{4} = 1.75 \text{ in.}$$

$$I_x = 1.817 \text{ in.}^4 \quad (I_x = I_U)$$

$$I_y = 5.66 \text{ in.}^4 \quad (I_y = I_V)$$

$$I_{xy} = 0 \quad y, V \text{ are axis of symmetry}$$

$$\tan \alpha = \frac{18.17}{5.66} \tan 30°$$

$$\alpha = 61.64°$$

Bending Stresses

Normal stresses caused by unsymmetrical bending are determined from [23]

$$\sigma = \frac{M_u v}{I_u} + \frac{M_v u}{I_v} \tag{8.302}$$

where

I_u, I_v – moments of internal with respect to principal axis U and V
u, v – coordinates to point where stress is to be determined
$M_u = M \cos \phi$
$M_v = M \sin \phi$

Example These equations will now be applied to the unsymmetrical beam shown in Figure 8.88. The bending moment diagram shows a maximum bending of 3200 ft lb along the middle portion. The bending components about the principal axes are shown below:

$$I_u = \frac{1}{12}bh^3 = \frac{1}{12}6(8)^3 = 256 \text{ in.}^4$$

$$I_v = \frac{8(6^3)}{12} = 144 \text{ in.}^4$$

Neutral axis is oriented per

$$\tan \alpha = \frac{256}{144} \tan 30 = 1.028$$

or $\alpha = 45.79°$ from the U principal axis.

The bending moment, M, is resolved into components about the U and V axis.

$$M_u = 3200(12) \cos 30 = 33\,300 \text{ in.-lb}$$

$$M_u = 3200(12) \sin 30 = 19\,200 \text{ in.-lb}$$

No sign has been attached to either of these components. It is somewhat simpler to observe that both bending components induce compressive stresses at point A.

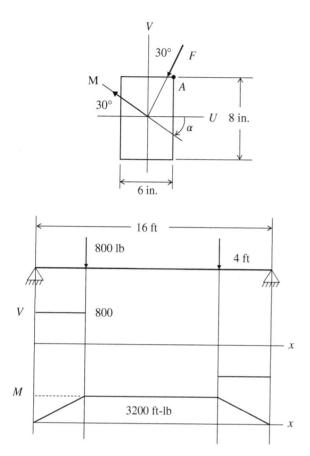

Figure 8.88 Shear and bending moment diagrams.

Substituting the numbers gives

$$\sigma_A = \frac{33\,000(4)}{256} + \frac{19\,200(3)}{144} = 916\,\text{psi (compression)}$$

The neutral axis angle, α, does not enter into this calculation directly.

Beams on Elastic Foundations

This analysis applies to long beams support by elastic foundations, such as a railroad track lying across a series of railroad ties. Winkler [9] developed the mathematics to this problem. This work was motivated by the need to understand bending stresses in railroads during the growth of steam locomotion in Europe. Loading in the beam can be concentrated or distributed as in regular beams.

Formulating the Problem

The equations of bending are developed with reference to the freebody diagram shown in Figure 8.89.

The shear forces and moments are shown in their positive directions. The elastic foundation force is $q\,dx$ with q being the foundation force per unit length, dx. From this diagram

$$\frac{dM}{dx} = V \tag{8.303}$$

$$\frac{dV}{dx} = q \tag{8.304}$$

$$EI\frac{d^2y}{dx^2} = -M \tag{8.305}$$

Combining these equations gives

$$EI\frac{d^4y}{dx^4} = -q \tag{8.306}$$

The magnitude of force, q (force/length) depends on deflection, y.

$$q = ky \tag{8.307}$$

where

k – distributed spring constant (force/length of beam per depth of deflection, force/length^2)

One way to experimentally determine k of a uniform support is to load a flat surface onto the uniform support with force, F, and measure displacement into support. The equation of equilibrium is

$$F = q\Delta x = ky\Delta x$$

$$k = \frac{F}{\Delta xy} = \frac{p\Delta xb}{\Delta xy} = \frac{pb}{y}\ \text{lb/in.}^2 \tag{8.308}$$

where

F – applied force
p – pressure applied to the top surface, psi
b – width of surface
Δx – incremental length
y – depth of penetration into surface

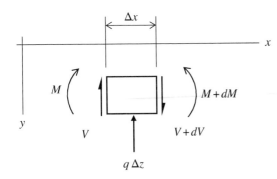

Figure 8.89 Differential element of beam on elastic foundation.

Another type of support is analogous to railroad tracks lying on ties. Assuming tie spacing is ℓ then

$$p\ell b = Ky$$

where K is the spring constant of each individual tie being pushed into the surface. Since $q = bp$, $\ell q = Ky$ and

$$q = \frac{K}{\ell}y \qquad (8.309)$$

So

$$k_{eq} = \frac{K}{\ell}.$$

By substitution for q in Eq. (8.306)

$$EI\frac{d^4y}{dx^4} = -ky \qquad (8.310)$$

$$EI\frac{d^4y}{dx^4} + ky = 0 \qquad (8.311)$$

$$\frac{d^4y}{dx^4} + 4\beta^4 y = 0 \qquad (8.312)$$

where

$$\beta^4 = \frac{k}{4EI}$$

Mathematical Solution

The solution to Eq. (8.312) is assumed to be $y(x) = Ce^{sx}$. From substitution, there are four values of "s" that satisfy this equation:

$$s^4 + 4\beta^4 = 0 \qquad (8.313)$$

$$s^4 = -4\beta^4$$

$$s^4 = 4\beta^4(-1) \qquad (8.314)$$

In the complex plane $e^{i\pi} = -1$ therefore

$$s^4 = 4\beta^4 e^{i\pi}$$

But since we need at least four roots

$$s^4 = 4\beta^4 e^{in\pi}$$

where $n = 1, 3, 5, 7$. Therefore

$$s = \sqrt{2}\beta e^{\frac{in\pi}{4}} \; n = 1, 3, 5, 7 \qquad (8.315)$$

These four roots are expressed in terms of complex coordinates as

$$s_1 = \beta(1 + i)$$
$$s_2 = \beta(-1 + i)$$
$$s_3 = \beta(-1 - i)$$
$$s_4 = \beta(1 - i)$$

All four values of "s" satisfy Eq. (8.312) and therefore the sum of each is also a solution. Therefore,

$$y(x) = C_1 e^{\beta(1+i)x} + C_2 e^{\beta(-1+i)x} + C_3 e^{\beta(-1-i)x} + C_4 e^{\beta(1-i)x}$$

Expanding further

$$y(x) = e^{\beta x}\left(C_1 e^{i\beta x} + C_4 e^{-i\beta x}\right) + e^{-\beta x}\left(C_2 e^{i\beta x} + C_3 e^{-i\beta x}\right) \tag{8.316}$$

Recall

$$e^{i\beta x} = \cos\beta x + i\sin\beta x$$

$$e^{-i\beta x} = \cos\beta x - i\sin\beta x$$

By substitution, the terms in the first and second brackets are

$$C_1 e^{i\beta x} + C_4 e^{-i\beta x} = (C_1 + C_4)\cos\beta x + i(C_1 - C_4)\sin\beta x \tag{8.317}$$

and

$$C_2 e^{i\beta x} + C_3 e^{-i\beta x} = (C_2 + C_3)\cos\beta x + i(C_2 - C_3)\sin\beta x \tag{8.318}$$

Since $y(x)$ is a real number, C_1 and C_4 must be complex conjugates as well as C_2 and C_3. Bringing all this together gives the general solution

$$y(x) = e^{\beta x}[A\cos\beta x + B\sin\beta x] + e^{-\beta x}[C\cos\beta x + D\sin\beta x] \tag{8.319}$$

Solution to Concentrated Force

The general solution is now applied to a beam loaded by a concentrated force. As in the case of open beams, the general solution does not apply across the concentrated force, so we look for the solution to the right side of load P. With respect to the x, y coordinates (Figure 8.90), positive values of x predict huge displacement at $+\infty$, therefor constants A and B must be zero, leaving

$$y(x) = e^{-\beta x}(C\cos\beta x + D\sin\beta x)$$

The problem is to determine the deflection, slope, and internal shear and moments caused by the concentrated force, P. The origin of the x axis is located at the point of application of the concentrated force. Beam deflection is measured positive in the downward direction.

The remaining equation has two arbitrary constants, C and D. These constants are determined by using the slope and shear at $x = 0$.

$$\left.\frac{dy}{dz}\right|_{x=0} = 0$$

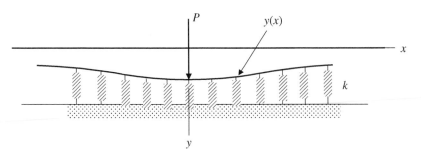

Figure 8.90 Deflection of a beam on an elastic foundation.

$$EI\frac{d^3y}{dx^3}\bigg|_{x=0} = V = -\frac{P}{2}$$

Using the slope at $x = 0$, $C = D$ leaving

$$y(x) = Ae^{-\beta x}(\sin \beta x + \cos \beta x) \quad (A \text{ is arbitrary}) \tag{8.320}$$

The shear condition at $x = 0$ gives

$$A = \frac{P\beta}{2k}$$

Collecting terms

$$y(x) = \frac{P\beta}{2k}e^{-\beta x}(\cos \beta x + \sin \beta x) \tag{8.321}$$

This equation shows that displacements diminish as x increases to $(+)$ infinity. Keep in mind that Eq. (8.321) was developed for displacement to the right of the concentrated load, P. It does not apply across the concentrated load, P, but we can apply it to the left side knowing that displacements are symmetrical about loading P.

Other parameters of interest are slope, shear, and bending moment along the beam. These parameters are determined from

$$y = \frac{P\beta}{2k}F_1 \quad \text{deflection} \tag{8.322}$$

$$y' = -\frac{P\beta^2}{k}F_2 \quad \text{slope} \tag{8.323}$$

$$EIy'' = -\frac{P}{4\beta}F_3 \quad \text{moment} \tag{8.324}$$

$$EIy'' = \frac{P}{2}F_4 \quad \text{shear} \tag{8.325}$$

where

$$F_1 = e^{-\beta x}(\cos \beta x + \sin \beta x) \tag{8.326}$$

$$F_2 = e^{-\beta x}\sin \beta x \tag{8.327}$$

$$F_3 = e^{-\beta x}(\cos \beta x - \sin \beta x) \tag{8.328}$$

$$F_4 = e^{-\beta x}\cos \beta x \tag{8.329}$$

In each case, displacement, slope, shear, and bending are symmetrical on each side of load P. If there are multiple concentrated loads applied to the beam, local parameters are determined by superposition. Other loading conditions are given.

Radial Deflection of Thin Wall Cylinders Due to Ring Loading

One of the most powerful applications of this theory is predicting bending and circumferential stresses in thin wall cylinders resulting from uniform circumferential loads. In this case, radial displacement, u, is the dependent variable and the distributed "spring" effect comes from the elastic circumferential strain.

Formulation of Spring Constant

From earlier discussions, hoop stress is related to external pressure surrounding a thin wall cylinder by

$$\sigma_\theta = \frac{a}{h} p \tag{8.330}$$

where

h – wall thickness
a – mean radius of cylinder
p – applied pressure (internal or external)

When external pressure is applied, thin wall cylinders shrink by radial displacement, u, where (+) displacement is in the inward direction (Figure 8.91). The hoop strain across the wall (assumed to be uniform) is

$$\varepsilon_\theta = \frac{2\pi(a-u) - 2\pi a}{2\pi a a} = -\frac{u}{a} \quad \text{(minus means compression)} \tag{8.331}$$

Noting that $\sigma_\theta = E\varepsilon_\theta$, then

$$\frac{a}{h} p = E \frac{u}{a} \tag{8.332}$$

then

$$u = \frac{a^2}{hE} p \tag{8.333}$$

As explained previously, the spring constant for elastic foundations is force per length of beam per deflection into the elastic foundation (k = lb/in. per inch of deflection with units of force/length^2). For the cylindrical strip (Figure 8.92), the elastic compression of the cylinder acts as the elastic foundation. The formulation of the expression for k is given below.

Consider a small arc of $b = a\Delta\theta$ and length L. When external pressure, p, is applied, the distributed load, w, is

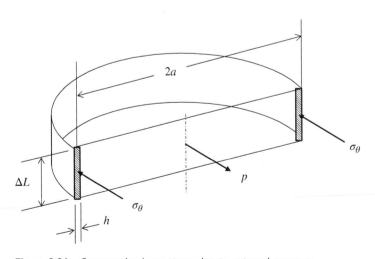

Figure 8.91 Compressive hoop stress due to external pressure.

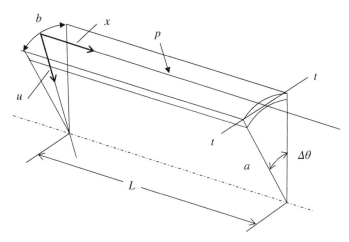

Figure 8.92 Arc section of a thin-walled cylinder.

$$wL = pbL \tag{8.334}$$

where

w – distributed load, lb/in.
L – length of strip, in.
p – pressure, psi
b – arc length

The distributed load produces an inward deflection of

$$u = \frac{a^2}{hE}p \tag{8.335}$$

The spring constant, k, is defined as

$$k = \frac{w}{u} \text{ lb/in.}^2 \tag{8.336}$$

By substitution

$$k = \frac{bhE}{a^2} \quad \text{or} \quad k = \frac{hE}{a^2} \text{ per unit arc length} \tag{8.337}$$

The arc distance, b, is a small arbitrary distance and cancels out later.

Equation of Bending for Cylindrical Arc Strip

The above formulation assumes zero bending and was developed for the sole purpose of developing the expression for the spring constant, k. We now examine deflections caused by nonuniform external loading, i.e. $w \neq$ constant.

The equation of bending for the strip beam (Figure 8.92) is developed from the differential element shown in Figure 8.93. Recall from beam the

$$M_t = EI_t \frac{d^2u}{dx^2} \tag{8.338}$$

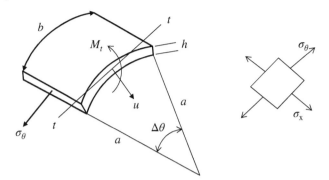

Figure 8.93 Element of cylindrical surface.

where

M_t – bending moment per length, b, about the t axis
I_t – moment of inertia per length, b, with respect to the t axis

In this case, EI_t is modified as follows. Each of the two sides of the strip beam lies in radial planes and hoop strains are not allowed when only bending is considered. Because of this,

$$\varepsilon_x = \frac{\sigma_x}{E} - \nu \frac{\sigma_\theta}{E} \tag{8.339}$$

$$\varepsilon_\theta = \frac{\sigma_\theta}{E} - \nu \frac{\sigma_x}{E} = 0 \tag{8.340}$$

From the second equation:

$$\sigma_\theta = \nu \sigma_x \tag{8.341}$$

Therefore, from the first equation:

$$\sigma_x = \frac{E}{1-\nu^2} \varepsilon_x = \frac{E}{1-\nu^2} \frac{z}{\rho} = -\frac{E}{1-\nu^2} z \frac{d^2 u}{dx^2} \tag{8.342}$$

where z is distance from neutral axis of bending of the cylinder wall.
The internal moment is

$$M_t = \int_{-\frac{h}{2}}^{\frac{h}{2}} \sigma_x z \, dz \quad \text{(internal moment per unit arc length)}$$

By substitution and integration:

$$M_t = \frac{Eh^3}{12(1-\nu^2)} \frac{d^2 u}{dx^2} = D \frac{d^2 u}{dx^2} \tag{8.343}$$

$$D = \frac{Eh^3}{12(1-\nu^2)} \tag{8.344}$$

Applying

$$\frac{dM}{dx} = V \quad \text{and} \quad \frac{dV}{dx} = q$$

gives

$$\frac{d^2M}{dx^2} = q \tag{8.345}$$

Turning now to the equation of bending in terms of cylinder dimension,

$$D\frac{d^4u}{dx^4} + \frac{hE}{a^2}u = 0 \tag{8.346}$$

$$D\frac{d^4u}{dx^4} + \frac{bhE}{h^2a^2}u = 0 \tag{8.347}$$

$$\frac{d^4u}{dx^4} + \frac{12(1-\nu^2)}{h^2a^2}u = 0 \tag{8.348}$$

$$\frac{d^4u}{dx^4} + 4\beta^4 u = 0 \tag{8.349}$$

where

$$\beta^4 = \frac{3(1-\nu^2)}{a^2h^2}$$

While Eq. (8.349) has the same mathematical form as Eq. (8.312), the mathematical expressions for β are different. All the previous examples of different loading apply directly to thin-walled cylinders, where loading is radial and symmetrical around the centroid axis of the cylinder as indicated in Figure 8.94. The circumferential loading (P) is analogous to a point load on a beam except in the pipe case P has units of force per circumferential length. The general solution is the same as in the previous case.

See Timoshenko and Woinowsky-Krieger [5] and Den Hartog [13] for a variety of other applications of this theory.

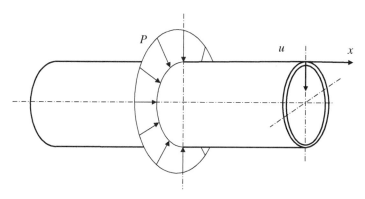

Figure 8.94 Thin cylinder with ring load (strap).

Reach of Bending Moment

The stiffness of a strip beam along the length of a thin-walled cylinder is typically much higher than for ordinary beams on an elastic foundation. As a result, wall deflections and bending stresses occur over a rather short distance away from the load. Both decay at a rate dictated by $e^{-\beta x}$. The value of this term becomes negligible when $\beta L = \frac{3\pi}{2}$. Assuming $\nu = 0.3$, the minimum length of the deflected surface is

$$L = \frac{3\pi}{2\beta}, \quad \text{where} \quad \beta^4 = \frac{3(1-\nu^2)}{a^2 h^2}$$

By substitution and assuming $\nu = 0.3$

$$L = \frac{3\pi}{2\beta} = 3.67a\sqrt{\frac{h}{a}} \tag{8.350}$$

which gives $L = 0.82a$ for $\frac{h}{a} = \frac{1}{20}$. If cylinder radius is $a = 8$ in., then the span of significant bending away from the load is 6.56 in. This observation is borne out by the following example.

Bending Stress in Wall of a Multi Banded Cylinder

Example Consider a situation where three collars are shrunk onto a cylinder as shown in Figure 8.95. Strain gauges on each collar indicate circumferential strains in each collar as

900μ (collar#1)

500μ (collar#2)

1000μ (collar#3)

Each collar has a rectangular cross section of ¼ in. thick by 1 in. wide. Dimensions of the cylinder are given in the figure. *Question:* What is the bending stress in the cylinder wall directly under collar #2?

Stress in each strap is relate to circumferential strain by

$$\sigma_\theta = \varepsilon_\theta E \tag{8.351}$$

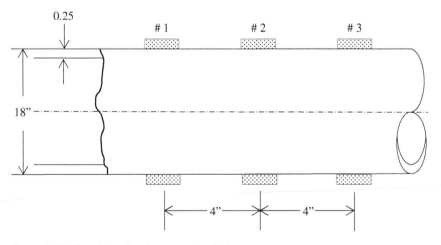

Figure 8.95 Local bending in strapped cylinder.

Contact pressure is related to stress by

$$\frac{r}{t}p = \varepsilon_\theta E \tag{8.352}$$

$$p = \frac{r}{t}\varepsilon_\theta E$$

Radial force per circumferential length is

$$P = pw = \frac{t}{r}\varepsilon_\theta E w \text{ lb/in. (} w \text{ is width of each band is 1 in.)}$$

where w is width of the strap.

$$P = \frac{0.25}{9} 30 \times 10^6 (1)\varepsilon_\theta = 0.833 \times 10^6 \varepsilon_\theta \text{ lb/in.}$$

Using this formula, the force per circumferential length is

Collar#1 $P_1 = 750$ lb/in.

Collar#2 $P_2 = 417$ lb/in.

Collar#3 $P_3 = 833$ lb/in.

The bending moment distribution caused by one collar is

$$M_t = -\frac{P}{4\beta}F_3 = -\frac{P}{4\beta}e^{-\beta x}(\cos\beta x - \sin\beta x) \text{ in.-lb/in.} \tag{8.353}$$

$$M_t = \frac{P}{4\beta}C_{\beta x} \text{ in.-lb/in.} \tag{8.354}$$

where

$$\beta^4 = \frac{3(1-\nu^2)}{r^2 t^2}$$

$$\beta^4 = \frac{3(1-0.25^2)}{9^2 0.25^2} = 0.5556$$

Therefore, $\beta = 0.8633$ 1/in. A plot of F_3 is shown in Figure 8.96. Note, $C_{\beta x} = -F_3$. The active distance of bending is

$$\frac{h}{a} = \frac{0.25}{9} = \frac{1}{36}$$

$$L = \frac{4.714}{1.3}a\sqrt{\frac{h}{a}} = 3.626a\sqrt{\frac{1}{36}} = 0.604a$$

$$L = 0.604(9) = 5.44 \text{ in.}$$

This number is consistent with Figure 8.96.

The total moment under strapped #2 is, by superposition

$$M_{total} = \frac{P_1}{4\beta}C_{\beta 4} + \frac{P_2}{4\beta}C_{\beta 0} + \frac{P_3}{4\beta}C_{\beta 4} \tag{8.355}$$

$$M_{total} = \frac{1}{4\beta}\left[(P_1 + P_3)C_{\beta 4} + P_2 C_{\beta 0}\right]$$

$$C_{\beta 0} = -1 \quad C_{\beta 4} = +0.0206 \quad 4\beta = 4(0.8569) = 3.43$$

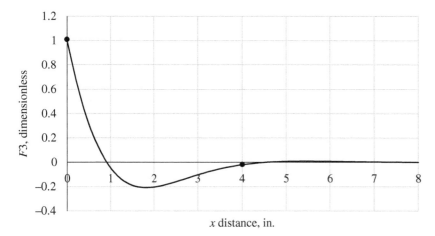

Figure 8.96 Distribution of bending from plane of application.

Substituting variables

$$M_{total} = \frac{1}{3.43}[(750 + 833)(0.0206) + 417(-1)] = -112 \text{ in.-lb per inch of circumference}$$

Notice the short reach of the bending moment (F_3 function).

$$\sigma_{bending} = \frac{M_x c}{I}$$

where

$c = 0.125$ in.

$I = \frac{0.25^3}{12} = 0.0013$ in.^{3} per inch circumference

Giving the total bending stress due to all three straps

$$\sigma_{bending} = \frac{112(0.125)}{0.0013} = 10\ 770 \text{ psi}$$

Criteria of Failure

Classic formulas are used to predict stress magnitudes associated with several different design configurations and loads. A design may be subjected to different combinations of these loads and stresses. In this case, local stress at a given point is the superposition of each stress type assuming stress levels are within the elastic limit. It is important to visualize each of these stress types and how to combine them. The combined stresses identify $\sigma_x \sigma_y \tau_{xy}$ in critical areas of a design. This information can then be used to determine principal stresses as previously discussed.

Combined Stresses

The stress condition at any point (biaxial or tri axial) in a design can be the combined effects of simple stress, direct pull, bending, torsion, pressure, etc.

Example   Consider the case where pipe is loaded by torsional shear, internal pressure plus bending. Using the following pipe data and loads:

- Pipe is 4-in. OD and 3.75 ID
- Internal pressure is 1000 psi
- Applied torque is 2000 ft lb
- Bending moment is 5000 ft lb
- Material yield is 60 000 psi

The challenge here is to determine the factor of safety against yielding. First, calculate the three stress components.

Moments of inertia are needed to determine torsion shear and bending moment stresses:

$$I = \frac{\pi}{4}\left(r_0^4 - r_i^4\right) = \frac{\pi}{4}\left(2^4 - 1.875^4\right) = 2.8591 \text{ in.}^4$$

$$J = 2I = 5.7183 \text{ in.}^4$$

$$A = \pi\left(r_0^2 - r_i^2\right) = \pi\left(2^2 - 1.875^2\right) = 1.5217 \text{ in.}^2$$

Internal Pressure

$$\frac{t}{r_0} = \frac{0.125}{2} = 0.0625 \text{ or } \frac{r_0}{t} = 16 \succ 10$$

The pipe can be considered a thin-walled cylinder:

$$r_{mean} = \frac{2 + 1.875}{2} = 1.94$$

$$\frac{r_{mean}}{0.125} = 15.5$$

$$\sigma_\theta = \frac{r_{mean}}{t}p = 15.5(1000) = 15\,500 \text{ psi}$$

Applied Torque

$$\tau = \frac{Tc}{J} = \frac{2000(12)2}{5.7183} = 8394 \text{ psi}$$

Bending Moment

$$\sigma_b = \frac{5000(12)2}{2.8591} = 41\,971 \text{ psi}$$

Assuming the pipe is support axially so that the internal pressure does not create axial stress, the state of stress is

$$\sigma_x = \sigma_b = 41\,971 \text{ psi}$$

$$\sigma_y = \sigma_\theta = 15\,500 \text{ psi}$$

$$\tau_{xy} = 8394 \text{ psi}$$

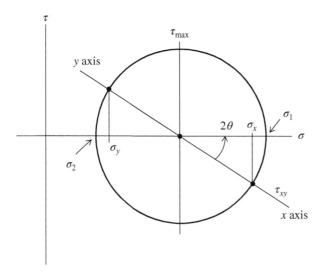

Figure 8.97 Mohr's circle for the example.

These stresses are shown in Mohr's circle (Figure 8.97), which also indicates principal stresses and maximum shear for the combined state of stress:

$$\sigma_{p1} = 44\,409 \text{ psi}$$

$$\sigma_{p2} = 13\,077 \text{ psi}$$

$$\tau_{max} = 15\,666 \text{ psi}$$

The orientation of the principal axis from the x axis is $\theta = 11.2°$ counterclockwise.

The combined effect of three different types of loading (internal pressure, bending, and torsion) produced a biaxial state of stress having principal stress as shown.

Failure of Ductile Materials

Many stress conditions, such as bending, truss analysis and simple tension or shear, allow the direct application of uniaxial stress data to determine load limits and determine factors of safety. However, many loading situations produce combinations of stresses which are superimposed to produce two or three-dimensional states of stress. Also, two and three-dimensional states of stress also may be produced simply due to complex geometry and these areas are often critical.

The *maximum normal stress criteria* states that material yielding occurs when the maximum principal stress equals the yield strength as determined from uniaxial test data. Experience shows that this theory is not accurate. An example is creatures that live in the deepest oceans. They experience extreme hydrostatic pressure but survive.

As per the *maximum shear stress criteria*, material yielding occurs when the maximum shear stress at a point is greater than the maximum shear yield of the material:

$$\tau_{max} \geq \tau_{yld} \tag{8.356}$$

Since the maximum shear stress at yield (per uniaxial test data) is ½ axial yield then

$$\tau_{yld} = 0.5\sigma_{yld} \tag{8.357}$$

and yielding occurs in a biaxial or tri-axial state of stress when

$$\tau_{max} \geq \tau_{yld} = 0.5\sigma_{yld} \tag{8.358}$$

This criterion is close to reality.

Von Mises' *criteria of energy distortion* more accurate and is the one most commonly used to determine material yielding. It states that material yielding occurs when the energy of distortion in a biaxial or triaxial stress state is equal to or greater than the energy of distortion of in a uniaxial test at yielding [23]. This criterion leads to the following set of equations commonly used in design.

$$(\sigma_1 - \sigma_2)^2 + (\sigma_2 - \sigma_3)^2 + (\sigma_3 - \sigma_1)^2 \geq 2\sigma_{yld}^2 \tag{8.359}$$

When the stress state is biaxial ($\sigma_3 = 0$), the criteria of failure can be expressed as

$$\sigma_1^2 - \sigma_1\sigma_2 + \sigma_2^2 \geq \sigma_{yld}^2 \tag{8.360}$$

Letting σ' represent von Mises stress

$$(\sigma') = \sigma_1^2 - \sigma_1\sigma_2 + \sigma_2^2 \tag{8.361}$$

Then yielding occurs when

$$\sigma' \geq \sigma_{yld} \tag{8.362}$$

This equation is often represented by ellipse on a stress diagram. The von Mises criteria is typically used to establish shear yield from normal yield strength determined from uniaxial test data. Consider a state of pure shear. The Mohr's circle has a radius of τ giving, $\sigma_1 = \tau$ and $\sigma_2 = -\tau$. By substitution into Eq. (8.361)

$$\sigma' = \sqrt{3}\tau \tag{8.363}$$

When yielding occurs

$$\sqrt{3}\tau_{yld} = \sigma_{yld} \tag{8.364}$$

Therefore,

$$\tau_{yld} = 0.577\sigma_{yld} \tag{8.365}$$

This formula is commonly used to establish shear yield strength. It is rarely a measured value. Appling the von Mises criteria to the previous example gives

$$\sigma' = \left[44.41^2 - 44.41(13.08 + 13.08^2)\right]^{\frac{1}{2}} = 39.53 \text{ ksi}$$

The factor of safety for this stress state is

$$FS = \frac{60}{39.53} = 1.518$$

Visualization of Stress at a Point

Example   A pipeline is subjected to the following loads (Figure 8.98):

- Internal fluid pressure of 150 000 psi
- Bending moment of 4000 ft lb
- Torque of 6000 ft lb

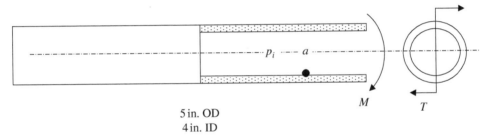

5 in. OD
4 in. ID

Figure 8.98 State of stress in pipe.

Draw the stress element for point "*a*" and show the stress components on the element caused by the applied loads:

$$I = \frac{\pi\left(r_o^4 - r_i^4\right)}{4} = \frac{\pi}{4}\left(2.5^4 - 2^4\right) = 18.11 \text{ in.}^4$$

$$J = 2I = 36.22 \text{ in.}^4$$

It is important to visualize stress components at any given point in an engineering structure. A stress situation is shown in the figure associated resulting from three types of loading: internal pressure, bending, and torque. The magnitude of stress from each load can be determined from classic formulas for each load:

$$\sigma_B = \frac{Mc}{I} = \frac{4000(12)2.5}{18.11} = 6626 \text{ psi} \quad \text{(bending)}$$

$$\sigma_\theta = \frac{R}{t}p = \frac{2.5}{0.5}1500 = 7500 \text{ psi} \quad \text{(pressure)}$$

$$\tau = \frac{Tc}{J} = \frac{6000(12)(2.5)}{36.22} = 4970 \text{ psi} \quad \text{(torque)}$$

The stress element at point "*a*" is shown in Figure 8.99.
Principal stresses and strains can be determined from this element.

Pressure Required to Yield a Cylindrical Vessel

Another application of von Mises criterion of failure is the determination of pressure limit on a given pressure vessels. Consider a cylinder vessel having an OD of 5.625 in. and a wall thickness of 0.04 in. (18-gauge sheet metal). The mean diameter is 5.587 in. The ends are capped with spherical shape caps having the same wall thickness as the cylinder. Assume a yield strength of $\sigma_{yld} = 60\,000$ psi and Poisson's ratio of $\nu = 0.29$.
Stress components in the body of the cylinder are

$$\sigma_z = \frac{r}{2t}p \quad \text{(axial stress)} \tag{8.366}$$

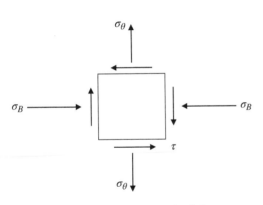

Figure 8.99 Stress element at point "*a*."

$$\sigma_\theta = \frac{r}{t}p \quad \text{(hoop stress)} \tag{8.367}$$

Substituting these expressions into Eq. (8.361) yields a von Mises stress of

$$\sigma' = 0.866\frac{r}{t}p \tag{8.368}$$

Equating to the yield strength of 60 000 psi gives a maximum allowable pressure of 9921.5 psi. The internal pressure also creates a third normal stress component, σ_r, at the inside surface, which is usually relative small by comparison.

The corresponding strains, ε_z, ε_θ, are determined Hooke's equations for biaxial stress.

Failure of Brittle Materials

Rock formations in the earth are subjected to complex biaxial or triaxial states of stress. Formation rocks are subjected to overburden as well as horizontal loads. When a well bore penetrates through rock formations, well bore pressure is applied to rock as well as overburden. Formation fluid pressures further complicate the analysis of rock stresses.

Commonly used criteria of failure for brittle materials are

- Maximum normal stress theory
- Coulomb–Mohr theory
- Modified Mohr theory

Each of these theories is illustrated in Figure 8.100.
Note

S_{ut} – ultimate tension strength as determined from uniaxial tests
S_{uc} – ultimate compressive strength as determined from uniaxial tests

The maximum normal stress criteria of failure theory is adequate for normal stress conditions in the first quadrant. In cases where there is a combination of tension and compression (second and

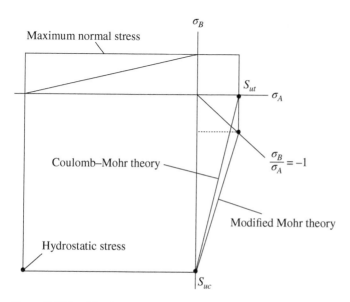

Figure 8.100 Criteria of failure of brittle materials.

fourth quadrant) the other two criteria are more appropriate. Consider the biaxial stress state for the fourth quadrant. The Coulomb–Mohr criteria is shown as a line between S_{ut}, ultimate strength in tension and S_{uc}, the ultimate strength in compression.

$$\tau = c + \sigma \tan \psi \tag{8.369}$$

where

τ – maximum shear stress at rupture
c – cohesive strength of rock
σ – compressive stress normal to the plane of rupture (compression is a + number)
ψ – angle of internal friction

If principal stresses are arranged so that $\sigma_1 \succ \sigma_2 \succ \sigma_3$, then the critical stress limits are σ_A and σ_B. These two applied stresses are related to the two strengths by

$$\frac{\sigma_A}{S_{ut}} - \frac{\sigma_B}{S_{uc}} = 1 \tag{8.370}$$

In this equation, S_{uc} is a positive number. For example, using numbers for sandstone:

$$S_{uc} = 20\,000 \text{ psi}, \quad S_{uc} = 500 \text{ psi}$$

$$\sigma_B = S_{uc}\left(\frac{\sigma_A}{S_{ut}} - 1\right) \tag{8.371}$$

This equation predicts failure in the fourth quadrant.

$$\sigma_B = 20\,000\left(\frac{\sigma_A}{500} - 1\right) \tag{8.372}$$

At $\sigma_A = 500$, $\sigma_B = 0$ and at $\sigma_B = -20\,000$, $\sigma_A = 0$. A biaxial state of stress (σ_1 and σ_3) located within this line does not produce failure; however, a biaxial state of stress outside this line does produce failure per this criterion.

Example Consider a design where Gray cast iron is used. Local stress in a critical area is defined by 40 ksi (compression) and 5 ksi (tension) (Figure 8.101). Determine the factor of safety against static failure using the modified Mohr criteria of failure:

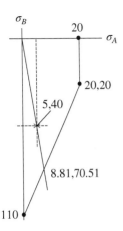

Figure 8.101 Factor of safety for gray cast iron application.

$$\sigma_{uc} = 110 \text{ ksi}$$

$$\sigma_{ut} = 20 \text{ ksi}$$

The factor of safety is the ratio of the line of intersection with the line of stress:

$$d = \sqrt{5^2 + 40^2} = 40.3$$

$$D = \sqrt{8.81^2 + 70.51^2} = 71.06$$

Factor of safety is

$$FS = \frac{71.06}{40.3} = 1.76$$

Experimental data show that actual points of failure fall outside of the Coulomb–Mohr criteria.

One point is the case, where $\frac{\sigma_B}{\sigma_A} = -1$. An example is the twisting of classroom chalk which causes failure on a 45° plane. Under this torsion loading, the principal normal stresses are equal but opposite and oriented on planes 45° from the axis of twist. The modified Mohr line is drawn from this point. The data of Coffin [25] and Grassi and Cornet [26] fall along this line. The modified Mohr criteria, then, appears to be more accurate than the Coulomb–Mohr criteria.

Mode of Failure in Third Quadrant

The mode of rock fracture due to compressive principal stresses (third quadrant) is different. The intrinsic line of failure is indicated in Figure 8.101. The Maximum Normal Stress Criteria of Failure does not apply in this case. Consider the case where all principal normal stresses are equal. Such a condition is indicated by the corner point identified by "hydrostatic stress." Experience shows that hydrostatic stress alone does not cause failure.

References

1 Ressler, S. (2011). Understanding the World's Greatest Structures, DVD Video, The Great Courses.
2 Mohr, O. (1882). Civil ingenieur, p. 113; also, see Mohr, O. (1906). Abhandlungen aus dem Gebiete der technischen Mechanik, Berlin, p. 219.
3 Perry, C.C. and Lissner, H.R. (1955). *The Strain Gage Primer*. McGraw – Hill Book Company, Inc.
4 Dareing, D.W. (2012). *Mechanics of Drillstrings and Marine Risers*. ASME Press.
5 Timoshenko, S. and Woinowsky-Krieger, S. (1959). *Theory of Plates and Shells*, 2e, 466. New York: McGraw-Hill.
6 Huang, T., Dareing, D.W., and Beran, W.T. (1980). Bending of tubular bundles attached to marine risers. *Trans. ASME J. Energy Resour. Technol.* 102: 24–29.
7 Dareing, D.W. and Ahlers, C.A. (1990). Tubular bending and pull out forces in high curvature well bores. *ASME J. Energy Resour. Technol.* 112: 84–89.
8 Rocheleau, D.N. and Dareing, D.W. (1992). Effect of drag forces on bit weight in high curvature well bores. *Trans. ASME J. Energy Resour. Technol.* 114 (3): 175–180.
9 Timoshenko, S.P. (1953). *History of Strength of Materials*. New York: McGraw-Hill; Also, see, Winkler, E. (1867). Die Lehre von der Elastizitat und Feitigheit, Prague, p 182.
10 (1989). *Manual of Steel Construction*, 9e. NY: American Institute of Steel Construction.
11 (1986). *Specifications for Aluminum Structures*. Washington, DC: Aluminum Association Inc.
12 American Institute of Timber Construction (1985). *Timber Construction Manual*. NY: Wiley.
13 Den Hartog, J.P. (1952). *Advanced Strength of Materials*. NY: McGraw – Hill.
14 Timoshenko, S. and Gere, J.M. (1961). *Theory of Elastic Stability*, 2e. New York: McGraw-Hill.
15 Dareing, D.W. (2019). *Oilwell Drilling Engineering*. ASME Press.
16 Lubinski, A. (1950). A study of the buckling of rotary drilling string. *API Drill. Prod. Prac.* 17: 178–214.
17 Huang, T. and Dareing, D.W. (1968). Buckling and lateral vibration of drill pipe. *Trans. ASME J. Eng. Ind.* 90, Series B (4): 613–619.
18 Huang, T. and Dareing, D.W. (1966). Predicting the stability of long vertical pipe transmitting torque in a viscous media. *Trans. ASME, J. Eng. Ind.* 88, Series B (2): 191–200.
19 Lame' and Clapeyron: "Memoire sur l'equilibre interieur des corps solides homogenes; Memoirs presents par dever sasvans" (1833) 4 (also see Timoshenko and Goodier, Theory of Elasticity)
20 Shigley, J.E. and Mitchell, L.D. (1983). *Mechanical Engineering Design*. McGraw-Hill Book Co.

21 Bresse, M. (1866). *Cours de mecanique applique*, Paris, Part I, p. 334 (also see Ref. 22).

22 Golovin, Trans. Inst. Tech. St. Petersburg, 1881 (also see Timoshenko and Goodier, Theory of Elasticity)

23 Seely, F.B. and Smith, J.O. (1956). *Advanced Mechanics of Materials*, 2e. New York: McGraw-Hill.

24 Riley, W., Sturges, L., and Morris, D. *Mechanics of Materials*, 6e. Wiley.

25 Coffin, L.F. (1950). The flow and fracture of brittle materials. *Trans. ASME J. Appl. Mech.* 17: 233–248.

26 Grassi, R.C. and Cornet, I. (1949). Fracture of grey cast iron tubes under biaxial stress. *Trans. ASME J. Appl. Mech.* 71: 178–182.

9

Modal Analysis of Mechanical Vibrations

Mechanical vibrations became an important design consideration during the early 1900s when machinery began to move at higher and higher speeds. Initially, mechanical vibrations were modeled as quasi-static events, i.e. structural responses were determined by treating dynamic forces as static even though they varied with time. The quasi-static approach is a practical and useful design tool provided the forcing frequency is much less than the fundamental frequency of the structure as discussed in Chapter 2. Fatigue analyses of many offshore structures are based on this assumption as ocean wave frequencies are much lower than natural frequencies of the structure. However, if the driving frequency is close to a natural frequency of a structure, the mass of the structure plays a significant role in dynamic response.

Vibrations, in general, are not good for machinery. They can produce noise and damage machine components through low-cycle or high-cycle fatigue. It is good practice to determine natural frequencies during the final phase of a design, and compare them with frequencies anticipated from engines, pumps, etc. Forcing frequencies are usually related to rotary speed. The final check on resonance comes from testing.

Many vibration problems can be solved by applying the fundamental results of single degree of freedom (SDOF) models. However, high-speed machinery may require consideration of higher modes of vibration. In these cases, modal analysis is most useful. This method develops solutions in terms of each vibration mode, thus making it easy to visualize the overall behavior in terms of the contribution of each mode.

Complex Variable Approach

The complex frequency response is a convenient way to solve differential equations of motion. It will be used throughout this chapter to explain vibration response of multiple degree of freedom systems. The complex variable approach will be first applied to the SDOF problem. In this case, the forcing function ($F_0 \cos \omega t$) is replaced with the imaginary function, $F_0 e^{i\omega t}$.

$$m \frac{d^2x}{dt^2} + c \frac{dx}{dt} + kx = F_0 e^{i\omega t} \tag{9.1}$$

The right side of the equation is shown below:

$$m \frac{d^2x}{dt^2} + c \frac{dx}{dt} + kx = F_0 (\cos \omega t + i \sin \omega t) \tag{9.2}$$

Engineering Practice with Oilfield and Drilling Applications, First Edition. Donald W. Dareing.

The complex force is introduced to expedite the solution.

Equation (9.2) is also written as

$$\ddot{x} + 2\zeta\omega_n\dot{x} + \omega_n^2 x = \omega_n^2 \frac{F_0}{k} e^{i\omega t} \tag{9.3}$$

where

$$\zeta = \frac{c}{c_{cr}}$$

$$c_{cr} = 2\sqrt{km}$$

The real part of the complex solution to Eq. (9.3) is the solution. The solution to Eq. (9.3) is assumed to be of the form

$$x(t) = \overline{X}e^{i\omega t} \tag{9.4}$$

where $x(t)$ is the complex response.

By substitution, Eq. (9.4) satisfies Eq. (9.1) provided

$$\overline{X} = \frac{\delta_{st}}{(1 - r^2) + i(2\zeta r)} \tag{9.5}$$

$$r = \frac{\omega}{\omega_n}$$

Therefore,

$$x(t) = \frac{\delta_{st}}{(1 - r^2) + i(2\zeta r)} e^{i\omega t} \tag{9.6}$$

which can also be written

$$x(t) = \delta_{st} \left[\frac{(1 - r^2) - i2\zeta r}{(1 - r^2)^2 + (2\zeta r)^2} \right] e^{i\omega t} \tag{9.7}$$

where $\delta_{st} = \frac{F_0}{k}$. The solution can also be written as

$$x(t) = \overline{X}e^{i\omega t} = (Xe^{-i\phi})e^{i\omega t} = Xe^{i(\omega t - \phi)} \tag{9.8}$$

which expands into

$$x(t) = X[\cos(\omega t - \phi) + i\sin(\omega t - \phi)] \tag{9.9}$$

The solution we seek is the real part of the complex displacement function:

$$x(t) = X\cos(\omega t - \phi) \tag{9.10}$$

The complex amplitude, $\overline{X}$, is conveniently represented in a Nyquist plot (Figure 9.1). The real and imaginary parts are identified in the complex plane.

It follows that

$$\text{Re}\left[\frac{\overline{X}}{\delta_{st}}\right] = \frac{(1 - r^2)}{(1 - r^2)^2 + (2\zeta r)^2} \tag{9.11}$$

$$\text{Im}\left[\frac{\overline{X}}{\delta_{st}}\right] = \frac{-2\zeta r}{(1 - r^2)^2 + (2\zeta r)^2} \tag{9.12}$$

Giving

$$X = \frac{\delta_{st}}{\left[(1-r^2)^2 + (2\zeta r)^2\right]^{\frac{1}{2}}} \qquad (9.13)$$

and

$$\tan \phi = \frac{2\zeta r}{1 - r^2} \qquad (9.14)$$

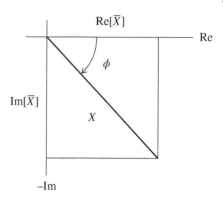

These expressions agree with earlier discussions of SDOF systems in Chapter 2.

Now, we have another way of expressing vibration response. Both pairs of equations give response amplitude, X, and phase angle ϕ.

Figure 9.1 Vibration response represented in complex plane.

Complex Transfer Function

The complex transfer function is defined as

$$\overline{H} = \frac{\overline{X}}{F_0} = \frac{1}{k} \frac{(1-r^2) - i(2\zeta r)}{(1-r^2)^2 + (2\zeta r)^2} \qquad (9.15)$$

which is the complex amplitude divided by the amplitude of the applied force.

This is a useful parameter because transfer functions can be determined experimentally. The calculus of the $\mathrm{Re}\left[\overline{H}\right]$ function, shows that its maximum occurs at

$$r = 1 - \zeta \qquad (9.16)$$

with a maximum value, of

$$\mathrm{Re}\left[\overline{H}\right]_{max} = \frac{1}{k} \frac{1}{4\zeta(1 - \zeta)} \qquad (9.17)$$

The minimum of $\mathrm{Re}\left[\overline{H}\right]$ occurs at

$$r = 1 + \zeta \qquad (9.18)$$

with a value, of

$$\mathrm{Re}\left[\overline{H}\right]_{min} = -\frac{1}{k} \frac{1}{4\zeta(1 + \zeta)} \qquad (9.19)$$

The frequency ratio difference between these two points is

$$\Delta r = 2\zeta \qquad (9.20)$$

giving a useful measure of the damping factor, ζ. Note that when $r = 0$, $\mathrm{Re}\left[\overline{H}\right] = \frac{1}{k}$, a useful measure of the stiffness parameter, k.

The maximum value of the $\mathrm{Im}\left[\overline{H}\right]$ function occurs at $r = 1$ and with a magnitude of minus $\frac{1}{2\zeta k}$.

Interpretation of Experimental Data

Consider the plot shown in Figure 9.2. Assume it is established experimentally. Steps for extracting damping factor, natural frequency, mass, and stiffness from this data are

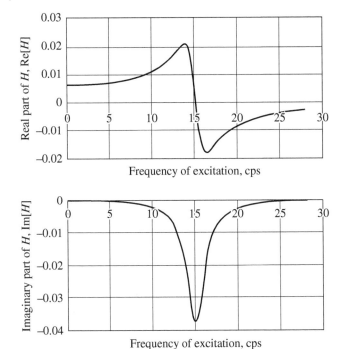

Figure 9.2 Experimentally determined complex transfer function.

Step 1 – $2\zeta = \Delta r$ (calculate ζ)

Step 2 – Determine $\dfrac{1}{2k\zeta}$ from imaginary part, (calculate k)

Step 3 – Determine natural frequency, $\omega_n = \sqrt{\dfrac{k}{m}}$, (calculate m)

Step 4 – Determine damping factor, $\zeta = \dfrac{c}{c_{cr}}$, (calculate c)

Natural Frequency

$$f_n = 15.2 \text{ cps}$$

as read directly from $Re[H]$.

Damping Factor

$$2\zeta = \Delta r$$

$$\Delta r = \frac{16.5 - 14}{15.5} = 0.1645$$

$$\zeta = \frac{0.1645}{2} = 0.0822$$

Spring Constant

$$\frac{1}{2\zeta k} = 0.038$$

$$k = \frac{1}{2(0.0822)(0.038)} = 160\,\text{lb/in.}$$

Mass

$$f_n = \frac{1}{2\pi}\sqrt{\frac{k}{m}}$$

$$[2\pi(15.2)]^2 = \frac{k}{m}$$

$$m = \frac{160}{9121} = 0.0175\,\text{lb-s}^2/\text{in.}$$

Thus,

$$W = mg = 0.0175(386) = 6.77\,\text{lb}$$

Damping Coefficient

The damping coefficient can now be determined:

$$c_{cr} = 2\sqrt{km}$$

$$c_{cr} = 2\sqrt{160(0.0175)} = 3.35\,\text{lb/ips}$$

Since

$$\zeta = \frac{c}{c_{cr}} = 0.0822$$

$$c = 0.0822(3.35) = 0.2751\,\text{lb/ips}$$

All basic parameters defining the SDOF system have now been quantified from the experimentally generated complex transfer function. The phase angle, ϕ, lag between the driving force, F, and displacement, $x(t)$, can be determined for any frequency ratio, r, by Eq. (9.12).

Two Degrees of Freedom

Systems with two degrees of freedom require two dependent variables to determine its vibration behavior. The response can be a free vibration or a forced vibration as in the case of SDOF systems. Two degrees of freedom systems have two distinct modes or shapes of vibration with each mode having its own natural frequency. The first step in analyzing vibrations is to determine these modal characteristics.

Natural Frequencies and Modes of Vibration

Consider the two-mass structure shown in Figure 9.3. Damping is not a consideration in determining mode shapes and natural frequencies.

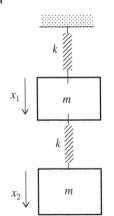

Figure 9.3 Schematic of a 2-DOF system.

For the sake of simplicity, both masses are assumed to be the same and both springs have the same elastic constant. The differential equations of motion for each mass are derived from separate freebody diagrams giving

$$m\ddot{x}_1 + 2kx_1 - kx_2 = 0 \tag{9.21}$$

$$m\ddot{x}_2 - kx_1 + kx_2 = 0 \tag{9.22}$$

These equations are put in matrix form for convenience:

$$\begin{bmatrix} m & 0 \\ 0 & m \end{bmatrix} \begin{Bmatrix} \ddot{x}_1 \\ \ddot{x}_2 \end{Bmatrix} + \begin{bmatrix} 2k & -k \\ -k & k \end{bmatrix} \begin{Bmatrix} x_1 \\ x_2 \end{Bmatrix} = \begin{Bmatrix} 0 \\ 0 \end{Bmatrix} \tag{9.23}$$

The solutions to these coupled differential equations are assumed to be of the form:

$$x_1 = X_1 \cos \omega t \tag{9.24}$$

$$x_2 = X_2 \cos \omega t \tag{9.25}$$

Upon substitution:

$$\begin{bmatrix} (2k - \omega^2 m) & -k \\ -k & (k - \omega^2 m) \end{bmatrix} \begin{Bmatrix} X_1 \\ X_2 \end{Bmatrix} = \begin{Bmatrix} 0 \\ 0 \end{Bmatrix} \tag{9.26}$$

Defining, $\xi = \frac{\omega^2 m}{k}$, then

$$\begin{bmatrix} (2 - \xi) & -1 \\ -1 & (1 - \xi) \end{bmatrix} \begin{Bmatrix} X_1 \\ X_2 \end{Bmatrix} = \begin{Bmatrix} 0 \\ 0 \end{Bmatrix} \tag{9.27}$$

This is called the *amplitude* equation. For a nontrivial solution,

$$\begin{vmatrix} (2 - \xi) & -1 \\ -1 & (1 - \xi) \end{vmatrix} = 0 \tag{9.28}$$

which expands into the *characteristic* equation:

$$\xi^2 - 3\xi - 1 = 0 \tag{9.29}$$

This equation has two roots, called eigenvalues: $\xi_1 = 0.382$ and $\xi_2 = 2.618$ giving

$$\omega_1 = 0.618 \sqrt{\frac{k}{m}} \quad \text{first mode} \tag{9.30}$$

$$\omega_2 = 1.618 \sqrt{\frac{k}{m}} \quad \text{second mode} \tag{9.31}$$

By substituting the eigenvalues into the amplitude equation:

$$\left(\frac{X_2}{X_1} \right)_1 = 1.618 \quad \text{(first mode shape)} \tag{9.32}$$

$$\left(\frac{X_2}{X_1} \right)_2 = -0.618 \quad \text{(second mode shape)} \tag{9.33}$$

Both mode shapes are displayed in Figure 9.4.

These results show the 2-DOF system has two natural modes of vibration. Each mode can respond separately or simultaneously. If the two masses are given initial displacements defining the first mode (or second mode), only that mode vibrates. However, in general both modes can participate in a vibration response simultaneously.

If the lower mass is also attached to ground through a spring of constant k, then both mode shapes and corresponding natural frequencies change to $\omega_1 = \sqrt{\frac{k}{m}}$ and $\omega_2 = 1.732\sqrt{\frac{k}{m}}$.

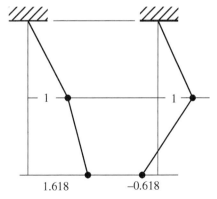

1.618   –0.618

Figure 9.4   Modes shapes and eigenvalues.

SDOF Converted to 2-DOF

Consider the system containing a rigid bar with mass distribution (μ), a discrete mass (m) and two springs of stiffness k_1 and k_2. As shown, the system is SDOF. However, if the pin connection is removed, the system becomes 2-DOF. Equations of motion are developed for both cases for analytical comparison. The mass of the bar is considered in both cases (Figure 9.5).

Single Degree of Freedom
Only one equation is required.

$$\sum M_0 = I_0 \alpha$$

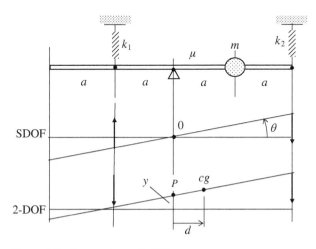

Figure 9.5   Conversion to 2-DOF system.

By expansion

$$a^2\theta k_1 - (2a)^2\theta k_2 = \left(I_b + a^2m\right)\ddot{\theta}$$

where the moment of inertia of the bar is with respect to point 0 is.

$$I_b = 2\int_0^{2a} x^2\mu\,dx = \frac{2}{3}\mu x^3\Big|_0^{2a} = \frac{16}{3}\mu a^3$$

and μ is mass per unit length of the bar.

The equation of motion for this case is

$$\ddot{\theta} + \frac{4a^2k_2 - a^2k_1}{I_b + a^2m}\theta = 0 \tag{9.34}$$

Natural circular frequency is

$$\omega_n^2 = \frac{4a^2k_2 - a^2k_1}{I_b + a^2m} \tag{9.35}$$

$$\omega_n^2 = \frac{4k_2 - k_1}{\frac{4}{3}M + m}$$

Two Degrees of Freedom

When the pin support is removed, the system becomes a 2-DOF problem and two equations of motion are required, a moment and a force equation. As explained in Chapter 7, there are two ways to set up the moment equation. One approach is

$$\sum M_P = I_{cg}\ddot{\theta} + d(m + \mu 4a)a_{cg} \tag{9.36}$$

The second

$$\sum M_P = I_P\ddot{\theta} + d(m + \mu 4a)a_P \tag{9.37}$$

The second moment equation is somewhat simpler when noting

$$am = d(4\mu a + m)$$
$$am = d(M + m) \ (M \text{ is mass of bar})$$

and

$$I_P = I_0 + a^2m$$

In this case, the moment equation becomes

$$\sum M_P = I_P\ddot{\theta} + am\ddot{y}$$

The moment equation expands into

$$ak_1(y - a\theta) - k_2(y + 2a\theta)2a = I_P\ddot{\theta} + am\ddot{y} \tag{9.38}$$

The force equation is

$$\sum F_y = (m + M)a_{cg} \tag{9.39}$$

$$-k_1(y - a\theta) - k_2(y + 2a\theta) = (m + M)(\ddot{y} + d\ddot{\theta}) \tag{9.40}$$

Two dependent variables, y and θ, are required. Rearranging terms the two equations of motion are

$$I_P\ddot\theta + am\ddot y + (k_1 + 4k_2)a^2\theta - (k_1 - 2k_2)ay = 0 \text{ (moment)}$$

and

$$am\ddot\theta + (m + M)\ddot y - (k_1 - 2k_2)a\theta + (k_1 + k_2)y = 0 \text{ (force)}$$

In this case the solutions are assumed to be in the form of

$$y(t) = Y\cos\omega t$$
$$\theta(t) = \Theta\cos\omega t$$

The procedure for finding natural modes and mode shapes is the same as given above.

Other 2-DOF Systems

There are many configurations of 2-DOF systems. The designer must envision an appropriate model for each situation. Typically, two dependent variables are required to describe the modes and their corresponding natural frequencies. Figure 9.6 illustrates several 2-DOF configurations.

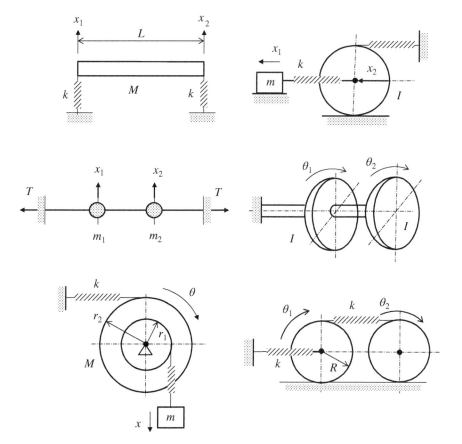

Figure 9.6 Examples of 2-DOF systems.

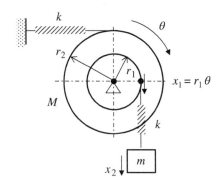

Figure 9.7 Combination of rotation and linear motion.

After setting up an appropriate math model, the equations of motion are developed based on the kinetics of rigid and discrete bodies as discussed earlier. As an example, consider the 2-DOF system in Figure 9.7.

Equations of motion are

$$\sum M = I_0 \ddot{\theta}$$

$$I_0 \ddot{\theta} = r_1(x_2 - r_1\theta)k - r_2(r_2\theta)k \quad \text{(disc)} \quad (9.41)$$

$$\sum F = m\ddot{x}_2$$

$$m\ddot{x}_2 = -(x_2 - r_1\theta)k \quad \text{(discrete mass)}$$

$$(9.42)$$

Since $\theta = \frac{x_1}{r_1}$, $I_0 = k_g^2 M$, k_g is radius of gyration, the two equations of motion become

$$M\left(\frac{k_g}{r_1}\right)^2 \ddot{x}_1 + \left[1 + \left(\frac{r_2}{r_1}\right)^2\right]kx_1 - kx_2 = 0 \tag{9.43}$$

and

$$m\ddot{x}_2 + kx_2 - kx_1 = 0 \tag{9.44}$$

Determination of the modal matrix and characteristic equation continues as before, starting with

$$x_1 = X_1 \cos \omega t$$
$$x_2 = X_2 \cos \omega t$$

The substitution of these two equations leads to amplitude equations and eigenvalues as before.

Undamped Forced Vibrations (2 DOF)

Consider a scenario (Figure 9.8) with a periodic force, $F(t)$, applied to the top mass. Assume the 2-DOF system is undamped.

The differential equations of motion are

$$m\ddot{x}_1 + 2kx_1 - kx_2 = F_0 \cos \omega t \tag{9.45}$$

and

$$m\ddot{x}_2 - kx_1 + kx_2 = 0 \tag{9.46}$$

For a solution to both equations, assume

$$x_1(t) = X_1 \cos \omega t \tag{9.47}$$
$$x_2(t) = X_2 \cos \omega t \tag{9.48}$$

These equations indicate both masses vibrate with the frequency of excitation (ω), but with different amplitudes. This frequency is not necessarily one of the natural frequencies (ω_1 or ω_2).

By substitution

$$\begin{bmatrix} (2 - \xi) & -1 \\ -1 & (1 - \xi) \end{bmatrix} \begin{Bmatrix} X_1 \\ X_2 \end{Bmatrix} = \begin{Bmatrix} \delta_{st} \\ 0 \end{Bmatrix} \tag{9.49}$$

where

$$\delta_{st} = \frac{F_0}{k} \quad \text{(static displacement)}$$

$$\xi = \frac{\omega^2 m}{k} \quad (\omega \text{ is driving frequency in this case})$$

By Cramer's rule the amplitudes of the responses are

$$X_1 = \frac{\begin{vmatrix} \delta_{st} & -1 \\ 0 & (1-\xi) \end{vmatrix}}{\begin{vmatrix} (2-\xi) & -1 \\ -1 & (1-\xi) \end{vmatrix}} \quad (9.50)$$

$$X_2 = \frac{\begin{vmatrix} (2-\xi) & \delta_{st} \\ -1 & 0 \end{vmatrix}}{\begin{vmatrix} (2-\xi) & -1 \\ -1 & (1-\xi) \end{vmatrix}} \quad (9.51)$$

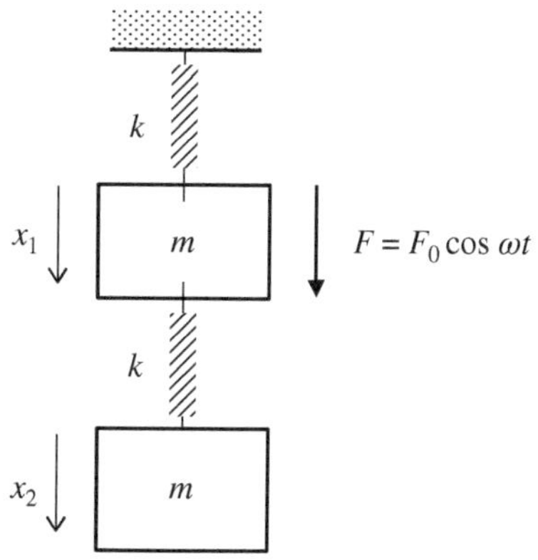

Figure 9.8 Forced vibration for 2-DOF system.

The solutions to Eqs. (9.45) and (9.46) are thus defined by substituting X_1 and X_2 into Eqs. (9.47) and (9.48). The amplitudes of each mass vary with the frequency of the force of excitation. The vibration amplitude of each mass vs. driving frequency is shown in Figures 9.9 and 9.10. Resonance occurs when the denominator is zero.

The denominator of both X_1 and X_2 expressions is zero when $\xi = \xi_1$ or $\xi = \xi_2$ per the characteristic equation given earlier for this 2 DOF system. Large vibration amplitudes are anticipated when

$$\frac{\omega^2 m}{k} = 0.382$$

and

$$\frac{\omega^2 m}{k} = 2.618$$

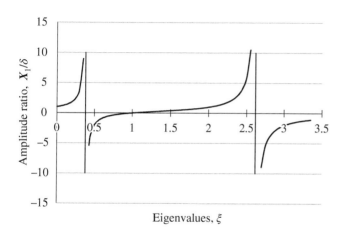

Figure 9.9 Frequency response of mass 1.

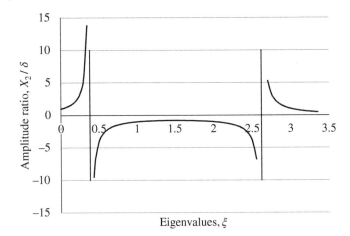

Figure 9.10 Frequency response of mass 2.

or

$$\omega = 0.618\sqrt{\frac{k}{m}} = \omega_1 \quad \text{(frequency of first mode)} \tag{9.52}$$

$$\omega = 1.618\sqrt{\frac{k}{m}} = \omega_2 \quad \text{(frequency of second mode)} \tag{9.53}$$

Either mode of vibration, defined earlier, can be excited provided the exciting frequency is tuned to one of the natural frequencies.

Undamped Dynamic Vibration Absorber

The discussion of dynamic absorbers is an extension of the previous model with slight modifications to the parameters [1]. The vibration model is shown in Figure 9.11.

The model depicts a base SDOF system (M and K), which is being excited by $F(t)$; its natural frequency is

$$\Omega_n = \sqrt{\frac{K}{M}} \tag{9.54}$$

From our previous discussion of forced vibrations of SDOF systems, resonance is expected when $\omega = \Omega_n$. The attachment of (m and k), which is the dynamic absorber, brings about a very interesting result.

If a machine or a component is operated at a constant speed and resonant vibrations exists, response vibrations can be suppressed or eliminated by use of a dynamic absorber. The added spring, k and mass, m, represent the dynamic absorber.

The dynamic absorber has a natural frequency of

$$\omega_n = \sqrt{\frac{k}{m}} \tag{9.55}$$

These are not modal frequencies but are frequencies of the two-separate spring-mass subparts.

The analysis below shows how the absorber eliminates the resonant vibration of mass, M, and absorbs the applied dynamic force. Note that the absorber is an attachment to mass, M.

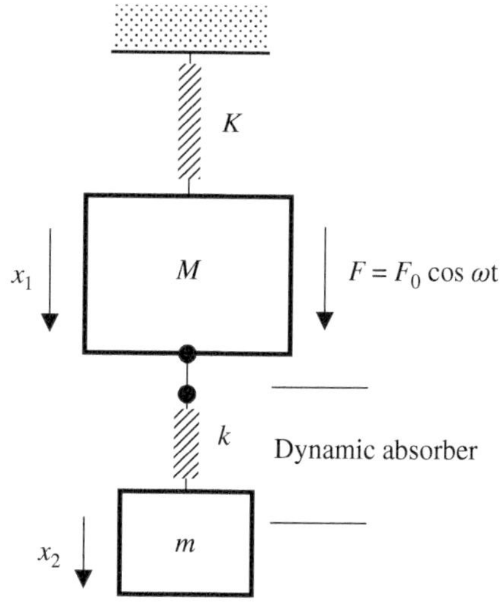

Base and Absorber Pinned Together

When the absorber is pinned to the base mass, the differential equation of motion for the base mass, M, is

$$M\ddot{x}_1 + (K + k)x_1 + kx_2 = F_0 \cos \omega t \qquad (9.56)$$

The differential equation of motion for the attached absorber is

$$m\ddot{x}_2 + k(x_2 - x_1) = 0 \qquad (9.57)$$

Vibration response of the two-mass system is assumed to be of the form:

$$x_1 = X_1 \cos \omega t \qquad (9.58)$$

$$x_2 = X_2 \cos \omega t \qquad (9.59)$$

By substitution and collecting terms

$$X_1 \left(1 + \frac{k}{K} - \frac{\omega^2}{\Omega_n^2}\right) - X_2 \frac{k}{K} = \delta_{st} \qquad (9.60)$$

$$X_1 = X_2 \left(1 - \frac{\omega^2}{\omega_n^2}\right) \qquad (9.61)$$

where

$$\delta_{st} = \frac{F_0}{K}$$

Simultaneous solution to these two equations is

$$\frac{X_1}{\delta_{st}} = \frac{1 - \frac{\omega^2}{\omega_a^2}}{\left(1 - \frac{\omega^2}{\omega_a^2}\right)\left(1 + \frac{k}{K} - \frac{\omega^2}{\Omega_n^2}\right) - \frac{k}{K}} \qquad (9.62)$$

$$\frac{X_2}{\delta_{st}} = \frac{1}{\left(1 - \frac{\omega^2}{\omega_a^2}\right)\left(1 + \frac{k}{K} - \frac{\omega^2}{\Omega_n^2}\right) - \frac{k}{K}} \qquad (9.63)$$

Figure 9.11 Math model of dynamic absorber.

From these equations, mass M has zero response, $\dfrac{X_1}{\delta_{st}} = 0$, when $\omega_a = \omega$ but mass, m, responds with amplitude

$$X_2 = -\frac{K}{k}\delta_{st} = -\frac{F_0}{k} \qquad (9.64)$$

The main mass, M, has zero response, $x_1(t) = 0$. The force in the damper spring varies per, $-F_0 \cos \omega t$, which is equal and opposite to the external force, F_0. This response is true for any value of $\dfrac{\omega}{\Omega_n}$. Since we are interested in suppressing resonance of the main mass, M, consider the case of $\omega_a = \Omega_n$ for which

$$\frac{k}{m} = \frac{K}{M} \quad \text{and} \quad \frac{k}{K} = \frac{m}{M} \qquad (9.65)$$

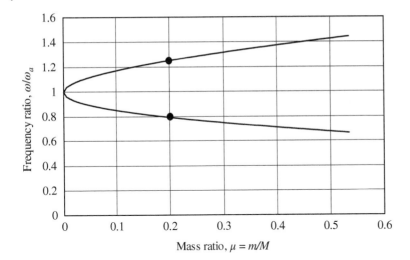

Figure 9.12 Separation of the natural frequencies.

Letting

$$\mu = \frac{k}{K}$$

Then

$$\frac{X_1}{\delta_{st}} = \frac{1 - \left(\frac{\omega}{\omega_a}\right)^2}{\left(1 - \left(\frac{\omega}{\omega_a}\right)^2\right)\left(1 + \mu - \left(\frac{\omega}{\Omega_n}\right)^2\right) - \mu} \tag{9.66}$$

$$\frac{X_2}{\delta_{st}} = \frac{1}{\left(1 - \left(\frac{\omega}{\omega_a}\right)^2\right)\left(1 + \mu - \left(\frac{\omega}{\Omega_n}\right)^2\right) - \mu} \tag{9.67}$$

To summarize, if the main mass is vibrating at resonance, $\Omega_n = \omega$. The absorber is chosen such that $\omega_a = \omega = \Omega_n$ (resonant condition). Question: what should be the size of k and m of the absorber? As a guide, consider the frequency spacing between the two natural frequencies of the *now* 2-DOF system. The denominator in Eqs. (9.66) and (9.67) will help with the answer.

The two resonant frequencies surrounding the null response are found by setting the denominator of Eq. (9.67) equal to zero.

$$\left(1 - \left(\frac{\omega}{\omega_a}\right)^2\right)\left(1 + \mu - \left(\frac{\omega}{\omega_a}\right)^2\right) - \mu = 0 \tag{9.68}$$

Expanding and collecting terms,

$$\left(\frac{\omega}{\omega_a}\right)^4 - \left(\frac{\omega}{\omega_a}\right)^2 (2 + \mu) + 1 = 0 \tag{9.69}$$

The two roots to this equation are

$$\left(\frac{\omega}{\omega_a}\right)^2 = \left(1 + \frac{\mu}{2}\right) \pm \sqrt{\mu + \frac{\mu^2}{2}} \tag{9.70}$$

The plot of this equation is shown in Figure 9.12. If, for example the ratio $\mu = \frac{k}{K} = \frac{m}{M}$ is taken as 0.2, then the new 2 DOF will resonate at $\frac{\omega}{\omega_a} = 0.8$ and $\frac{\omega}{\omega_a} = 1.25$. These two frequency ratios are

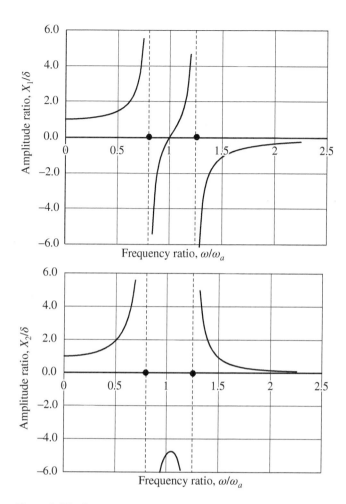

Figure 9.13 Frequency responses with dynamic absorber.

also shown in Figure 9.13 by the dashed lines. Keep in mind that dynamic absorbers are designed for one operating frequency, $\dfrac{\omega}{\omega_a} = 1.0$.

These diagrams show that the separation of resonance points depends on the mass or stiffness ratio.

Example Steps for determining the magnitude of k and m are illustrated an example with the parameter as follows:

$F_0 = 50\,\text{N}$

$M = 2\,\text{kg}$

$\omega = 100\,\text{rad/s}$

$X_2 \leq 1\,\text{cm}$ (requirement)

The problem is to determine k and m of the absorber.

Step 1 $\quad X_2 = \dfrac{F_0}{k}$

$\quad\quad k = \dfrac{50\text{ N}}{1\text{ cm}}\left|\dfrac{100\text{ cm}}{1\text{ m}}\right| = 5000\text{ N/m}$

Step 2 $\quad \omega_a = \omega$

$\quad\quad \dfrac{k}{m} = \omega^2$

$\quad\quad m = \dfrac{k}{\omega^2} = \dfrac{5000}{10\,000} = 0.5\text{ kg}$

Since $\mu = \dfrac{k}{m} = \dfrac{50}{200} = 0.25$. According to Eq. (9.70), resonant frequencies are separated by $\dfrac{\omega}{\omega_a} = 1.29$ and $\dfrac{\omega}{\omega_a} = 0.77$.

Dynamic absorber theory was explained in terms of a linear spring-mass model. This model, however, applies to many practical geometric configurations. The torsion model is like the discrete mass linear model discussed above so the results in the above apply directly when linear parameters are converted to angular parameters. The conversion to other configurations is not as obvious. The goal is to put the model of any SDOF system into the form of a simple spring-mass system.

Example Consider the system shown in Figure 9.14. The primary system is the disc, while the linear spring-mass is the absorber.

$$I_P\ddot\theta = -(2r\theta k_1)2r + 2rF_0\cos\omega t \tag{9.71}$$

$$\frac{3}{2}r^2 m_1\left(\frac{\ddot x_1}{r}\right) + 2r\left(\frac{x_1}{r}\right)k_1 2r = 2rF_0\cos\omega t$$

$$\frac{3}{4}m_1\ddot x_1 + 2k_1 x_1 = F_0\cos\omega t$$

Its linear equivalence is $M = \dfrac{3}{4}m_1$ and $K = 2k_1$.

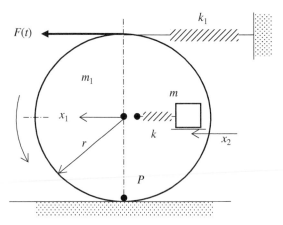

Figure 9.14 Dynamic absorber applied to a rolling disc.

$F(t)$

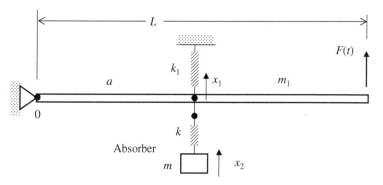

Figure 9.15 Damper applied to pivoted rod.

Example Consider the system shown in Figure 9.15. The equation of motion for the primary system (the bar) is

$$I_0\ddot{\theta} = -ak_1(a\theta) + 2aF_0 \cos \omega t \tag{9.72}$$

$$\left(\frac{L^2 m_1}{3}\right)\frac{\ddot{x}_1}{a} + ak_1 x_1 = 2aF_0 \cos \omega t$$

which translates into

$$\frac{2}{3}m_1\ddot{x}_1 + \frac{1}{2}k_1 x_1 = F_0 \cos \omega t \tag{9.73}$$

Therefore, the linear equivalences are $M = \dfrac{2m_1}{3}$ and $K = \dfrac{k_1}{2}$.

The equations of motion for the vibration absorber system are

$$M\ddot{x}_2 + (K + k)x_1 - kx_2 = 2F(t) \tag{9.74}$$

and

$$m\ddot{x}_2 + k(x_2 - x_1) = 0 \tag{9.75}$$

The math model of Figure 9.11 and results as explained earlier apply directly to this case.

Multi-DOF Systems – Eigenvalues and Mode Shapes

In general, free vibration equations of motion for a multi-DOF system in are expressed in matrix form by

$$[m]\{\ddot{x}\} + [k]\{x\} = \{0\} \tag{9.76}$$

Assuming the solution has the form of

$$x(t) = X \cos \omega t$$

gives

$$[[k] - \omega^2[m]]\{X\} = \{0\} \quad \text{(amplitude equation)} \tag{9.77}$$

For a nontrivial solution,

$$\left| [k] - \omega^2 [m] \right| = 0 \quad \text{(characteristic equation)} \tag{9.78}$$

The characteristic equation gives natural frequencies of each mode in the system. Mode shapes are determined from the amplitude equation.

The result from each equation depends on the stiffness $[k]$ matric and the mass matric $[m]$. Each of these matrices is a by-product of the equations of motion for each mass as illustrated with the 2 DOF systems. However, both matrices can be determined directly without developing the differential equations of motion as explained below.

Flexibility Matrix – Stiffness Matrix

Displacements in engineering structures can be determined through classic bean analysis or using computer software. These tools provide a convenient way to build a flexibility matrix. Consider a cantilever beam supporting two concentrated forces. Deflections, x_1 and x_2 can be determined by use of a flexibility matrix.

$$\left\{ \begin{matrix} x_1 \\ x_2 \end{matrix} \right\} = \begin{bmatrix} \alpha_{11} & \alpha_{12} \\ \alpha_{21} & \alpha_{22} \end{bmatrix} \left\{ \begin{matrix} F_1 \\ F_2 \end{matrix} \right\} \tag{9.79}$$

By inversion, the stiffness matrix is

$$\left\{ \begin{matrix} F_1 \\ F_2 \end{matrix} \right\} = \begin{bmatrix} k_{11} & k_{12} \\ k_{21} & k_{22} \end{bmatrix} \left\{ \begin{matrix} x_1 \\ x_2 \end{matrix} \right\}$$

where α_{ij} is the displacement at location "i" due to a unit force at location "j." As per Maxwell's law of reciprocity [2], $\alpha_{ij} = \alpha_{ji}$. The flexibility elements can easily be determined from beam deflection theory.

The relationship between the flexibility matrix and the stiffness matrix is determined as follows:

$$\{F\} = [k]\{x\} \tag{9.80}$$

$$\{x\} = [\alpha]\{F\} \tag{9.81}$$

Substituting for $\{F\}$ gives

$$\{x\} = [\alpha][k]\{x\} \tag{9.82}$$

Therefore,

$$[\alpha][k] = [I] \tag{9.83}$$

or

$$[k] = [\alpha]^{-1} \tag{9.84}$$

showing the stiffness matrix is the inverse of the flexibility matrix. In consideration of the top drawing and Eq. (9.81), if $F_2 = F_3 = 0$, then

$$\left\{ \begin{matrix} x_1 \\ x_2 \\ x_3 \end{matrix} \right\} = \left\{ \begin{matrix} \alpha_{11} \\ \alpha_{21} \\ \alpha_{31} \end{matrix} \right\} F_1$$

Figure 9.16 Flexibility (α_{ij}) and stiffness (k_{ij}) elements.

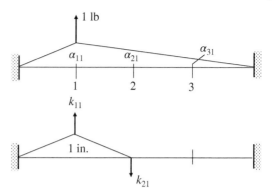

Letting $F_1 = 1$ (Figure 9.16)

$$x_1 = \alpha_{11} \quad x_2 = \alpha_{21} \quad x_3 = \alpha_{31}$$

According to Eq. (9.80), elements in the stiffness matrix are developed as follows:

$$\left\{ \begin{array}{c} F_1 \\ F_2 \\ F_3 \end{array} \right\} = \left[\begin{array}{ccc} k_{11} & k_{12} & k_{13} \\ k_{21} & k_{22} & k_{23} \\ k_{31} & k_{32} & k_{33} \end{array} \right] \left\{ \begin{array}{c} x_1 \\ x_2 \\ x_3 \end{array} \right\} \tag{9.85}$$

If locations x_2 and x_3 are fixed then Eq. (9.85) reduces to

$$\left\{ \begin{array}{c} F_1 \\ F_2 \\ F_3 \end{array} \right\} = \left\{ \begin{array}{c} k_{11} \\ k_{21} \\ k_{31} \end{array} \right\} x_1$$

Furthermore if x_1 is given a unit displacement the applied force at location 1 is

$$F_1 = k_{11}$$

and

$$F_2 = k_{21} \quad \text{(reaction force)}$$

$$F_3 = k_{31} = 0 \quad \text{(reaction force)}$$

Differential equations of motion may be developed using the flexibility matrix directly:

$$\{x\} = -[\alpha][m]\{x\} \tag{9.86}$$

$$[k]\{x\} = -[k][\alpha][m]\{\ddot{x}\}$$

Since $[k][\alpha] = [I]$

$$[m]\{\ddot{x}\} + [k][x] = \{0\} \tag{9.87}$$

Example   Consider the flexibility and stiffness matrix for the system shown in Figure 9.17. Following the above procedure

$$[\alpha] = \frac{1}{k} \left[\begin{array}{cc} 5 & -1 \\ -1 & 1 \end{array} \right] \tag{9.88}$$

$$[k] = \frac{k}{4} \left[\begin{array}{cc} 1 & 1 \\ 1 & 5 \end{array} \right] \tag{9.89}$$

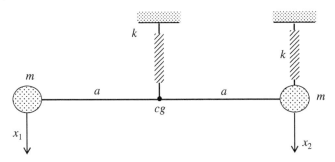

Figure 9.17 Two degree of freedom system.

The differential equations of motion are given in matrix form as

$$m\begin{bmatrix} 1 & 0 \\ 0 & 1 \end{bmatrix}\begin{Bmatrix} \ddot{x}_1 \\ \ddot{x}_2 \end{Bmatrix} + \frac{k}{4}\begin{bmatrix} 1 & 1 \\ 1 & 5 \end{bmatrix}\begin{Bmatrix} x_1 \\ x_2 \end{Bmatrix} = \begin{Bmatrix} 0 \\ 0 \end{Bmatrix} \tag{9.90}$$

yielding

$$m\ddot{x}_1 + \frac{k}{4}(x_1 + x_2) = 0$$

and

$$m\ddot{x}_2 + \frac{k}{4}(x_1 + 5x_2) = 0$$

For comparison, we now develop the differential equations of motion from rigid body analysis from Chapter 7 (see Figure 9.18).

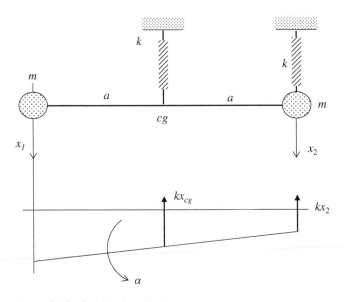

Figure 9.18 Rigid body analysis.

From force equation

$$\sum \bar{F} = m\bar{a}_G$$

$$-\frac{1}{2}(x_1 + x_2)k - kx_2 = 2m\frac{1}{2}(\ddot{x}_1 + \ddot{x}_2)$$

$$m(\ddot{x}_1 + \ddot{x}_2) + \frac{1}{2}kx_1 + \frac{3}{2}kx_2 = 0 \tag{9.91}$$

From moment equation

$$\sum M_G = I_G\alpha \quad (G \text{ is mass center})$$

$$+ akx_2 = 2a^2 m\left(\frac{\ddot{x}_1 - \ddot{x}_2}{2a}\right)$$

$$m(\ddot{x}_1 - \ddot{x}_2) = -kx_2 = 0 \tag{9.92}$$

Notice that equations 9.91 and 9.92 look completely different from the ones developed by the direct method. However, if the two are added and then subtracted, we get the same two equations.

Direct Determination of the Stiffness Matrix

In general, the stiffness matrix is defined in general by

$$[k] = \begin{bmatrix} k_{11} & k_{12} & k_{13} & \text{etc} \\ k_{21} & k_{22} & & \\ k_{31} & k_{32} & & \\ & & k_{ij} & \text{etc} \end{bmatrix} \tag{9.93}$$

where k_{ij} is the force at point "i" as a result of a *unit* displacement at point "j." Similar, m_{ij} is the force at point "i" due to a force required to produce a unit acceleration at point "j."

Example A tight cable with three masses equally spaced by "a" distance as shown in Figure 9.19. The cable is tensioned by force T.
From the diagram

$$k_{11} = 2\frac{T}{a}, \quad k_{21} = -\frac{T}{a}, \quad k_{31} = 0$$

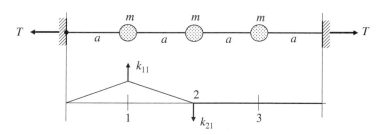

Figure 9.19 Stiffness matrix of 3-DOF system.

Accordingly, the stiffness matrix for the system is

$$[k] = \frac{T}{a} \begin{bmatrix} 2 & -1 & 0 \\ -1 & 2 & -1 \\ 0 & -1 & 2 \end{bmatrix} \tag{9.94}$$

Direct Determination of the Mass Matrix

The mass matrix can also be established in a similar manner. Consider the matrix equation:

$$\begin{Bmatrix} F_1 \\ F_2 \\ F_3 \end{Bmatrix} = \begin{bmatrix} m_{11} & m_{12} & m_{13} \\ m_{21} & m_{22} & m_{23} \\ m_{31} & m_{32} & m_{33} \end{bmatrix} \begin{Bmatrix} \ddot{x}_1 \\ \ddot{x}_2 \\ \ddot{x}_3 \end{Bmatrix}$$

If masses 2 and 3 are fixed and mass 1 given a unit acceleration

$$\begin{Bmatrix} f_1 \\ 0 \\ 0 \end{Bmatrix} = \begin{bmatrix} m_{11} & m_{12} & m_{13} \\ m_{21} & m_{22} & m_{23} \\ m_{31} & m_{32} & m_{33} \end{bmatrix} \begin{Bmatrix} 1 \\ 0 \\ 0 \end{Bmatrix}$$

$$f_1 = m_{11}(1) \quad (f_2 = 0, f_3 = 0)$$

but $f_1 = m(1)$, therefore $m(1) = m_{11}(1)$ and $m_{11} = m$.
Following this procedure, the mass matrix fills out as follows:

$$[m] = m \begin{bmatrix} 1 & 0 & 0 \\ 0 & 1 & 0 \\ 0 & 0 & 1 \end{bmatrix} \tag{9.95}$$

Amplitude and Characteristic Equations

The amplitude equation is

$$\begin{bmatrix} (2-\xi) & -1 & 0 \\ -1 & (2-\xi) & -1 \\ 0 & -1 & (2-\xi) \end{bmatrix} \begin{Bmatrix} X_1 \\ X_2 \\ X_3 \end{Bmatrix} = \begin{Bmatrix} 0 \\ 0 \\ 0 \end{Bmatrix} \tag{9.96}$$

where $\xi = \frac{m \omega^2 a}{T}$. The characteristic equation is

$$(2-\xi)\left[(2-\xi)^2 - 2\right] = 0 \tag{9.97}$$

The eigenvalues from this equation are

$$\xi_1 = 2 - \sqrt{2}, \quad \omega_1^2 = \frac{T}{ma}\left(2 - \sqrt{2}\right) \tag{9.98}$$

$$\xi_1 = 2, \qquad \omega_1^2 = \frac{T}{ma}(2) \tag{9.99}$$

$$\xi_1 = 2 + \sqrt{2}, \quad \omega_1^2 = \frac{T}{ma}\left(2 + \sqrt{2}\right) \tag{9.100}$$

The modal matrix is

$$\{X_1 \ X_2 \ X_3\} = \begin{bmatrix} 1 & 1 & 1 \\ \sqrt{2} & 0 & -\sqrt{2} \\ 1 & -1 & 1 \end{bmatrix} \tag{9.101}$$

The first column is the shape of the first mode, etc. Mode shapes are useful for visualizing the total response to various dynamic loads applied to a spring mass system. In general, total motion involves all modes, however, any one mode can stand out especially, when the frequency of excitation is close to the natural frequency of a given mode.

Example   Consider the system shown in Figure 9.20.

$$\{X_1, X_2 X_3\} = \begin{bmatrix} 1 & 1 & 1 \\ 1.83 & 0.445 & -1.25 \\ 2.247 & -0.802 & 0.555 \end{bmatrix} \tag{9.102}$$

The corresponding eigenvalues are

$$\xi_1 = 0.198$$
$$\xi_2 = 1.55$$
$$\xi_3 = 3.25$$

where

$$\xi = \frac{m\omega^2}{k}$$

Therefore,

$$\omega_1 = 0.445\sqrt{\frac{k}{m}}, \quad \omega_2 = 1.072\sqrt{\frac{k}{m}}, \quad \omega_3 = 1.803\sqrt{\frac{k}{m}}$$

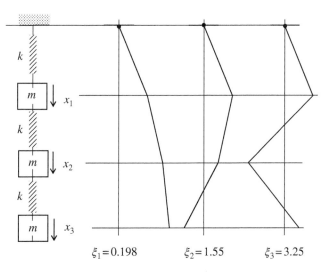

$$\xi_1 = 0.198 \qquad \xi_2 = 1.55 \qquad \xi_3 = 3.25$$

Figure 9.20   Mode shapes and frequencies.

Parameters Not Chosen at Discrete Masses

In some 2-DOF systems, there may be more than one discrete mass and only two dependent variables are required such as shown in Figure 9.21. In this example, two of the discrete masses are not located at the points of displacements (x_1 and x_2). This system is viewed as a rigid body composed of a rigid weightless bar containing three discrete masses.

The stiffness matrix can be determined as in the previous example. In this case, give point 1 a unit displacement and determine the required force, f_1. At the same time determine the reaction force, f_2, at point 2. The results are

$$f_1 = k_{11} = \frac{10}{9}k \tag{9.103}$$

$$f_2 = k_{21} = \frac{2}{9}k \tag{9.104}$$

Next fix point 1 and give point 2 a unit displacement:

$$f_2 = k_{22} = \frac{13}{9}k \tag{9.105}$$

$$f_1 = k_{12} = \frac{2}{9}k \tag{9.106}$$

The stiffness matrix thus becomes

$$[k] = \frac{k}{9}\begin{bmatrix} 10 & 2 \\ 2 & 13 \end{bmatrix} \tag{9.107}$$

The mass matrix is determined in a similar manner (Figure 9.22). First fix point 2 and give point 1 a unit acceleration to determine m_{11} and m_{21}. This assessment requires dynamic principles as follows.

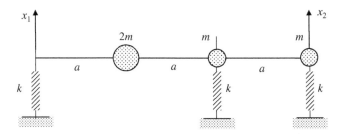

Figure 9.21 Two degree of freedom system.

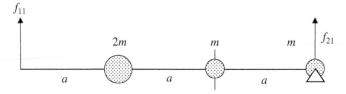

Figure 9.22 Model for mass matrix.

With point 2 fixed and applying $T = \sum m_i a_i r_i$

$$f_{11} 3\,a = 2m \frac{2}{3} (2a) + m \frac{1}{3} a \tag{9.108}$$

$$f_{11} = m \tag{9.109}$$

Applying $\sum F = \sum m_i a_1$ to the freebody gives

$$f_{11} + f_{21} = 2m \frac{2}{3} + m \frac{1}{3} = \frac{5}{3} m \tag{9.110}$$

$$f_{21} = \frac{5}{3} m - m = \frac{2}{3} m \tag{9.111}$$

Therefore,

$$m_{11} = m \quad \text{and} \quad m_{21} = \frac{2}{3} m \tag{9.112}$$

Next fix point 1 and give point 2 a unit acceleration. Reactions are

$$f_{22} = m_{22} = \frac{5}{3} m \tag{9.113}$$

$$f_{12} = m_{12} = \frac{2}{3} m \tag{9.114}$$

The mass matrix thus becomes

$$[m] = \frac{m}{3} \begin{bmatrix} 3 & 2 \\ 2 & 5 \end{bmatrix} \tag{9.115}$$

Lateral Stiffness of a Vertical Cable

A long vertical cable containing three discrete masses is shown in Figure 9.23. The tension at the top is greater than the tension at the bottom due to the distribution of the three masses. Determine the stiffness matrix for lateral modes of vibration.

The difference between this example and the example in Figure 9.19 is variable tension between the masses. There is some similarity between this example and vertical drill pipe, but in this case bending flexibility included. Bending stiffness, however, could be modeled as torsion springs at the connection points. The mass matrix is the same, however, the stiffness matrix is modified as explained below. Tensions between the three masses will be represented by

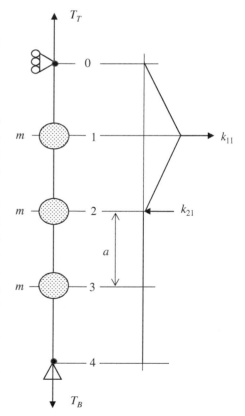

T_0	tension between points 0 and 1	Top tension
T_1	tension between points 1 and 2	$T_1 = T_0 - m/g$
T_2	tension between points 2 and 3	$T_2 = T_0 - 2m/g$
T_3	tension between points 3 and 4	$T_3 = T_0 - 3m/g$

Figure 9.23 Lateral vibration of hanging pipe.

Tension in the fourth member is also the pull at the bottom end. If it is desired to maintain a bottom pull of T_B the top pull force has to be $T_T = T_B + 3m/g$.

$$k_{11} = \frac{1}{a}(T_0 + T_1) \tag{9.116}$$

$$k_{21} = -\frac{1}{a}T_1 \tag{9.117}$$

$$k_{31} = 0 \tag{9.118}$$

$$k_{22} = \frac{1}{a}(T_1 + T_2) \tag{9.119}$$

$$k_{32} = -\frac{1}{a}T_2 \tag{9.120}$$

$$k_{33} = \frac{1}{a}(T_3 + T_4) \quad k_{13} = 0 \tag{9.121}$$

Building the Damping Matrix

In the general case, damping must be included in the equations of motion:

$$[m]\{\ddot{x}\} + [c]\{\dot{x}\} + [k]\{x\} = \{F\} \tag{9.122}$$

The damping matrix is represented by

$$[c] = \begin{bmatrix} c_{11} & c_{12} & c_{13} & c_{14} \\ c_{21} & c_{22} & c_{23} & \\ c_{31} & c_{32} & c_{ij} & \\ c_{41} & & & etc \end{bmatrix} \tag{9.123}$$

The damping matrix can also be developed in a manner similar to the stiffness and mass matrices. In this case, c_{ij} is the force at point "i" due to a force required to produce a unit velocity at point "j."

Modal Analysis of Discrete Systems

The modal analysis method uncouples the differential equations of motion and puts them into the form of a SDOF equation, whose solution was discussed earlier [3]. The transformation is based on

$$\{x(t)\} = [X]\{\eta(t)\} \tag{9.124}$$

where

$\{x(t)\}$ – local coordinates
$[X]$ – modal matrix
$\{\eta(t)\}$ – modal coordinates

The expansion of Eq. (9.124) for a 3-DOF system is

$$\begin{Bmatrix} x_1 \\ x_2 \\ x_3 \end{Bmatrix} = \begin{bmatrix} X_{11} & X_{12} & X_{13} \\ X_{21} & X_{22} & X_{23} \\ X_{31} & X_{32} & X_{33} \end{bmatrix} \begin{Bmatrix} \eta_1 \\ \eta_2 \\ \eta_3 \end{Bmatrix} \tag{9.125}$$

$$\begin{Bmatrix} x_1 \\ x_2 \\ x_3 \end{Bmatrix} = \begin{Bmatrix} X_{11} \\ X_{21} \\ X_{31} \end{Bmatrix} \eta_1(t) + \begin{Bmatrix} X_{12} \\ X_{22} \\ X_{32} \end{Bmatrix} \eta_2(t) + \begin{Bmatrix} X_{13} \\ X_{23} \\ X_{33} \end{Bmatrix} \eta_3(t) \tag{9.126}$$

$$\begin{Bmatrix} x_1 \\ x_2 \\ x_3 \end{Bmatrix} = X_{(1)}\eta_1(t) + X_{(2)}\eta_2(t) + X_{(3)}\eta_3(t) \tag{9.127}$$

where $X_{(j)}$ represents the jth mode. The displacement of the ith coordinate expressed in terms of contributions from each mode is

$$x_i = \sum_{j=1}^{N} X_{ij}\eta_j(t) \tag{9.128}$$

where

i – ith location
j – jth mode
X_{ij} – mode amplitude at ith location within the jth mode

The overall vibration of a multisystem can be visualized in terms of each modal responses.

Orthogonal Properties of Natural Modes

Consider a discrete mass system whose displacements are defined by $\{x(t)\}$. The equations of motion for each mass in terms of local coordinates are

$$[m]\{\ddot{x}\} + [c]\{\dot{x}\} + [k]\{x\} = \{F\} \tag{9.129}$$

Using Eq. (9.128),

$$[m][X]\{\ddot{\eta}\} + [c][X]\{\dot{\eta}\} + [k][X]\{\eta\} = \{F\} \tag{9.130}$$

Multiplying each term by the transposed modal matrix gives

$$[X]^T[m][X]\{\ddot{\eta}\} + [X]^T[c][X]\{\dot{\eta}\} + [X]^T[k][X]\{\eta\} = [X]^T\{F\} \tag{9.131}$$

The orthogonality property of the modes with respect to mass and stiffness matrices gives

$$X_{(r)}^T[m]X_{(s)} = 0 \tag{9.132}$$

$$X_{(r)}^T[m]X_{(r)} = M_r \quad \text{(modal mass)} \tag{9.133}$$

$$X_{(r)}^T[k]X_{(s)} = 0 \tag{9.134}$$

$$X_{(r)}^T[k]X_{(r)} = K_r \quad \text{(modal stiffness)} \tag{9.135}$$

The modes are not necessarily orthogonal with respect to the damping matrix. This is true only if damping is proportional. Assuming proportional damping

$$X_{(r)}^T[c]X_{(s)} = 0 \tag{9.136}$$

$$X_{(r)}^T[c]X_{(r)} = C_r \quad \text{(modal damping)} \tag{9.137}$$

In each case, modal mass, damping, and stiffness are diagonal matrices. This orthogonality property of the modes allows the equations of motion to be converted into modal coordinates, $\eta_i(t)$. The converted equation becomes

$$[M]\{\ddot{\eta}\} + [C]\{\dot{\eta}\} + [K]\{\eta\} = \{Q\} \tag{9.138}$$

where modal mass, damping, and stiffness matrices are diagonal, and the modal force is

$$\{Q\} = \{X^T\}\{F\} \tag{9.139}$$

Because of the orthogonality properties of the natural modes, the modal mass and stiffness matrices are diagonal. If damping is proportional, the modal damping matrix is also diagonal. The differential equation defining the response of each mode is

$$M_i\ddot{\eta}_i + C_i\dot{\eta}_i + K_i\eta = Q_i \quad i = 1, 2, 3, 4, \ldots \tag{9.140}$$

where

$$Q_i = X_{(i)}{}^T\{F\} \tag{9.141}$$

The solution to Eq. (9.140) gives the time history of each mode. Theory discussed earlier for SDOF problems applies directly to modal responses to various forces, whether periodic or not.

The results in terms of modal coordinates transfer directly back to local coordinates according to Eq. (9.128).

Proportional Damping

The local damping matrix converts into a diagonal matrix only if $[c]$ is proportional to $[m]$ and $[k]$:

$$[c] = \alpha[m] + \beta[k] \tag{9.142}$$

The proof that this type of local damping leads to a diagonal matrix is as follows. For the off-diagonal terms in the modal damping matrix to be zero:

$$X_s^T[c]X_r = 0 \tag{9.143}$$

By substitution

$$X_s^T[\alpha[m] + \beta[k]]X_r = \alpha X_s^T[m]X_r + \beta X_s^T[k]X_r \tag{9.144}$$

Since

$$X_s^T[m]X_r = 0 \tag{9.145a}$$
$$X_s^T[k]X_r = 0 \tag{9.145b}$$

then

$$X_s^T[c]X_r = 0 \tag{9.146}$$

Therefore,

$$X_r^T[c]X_r = C_r \tag{9.147}$$

Examples are illustrated in Figure 9.24.

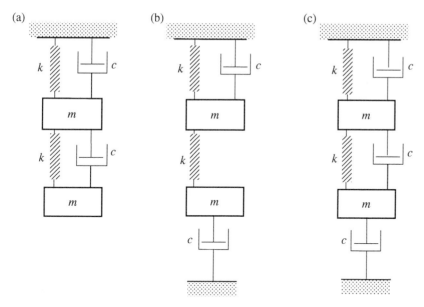

Figure 9.24 Examples of damped systems.

Case a

$$[c] = c \begin{bmatrix} 2 & -1 \\ -1 & 1 \end{bmatrix}, \quad \alpha = 0, \quad \beta = \frac{c}{k} \quad \text{(proportional damping)}$$

Case b

$$[c] = c \begin{bmatrix} 1 & 0 \\ 0 & 1 \end{bmatrix}, \quad \alpha = \frac{c}{m}, \quad \beta = 0 \quad \text{(proportional damping)}$$

Case c

$$[c] = c \begin{bmatrix} 2 & -1 \\ -1 & 2 \end{bmatrix} \quad \text{(not proportional damping)}$$

Transforming Modal Solution to Local Coordinates

The transformation of local coordinates into modal coordinates resolves multidegree of freedom systems into the form:

$$[M]\{\ddot{\eta}\} + [C]\{\dot{\eta}\} + [K]\{\eta\} = \{0\} \tag{9.148}$$

as explained above and greatly simplifies the mathematics. The solution to this equation is well documented [1]. Once a solution is obtained in terms of modal coordinates, it is transformed back to local coordinates according to

$$\{x(t)\} = [X]\{\eta(t)\} \tag{9.149}$$

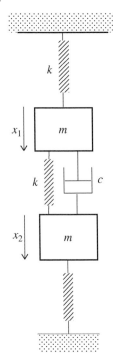

where

{$x(t)$} – local coordinates
$[X]$ – modal matrix
{$\eta(t)$} – modal coordinates

This equation shows how each mode contributes to the total response, a useful aspect of modal analysis.

For the 2-D example shown in Figure 9.25, the reverse transformation requires the following matrices:

$$[X] = \begin{bmatrix} 1 & 1 \\ 1 & -1 \end{bmatrix} \quad \text{modal matrix}$$

$$[m] = m\begin{bmatrix} 1 & 0 \\ 0 & 1 \end{bmatrix} \quad [c] = c\begin{bmatrix} 1 & -1 \\ -1 & 1 \end{bmatrix} \quad [k] = k\begin{bmatrix} 2 & -1 \\ -1 & 2 \end{bmatrix}$$

$$[M] = m\begin{bmatrix} 2 & 0 \\ 0 & 2 \end{bmatrix} \quad [C] = c\begin{bmatrix} 0 & 0 \\ 0 & 4 \end{bmatrix} \quad [K] = k\begin{bmatrix} 2 & 0 \\ 0 & 6 \end{bmatrix}$$

$$\alpha = -\frac{c}{m} \quad \beta = \frac{c}{k}$$

All modal matrices are diagonal, and damping is proportional. In some cases, off diagonal terms are ignored for the sake of simplicity. In many cases, this assumption gives reasonable engineering solutions.

Figure 9.25 Damping between two masses.

Free Vibration of Multiple DOF Systems

Free vibration of multidegrees of freedom systems can be visualized as the summation of all vibration modes. The level of participation of each mode depends on initial conditions. The approach to free vibration response follows the same approach outlined above. Initial conditions are expressed in terms of local coordinates, which must be transformed into modal coordinates.

The modal equation when there are no external forces is

$$[M]\{\ddot{\eta}\} + [C]\{\dot{\eta}\} + [K]\{\eta\} = \{0\} \tag{9.150}$$

giving

$$M_i\ddot{\eta}_i + C_i\dot{\eta}_i + K_i\eta_i = 0 \tag{9.151}$$

Assuming zero damping the solution to each modal response is

$$\eta_i(t) = A_i\cos\omega_i t + B_i\sin\omega_i t \tag{9.152}$$

The arbitrary constants are determined from initial conditions given in terms of local coordinates and transformed to modal coordinates per

$$\{x(0)\} = [X]\{\eta(0)\} \tag{9.153}$$
$$\{\dot{x}(0)\} = [X]\{\dot{\eta}(0)\} \tag{9.154}$$

Transferring to modal coordinates,

$$\eta_i(0) = \frac{X_i^T[m]\{x(0)\}}{M_i}; \quad [k] \text{ could also be used} \tag{9.155}$$

$$\dot{\eta}_i(0) = \frac{X_i^T[m]\{\dot{x}(0)\}}{M_i}; \quad [k] \text{ could also be used} \tag{9.156}$$

Once the solution is obtained in modal coordinates, it is transformed back into local coordinates for the true response.

The overall response to a given set of initial conditions is defined by Eq. (9.128).

Free Vibration of 2 DOF Systems

Example Consider the free vibrations of the 2-degree system in Figure 9.3. Assume the following initial conditions for x_1 and x_2. The initial position of $x_1(0) = x_{10}$, while $x_2(0) = \dot{x}_1(0) = \dot{x}_2(0) = 0$.

$$\{x(0)\} = \begin{Bmatrix} x_{10} \\ 0 \end{Bmatrix} \tag{9.157}$$

$$\{\dot{x}(0)\} = \begin{Bmatrix} 0 \\ 0 \end{Bmatrix} \tag{9.158}$$

Modal mass and modal stiffness matrics are

$$[M] = m \begin{bmatrix} 3.618 & 0 \\ 0 & 1.382 \end{bmatrix} \tag{9.159}$$

$$[K] = k \begin{bmatrix} 1.386 & 0 \\ 0 & 3.618 \end{bmatrix} \tag{9.160}$$

Modal equations are

$$M_1\ddot{\eta}_1 + K_1\eta_1 = 0$$
$$M_2\ddot{\eta}_2 + K_2\eta_2 = 0$$

The solutions to both equations are

$$\eta_1(t) = \eta_1(0)\cos\omega_1 t + \frac{\dot{\eta}_1(0)}{\omega_1}\sin\omega_1 t \tag{9.161}$$

$$\eta_2(t) = \eta_2(0)\cos\omega_2 t + \frac{\dot{\eta}_2(0)}{\omega_2}\sin\omega_2 t \tag{9.162}$$

From Eqs. (9.155) and (9.156)

$$\eta_1(0) = \frac{X_1^T[m]\{x(0)\}}{M_1} = \frac{x_{10}}{3.618}$$

$$\eta_2(0) = \frac{X_2^T[m]\{x(0)\}}{M_2} = \frac{x_{10}}{1.382}$$

$$\dot{\eta}_1(0) = \frac{X_1^T[m]\{\dot{x}(0)\}}{M_1} = 0$$

$$\dot{\eta}_2(0) = \frac{X_2^T[m]\{\dot{x}(0)\}}{M_2} = 0$$

Therefore,

$$\eta_1(t) = \frac{x_{10}}{3.618}\cos\omega_1 t$$

$$\eta_2(t) = \frac{x_{10}}{1.382}\cos\omega_2 t$$

The response of each mass is the modal summation:

$$x_1(t) = \eta_1(t) + \eta_2(t) \tag{9.163}$$

$$x_2(t) = 1.618\eta_1(t) - 0.618\,\eta_2(t) \tag{9.164}$$

If the masses are displaced initially in a mode shape, only that mode vibrates in free vibration.

Suddenly Stopping Drill Pipe with the Slips

Drill pipe is lowered into a well bore by block and tackle. Downward movement is controlled by a band brake. The rate of stopping affects magnitude of stress waves set up in the pipe. Sometimes the slips are clamped around the pipe causing an instantaneous stop. Axial vibration response resulting from setting the slips too early is discussed below. The problem is to determine the response of pipe (modeled as multiple springs and masses) moving downward with velocity v_0 after point 0 is suddenly stopped (Figure 9.26).

Ignoring damping, the equation of motion of the system is

$$[m_i]\{\ddot{x}_i\} + [k]\{x_i\}0 = \{0\} \tag{9.165}$$

where

$$[m] = m \begin{bmatrix} 1 & 0 & 0 & 0 \\ 0 & 1 & 0 & 0 \\ 0 & 0 & 1 & 0 \\ 0 & 0 & 0 & 0 \end{bmatrix} \quad \text{and} \quad [k] = k \begin{bmatrix} 2 & -1 & 0 & 0 \\ -1 & 2 & -1 & 0 \\ 0 & -1 & 2 & -1 \\ 0 & 0 & -1 & 1 \end{bmatrix}$$

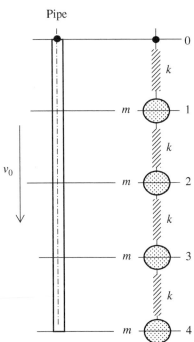

Figure 9.26 Suddenly stopping pipe with the slips.

Values of m and k according to the approximation are

$$m = \frac{wl}{g} \quad \text{and} \quad k = \frac{EA}{l}$$

The modal matrix, X, is determined as before along with the natural frequency, ω_i of each mode.

Transforming from local coordinates to modal coordinates using

$$\{x_i\} = [X]\{\eta_i\}$$

uncouples the modal differential equation:

$$[M]\{\ddot{\eta}_i\} + [K]\{\eta_i\} = \{0\} \tag{9.166}$$

Solutions to this equation, which describes the time history of each mode separately, are of the form:

$$\eta_i(t) = A_i \cos \omega_i t + B_i \sin \omega_i t \tag{9.167}$$

In terms of initial conditions

$$\eta_i(t) = \eta_i(0) \cos \omega t + \frac{\dot{\eta}_i(0)}{\omega_i} \sin \omega t$$

Initial conditions in terms of local coordinates are

$$\{x_i(0)\} = \{0\} \tag{9.168}$$
$$\{\dot{x}_i(0)\} = v_0\{1\} \tag{9.169}$$

These initial conditions transform into modal coordinates by reverse transformation of

$$\{x_i\} = [X]\{\eta_i\}$$
$$[X]^T[m]\{x\} = [X^T][m][X]\{\eta\} = [M]\{\eta\}$$

This equation yields

$$\eta_i(0) = \frac{X_i^T[m]\{x(o)\}}{M_i} = 0 \tag{9.170}$$

and

$$\dot{\eta}_i(0) = \frac{X_i^T[m]\{\dot{x}(o)\}}{M_i} = \frac{X_i^T[m]\{v_0\}}{M_i} \tag{9.171}$$

By substitution

$$\eta_i(t) = \frac{\dot{\eta}_i(0)}{\omega_i} \sin \omega_i t \tag{9.172}$$

Note: X_i^T is a horizontal 1×4 matrix. When multiplied into the square matrix gives a 1×4 flat matrix.

Response in terms of local coordinates is

$$\{x_i\} = \sum_{i = 1, 2, 3, 4} X_i \eta_i(t) \tag{9.173}$$

Example   Consider the 3-DOF system of Figure 9.20. The modal matrix, $[X]$ and modal frequencies are known. Using this model:

$$\dot{\eta}_1(0) = \frac{X_1^T[m]\{v_0\}}{M_i} = \frac{[1 \quad 1.83 \quad 2.247\,]m\begin{bmatrix} 1 & 0 & 0 \\ 0 & 1 & 0 \\ 0 & 0 & 1 \end{bmatrix} v_0 \begin{Bmatrix} 1 \\ 1 \\ 1 \end{Bmatrix}}{M_1} \tag{9.174}$$

$$\dot{\eta}_1(0) = \frac{5.077mv_o}{M_1} = \frac{5.077mv_0}{9.398m} = 0.54v_0 \tag{9.175}$$

$$\dot{\eta}_2(0) = \frac{X_2^T[m]\{v_0\}}{M_2} = \frac{[1 \quad 0.445 \quad -0.802\,]m\begin{bmatrix} 1 & 0 & 0 \\ 0 & 1 & 0 \\ 0 & 0 & 1 \end{bmatrix} v_0 \begin{Bmatrix} 1 \\ 1 \\ 1 \end{Bmatrix}}{M_2} \tag{9.176}$$

$$\dot{\eta}_2(0) = \frac{0.652mv_o}{M_2} = \frac{0.652mv_0}{1.84m} = 0.354v_0 \tag{9.177}$$

$$\dot{\eta}_3(0) = \frac{X_3^T[m]\{v_0\}}{M_3} = \frac{[1 \quad -1.25 \quad 0.555]m \begin{bmatrix} 1 & 0 & 0 \\ 0 & 1 & 0 \\ 0 & 0 & 1 \end{bmatrix} v_0 \begin{Bmatrix} 1 \\ 1 \\ 1 \end{Bmatrix}}{M_3} \tag{9.178}$$

$$\dot{\eta}_3(0) = \frac{0.305mv_o}{M_3} = \frac{0.305mv_0}{2.87m} = 0.106v_0 \tag{9.179}$$

Therefore

$$\eta_1(t) = \frac{0.54v_0}{\omega_1} \sin \omega_1 t$$

$$\eta_2(t) = \frac{0.354v_0}{\omega_2} \sin \omega_2 t$$

$$\eta_3(t) = \frac{0.106v_0}{\omega_3} \sin \omega_3 t$$

Recall model mass, M_i is

$$M_i = [X_i]^T[m][X_i]$$

$$M_1 = [1 \quad 1.83 \quad 2.247] \begin{bmatrix} m & 0 & 0 \\ 0 & m & 0 \\ 0 & 0 & m \end{bmatrix} \begin{Bmatrix} 1 \\ 1.83 \\ 2.247 \end{Bmatrix} = m[1 \quad 1.83 \quad 2.247] \begin{Bmatrix} 1 \\ 1.83 \\ 2.247 \end{Bmatrix}$$

$$M_1 = 9.398m$$
$$M_2 = 1.841m$$
$$M_3 = 2.87m$$

The response of each mass will contain contributions from each of the three modes. The final solution is

$$\begin{Bmatrix} x_1 \\ x_2 \\ x_3 \end{Bmatrix} = \begin{Bmatrix} 1 \\ 1.83 \\ 2.247 \end{Bmatrix} \eta_1(t) + \begin{Bmatrix} 1 \\ 0.445 \\ -0.802 \end{Bmatrix} \eta_2(t) + \begin{Bmatrix} 1 \\ -1.25 \\ 0.555 \end{Bmatrix} \eta_3(t) \tag{9.180}$$

Critical Damping of Vibration Modes

Damping characteristics of each mode are determined from the free vibration equation:

$$M_i \ddot{\eta}_i + C_i \dot{\eta}_i + K_i \eta_i = 0 \tag{9.181}$$

Similar to the SDOF equation in local coordinates:

$$\zeta_i = \frac{C_i}{C_{i,cr}} = \frac{C_i}{2M_i\omega_i} \tag{9.182}$$

$$C_{i,cr} = 2\sqrt{K_iM_i} \tag{9.183}$$

Consider the model in Figure 9.27.

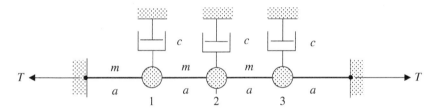

Figure 9.27 Damped vibration model.

As given earlier the modal matrix for this system is

$$\{X_1 \ \ X_2 \ \ X_3\} = \begin{bmatrix} 1 & 1 & 1 \\ \sqrt{2} & 0 & -\sqrt{2} \\ 1 & -1 & 1 \end{bmatrix} \tag{9.184}$$

The eigenvalues and natural frequencies are

$$\xi_1 = 2 - \sqrt{2}, \quad \omega_1^2 = \frac{T}{ma}\left(2 - \sqrt{2}\right)$$

$$\xi_1 = 2, \quad \omega_1^2 = \frac{T}{ma}(2)$$

$$\xi_1 = 2 + \sqrt{2}, \quad \omega_1^2 = \frac{T}{ma}\left(2 + \sqrt{2}\right)$$

The damping matrix is

$$[C] = c \begin{bmatrix} 1 & \sqrt{2} & 1 \\ 1 & 0 & -1 \\ 1 & -\sqrt{2} & 1 \end{bmatrix} \begin{bmatrix} 1 & 0 & 0 \\ 0 & 1 & 0 \\ 0 & 0 & 1 \end{bmatrix} \begin{bmatrix} 1 & 1 & 1 \\ \sqrt{2} & 0 & -\sqrt{2} \\ 1 & -1 & 1 \end{bmatrix} \tag{9.185}$$

$$[C] = c \begin{bmatrix} 4 & 0 & 0 \\ 0 & 2 & 0 \\ 0 & 0 & 4 \end{bmatrix} \tag{9.186}$$

Similarly, the modal mass matrix is

$$[M] = m \begin{bmatrix} 4 & 0 & 0 \\ 0 & 2 & 0 \\ 0 & 0 & 4 \end{bmatrix} \tag{9.187}$$

The damping factor for each of the three modes is

$$\zeta_i = \frac{C_i}{C_{i,cr}} = \frac{C_i}{2M_i\omega_i} \tag{9.188}$$

$$\zeta_1 = \frac{c}{2m}\frac{1}{\omega_1}, \quad \zeta_2 = \frac{c}{2m}\frac{1}{\omega_2}, \quad \zeta_3 = \frac{c}{2m}\frac{1}{\omega_3} \tag{9.189}$$

Showing for this example $\zeta_1 \succ \zeta_2 \succ \zeta_3$. The maximum amplitude for each modal response occurs at $r_r = 1$ and is defined by

$$|\eta_r| = \frac{Q_{r0}}{K_r} \frac{1}{2\zeta_r} \tag{9.190}$$

Forced Vibration by Harmonic Excitation

Consider separate forces are applied simultaneously to each mass. Modal responses are obtained from the solution to

$$\ddot{\eta}_r + 2\zeta_r \omega_r \dot{\eta}_r + \omega_r^2 \eta_r = \frac{Q_r}{M_r} \tag{9.191}$$

where

ζ_r – mode damping factor, $\left(= \frac{C_r}{C_{r,cr}} \right)$

$C_{r,cr} = 2M_r\omega_r$

$Q_r = X_{(r)}{}^T\{F\}$

and assuming the force amplitude at each point is periodic:

$$\{F\} = \{F_0\} \cos \omega t \tag{9.192}$$

By substitution

$$Q_r = Q_{r0} \cos \omega t \tag{9.193}$$

where

$$Q_{r0} = X_{(r)}{}^T\{F_0\}$$

Following the solution of a SDOF system,

$$\eta_r(t) = \frac{\frac{Q_{r0}}{K_r}}{\left[\left(1 - r_r^2\right)^2 + \left(2\zeta_r r_r\right)^2 \right]^{\frac{1}{2}}} \cos(\omega t - \phi_r) \tag{9.194}$$

where

$$\tan \phi_r = \frac{2\zeta_r r_r}{1 - r_r^2}$$

$$r_r = \frac{\omega}{\omega_r}$$

ϕ_r represents the phase lag of each mode with respect to the modal force, Q_r.

Complex Variable Approach

In terms of complex solution, the complex modal force is

$$Q_r = X_{(r)}{}^T\{F_0\}e^{i\omega t} \tag{9.195}$$

The complex variable solution to Eq. (9.191) is

$$\eta_r(t) = \frac{Q_{r0}/K_r}{(1 - r_r^2) + i(2\zeta_r r_r)} e^{i\omega t} \tag{9.196}$$

In terms of local coordinates

$$\{x\} = [X][\eta] = \sum_{r=1}^{N} X_{(r)} \eta_r \tag{9.197}$$

For example, the expression for a 3-DOF system is

$$\begin{Bmatrix} x_1 \\ x_2 \\ x_3 \end{Bmatrix} = \{X_{(1)}\}\eta_1 + \{X_{(2)}\}\eta_2 + \{X_{(3)}\}\eta_3 \tag{9.198}$$

The response at point i is

$$x_i = \sum_{r=1}^{N} X_{ir} \eta_r \tag{9.199}$$

where $X_{(r)}$ is the shape of the rth mode. X_{ir} is mode amplitude at location i of the rth mode. By substitution

$$\{x_i(t)\} = \sum_{r=1}^{N} X_{(r)} \frac{Q_{r0}/K_r}{(1 - r_r^2) + i(2\zeta_r r_r)} e^{i\omega t} \tag{9.200}$$

$$\{x_i(t)\} = \sum_{r=1}^{N} X_{(r)} \frac{X_{(r)}^T \{F_0\}}{K_r} \frac{1}{(1 - r_r^2) + i(2\zeta_r r_r)} e^{i\omega t} \tag{9.201}$$

The *complex amplitude* of the response is

$$\{X_i\} = \sum_{r=1}^{N} \frac{X_{(r)} X_{(r)}^T \{F_0\}}{K_r} \frac{1}{(1 - r_r^2) + i(2\zeta_r r_r)} \tag{9.202}$$

From which the *complex amplitude* at each location is

$$\bar{X}_1 = \sum_{r=1}^{N} \frac{X_{(1r)} X_{(r)}^T \{F_0\}}{K_r} \frac{1}{(1 - r_r^2) + i(2\zeta_r r_r)}, \quad \text{for } i = 1 \tag{9.203}$$

$$\bar{X}_2 = \sum_{r=1}^{N} \frac{X_{(2r)} X_{(r)}^T \{F_0\}}{K_r} \frac{1}{(1 - r_r^2) + i(2\zeta_r r_r)}, \quad \text{for } i = 2, \text{ etc.} \tag{9.204}$$

Note that when $r_r = 1$ for any mode, that mode responds with resonance.

Harmonic Excitation of 3 DOF Systems

Consider the 3-DOF system of Figure 9.28. Assume separate forces applied simultaneously to each mass and linear damping is parallel to each spring. Modal responses are obtained from the solution to

$$\ddot{\eta} + 2\zeta_r \omega_r \dot{\eta}_r + \omega_r^2 \eta_r = \frac{Q_r}{M_r} \tag{9.205}$$

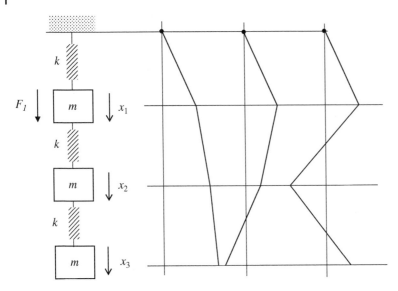

Figure 9.28 Mode shapes.

where

ζ_r – mode damping factor, $\left(= \frac{C_r}{C_{r,cr}} \right)$

$C_{r,\ cr} = 2M_r\omega_r$

$Q_r = X_{(r)}{}^T\{F\}$

and assuming the force amplitude at each point varies according to

$$\{F\} = \{F_{r0}\} \cos \omega t \tag{9.206}$$

By substitution

$$Q_r = Q_{r0} \cos \omega t \tag{9.207}$$

where

$$Q_{r0} = X_{(r)}{}^T\{F_0\}$$

Following the solution of SDOF differential,

$$\eta_r(t) = \frac{\frac{Q_{r0}}{K_r}}{\left[\left(1 - r_r^2\right)^2 + \left(2\zeta_r r_r\right)^2 \right]^{\frac{1}{2}}} \cos\left(\omega t - \phi_r\right) \tag{9.208}$$

where

$$\tan \phi_r = \frac{2\zeta_r r_r}{1 - r_r^2}$$

$$r_r = \frac{\omega}{\omega_r}$$

ϕ_r represents the phase lag of each mode with respect to the modal force, Q_r.

Modal Solution of a Damped 2-DOF System

The following damped system is used to illustrate the modal analysis approach (Figure 9.29). The modal matrix for this arrangement is

$$[X] = \begin{bmatrix} 1 & 1 \\ 1 & -1 \end{bmatrix}$$

The local mass, stiffness and damping matrices are given below:

$$[m] = m \begin{bmatrix} 1 & 0 \\ 0 & 1 \end{bmatrix} \quad [c] = c \begin{bmatrix} 1 & -1 \\ -1 & 1 \end{bmatrix} \quad [k] = k \begin{bmatrix} 1 & -1 \\ -1 & 1 \end{bmatrix}$$

The modal matrices are

$$[M] = m \begin{bmatrix} 1 & 1 \\ 1 & -1 \end{bmatrix} \begin{bmatrix} 1 & 0 \\ 0 & 1 \end{bmatrix} \begin{bmatrix} 1 & 1 \\ 1 & -1 \end{bmatrix} = m \begin{bmatrix} 2 & 0 \\ 0 & 2 \end{bmatrix}$$

$$[K] = k \begin{bmatrix} 1 & 1 \\ 1 & -1 \end{bmatrix} \begin{bmatrix} 2 & -1 \\ -1 & 2 \end{bmatrix} \begin{bmatrix} 1 & 1 \\ 1 & -1 \end{bmatrix} = k \begin{bmatrix} 2 & 0 \\ 0 & 6 \end{bmatrix}$$

$$[C] = c \begin{bmatrix} 1 & 1 \\ 1 & -1 \end{bmatrix} \begin{bmatrix} 1 & -1 \\ -1 & 1 \end{bmatrix} \begin{bmatrix} 1 & 1 \\ 1 & -1 \end{bmatrix} = c \begin{bmatrix} 0 & 0 \\ 0 & 4 \end{bmatrix}$$

Modal force is

$$\{Q\} = \begin{bmatrix} 1 & 1 \\ 1 & -1 \end{bmatrix} \begin{Bmatrix} 1 \\ 0 \end{Bmatrix} F_0 \cos \omega t$$

Modal equations are

$$M_1 \ddot{\eta}_1 + C_1 \dot{\eta}_1 + K_1 \eta_1 = F_0 \cos \omega t$$
$$M_2 \ddot{\eta}_2 + C_2 \dot{\eta}_2 + K_2 \eta_2 = F_0 \cos \omega t$$

Modal solutions then become

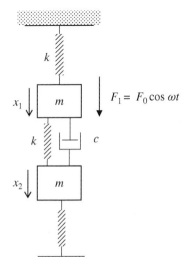

$$\eta_1(t) = \frac{F_0/K_1}{\left[(1 - r_1^2)^2 + \left[(2\zeta_1 r_1)^2\right]\right]^{\frac{1}{2}}} \cos(\omega t - \phi_1) \tag{9.209}$$

$$\eta_2(t) = \frac{F_0/K_2}{\left[(1 - r_2^2)^2 + \left[(2\zeta_2 r_2)^2\right]\right]^{\frac{1}{2}}} \cos(\omega t - \phi_2) \tag{9.210}$$

Bring it all together

$$x_1(t) = X_{11}\eta_1(t) + X_{12}\eta_2(t)$$
$$x_1(t) = \eta_1(t) + \eta_2(t) \tag{9.211}$$

and

$$x_2(t) = X_{21}\eta_1(t) + X_{22}\eta_2(t)$$
$$x_2(t) = \eta_1(t) - \eta_2(t) \tag{9.212}$$

Figure 9.29 Damped forced vibration.

General Complex Variable Solution

In terms of complex solution, the complex modal force is

$$Q_r = X_{(r)}{}^T \{F_0\} e^{i\omega t} \tag{9.213}$$

The complex variable solution to Eq. (9.205) is

$$\eta_r(t) = \frac{Q_{r0}/K_r}{(1 - r_r^2) + i(2\zeta_r r_r)} e^{i\omega t} \tag{9.214}$$

In terms of local coordinates

$$\{x\} = [X]\{\eta\} = \sum_{r=1}^{N} X_{(r)} \eta_r \tag{9.215}$$

For example, the expression for a 3-DOF system is

$$\begin{Bmatrix} x_1 \\ x_2 \\ x_3 \end{Bmatrix} = \{X_{(1)}\} \eta_1 + \{X_{(2)}\} \eta_2 + \{X_{(3)}\} \eta_3 \tag{9.216}$$

The response at point i is

$$\{x\} = [X]\{\eta\} = \sum_{r=1}^{N} X_{(r)} \eta_r$$

$$x_i = \sum_{r=1}^{N} X_{ir} \eta_r \tag{9.217}$$

where $X_{(r)}$ is the shape of the rth mode. X_{ir} is mode amplitude at location i of the rth mode. By substitution [4]

$$\{x_i(t)\} = \sum_{r=1}^{N} X_{(r)} \frac{Q_{r0}/K_r}{(1 - r_r^2) + i(2\zeta_r r_r)} e^{i\omega t} \tag{9.218}$$

$$\{x_i(t)\} = \sum_{r=1}^{N} X_{(r)} \frac{X_{(r)}{}^T \{F_0\}}{K_r} \frac{1}{(1 - r_r^2) + i(2\zeta_r r_r)} e^{i\omega t} \tag{9.219}$$

The *complex amplitude* of the response is

$$\{X_i\} = \sum_{r=1}^{N} \frac{X_{(r)} X_{(r)}{}^T \{F_0\}}{K_r} \frac{1}{(1 - r_r^2) + i(2\zeta_r r_r)} \tag{9.220}$$

From which the *complex amplitude* of each mass is

$$\overline{X}_1 = \sum_{r=1}^{N} \frac{X_{(1r)} X_{(r)}{}^T \{F_0\}}{K_r} \frac{1}{(1 - r_r^2) + i(2\zeta_r r_r)}, \quad \text{for } i = 1 \tag{9.221}$$

$$\overline{X}_2 = \sum_{r=1}^{N} \frac{X_{(2r)} X_{(r)}{}^T \{F_0\}}{K_r} \frac{1}{(1 - r_r^2) + i(2\zeta_r r_r)}, \quad \text{for } i = 2, \text{ etc.} \tag{9.222}$$

Note that when $r_r = 1$ for any mode, that mode responds with resonance.

Consider, for example a 3-DOF system where a periodic force is applied to only one the mass at location number one (1):

$$F_1 = F_{01} \cos \omega t \tag{9.223}$$

Using the complex notation, the force matrix becomes

$$\{F\} = \begin{Bmatrix} F_{01} \\ 0 \\ 0 \end{Bmatrix} e^{i\omega t} \tag{9.224}$$

Since

$$\{Q\} = [X]^T \{F\} \tag{9.225}$$

By Eq. (9.219)

$$x_i(t) = \sum_{r=1}^{N} \frac{X_{ir} X_{(r)}{}^T \{F_0\}}{K_r} \frac{1}{(1 - r_r^2) + i(2\zeta_r r_r)} e^{i\omega t} \tag{9.226}$$

Assuming a harmonic force is applied only to mass number one (1)

$$\{Q\} = [X]^T \begin{Bmatrix} F_{01} \\ 0 \\ 0 \end{Bmatrix} e^{i\omega t} \tag{9.227}$$

then

$$Q_r = X_{(r)}^T F_{01} e^{i\omega t}$$

If each mode is normalized to the amplitude at location number on (1), then [4]

$$x_i(t) = \sum_{r=1}^{N} \frac{X_{ir} F_{01}}{K_r} \frac{1}{(1 - r_r^2) + i(2\zeta_r r_r)} e^{i\omega t} \tag{9.228}$$

Maximum displacement amplitudes are defined by

$$\frac{X_1}{F_{01}} = \sum_{r=1}^{N} X_{1r} \frac{1}{K_r} \frac{1}{(1 - r_r^2) + i(2\zeta_r r_r)} \quad \text{(DTF)} \tag{9.229}$$

$$\frac{X_2}{F_{01}} = \sum_{r=1}^{N} X_{2r} \frac{1}{K_r} \frac{1}{(1 - r_r^2) + i(2\zeta_r r_r)} \quad \text{(CTF)} \tag{9.230}$$

$$\frac{X_3}{F_{01}} = \sum_{r=1}^{N} X_{3r} \frac{1}{K_r} \frac{1}{(1 - r_r^2) + i(2\zeta_r r_r)} \quad \text{(CTF)} \tag{9.231}$$

Expanding each of these equations gives

$$\frac{X_1}{F_{01}} = X_{11} \frac{1}{K_1} \frac{1}{(1 - r_1^2) + i(2\zeta_1 r_1)} + X_{12} \frac{1}{K_2} \frac{1}{(1 - r_2^2) + i(2\zeta_2 r_2)} + X_{13} \frac{1}{K_3} \frac{1}{(1 - r_3^2) + i(2\zeta_3 r_3)} \tag{9.232}$$

$$\frac{X_2}{F_{01}} = X_{21}\frac{1}{K_1}\frac{1}{(1-r_1^2) + i(2\zeta_1 r_1)} + X_{22}\frac{1}{K_2}\frac{1}{(1-r_2^2) + i(2\zeta_2 r_2)} + X_{23}\frac{1}{K_3}\frac{1}{(1-r_3^2) + i(2\zeta_3 r_3)}$$

$$(9.233)$$

$$\frac{X_3}{F_{01}} = X_{31}\frac{1}{K_1}\frac{1}{(1-r_1^2) + i(2\zeta_1 r_1)} + X_{32}\frac{1}{K_2}\frac{1}{(1-r_2^2) + i(2\zeta_2 r_2)} + X_{33}\frac{1}{K_3}\frac{1}{(1-r_3^2) + i(2\zeta_3 r_3)}$$

$$(9.234)$$

Each of these transfer functions has real and imaginary parts.

Experimental Modal Analysis

Sometimes it is useful to determine modal mass, damping, and stiffness as well as modal frequencies and mode shapes experimentally. This is especially useful when shape and geometry of an elastic system is complex and difficult to model mathematically. The analytical basis for obtaining vibration parameters experimentally is explained below.

If this information is established experimentally, then it is possible to extract mode shapes, modal parameters (mass, stiffness, and damping), and natural frequencies of each mode. This is possible because the reach of each term in each of the three equations as a short reach as shown in Figures 9.30 and 9.31. Modal information can be transferred back to local coordinates as discussed earlier.

For simplicity, consider the 2-DOF system in Figure 9.8 with the addition of dampers parallel to the springs. Specific parameters for this example are

$m_1 = m = 10/386$ lb s^2/in.
$m_2 = m = 10/386$ lb s^2/in.
$k_1 = k = 50$ lb/in.
$k_2 = k = 50$ lb/in.
$c_1 = c = 0.05$ lb/ips
$c_2 = c = 0.05$ lb/ips

The modal matrix from earlier discussion is

$$[X] = \begin{bmatrix} 1 & 1 \\ 1.618 & -0.618 \end{bmatrix}$$

The response to a periodic force applied to mass #1 is

$$\frac{X_1}{F_{01}} = (1)\frac{1}{K_1}\frac{1}{(1-r_1^2) + i(2\zeta_1 r_1)} + (1)\frac{1}{K_2}\frac{1}{(1-r_2^2) + i(2\zeta_2 r_2)}$$

$$(9.235)$$

$$\frac{X_2}{F_{01}} = 1.618\frac{1}{K_1}\frac{1}{(1-r_1^2) + i(2\zeta_1 r_1)} - 0.618\frac{1}{K_2}\frac{1}{(1-r_2^2) + i(2\zeta_2 r_2)}$$

$$(9.236)$$

Local mass, stiffness, and damping matrices are

$$[m] = m\begin{bmatrix} 1 & 0 \\ 0 & 1 \end{bmatrix}, \quad [k] = k\begin{bmatrix} 2 & -1 \\ -1 & 1 \end{bmatrix}, \quad [c] = c\begin{bmatrix} 2 & -1 \\ -1 & 1 \end{bmatrix}$$

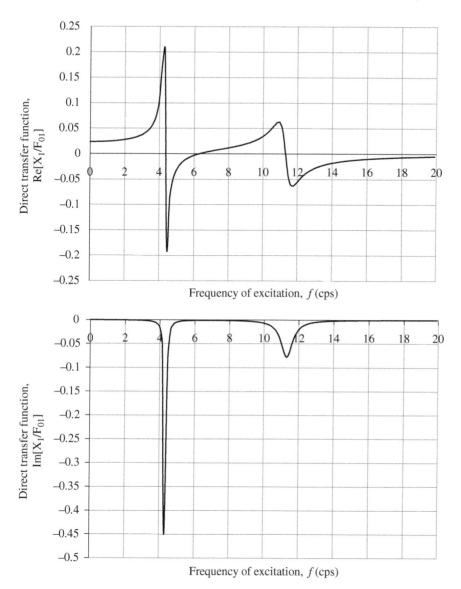

Figure 9.30   Direct transfer function.

Elements in the *modal* mass, stiffness, and damping matrices are determined from

$$[M] = [X]^T[m][X] = m\begin{bmatrix} 3.618 & 0 \\ 0 & 1.382 \end{bmatrix}; \quad M_1 = 0.094 \text{ and } M_2 = 0.0358$$

$$[K] = [X]^T[k][X] = k\begin{bmatrix} 1.382 & 0 \\ 0 & 3.618 \end{bmatrix}; \quad K_1 = 69.1 \text{ and } K_2 = 181$$

$$[C] = [X]^T[c][X] = c\begin{bmatrix} 1.382 & 0 \\ 0 & 3.618 \end{bmatrix}; \quad C_1 = 0.07 \text{ and } C_2 = 0.18$$

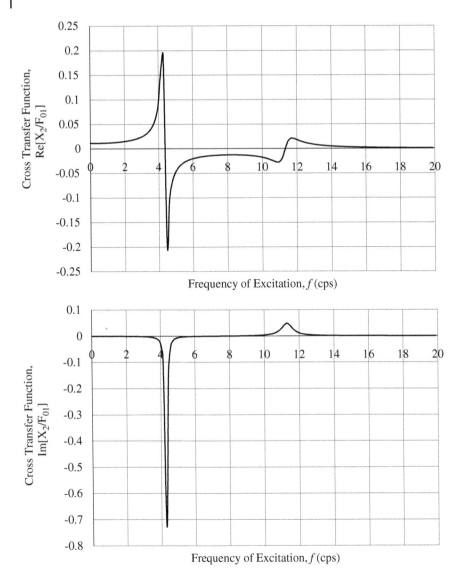

Figure 9.31 Cross transfer function.

The critical damping factor for the first and second mode are

$$C_{cr1} = 2M_1\omega_1 = 5.09 \text{ lb/ips} \quad \zeta_1 = \frac{C_1}{C_{cr1}} = 0.0136$$

$$C_{cr2} = 2M_2\omega_2 = 5.09 \text{ lb/ips} \quad \zeta_2 = \frac{C_2}{C_{cr2}} = 0.0355$$

The natural frequencies of the first and second modes are

$$\omega_1 = 0.618\sqrt{\frac{k}{m}} = \sqrt{\frac{K_1}{M_1}} = 27.15 \text{ rad/s} \quad \text{or} \quad f_1 = 4.32 \text{ cps}$$

$$\omega_2 = 1.618\sqrt{\frac{k}{m}} = \sqrt{\frac{K_2}{M_2}} = 71.08 \text{ rad/s} \quad \text{or} \quad f_2 = 11.3 \text{ cps}$$

Modal Response to Nonperiodic Forces

The modal analysis method is a powerful tool in analyzing the response of multi-DOF systems to a variety of applied forces such as

1) Nonperiodic forces
2) Step, ramp functions
3) Base motion
4) Imbalance rotating mass

These applications show the true value of the modal analysis method.

A tight cable contains two discrete masses that is stretched by force T as depicted in Figure 9.32. The objective is to determine system response to a nonperiodic force. An assumed force history is also shown in the figure. One of the special features of modal analysis is its ability to adapt SDOD solutions to multidegrees of freedom problems, which in this case is a 2-DOF system.

In this case, the amplitude and characteristic equations are

$$[X] = \begin{bmatrix} 1 & 1 \\ 1 & -1 \end{bmatrix} \quad \text{and} \quad (2-\xi)^2 - 1 = 0$$

The eigenvalues are $\xi_1 = 1$ and $\xi_2 = 3$, where $\xi = \frac{m\omega^2 a}{T}$.

As before, modal response is determined from

$$\ddot{\eta}_r + 2\zeta_r \omega_r \dot{\eta}_r + \omega_r^2 \eta_r = \frac{Q_r}{M_r}$$

However, in this case, the external force is not periodic, but defined by the graph. For the sake of simplicity, we consider an undamped system such that

$$\ddot{\eta}_r + \omega_r^2 \eta_r = \frac{Q_r}{M_r} \tag{9.237}$$

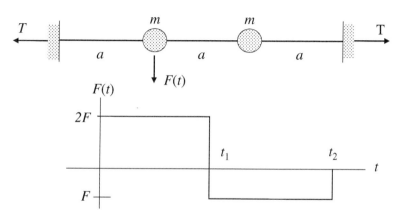

Figure 9.32 Response of 2-DOF to non-periodic force.

Applying Duhamel's method [5], modal response is determined from

$$\eta_r(t) = \frac{1}{M_r\omega_r} \int_0^t Q_r(\tau)(-i)e^{i\omega_r(t-\tau)}\, d\tau \qquad (9.238)$$

where the modal force, is

$$Q_r = X_{(r)}^T \left\{ \begin{matrix} F_1(t) \\ 0 \end{matrix} \right\} \quad \text{(applies for any } F_1(t)) \qquad (9.239)$$

In this case, F_1 is a square function defined by the diagram in Figure 9.32. By substitution,

$$\eta_r(t) = \frac{1}{M_r\omega_r} X_r^T \int_0^t \left\{ \begin{matrix} F(t) \\ 0 \end{matrix} \right\}(-i)e^{i\omega_r(t-\tau)}d\tau \qquad (9.240)$$

Modal response by both vibration modes is defined by $\eta_1(t)$ and $\eta_2(t)$. Local response is defined by

$$x_i = \sum_{r=1}^N X_{ir}\eta_r \qquad (9.241)$$

Natural Frequencies of Drillstrings

Natural frequencies of drill strings can be determined by solving the differential equation of motion for the continuous pipe system. The differential equation of motion for axial vibration is

$$AE\frac{\partial^2 u}{\partial x^2} = m\frac{\partial^2 u}{\partial t^2} \qquad (9.242)$$

$$\frac{\partial^2 u}{\partial x^2} = \frac{1}{c^2}\frac{\partial^2 u}{\partial t^2} \qquad (9.243)$$

where

$u(x,t)$ – axial displacement of any point within the pipe
A – cross-section area of pipe
E – modulus of elasticity
m – mass per unit length of pipe
c – acoustic velocity of compression(tension) wave (16 850 ft/s in steel pipe)

This equation applies over both drill pipe and drill collar sections.

$u_1(x, t)$ – drill collar displacements
$u_2(x, t)$ – drill pipe displacements

The mathematical modal considered here is shown in Figure 9.33.
The determination of mode shapes and corresponding natural frequencies proceeds as follows. Boundary conditions are

$$u_1(0,t) = 0 \quad \text{(drill bit)} \qquad (9.244a)$$

$$u_1(L_1,t) = u_2(L_1,t) \quad \text{(interface; pipe/collars)} \qquad (9.244b)$$

$$A_1E\frac{du_1}{dx}(L_1,t) = A_2E\frac{du_2}{dx}(L_1,t) \qquad (9.244c)$$

(interface; pipe/collars)

$$M\frac{\partial^2 u_2}{\partial t^2} + A_2E\frac{\partial u_2}{\partial x} + ku_2 = 0 \quad \text{(drawworks)} \qquad (9.244d)$$

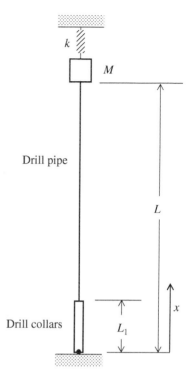

Figure 9.33 Axial vibration models of drillstrings.

The fourth boundary condition is derived by direct application of the second law to the top mass.

Natural frequencies are determined, following the method of separation of variables as applied in Ref. [6] starting with

$$u(x,t) = X(x)\sin \omega t \qquad (9.245)$$

Figure 9.34 shows how natural frequencies change with hole depth for the following set of drillstring parameters. Multiple roots to the characteristic equation give the frequencies of each of the modes of vibration:

Drill pipe size (5 in. OD by 4.276 in. ID)
Drill collar size (7 in. OD by 2.937 in. ID)
Length $(L_1) = 550$ ft
Acoustic velocity = 16 850 ft/s
Spring constant of drawwork cables $(k) = 640\,000$ lb/ft (53 333 lb/in.)
Mass of swivel and traveling block $(M) = 466$ slugs (15 000 lb)
Drillstring length (L) is a variable.

Considering a well depth of 5000 ft, the natural frequency of the fourth mode is 6.13 cps and a critical rotary speed of 122.6 rpm (Table 9.1). The fourth mode was resonant with a roller cone bit at this speed. Note also that the frequency band width of the response is quite large so any of the axial modes can easily be excited.

On the other hand, a rotary speed of 80 rpm would fall between critical speeds of the second and third axial modes so a roller cone bit should operate relatively smooth at this speed.

According to Figure 9.34, the frequency bandwidth of any mode is in the order of $\Delta f \sim 0.1$ cps ($\Delta N \sim 2$ rpm). Since the exciting frequency is related directly to bit rotation, rotary speed has to be maintained within a ± 1.0 rpm variance in order for energy to be fed into the mode. It takes time to put energy into a mode because of drill string length. On the other hand, the frequency band width of drill collar response is much larger, plus the drill collars are much shorter making it easier to excite drill collar modes as demonstrated in shallow test wells. The frequency band width at 20 000 ft is much smaller, and it is more difficult for bit speeds to latch onto these modes.

Figure 9.34 also shows how natural frequency changes with depth. For example, at a rotary speed of 120 rpm (6 cps), mode number changes about every 1500 ft.

Natural frequencies of marine risers and modal responses are discussed in Refs. [7, 8].

Table 9.1 Natural frequencies of axial modes [6].

| Mode | Length – 5000 ft | | Length – 20 000 ft | |
	Natural frequency (cps)	Critical rotary speed[a] (rpm)	Natural frequency (cps)	Critical rotary speed[a] (rpm)
1	1.72	34.4	0.414	8.3
2	3.39	67.8	0.84	16.8
3	4.87	97.4	1.26	25.2
4	6.13	122.6	1.67	33.4
5	7.26	145.2	2.09	41.8
6	8.30	166.0	2.50	50.0
7	9.77	195.4	2.92	58.4

[a] Based on three cycles per bit rotation.
Source: Based on Dareing [6].

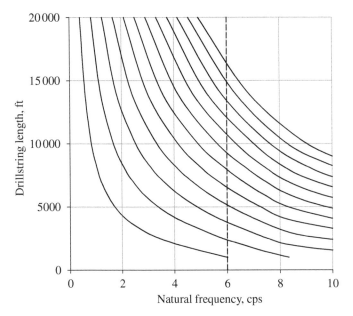

Figure 9.34 Natural frequencies of axial modes (critical rotary speed is 20 times frequency). *Source:* Based on Dareing [6].

References

1 Den Hartog, J.P. (1953). *Mechanical Vibrations*, 4e. New York: McGraw-Hill.
2 Langhaar, H.L. (1962). *Energy Methods in Applied Mechanics*. New York: Wiley.
3 Timoshenko, S., Young, D.H., and Weaver, W. Jr. (1974). *Vibration Problems in Engineering*, 4e. Wiley.
4 Craig, R.R. Jr. (1981). *Structural Dynamics*. New York: Wiley.

5 Thomson, W.T. and Dahleh, M.D. (1998). *Theory of Vibrations with Applications*. New York: Prentice Hall.

6 Dareing, D.W. (2019). *Oilwell Drilling Engineering*. ASME Press.

7 Dareing, D.W. and Huang, T. (1976). Natural frequencies of marine risers. *J. Pet. Technol.* 28: 813–818.

8 Dareing, D.W. and Huang, T. (1979). Marine riser vibration response determined by modal analysis. *ASME J. Energy Resour. Technol.* 101: 159–166.

10

Fluid Mechanics

Two types of fluid flow are turbulent and laminar. Turbulent fluid flow problems require the use of empirical formulas based on experimental data. Laminar flow is more amenable to mathematical predictions, and there are many problems where this theory applies. Two areas of flow of fluids where laminar theory applies are thin films and capillary tubes.

Laminar Flow

Laminar flow predictions depend on rheology of the fluid. A Newtonian fluid is one in which shear stress is proportional to shear rate as expressed by

$$\tau = \mu \frac{du}{dy} \tag{10.1}$$

where the constant, μ, defines the viscosity of the fluid. The rheology of non-Newtonian fluids can often be modeled by Bingham or Power law models [1, 2].

Viscous Pumps

Viscous pumps move fluids between parallel surfaces by simple shear. Fluid flow is induced by one surface moving parallel to the other. The pump is usually working against back pressure, so there is a negative pressure-induced effect to be considered. A simple pump configuration and flow patterns are illustrated in Figure 10.1. In this case

$$\tau = \mu \frac{du}{dy} \approx \mu \frac{U}{h} \tag{10.2}$$

The velocity gradient across a thin film produced by the runner, is $\frac{U}{h}$. The shear stress is constant across the film.

Back pressure also affects the velocity profile. The freebody diagram of a fluid element within the thin film gives

$$\mu \frac{d^2 u}{dy^2} = \frac{dp}{dx}$$

$$u(y) = \frac{1}{2\mu} \frac{dp}{dx} y^2 + C_1 y + C_2$$

Engineering Practice with Oilfield and Drilling Applications, First Edition. Donald W. Dareing.
© 2022 John Wiley & Sons, Inc. Published 2022 by John Wiley & Sons, Inc.

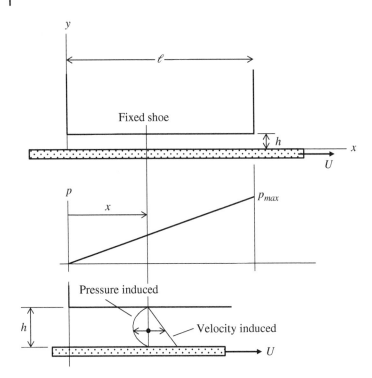

Figure 10.1 Mechanics of viscous pumps.

By integration and applying boundary conditions:

$$u(0) = U \quad \text{and} \quad u(h) = 0$$

gives

$$u(y) = \frac{1}{2\mu}\frac{dp}{dx}(y^2 - hy) - U\left(\frac{y}{h} - 1\right) \tag{10.3}$$

For fluid pumps, film thickness, h, is constant [3]. The overall flow through the film from left to right is the sum of both velocity and pressure-induced flows. Flow through the film (from left to right) is determined by integration of the velocity profile. At any location x

$$Q = b\int_0^h u(y)dy$$

$$Q = \frac{1}{2}Uh - \frac{h^3}{12\mu}\frac{dp}{dx} \quad \text{(flow rate per unit width)} \tag{10.4}$$

It is useful to define

$$Q = \alpha\frac{1}{2}hU \tag{10.5}$$

where "α" is a measure of the effect of back pressure on positive flow. It is also a measure of the volumetric efficiency of the pump.

Noting that

$$\frac{dp}{dx} = \frac{p_{max}}{\ell} \tag{10.6}$$

Equating Eqs. (10.4) and (10.5)

$$\alpha \frac{1}{2} hU = \frac{1}{2} hU - \frac{h^3}{12\mu} \frac{p_{max}}{\ell} \tag{10.7}$$

$$\frac{p_{max}}{\ell} = \frac{6\mu U}{h^2}(1 - \alpha) \tag{10.8}$$

giving

$$\alpha = 1 - \frac{p_{max}}{\ell} \frac{h^2}{6\mu U} \tag{10.9}$$

Again, α is used for the convenience of calculating flow, Q.
When

$$p_{max} = 0, \quad \alpha = 1$$

$$\frac{p_{max}}{\ell} = \frac{6\mu U}{h^2}, \quad \alpha = 0$$

Example   Consider the following operating conditions.

$$p_{max} = 200 \text{ psi}$$

$$\ell = 5 \text{ in.}$$

$$U = 10 \text{ ips}$$
$$\mu = 100 \text{ cp}(100 \times 1.45 \times 10^{-7} = 1.45 \times 10^{-5} \text{ reyn})$$

$$h = 0.002 \text{ in.}$$

Without back pressure, flow rate is 0.01 in.^{3}/sec. With back pressure flow rate delivered by the viscous pump is

$$Q = \frac{1}{2}(10)0.002 - \frac{(0.002)^3}{12(1.45 \times 10^{-5})} \frac{200}{5}$$

$$Q = 0.01 - 18.39 \times 10^{-4} = 0.008\,16 \text{ in.}^3/\text{s}$$

For this case,

$$\alpha = \frac{0.008\,16}{0.01} = 0.816$$

Force to Move Runner
The force, F_r, required to move the runner to the right is

$$F_r = \int_0^\ell \mu \frac{du}{dy}\bigg|_{y=0} dx \tag{10.10}$$

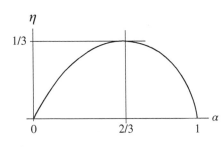

Figure 10.2 Mechanical efficiency vs. flow rate.

$$\left.\frac{du}{dy}\right|_{y=0} = -\frac{1}{2\mu}\frac{dp}{dx}h - \frac{U}{h} \tag{10.11}$$

By Eq. (10.10)

$$F_r = p_{max}\frac{h}{2} + \mu U \frac{\ell}{h}$$

The *mechanical power* to move the runner at velocity, U, is

$$P_M = F_r\,U \tag{10.12}$$

$$P_M = (4 - 3\alpha)\mu U^2 \frac{\ell}{h} \tag{10.13}$$

The *hydraulic power* delivered by the pump is

$$P_H = p_{max}Q \tag{10.14}$$

$$P_H = \mu U^2 \frac{\ell}{h} 3\alpha(1 - \alpha) \tag{10.15}$$

Pump efficiency [3]

$$\eta = \frac{P_H}{P_M} = \frac{p_{max}Q}{P_M} \tag{10.16}$$

$$\eta = \frac{3\alpha(1 - \alpha)}{(4 - 3\alpha)} \tag{10.17}$$

Figure 10.2 shows how mechanical efficiency varies with flow rate indicated by α.

Maximum pump efficiency (33.3%) occurs at $\alpha = \frac{2}{3}$. Recall that α is the volumetric efficiency of the pump, $\frac{Q}{1/2Uh} = \alpha$, with a value of 66.6% at maximum pump efficiency.

For the above example $\alpha = 0.816$ and $\eta = 29\%$, which is less-than optimum pump efficiency of 33.3%.

A useful configuration of a viscous pump is shown in Figure 10.3.

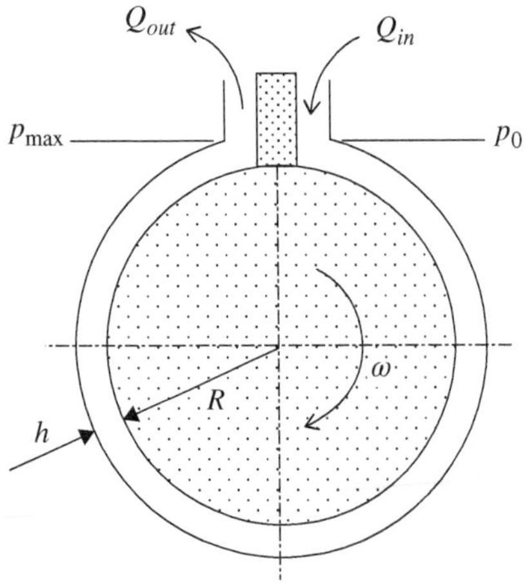

Capillary Tubes

Summing forces on the fluid plug (Figure 10.4) and assuming a Newtonian fluid gives

$$dp\pi r^2 = -2\pi r\mu \frac{du}{dr}dx \tag{10.18}$$

$$\frac{du}{dr} = -\frac{r}{2\mu}\frac{dp}{dx} \tag{10.19}$$

The velocity distribution of fluid moving within pipe follows:

$$u(r) = \frac{1}{4\mu}\frac{dp}{dx}\left(R^2 - r^2\right) \tag{10.20}$$

Figure 10.3 Viscous pump.

which shows that fluid velocity is maximum in the center ($r = 0$) and zero at the inside wall of the pipe ($r = R$). The derivation assumes Newtonian fluid having viscosity of μ. Using this expression for the velocity profile and integrating across the inside pipe area gives the Hagen–Poiseuille equation [4, 5].

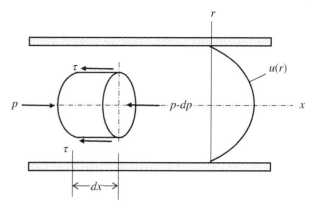

$$Q = \frac{\pi}{8\mu} \frac{dp}{dx} R^4 \qquad (10.21)$$

Rearranging and letting $Q = VA$

Figure 10.4 Freebody diagram of fluid plug inside pipe.

$$\frac{dp}{dx} = \frac{8\mu}{\pi R^4} AV \qquad (10.22)$$

The pressure drop in pipe for laminar flow conditions is

$$\frac{\Delta p}{L} = \frac{32\mu}{D^2} V \qquad (10.23)$$

Flow equations that relate flow rate to pressure gradient for non-Newtonian fluids (Bingham, Power law models) are given in Ref. [1].

Flow Through Noncircular Conduits

The Navier–Stokes equations for flow through noncircular conduits reduces to

$$\frac{\partial^2 w}{\partial x^2} + \frac{\partial^2 w}{\partial y^2} = \frac{1}{\mu} \frac{dp}{dz} \qquad (10.24)$$

This is Poisson's equation which applies to a variety of problems, such as twisting of noncircular shafts [6], squeeze films, flow through porous media, and others. There are many solutions readily available.

Elliptical Conduit

For example, consider flow through an elliptical conduit (Figure 10.5). Some solutions fall into place with ease. For example

$$w(x,y) = \frac{1}{2\mu} \frac{dp}{dz} \frac{a^2 b^2}{(a^2 + b^2)} \left(\frac{x^2}{a^2} + \frac{y^2}{b^2} - 1 \right) \qquad (10.25)$$

satisfies Eq. (10.24) and boundary conditions, $w = 0$, around the outer edges. The flow rate through the elliptical conduit is determined from

$$Q = \int\int w(x,y) dx \, dy$$

While the fluid friction at the conduit walls is found by

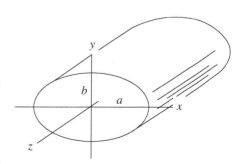

Figure 10.5 Elliptical conduits.

$$\tau = \mu \frac{dw}{dn}$$

where

$$\frac{dw}{dn} = \frac{\partial w}{\partial x}\frac{dx}{dn} + \frac{\partial w}{\partial y}\frac{dy}{dn}$$

$$\frac{dw}{dn} = \frac{\partial w}{\partial x}\cos\phi + \frac{\partial w}{\partial y}\sin\phi$$

$$\tan(90-\phi) = -\frac{dy}{dx}$$

Also, when $a = b$, Eq. (10.25) simplifies to

$$w(r) = \frac{-1}{4\mu}\frac{dp}{dz}\left(R^2 - r^2\right) \tag{10.26}$$

which agrees with Eq. (10.20).

Rectangular Conduit

The velocity distribution in a *rectangular conduit* (Figure 10.6) does not have a direct solution as described above. The solution requires the method of separation of variables. Poisson's equation (Eq. (10.24)) still applies.

Observing the velocity $w(x, y)$ will be symmetrical with respect to the y axis, we assume the solution to be of the form

$$w(x,y) = \sum_{n=1,3,5}^{\infty} \cos\frac{n\pi x}{2a} Y_n(y) \tag{10.27}$$

This function automatically satisfies boundary conditions, $w(a, y) = 0$ and $w(-a, y) = 0$. $Y_n(y)$ defines how $w(x, y)$ varies with y.

The constant, F, on the right side of the Eq. (10.24) is expressed in terms of a Fourier series to factor out the cosine term (see Figure 10.7).

Since this function is even only, the cosine part the series is needed:

$$F(x) = \sum a_n \cos\frac{n\pi x}{2a} \quad n = 1, 3, 5, \ldots \tag{10.28}$$

where

$$a_n = \frac{1}{2a}\int_{-2a}^{2a} F\cos\frac{n\pi x}{2a}dx \tag{10.29}$$

and

$$F = \frac{1}{\mu}\frac{dp}{dz} \tag{10.30}$$

This integration gives

$$a_n = \frac{2F}{2a}\frac{2a}{n\pi}\left[\sin\frac{n\pi x}{2a}\Big|_0^a - \sin\frac{n\pi x}{2a}\Big|_a^{2a}\right]$$

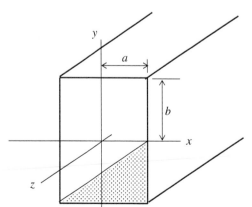

Figure 10.6 Flow through rectangular conduits.

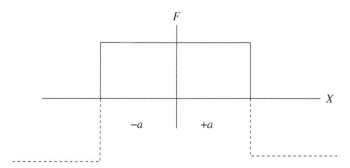

Figure 10.7 Even function.

$$a_n = \frac{4F}{n\pi} \sin \frac{n\pi}{2} \quad n = 1, 3, 5$$

$$a_n = \frac{4F}{n\pi}(-1)^{\frac{n-1}{2}} \quad (10.31)$$

Therefore,

$$F(x) = \sum_{1,3,5}^{\infty} F\frac{4}{n\pi}(-1)^{\frac{n-1}{2}} \cos \frac{n\pi x}{2a} \quad (10.32)$$

Substituting Eqs. (10.27) and (10.28) into Eq. (10.24) gives

$$Y''_n - \frac{n^2\pi^2}{4a^2}Y_n = F\frac{4}{n\pi}(-1)^{\frac{n-1}{2}} \quad (10.33)$$

The solution to this equation is

$$Y_n(y) = A_n \sinh \frac{n\pi y}{2a} + B_n \cosh \frac{n\pi y}{2a} - \frac{16Fa^2}{n^3\pi^3}(-1)^{\frac{n-1}{2}} \quad (10.34)$$

Observing that $w(x, y)$ is symmetrical with respect to the x axis, A_n must be zero. B_n is determined from $w(x, b) = 0$ and $w(x, -b) = 0$.

$$B_n = \frac{1}{\cosh \frac{n\pi b}{2a}} \frac{16Fa^2}{n^3\pi^3}(-1)^{\frac{n-1}{2}}$$

Keep in mind that $F = \frac{1}{\mu}\frac{dp}{dz}$. Bringing everything together

$$w(x, y) = -\frac{16a^2}{\pi^3\mu}\frac{dp}{dz}\sum \frac{1}{n^3}(-1)^{\frac{n-1}{2}}\left[1 - \frac{\cosh \frac{n\pi y}{2a}}{\cosh \frac{n\pi b}{2a}}\right]\cos \frac{n\pi x}{2a} \quad n = 1, 3, 5, \dots \quad (10.35)$$

When $a = b$, the solution gives the velocity distribution across a square conduit.

Unsteady Flow Through Pipe

Consider a situation where a pressure gradient is suddenly applied between entrance and exit of a pipe. Every location within the pipe experiences the same pressure gradient. The fluid starts at rest and time is required to develop the steady-state velocity profile given by Eq. (10.20).

The Navier–Stokes equation [7] for this case, describing start-up flow in the z direction is

$$\rho \frac{\partial v}{\partial t} = -\frac{\partial p}{\partial z} + \mu \frac{1}{r} \frac{\partial}{\partial r}\left(r\frac{\partial v}{\partial r}\right) \tag{10.36}$$

The equation is expressed in terms of polar coordinates. Following the work of Szymanski [8], the solution is expressed as the sum of a steady state apart and an unsteady part.

$$v(r,t) = v_s(r) + v_u(r,t) \tag{10.37}$$

The steady-state solution comes from the solution to

$$\frac{1}{\mu}\frac{dp}{dz} = \frac{1}{r}\frac{d}{dr}\left(r\frac{dv}{dr}\right) \tag{10.38}$$

Using boundary conditions of

$$\left.\frac{dv}{dr}\right|_{r=0} = 0 \quad C_1 = 0$$

$$v(R) = 0 \quad C_2 = -\frac{1}{4\mu}\frac{dp}{dz}R^2$$

Therefore,

$$v_s(r) = -\frac{1}{4\mu}\frac{dp}{dz}\left(R^2 - r^2\right) \tag{10.39}$$

The unsteady part of the solution is determined from

$$\rho \frac{\partial v_u}{\partial t} = \mu \frac{1}{r}\frac{\partial}{\partial r}\left(r\frac{\partial v_u}{\partial r}\right) \tag{10.40}$$

Using separation of variables and letting $v_u(r, t) = R(r)T(t)$, Eq. (10.40) becomes

$$\rho R T' = \mu \frac{1}{R}\frac{d}{dr}\left(r\frac{dR}{dr}\right)T \tag{10.41}$$

Separating the variable and letting $\nu = \frac{\mu}{\rho}$

$$\frac{1}{\nu}\frac{T'}{T} = \frac{\frac{1}{r}\frac{d}{dr}\left(r\frac{dR}{dr}\right)}{R} = \begin{cases} \lambda^2 \\ 0 \\ -\lambda^2 \end{cases} \tag{10.42}$$

Choosing $(-\lambda^2)$

$$T' + \nu\lambda^2 T = 0 \tag{10.43}$$

and

$$\frac{1}{r}\frac{d}{dr}\left(r\frac{dR}{dr}\right) + \lambda^2 R = 0 \tag{10.44}$$

$$r^2\frac{d^2R}{dr^2} + r\frac{dR}{dr} + \lambda^2 r^2 R = 0 \tag{10.45}$$

Let, $x = \lambda r$ and by substitution

$$x^2 \frac{d^2R}{dx^2} + x\frac{dR}{dx} + x^2R = 0 \tag{10.46}$$

The general expression for the Bessel equation is

$$x^2 \frac{d^2R}{dx^2} + x\frac{dR}{dx} + \left(x^2 - n^2\right)R = 0$$

In our case, $n = 0$ and Eq. (10.46) is a Bessel equation of order 0. Its solution is the Bessel function:

$$R(r) = AJ_0(x) \tag{10.47}$$

where $J_0(x)$ is a Bessel function of the first kind, whose expression is [9]

$$J_0(x) = 1 - \frac{x^2}{2^2} + \frac{x^4}{2^4(2!)^2} - \frac{x^6}{2^6(3!)^2} + \cdots(-1)^k \frac{x^{2k}}{2^{2k}(k!)^2} + \cdots \quad (k = 1, 2, 3, \ldots) \tag{10.48}$$

The plot of this function is shown in Figure 10.8.
The solution to Eq. (10.43) is

$$\frac{dT}{dt} + \nu\lambda^2T = 0 \tag{10.49}$$

$$\frac{dT}{T} = -\nu\lambda^2dt \tag{10.50}$$

$$\ln T = -\nu\lambda^2t + C' \tag{10.51}$$

$$T(t) = Ce^{-\nu\lambda^2t} \tag{10.52}$$

Bringing everything together, the total solution becomes

$$u(x, t) = -\frac{1}{4\mu}\frac{dp}{dx}\left(R^2 - r^2\right) + AJ_0(\lambda r)e^{-\nu\lambda^2t} \tag{10.53}$$

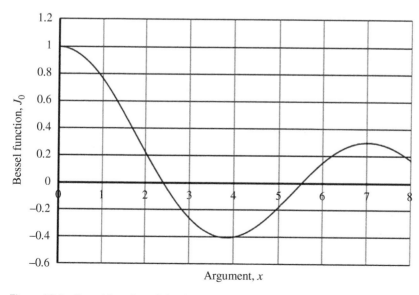

Figure 10.8 Bessel function of the first kind.

Two unknowns in this equation are the arbitrary constant, A, and the eigenvalue, λ. These constants are determined as follows:

$$u_u(R, t) = 0 \quad \text{(velocity is zero at the surface always)}$$

This means that $J_0(\lambda R) = 0$. The first root of this function is 2.4048; therefore, $\lambda R = 2.4048$ or $\lambda = \frac{2.4048}{R}$. The second requirement is the initial condition $u_u(R, 0) = 0$.

$$0 = -\frac{1}{4\mu}\frac{dp}{dx}R^2 + AJ_0(0) \tag{10.54}$$

Since $J_0(0) = 1$

$$A = \frac{1}{4\mu}\frac{dp}{dx}R^2 \tag{10.55}$$

The total solution is

$$u(r, t) = -\frac{1}{4\mu}\frac{dp}{dx}\left[(R^2 - r^2) - R^2 J_0(\lambda r)e^{-\nu\lambda^2 t}\right] \tag{10.56}$$

This equation is put in dimensionless form

$$\frac{u}{u_m} = 1 - \left(\frac{r}{R}\right)^2 - J_0\left(2.4048\frac{r}{R}\right)e^{-(2.4048)^2\tau} \tag{10.57}$$

where $\tau = \frac{\nu}{R^2}t$ and $u_m = -\frac{R^2}{4\mu}\frac{dp}{dx}$.

The velocity profiles shown in Figure 10.9 correspond to the following values of τ (∞, 0.5, 0.3, 0.2, 0.1, 0.05). The velocity profile at $\tau = \infty$, corresponds to the steady-state condition discussed earlier. At $\tau = 0.05$, the velocity profile shows the greatest velocity gradient and shear stress being greatest at $r = R$, the boundary radius, with very little velocity gradient over the center portion.

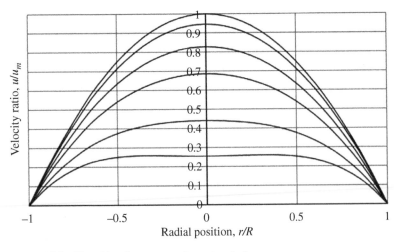

Figure 10.9 Transition from unsteady to steady flow.

Hydraulics of Non-Newtonian Fluids

The hydraulics of oil well drilling is central to the rotary drilling method. A special drilling fluid is pumped down drilling pipe to supply jetting action under drill bits to clean and cool them and then carry the cuttings back to the surface for analysis and disposal. Drilling hydraulics is designed to optimize hydraulic horsepower under the drill bit for best cleaning and drilling. Drilling cost in terms of feet drilled per hour (rate of penetration, ROP).

Drilling hydraulics offers several aspects of fluid flow including non-Newtonian fluid flow, optimizing hydraulics for maximum available hydraulic horsepower at the end of several thousand feet of pipe, and the application of downhole drilling motors and turbines [2].

Hydraulics of Drilling Fluids

Drilling fluids, called drilling mud, are specially designed for variable density and gel strength properties. Mud density is needed to provide sufficient downhole pressure to control formation fluid; barite is added for this purpose. Bentonite is added to give gel strength for suspending cutting produced by drill bits. These additives produce non-Newtonian fluids whose rheology are typically describes by the Bingham Plastic model or the Power Law model. Much work has been done over the years to relate the drilling variables into a mathematical model. It is essential to have the technology to predict fluid losses, which occur mainly inside drill pipe, which can be several thousand feet long.

Pressure Loss Inside Drill Pipe

The equation for predicting pressure losses inside drill pipe (or any tubular) is [10, 11] becomes

$$\frac{\Delta p}{L} = 7.7(10^{-5}) \frac{PV^{0.2} \gamma^{0.8} Q^{1.8}}{D^{4.8}} \qquad (10.58)$$

where

D	–	ID of pipe, in.
PV	–	plastic viscosity, cp
γ	–	mud weight, ppg
Q	–	flow rate, gpm
L	–	pipe length, ft
Δp	–	pressure drop, psi

This equation is commonly used within the petroleum industry.

Example   Assume the following:

Mud weight – 11 ppg
Plastic viscosity – 10 cp
Drill pipe size – 4½ in. (16.6 lb/ft) (ID = 3.826 in.)
Flow rate, Q – 400 gpm

Substituting directly into Eq. (10.58) gives

$$\frac{\Delta p}{L} = 0.0634 \text{ psi/ft}$$

Pressure loss over 10 000 ft of pipe is predicted to be 634 psi.

Pressure Loss in Annulus

Pressure loss in the annulus is predicted by [10, 11]

$$\frac{\Delta p}{L} = 7.7(10^{-5}) \frac{PV^{0.2}\gamma^{0.8}Q^{1.8}}{(D_h - D_p)^3 (D_h + D_p)^{1.8}} \tag{10.59}$$

Oil Well Drilling Pumps

Mud pumps have two distinct parts: (i) fluid end and (ii) power end (Figure 10.10). The fluid end contains fluid manifold, intake and exhaust valves, pistons, and liner. The power end is a mechanical drive containing a slider crank mechanism, which converts rotary motion and torque to linear piston motion and force.

Mud pumps are positive displacement pumps. Volume output is directly related to the number of strokes of the crank. Output capacity is normally expressed in terms of volume per stroke. Volume output depends on liner size and stroke length. Actual flow rate output depends on volumetric efficiency of the pump and suction intake.

Fluid ends are limited by fluid pressure in the manifold. Triplex pumps are limited to a maximum discharge pressure of approximately 5500 psi. Higher manifold pressures can cause fatigue cracks and premature fatigue failure.

Power ends are limited by force transmitted from the piston through the slider crank mechanism. Bearings usually limit the maximum allowable force on the piston, and this force depends on fluid pressure and piston area. Allowable fluid pressure in large liners is low (high volume). Allowable pressure in small liners is typically high (small volume).

While pressure and volume output change with liner size, power input remains the same. Hydraulic horsepower is related mathematically to pressure and flow rate by

$$\text{Power} = pQ$$

which is adjusted to horsepower units by

$$HHP = pQ \left| \frac{144\ \text{in.}^2}{1\ \text{ft}^2} \right| \left| \frac{1\ \text{ft}^3}{7.48\ \text{gal}} \right| \left| \frac{1\ \text{min}}{60\ \text{s}} \right| \left| \frac{1\ \text{hp}}{550\ \text{ft-lb/s}} \right| \text{hp}$$

$$HHP = \frac{pQ}{1714}\ \text{hp}$$

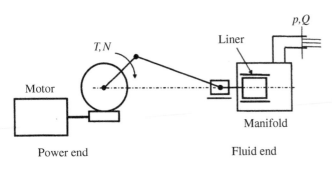

Figure 10.10 Schematic of mud pumps.

where

HHP	–	hydraulic horsepower, hp.
p	–	fluid pressure, psi
Q	–	volume flow rate, gpm

Performance of mud pumps is measured in terms of hydraulic horsepower output. Mud pumps produce maximum hydraulic horsepower with different combinations of output pressure and flow rate (Table 10.1). The constant horsepower curve is shown in Figure 10.11.

The rectangle that touches the horsepower curve represents pressure – flow rate capability the pump. The horizontal line indicates the maximum operating pressure set by the manufacturer. Drilling contractors usually operate below the maximum pressure. Flow rates listed in Table 10.1 are limits for each of the liner sizes.

Table 10.1 Typical triplex mud pump performance features.

Liner (in.)	p_{max} (psi)	F_{max} (lb)	Flow rate (gpm)	HHP (hp)
5½	5558	132 048	481	1560
6	4665	131 879	573	1560
6½	3981	132 090	681	1571
7	3423	131 717	778	1554
7½	2988	132 010	894	1559

Figure 10.11 Pump performance vs. liner size.

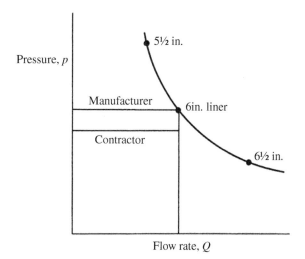

Drilling Hydraulics

The purpose of a hydraulic analysis is to establish available hydraulic horsepower at the bottom of a drillstring to make best use of what is available to drive motors (or turbines), clean drill bits, and carry cutting out of the well bore.

Example   Consider the follow operating conditions [2].

Hole size – 9⅞in.

Hole depth – 8500 ft
Mud weight – 11 ppg
Plastic viscosity – 10 cp
Drill pipe size – 4½ in. (16.6 lb/ft); ID = 3.826 in.
Drill pipe length – 7900 ft
Drill collar size – 8 × 3 in.
Drill collar length – 600 ft
Pump data:
  6-in. liner
  12-in. stroke
  Liner pressure rating is 4670 psi
  Operating pressure limit is 3000 psi

Surface equipment – Type 4
Step # 1 Calculate parasitic pressure losses for a flow rate of 400 gpm.
Step # 2 Calculate parasitic pressure losses for a range of flow rates (Figure 10.12).
Step # 3 Calculate and plot the hydraulic horsepower loss curve for each flow rate (Table 10.2).
Step # 4 Superimpose the three hydraulic horsepower curves to obtain the available power curve (Figure 10.13).

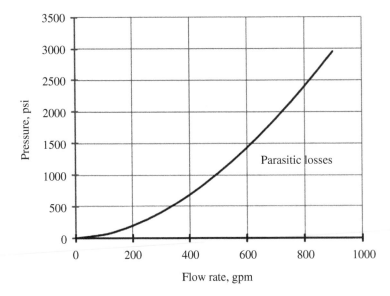

Figure 10.12   Effect of flow rate on parasitic losses.

Table 10.2 Pressure losses and available hydraulic horsepower.

Flow rate, Q (gpm)	p, pipe ID (psi)	p, collar ID (psi)	p, pipe annulus (psi)	p, collar annulus (psi)	Surface equipment (psi)	System losses (psi)	Pump HHP	System HHP	Available HHP
100	42	10	1.4	1.7	1.7	57	175	3	172
150	87	21	2.9	3.5	3.4	117	263	10	252
200	145	35	4.8	5.8	5.7	197	350	23	327
250	217	53	7.2	8.7	8.6	294	438	43	395
300	301	74	10.0	12.1	11.9	409	525	72	454
350	398	97	13.2	16.0	15.7	540	613	110	502
400	506	123	16.8	20.3	20.0	686	700	160	540
450	625	153	20.8	25.1	24.7	848	788	223	565
500	755	184	25.2	30.4	29.9	1025	875	299	576
550	897	219	29.9	36.1	35.5	1217	963	391	572
600	1049	256	34.9	42.2	41.5	1424	1050	498	552
650	1211	296	40.3	48.7	47.9	1644	1138	624	514
700	1384	338	46.1	55.7	54.8	1879	1225	767	458
750	1567	383	52.2	63.1	62.0	2127	1313	931	382
800	1760	430	58.6	70.8	69.7	2389	1400	1115	285
850	1963	479	65.4	79.0	77.7	2665	1488	1321	166
900	2176	531	72.5	87.6	86.1	2953	1575	1551	24
950	2399	585	79.9	96.5	94.9	3255	1663	1804	−142
1000	2631	642	87.6	105.9	104.1	3570	1750	2083	−333

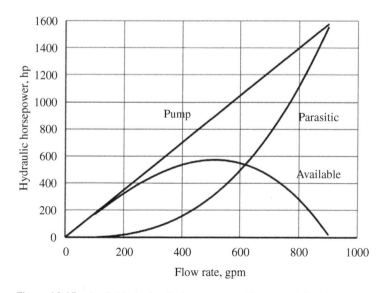

Figure 10.13 Available hydraulic horsepower at bottom of drill string.

Note the difference between the maximum operating pump pressure (3000 psi) and system losses at a given flow rate represents the pressure available at the bottom of the drill bit.

Pressure drop predictions listed in Table 10.2 were obtained using Eqs. (10.58) and (10.59).

The straight line shows how the hydraulic horsepower supplied by the pump varies with flow rate. This line is based on the maximum operating pump pressure of 3000 psi. For example, the fluid power supplied by the pump at 400 gpm is

$$HHP_P = \frac{3000(400)}{1714} = 700 \text{ hp}$$

while the fluid power consumed by friction at 400 gpm is

$$HHP_f = \frac{686(400)}{1714} = 160 \text{ hp}$$

which agrees with Table 10.2.

Note that the flow rate limit of one pump is 573 gpm, which is slightly higher than the optimum flow rate.

The example shows that the maximum available hydraulic power of 590.85 hp is achieved at a flow rate of 525 gpm (see Figure 10.13). The system pressure loss at 525 gpm is 1071 psi, which is 1/2.8 of the maximum allowable operating pressure of 3000 psi. This also means that 589 HHP (1929 psi at 525 gpm) is available for motor, turbine, or drill bit.

The annular velocity between drill pipe and well bore is determined from

$$V_{an} = \frac{Q}{\frac{\pi}{4}\left(D_h^2 - D_{dp}^2\right)} \left|\frac{144}{7.48}\right|$$

$$V_{an} = 24.51 \frac{Q}{\left(D_h^2 - D_{dp}^2\right)} \tag{10.60}$$

where

Q	–	flow rate, gpm
D_h	–	well bore diameter, in.
D_{dp}	–	OD of drill pipe, in.

Both HHP_a and V_{an} are shown in a composite diagram (Figure 10.14). The selection of the best flow rate is a compromise between highest HHP_a and annular velocity requirements.

Power Demands of Downhole Motors

Analytical tools are useful for configuring a motor for a given set of operational specifications. However, after a tool is built its true performance is determined by laboratory testing. Test results establish how a given motor will perform in the field. A typical test arrangement is shown schematically in Figure 10.15.

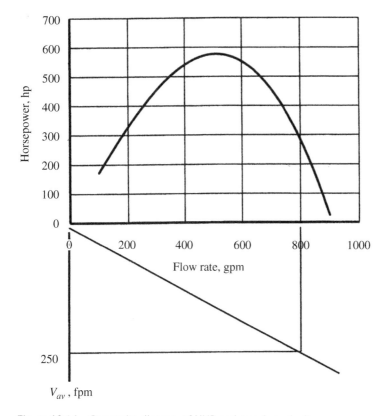

Figure 10.14 Composite diagram of HHP and annular velocity.

Fluid, usually water, is pumped through the tool and each of the above parameters are measured. The dynamometer generates torque on the drive shaft and sensors measures each parameter for several torque settings. Performance tests are made for each size and model.

Performance of Positive Displacement Motors (PDM)

Typical performance data for positive displacement motors (PDMs) is shown in Figure 10.16. Pressure differentials for various dynamometer torque settings are recorded at constant flow rates. Applied torque and rotational speeds are recorded. When torque is zero, rotational speed is a maximum. Pressure differential is required to overcome internal friction but is lowest under no torque conditions. When torque is increased, rotational speed drops off slightly because of fluid leakage through each stage. Higher torque requires higher differential pressures and causes a reduction in rotational speed due to leakage.

When applied torque is taken to the extreme, PDMs stall out. This means that the pressure level is great enough to force fluid through motors without rotation. Fluid leakage is 100%. Even though torque is a maximum under stall conditions, power output is zero because the output shaft does not turn. Therefore, operating a PDM near stall conditions produces little or no power to the drill bit.

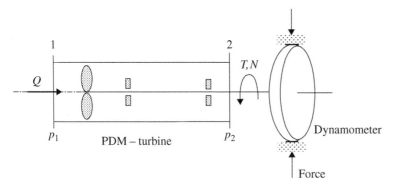

Figure 10.15 Performance testing parameters.

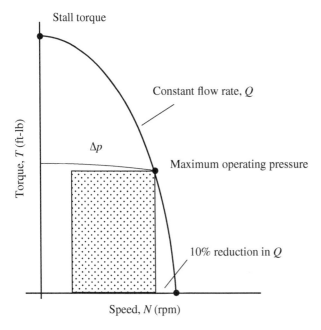

Figure 10.16 Typical performance diagram for Moineau type motors.

PDMs are normally rated at torque and pressure levels corresponding to about a 10% leakage through the motor.

PDMs operate between recommended maximum and minimum flow rates. The maximum flow rate is based on a maximum allowable whirl velocity of the rotor. The rotor is constantly undergoing accelerations which generate forces on rubber undulations in stators. This could cause premature failure of the rubber. Minimum flow rate is based on inefficiency, which is typical at low flow rates. Based on these guidelines, laboratory performance data can be approximated as shown in

Figure 10.17. The numbers are typical of a 1 : 2 PDM. The performance of multiple lobed PDMs deviates somewhat from this model. Performance data for a variety of motor sizes are given in Table 10.3.

It is important to note that maximum recommended operating torque and pressure levels are not stall conditions. Stall torque is approximately 50% higher than recommended operating torque.

The capability to monitor performance of PDMs from the surface is a distinct advantage. Pressure drop across these motors can be monitored by standpipe pressure and output speed can be monitored by pump flow rate.

Figure 10.17 Performance diagrams for a 6¾ Moineau motor.

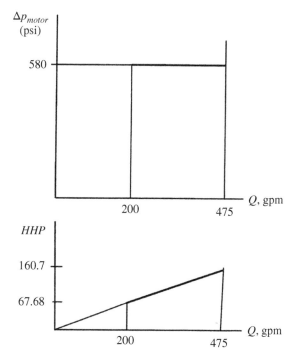

Table 10.3 Positive displacement motor performance data.

Tool size	Pump rate (gpm) (min)	Pump rate (gpm) (max)	Bit speed (rpm) (min)	Bit speed (rpm) (max)	Max diff. (psi)	Torque (ft-lb)	Power range (HP)	HHP
3¾	75	85	280	700	580	385	20–51	
4¾	100	240	245	600	580	585	27–67	
6¾	200	475	205	485	580	1500	59–138	(67.7–160.7)
8	245	635	145	380	465	2090	59–152	
9½	395	740	195	365	695	3890	145–271	

PDMs are a spin-off from the Moineau pump [12]. This pump, sometimes called a progressive cavity pump, is still in use today for transporting: food, oil, sewage, and viscous chemicals. Fluid is trapped within cavities (characteristic of the Moineau pump), which progress axially from the intake to the outlet end. Pressure within each cavity may vary because of leakage between cavities

PDMs have a direct relationship between output speed and through-put flow rate. This characteristic makes it possible to monitor output speed by monitoring flow rate:

$$N = CQ \tag{10.61}$$

where N is rotational speed (rpm), Q is flow rate through the motor (gpm), and C is a constant depending on geometry of rotor and stator. The proportionality constant, C, is determined from maximum rotational speed and corresponding flow rate data from the manufacture's performance information:

$$C = \frac{N}{Q} = \frac{N_{max}}{Q_{max}} \tag{10.62}$$

For the 6¾ motor performance data given in Table 10.3,

$$C = \frac{485}{475} = 1.02$$

Rotational speed and flow rate are related by

$$N = 1.02Q$$

The motor's rotational speed, therefore, can be monitored by the flow rate.

Application of Drilling Turbines

The successful application of downhole turbines depends on the amount of hydraulic horsepower available to drive a turbine. It is therefore important to establish the amount of available hydraulic horsepower at the bottom of the drillstring. The available hydraulic horsepower varies with flow rate. Hydraulic horsepower consumed by turbines also varies with flow rate and increases with the third power of flow rate.

For example, consider the turbine having performance data as listed in Table 10.4. The amount of hydraulic horsepower consumed by this turbine depends on both flow rate, Q, and pressure drop, Δp, according to

Table 10.4 Hydraulic horsepower consumed 6¾ in. turbine (12 ppg) (HHP is developed from Eq. (10.64)).

Flow rate (gpm)	Pressure drop (psi)	Hydraulic horsepower (hp)
250	621	90.6
300	894	156.5
350	1217	248.5
400	1589	370.8
450	2011	527.9
500	2483	724.3

$$HHP = \frac{Q\Delta p}{1714} \qquad (10.63)$$

Since pressure drop across turbines is independent of the amount of mechanical power delivered, turbines consume the calculated hydraulic horsepower level, whether the drill bit is off or on bottom. Mechanical output power varies considerably with bit torque, however.

When this type of data is plotted, the power consumption curve for turbines can be compared with the available hydraulic horsepower to see if there is enough power to drive the turbine. To work turbines to the fullest, turbines should be run at high flow rates. This usually means that all the available horsepower is consumed by turbines, and there is little or no hydraulic horsepower left to clean the drill bit. In some cases, open flow diamond bits can be cleaned and cooled by high flow rates without regard to HHP when used with turbines.

Hydraulic Demands of Drilling Motors – Turbines

Typical PDM power consumption is shown in Figure 10.18 along with the available HHP curve to show how much power is consumed by a PDM. The difference between to two curves represents the HHP available for bit cleaning. This difference is the basis for selecting nozzle size for best bottom-hole cleaning. This type of diagram is useful for determining best flow rate, too. Flow rate selection also should consider annular velocity requirements.

Successful application of downhole turbines depends on the amount of hydraulic horsepower available at the bottom of the drillstring. Hydraulic analysis of the total circulating system is critical. Figure 10.19 shows an available hydraulic power curve and turbine power consumption curve, assuming the turbine operates at maximum power output. Note that the power consumption curve is basically different from the PDM power curve.

Turbines usually operate at high flow rates to make best use of the turbine's power capability. This usually means that all the available hydraulic horsepower is consumed by turbines, and there is little or no hydraulic horsepower left for bit cleaning. This condition is represented by the intersection of the turbine power with the available power curve. For this reason, open drill bits (no nozzles) are often used with turbines. In this case, bit cleaning and cooling are accomplished by high flow rates.

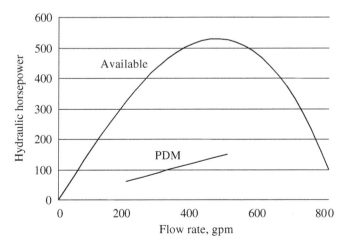

Figure 10.18 Typical PDM power consumption.

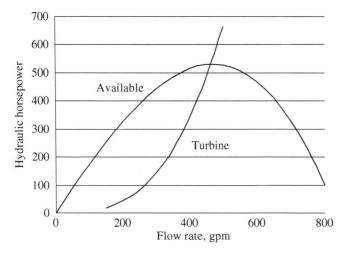

Figure 10.19 Power consumption of drilling turbine.

Fluid Flow Around Vibrating Micro Cantilevers

Measurements of fluid viscosity have enormous applications in areas ranging from clinical diagnostics to industrial and engineering applications. Since determination of absolute viscosities is a challenging problem, most applications exploit relative variation in viscosities. Viscosity measurements, quite often, require bulky experimental apparatus and relatively large volumes of samples. Measuring viscosities of very small volume samples using miniature devices are very attractive. Although resonance responses of cantilevers, the size of a meter, is insensitive to viscosity of surrounding medium, fluids significantly influence the resonance response of a microfabricated cantilever.

Consider the Navier–Stokes equations for flow through a control volume located above a cantilever surface (Figure 10.20). The solution is given in closed form and links viscosity directly to the damping factor, which can easily be extracted from experimental frequency response data [13].

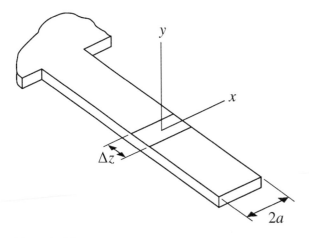

Figure 10.20 Microcantilever with coordinate system.

Mathematical Model

Key assumptions made in the mathematical development of fluid flow adjacent to microcantilevers surfaces are

1) Fluid flow is laminar.
2) Inertia forces within the flow pattern are small in comparison with viscous forces and are therefore not included in the hydrodynamic analysis. This is consistent with Reynolds basic equation of lubrication and squeeze film vibration damping theory. The mass of the fluid on and near the surfaces of microcantilevers does, however, affect the natural frequency of cantilevers because of the small dimensions and mass density of microcantilevers. The effective mass is best determined experimentally. It is a separate determination and not an integral factor in our hydrodynamic analysis.
3) Fluid flow pattern depends only on the instantaneous velocity of the cantilever surface. This is consistent with assumption (2). This assumption allows viscous forces and viscosity to be related to surface velocity, damping coefficients, and damping factors in a simple and easy to use formula.
4) Actual hydrodynamic fluid flow near a cantilever surface is approximated with flow within a rectangular control volume.
5) Fluid flow is two-dimensional. There is no fluid flow along the length of the cantilever.

The coordinate system used in the development is illustrated in Figure 10.21. A typical surface section of length, Δz, and width, $2a$ is taken as the base of the control volume. The pressure function, $p(x, y)$, and velocity function, $u(x, y)$ vary with x and y and are assumed constant over the unit distance, Δz.

The predicted damping coefficient, c, will have units of force per velocity per unit length of the microcantilever. The control volume under consideration is defined by a rectangle have dimensions of $(2a \times h \times \Delta z)$. Fluid flows into and out of this control volume to maintain continuity of flow. Fluid is assumed to be incompressible.

The simplified Navier–Stokes equations governing fluid flow within the control volume are

$$\frac{\partial p}{\partial x} = \mu \left(\frac{\partial^2 u}{\partial x^2} + \frac{\partial^2 u}{2y^2} \right) \tag{10.64}$$

$$\frac{\partial p}{\partial y} = \mu \left(\frac{\partial^2 v}{\partial x^2} + \frac{\partial^2 v}{2y^2} \right) \tag{10.65}$$

Figure 10.21   Cross section of micro cantilever with control volume.

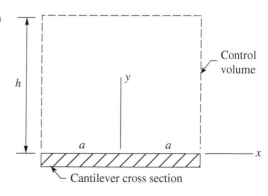

These equations, plus the continuity equation,

$$\frac{\partial u}{\partial x} + \frac{\partial v}{\partial y} = 0 \qquad (10.66)$$

are satisfied along with appropriate boundary conditions, which are assumed to be

$$p(a,y) = 0$$
$$p(-a,y) = 0$$
$$p(x,h) = 0$$

$$p(x,0) = p_s \cos \frac{\pi x}{2a} \qquad (p_s \text{ is stagnation pressure})$$

Velocity related boundary conditions are

$$u(x,h) = 0$$
$$u(x,0) = 0$$

$$u(0,y) = 0$$

These boundary conditions approximate the actual flow pattern above and around the top surface. They give reasonable results as will be shown.

Fluid Pressure Formulation

The simplified Navier–Stokes equations and continuity equation can be combined to produce the Laplace equation with pressure, $p(x, y)$, as the dependent variable:

$$\frac{\partial^2 p}{\partial x^2} + \frac{\partial^2 p}{\partial y^2} = 0 \qquad (10.67)$$

Using the separation of variables technique and applying the first two pressure boundary conditions, gives

$$p(x,y) = \cos \frac{\pi x}{2a} \left[C \cosh \frac{\pi y}{2a} + D \sinh \frac{\pi y}{2a} \right] \qquad (10.68)$$

The remaining two boundary conditions at $y = 0$ and $y = h$ yield

$$p(x,y) = p_s \cos \frac{\pi x}{2a} \left[\cosh \frac{\pi y}{2a} - \left(\frac{\cosh \dfrac{\pi h}{2a}}{\sinh \dfrac{\pi h}{2a}} \right) \sinh \frac{\pi y}{2a} \right] \qquad (10.69)$$

The shape of the predicted pressure profile depends on the ratio, h/a. By trying different values of h/a, the pressure slope at $y = h$ becomes essentially zero when $h/a = 5$. This means that the pressure above the cantilever's surface blends into the atmospheric pressure at this height ($h = 5a$) above the cantilever. For example, if the width of a microcantilever is 30 μm, then film pressure extends up to a height of about 5×15 or 75 μm with significant pressure occurring out to about $y = \frac{1}{2}h$ or 37.5 μm (see Figure 10.22).

Fluid Velocity Formulation

When the pressure function is applied to the x component of the Navier–Stokes equations, we obtain

$$\frac{\partial^2 u}{\partial x^2} + \frac{\partial^2 u}{\partial y^2} = -\frac{p_s}{\mu}\left(\frac{\pi}{2a}\right)\sin\frac{\pi x}{2a}\left[\cosh\frac{\pi y}{2a} - D\sinh\frac{\pi y}{2a}\right] \qquad (10.70)$$

where

$$D = \frac{\cosh\frac{\pi h}{2a}}{\sinh\frac{\pi h}{2a}}$$

To solve Eq. (10.70), we start by assuming

$$u(x,y) = \sin\frac{\pi x}{2a}g(y) \qquad (10.71)$$

The substitution of Eq. (10.71) into Eq. (10.70) produces the following ordinary differential equation:

$$\frac{d^2 g}{dy^2} - \left(\frac{\pi}{2a}\right)^2 g = -\frac{p_s}{\mu}\frac{\pi}{2a}\left[\cosh\frac{\pi y}{2a} - D\sinh\frac{\pi y}{2a}\right] \qquad (10.72)$$

Through trial and error, we found that the following function satisfies Eq. (10.72) and is therefore a particular solution:

$$g_p(y) = -\frac{1}{2}\frac{p_s}{\mu}y\left[\sinh\frac{\pi y}{2a} - D\cosh\frac{\pi y}{2a}\right] \qquad (10.73)$$

The complimentary solution to

$$\frac{d^2 g}{dy^2} - \left(\frac{\pi}{2a}\right)^2 g = 0 \qquad (10.74)$$

is

$$g_c(y) = E\cosh\frac{\pi y}{2a} + F\sinh\frac{\pi y}{2a} \qquad (10.75)$$

The total solution to Eq. (10.72) is the sum of g_p and g_c or

$$g(y) = E\cosh\frac{\pi y}{2a} + F\sinh\frac{\pi y}{2a} - \frac{1}{2}\frac{p_s}{\mu}y\left[\sinh\frac{\pi y}{2a} - D\cosh\frac{\pi y}{2a}\right] \qquad (10.76)$$

The constants of integration, E and F, are determined by applying the following boundary conditions:

$$g(0) = 0$$

$$g(h) = 0$$

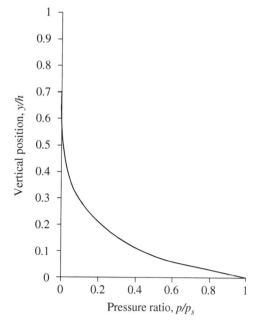

Figure 10.22 Pressure profile at $x = 0$ ($h/a = 5$).

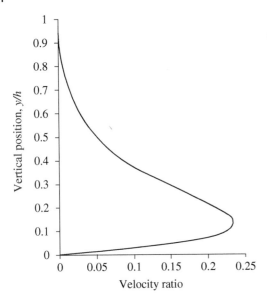

Figure 10.23 Velocity profile at $x = a$ ($h/a = 5$).

The first boundary condition gives, $E = 0$. The second boundary condition gives,

$$F = \frac{1}{2}\frac{P_s}{\mu}h(1 - D^2)$$

Applying these results to Eqs. (10.76) and (10.71) gives

$$u(x,y) = \frac{1}{2}\frac{P_s}{\mu}\sin\frac{\pi x}{2a}\left\{h(1 - D^2)\sinh\frac{\pi y}{2a}\right.$$

$$\left. -y\left[\sinh\frac{\pi y}{2a} - D\cosh\frac{\pi y}{2a}\right]\right\} \quad (10.77)$$

This equation can be formatted in terms of useful dimensionless groupings as follows:

$$u(x,y) = \frac{1}{2}\frac{P_s a}{\mu}\sin\frac{\pi x}{2a}\left\{\frac{h}{a}(1 - D^2)\sinh\frac{\pi}{2}\frac{h}{a}\frac{y}{h}\right.$$

$$\left. -\frac{h}{a}\frac{y}{h}\left[\sinh\frac{\pi}{2}\frac{h}{a}\frac{y}{h} - D\cosh\frac{\pi}{2}\frac{h}{a}\frac{y}{h}\right]\right\}$$

$$(10.78)$$

Equation (10.78) is plotted in terms of dimensionless parameters (see Figure 10.23), where the velocity ratio is defined by

$$\text{velocity ratio} = \frac{2\mu u}{P_s a} \quad (10.79)$$

This graph shows how the x component of the velocity varies from the cantilever surface ($y = 0$) to the top edge of the control volume ($y = h$) along the line, $x = a$. A value of $h/a = 5$ was also chosen for this plot, which shows the velocity gradient at $y = h$ to be essentially zero as expected. Also, we note that the maximum horizontal velocity occurs at y/h equal to 0.125 or ⅛th of h. If, for example, the cantilever width is 50 µm, the maximum velocity occurs at $0.125(5 \times 25) = 16.625$ µm above the cantilever surface.

References

1 Dareing, D.W. and Dayton, R.D. (1992). Non-Newtonian behavior of power lubricants mixed with ethylene glycol. *STLE Tribol. Trans.* 35 (1): 114–120.

2 Dareing, D.W. (2019). *Oilwell Drilling Engineering.* ASME Press.

3 Cameron, A. (1966). *The Principles of Lubrication.* Longmans Green and Co., Ltd.

4 Hagen, G. (1839). On the motion of water in narrow cylindrical tubes. *Poggendorff's Ann. Phys. Chem.* 46: 423.

5 Poiseuille, J.L.M. (1843). Recherches Experimentales sur le Mouvement des Liquides dans les Tubes de Tres Petits Diametres. *Acad. Sci. Savants Etrangers* 9.

6 Timoshenko, S. and Goodier, J.N. (1951). *Theory of Elasticity,* 2e. New York: McGraw-Hill.

7 Schlichting, H. (1960). *Boundary Layer Theory.* McGraw-Hill Book Company.

8 Szymanski, F. (1932). Quelques solutions exactes des equations de l'hydrodynamicque de fluide visqueux dans le cas d'um tube cylindrique. *J. de math. pures et appliquees, Series 9* 11: 67; see also *Intern. Congr. Appl. Mech. Stockholm* **I**, 249 (1930).

9 Reddick, H.W. and Miller, F.H. (1956). *Advanced Mathematics for Engineers*, 3e. NY: Wiley.

10 Moore, P.L. (1974). *Drilling Practice Manual*. Tulsa, Oklahoma: Penn Well Books.

11 Kendall, H.A. and Goins, W.C. (1960). Design and operation of jet-bit programs for maximum horsepower, maximum impact force and maximum jet velocity. *Trans. AIME* 219.

12 Moineau, R.J.L. (1932). Gear mechanism. US Patent 1,892,217, 27 December.

13 Dareing, D.W., Yi, D., and Thundat, T. (2007). Vibration response of microcantilevers bounded by confined fluid. *Ultramicroscopy* 107: 1105–1110.

11

Energy Methods

There are two approaches to engineering mechanics: (i) Newtonian mechanics and (ii) energy methods. Both approaches lead to the same results; however, in some cases, energy methods can give a more direct solution. Earlier chapters focused on Newtonian mechanics. Newtonian mechanics requires the use of force and acceleration vectors. Energy methods requires only the tracking of energy, a scalar quantity.

Principle of Minimum Potential Energy

This principle states that: "A motionless conservative system is in stable equilibrium if its potential energy is a relative minimum" [1]. A conservative system is one in which the change in potential energy is independent of the path taken in going from on configuration to another.

Total potential energy is defined as

$$V = U + \Omega \tag{11.1}$$

where

V	–	total potential energy
U	–	potential of internal forces (elastic energy)
Ω	–	potential of external forces (applied forces)

Both potentials can be viewed as work one performs to achieve the virtual displacement.

The concept of this principle is illustrated in Figure 11.1, which depicts a bead on a curved wire in three different positions. In each of the three positions, intuitively we note the bead is in equilibrium. However, only the first position represents stable equilibrium. In this case, $V(x)$, is a function of one independent variable; the system has one degree of freedom.

Stable and Unstable Equilibrium

Let us examine the change in potential about $x = a$ when the bead is given a virtual displacement, h. Expanding $V(x)$ in a Taylor Series about point "a" gives

$$V(a + h) = V(a) + hV'(a) + \frac{h^2}{2!}V''(a) + \frac{h^3}{3!}V'''(a) + \cdots \tag{11.2}$$

Engineering Practice with Oilfield and Drilling Applications, First Edition. Donald W. Dareing.
© 2022 John Wiley & Sons, Inc. Published 2022 by John Wiley & Sons, Inc.

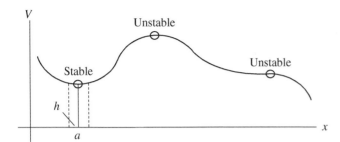

Figure 11.1 Bead on a curved wire.

The change in potential energy, ΔV, due to the virtual displacement, h, is

$$\Delta V = V(a + h) - V(a)$$

$$\Delta V = hV'(a) + \frac{h^2}{2!}V''(a) + \frac{h^3}{3!}V'''(a) + \cdots + \frac{h^{m-1}}{(m-1)!}V^{m-1}(a) + \tag{11.3}$$

which may be written

$$\Delta V = \delta V + \frac{1}{2!}\delta^2 V + \cdots + \frac{1}{(m-1)!}\delta^{(m-1)}V + \cdots \tag{11.4}$$

where δ^k is the kth variation of the total potential energy, V.

$$\delta^k V = h^k V^k(a)$$

If h is sufficiently small, ΔV has the same sign as the first nonzero term in the series. If the value of V is a relative minimum, the first nonzero term is an even term, because an odd term reverses its sign when h is reversed. Therefore, for V to have relative minimum at $x = a$.

1) The first nonzero variation $\delta^k V$ is of even order
2) $\delta^k V \succ 0$, i.e. $V^k(a) \succ 0$

Condition (1) implies that $\delta V = 0$. In other words, for the illustration (Figure 11.1)

$$\text{First location} - \frac{dV}{dx} = 0 \quad \text{and} \quad \frac{d^2 V}{dx^2} \succ 0 \quad \text{(stable equilibrium)}$$

$$\text{Second location} - \frac{dV}{dx} = 0 \quad \text{and} \quad \frac{d^2 V}{dx^2} < 0 \quad \text{(unstable equilibrium)}$$

$$\text{Third location} - \frac{dV}{dx} = 0 \quad \text{and} \quad \frac{d^2 V}{dx^2} = 0 \quad \text{(unstable equilibrium)}$$

Stability of Floating Objects

Consider a wood block floating in water (Figure 11.2). Since the density of wood is less than the density of water, part of the block is above the water line and part below. The center of gravity

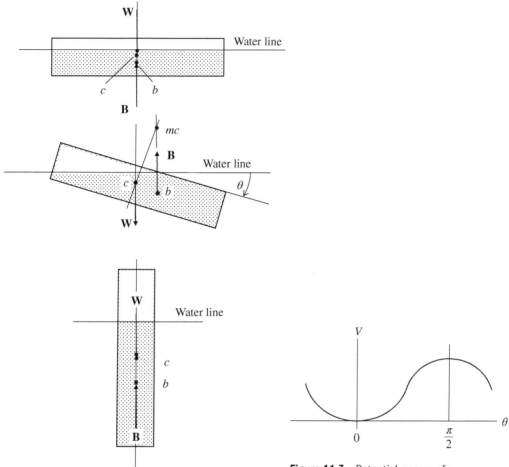

Figure 11.2 Stabilizing moment on a floating block.

Figure 11.3 Potential energy of a floating block.

of the block is indicated by point "*c*," while the center of buoyancy is indicated by point "*b*." From experience and intuition, we expect the block to float horizontally as shown. Any effort to rotate the block brings about a corrective action that tends to put the block horizontal again. What causes this to happen is the shifting of vector **B** away from vector **W**, which brings about a corrective couple (torque) as shown.

Another parameter shown in the rotated figure is the metacenter, "*mc*." The distance between *c* and *mc* is the metacentric height. The metacentric height varies with rotation angle, but over small angles of θ its change is small and often assumed constant. The height is a measure of how fast the block (or ship) will return to its horizontal position. Floating objects with long metacentric heights return faster.

The potential energy of this floating object is illustrated in Figure 11.3. When the block is flat, it is stable. When the block is vertical, it is unstable.

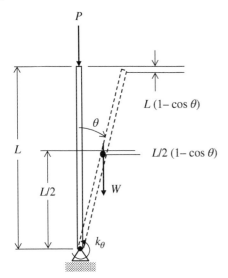

Figure 11.4 Stability of a vertical column.

Stability of a Vertical Rod

As another example, consider the problem shown in Figure 11.4 pinned at the lower end but constrained by a torsional spring, k_θ. The bar of length, L, is rigid.

$$V(\theta) = U(\theta) + \Omega(\theta) \tag{11.5}$$

where

$$U = \frac{1}{2}k_\theta\theta^2 \tag{11.6}$$

$$\Omega = -(P + W/2)L(1 - \cos\theta) \tag{11.7}$$

Note that in this example "you" do plus work on the elastic spring and minus work on gravity and concentrated force, P.

Expanding

$$\cos\theta = 1 - \frac{\theta^2}{2} + \frac{\theta^4}{4} - \frac{\theta^6}{6} + \cdots \tag{11.8}$$

And combining terms

$$V(\theta) = (k - PL - WL/2)\frac{\theta^2}{2} + (PL + WL/2)\left[\frac{\theta^4}{4} - \frac{\theta^6}{6} + \cdots\right] \tag{11.9}$$

$$V'(\theta) = (k - PL - WL/2)\theta + (PL + WL/2)\left[\theta^3 - \theta^5 + \cdots\right] \tag{11.10}$$

$$V''(\theta) = (k - PL - WL/2) + (PL + WL/2)\left[3\theta^2 - 5\theta^4 + \cdots\right] \tag{11.11}$$

$$\delta V(0) = 0 \tag{11.12}$$

$$\delta^2 V(0) = k - (PL + WL/2) \quad \text{(first nonzero term is even)} \tag{11.13}$$

For stability, $V''(0) \succ 0$, therefore, $P_{cr} = \dfrac{k_\theta}{L} - \dfrac{W}{2}$.

The plot in Figure 11.5 shows the effects of the magnitude of applied force, P, on total potential energy, V, is calculated from Eq. (11.5).

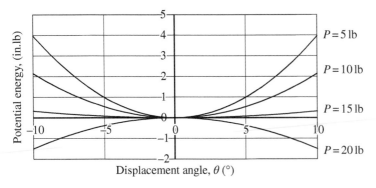

Figure 11.5 Potential energy of vertical column.

Input data for the potential energy calculations are

$$L = 24 \text{ in.}$$

$$W = 10 \text{ lb}$$

$$k_\theta = 500 \text{ in.-lb/rad}$$

The potential energy curves show a strong return to the neutral position for the lower end loads. When $P = 20$ lb, the potential curvature is negative, indicating unstable equilibrium at $\theta = 0$.
  The predicted critical force for this set of numbers is

$$P_{cr} = \frac{k_\theta}{L} - \frac{W}{2} = \frac{500}{24} - \frac{10}{2} = 15.83 \text{ lb} \tag{11.14}$$

This condition is represented by the horizontal line.

Rayleigh's Method

This approach is based on assuming a reasonable displacement function, $y(x)$. Consider, for example, a cantilever beam supported transversely at the top by a spring. A force of magnitude P is applied at the top as shown (Figure 11.6).
  Following the Rayleigh method, we assume the cantilever will buckle into a configuration approximated by

$$y = y_0 \left(1 - \cos \frac{\pi x}{2L}\right) \tag{11.15}$$

The unknown parameter is y_0. It is important that the assumed displacement satisfies the boundary conditions.
  The total internal energy of the system is

$$U = \frac{1}{2} EI \int_0^L (y'')^2 dx + \frac{1}{2} k y_0^2 \tag{11.16}$$

where

$$y'' = y_0 \left(\frac{\pi}{2L}\right)^2 \cos \frac{\pi x}{2L}$$

The potential energy of the external force, P, is

$$\Omega = -\frac{1}{2} P \int_0^L (y')^2 dx \tag{11.17}$$

$$y' = y_0 \frac{\pi}{2L} \sin \frac{\pi x}{2L}$$

$$\Omega = -\frac{1}{2} P \int_0^L \left[\frac{\pi}{2L} y_0 \sin \frac{\pi x}{2L}\right]^2 dx$$

$$\Omega \sim \frac{1}{2} P y_0^2 \frac{\pi}{2L} \frac{\pi}{4}$$

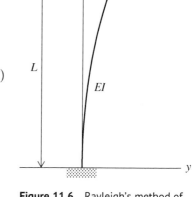

Figure 11.6 Rayleigh's method of predicting buckling.

Total potential energy then is

$$V = U + \Omega$$

$$V = \frac{1}{2}EIy_0^2\left(\frac{\pi}{2L}\right)^3\frac{\pi}{4} + \frac{1}{2}ky_0^2 - \frac{1}{2}Py_0^2\frac{\pi}{2L}\frac{\pi}{4} \tag{11.18}$$

$$\frac{dV}{dy_0} = 0 \text{ at } y_0 = 0; \text{ therefore, the system is in equilibrium at } y_0 = 0$$

$$\frac{d^2V}{dy_0^2} \text{ is plus only when}$$

$$P < EI\left(\frac{\pi}{2L}\right)^2 + k\frac{8L}{\pi^2}$$

Therefore, the system is unstable when

$$P \geq EI\left(\frac{\pi}{2L}\right)^2 + k\frac{8L}{\pi^2}$$

So, the critical value of P is

$$P_{cr} = EI\left(\frac{\pi}{2L}\right)^2 + k\frac{8L}{\pi^2} \tag{11.19}$$

If the spring is removed, the buckling force becomes $P_{cr} = \left(\frac{\pi}{2L}\right)^2 EI$, which agrees with the earlier discussing of Euler buckling.

Multiple Degrees of Freedom

The application of the principle of minimum potential energy generally follows the steps below.

1) Set up generalized coordinates at joints, taking advantage of symmetry whenever possible.
2) Express elongation of each member in terms of the generalized coordinates.
3) Write expressions of U and Ω in terms of the generalized coordinates; $V = U + \Omega$.
4) $\frac{\partial V}{\partial x_i} = 0, x_i$ – generalized coordinate.
5) Solve for generalized coordinate.
6) Substitute to obtain elongation (e) of each member.
7) Axial force in each member is then determined from $T = \frac{EA}{L}e$.

Structure Having Two Degrees of Freedom

Example Consider a structure made of three members as shown in Figure 11.7. The problem is to determine displacement components u,v due to a horizontal force of 100 lb. The cross-section properties are:

$$(EA)_a = C$$

$$(EA)_b = \frac{1}{2}C$$

$$(EA)_c = \frac{3}{2}C$$

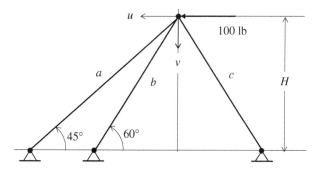

Figure 11.7 Deflection components.

This problem is a statically indeterminate but is solved conveniently with the energy method. Internal elastic energy is expressed in terms of the two displacement components.

$$U = \frac{1}{2}\sum e_i^2 k_i \tag{11.20}$$

where

$$e_i = \bar{q}\cdot\hat{n}_i$$

$$\bar{q} = -(u\hat{i} + v\hat{j})$$

$$\hat{n}_a = \frac{1}{\sqrt{2}}(\hat{i} + \hat{j}) \quad \hat{n}_b = -\frac{1}{2}(\hat{i} + \sqrt{3}\hat{j}) \quad \hat{n}_c = \frac{1}{2}(\hat{i} - \sqrt{3}\hat{j})$$

$$k_i = \frac{E_i A_i}{L_i}$$

The axial displacements of each of the three members are

$$e_a = -\frac{1}{\sqrt{2}}(u + v)$$

$$e_b = -\frac{1}{2}\left(u + \sqrt{3}v\right)$$

$$e_c = \frac{1}{2}\left(u - \sqrt{3}v\right)$$

By substitution

$$U = \frac{1}{2}k_a\left[-\frac{1}{\sqrt{2}}(u + v)\right]^2 + \frac{1}{2}k_b\left[-\frac{1}{2}\left(u + \sqrt{3}v\right)\right]^2 + \frac{1}{2}k_c\left[\frac{1}{2}\left(u - \sqrt{3}v\right)\right]^2 \tag{11.21}$$

The potential energy of the external force, 100 lb is

$$\Omega = -100u \tag{11.22}$$

Total potential energy is

$$V = U + \Omega \tag{11.23}$$

For equilibrium

$$\frac{\partial V}{\partial u} = \frac{1}{2}k_a(u + v) + \frac{1}{4}k_b\left(u + \sqrt{3}v\right) + \frac{1}{4}k_c\left(u - \sqrt{3}v\right) - 100 = 0 \tag{11.24}$$

$$\frac{\partial V}{\partial v} = \frac{1}{2}k_a(u + v) + \frac{1}{4}k_b\left(\sqrt{3}u + 3v\right) + \frac{1}{4}k_c\left(-\sqrt{3}u + 3v\right) = 0 \tag{11.25}$$

where

$$k_a = \frac{C}{L_a} = \frac{C}{H\sqrt{2}}$$

$$k_b = \frac{C}{2L_b} = \frac{C}{2\frac{2}{\sqrt{3}}H} = \frac{\sqrt{3}C}{4H}$$

$$k_c = \frac{3C}{2L_c} = \frac{3C}{2\frac{2}{\sqrt{3}}H} = \frac{3\sqrt{3}C}{4H}$$

Giving

$$0.785u - 0.021v = \frac{H}{C}100$$

$$-0.021u + 1.649v = 0$$

Simultaneous solutions give

$$u = 127.3\frac{H}{C} \quad \text{and} \quad v = 1.62\frac{H}{C}$$

By substitution

$$e_a = -91.1\frac{H}{C} \quad \text{and} \quad T_a = k_a e_a = -64.5\,\text{lb}$$

$$e_b = -64.9\frac{H}{C} \quad \text{and} \quad T_b = k_b e_b = -28.2\,\text{lb}$$

$$e_c = 62\frac{H}{C} \quad \text{and} \quad T_c = k_c e_c = +80.5\,\text{lb}$$

Analysis of Beam Deflection by Fourier Series

The principle of minimum potential energy along with the use of Fourier series will now be applied to beams. The first case will be a simply supported beam supporting a concentrated force (Figure 11.8). The pinned boundaries suggest the use of the sine (or odd function) in the series.

Recall the internal energy (strain) is expressed in terms of beam deflection by a Fourier series in the form of

$$U = \frac{1}{2}EI\int_0^L \left(\frac{d^2y}{dx^2}\right)^2 dx \tag{11.26}$$

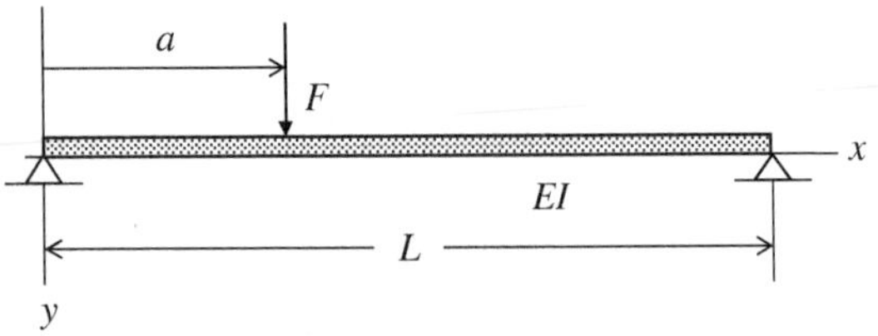

Figure 11.8 Fourier series solution for concentrated beam load.

$$y(x) = \sum_{n=1}^{\infty} a_n \sin \frac{n\pi x}{L} \tag{11.27}$$

$$y'' = -\sum_{n=1}^{\infty} \left(\frac{n\pi}{L}\right)^2 a_n \sin \frac{n\pi x}{L} \tag{11.28}$$

By substitution

$$U = \frac{1}{2} EI \left(\frac{\pi}{L}\right)^4 \int_0^L n^4 a_n^2 \sin^2 \frac{n\pi}{L} x \, dx \tag{11.29}$$

$$U = \frac{EI\pi^4}{4L^3} \sum_1^{\infty} n^4 a_n^2 \tag{11.30}$$

Concentrated Load

For the work done by the external force, F

$$\Omega = -Fy(a)$$

where

$$y(a) = \sum_1^{\infty} a_n \sin \frac{n\pi a}{L} \tag{11.31}$$

Plus (+) y is taken downward, so displacement y is (+). Total potential energy is

$$V = U + \Omega$$

$$V = \sum_1^{\infty} \left[\frac{EIL}{4} \left(\frac{n\pi}{L}\right)^4 a_n^2 - Fa_n \sin \frac{n\pi a}{L} \right] \tag{11.32}$$

For minimum potential energy $\dfrac{dV}{da_n} = 0$, which gives

$$a_n = \frac{2F}{EIL} \left(\frac{L}{n\pi}\right)^4 \sin \frac{n\pi a}{L} \tag{11.33}$$

and

$$y(x) = \frac{2F}{EIL} \sum_1^{\infty} \left(\frac{L}{n\pi}\right)^4 \sin \frac{n\pi a}{L} \sin \frac{n\pi x}{L} \tag{11.34}$$

Distributed Load

The approach in representing beam deflection with a Fourier series for distributed load follows the same process as for the concentrated load (Figure 11.9). The expression for the strain energy remains the same. However, the work done by the external forces differs. Again, based on pinned boundary conditions

$$y(x) = \sum_1^{\infty} a_n \sin \frac{n\pi x}{L}$$

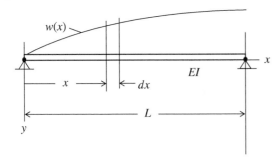

Figure 11.9 Fourier series solution for distributed loading.

The potential energy of the external distributed load is

$$d\Omega = -w(x)dxy(x)$$

which leads to

$$V = \sum_{1}^{\infty}\left[\frac{EIL}{4}\left(\frac{n\pi}{L}\right)^4 a_n^2 - a_n \int_0^L w(x)\sin\frac{n\pi x}{L}dx\right] \qquad (11.35)$$

For stationary potential energy, $\dfrac{dV}{da_n} = 0$, which gives

$$a_n = \frac{2}{LEI}\int_0^L \left(\frac{L}{n\pi}\right)^4 w(x)\sin\frac{n\pi x}{L}dx \qquad (11.36)$$

$$y(x) = \frac{2}{LEI}\sum_{1}^{\infty}\left(\frac{L}{n\pi}\right)^4\left[\int_0^L w(x)\sin\frac{n\pi x}{L}dx\right]\sin\frac{n\pi x}{L} \qquad (11.37)$$

Axially Loaded Beam (Column)

We now consider the Euler buckling problem as shown in Figure 11.10.

The post buckle displacement, e, of the applied force, P, is determined as follows.

$$ds^2 = dx^2 + dy^2$$

$$ds = \sqrt{1 + \left(\frac{dy}{dx}\right)^2}\,dx$$

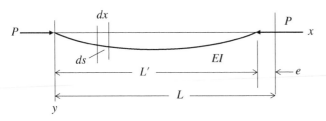

Figure 11.10 Axial buckling load.

$$L = \int_0^{L'} ds = \int_0^{L'} \sqrt{1 + \left(\frac{dy}{dx}\right)^2}\, dx \tag{11.38}$$

$$L \approx \int_0^{L'} \left[1 + \left(\frac{dy}{dx}\right)^2\right] dx \approx L' + \frac{1}{2}\int_0^{L'} (y')^2 dx$$

$$e = L - L' = \int_0^{L'} (y')^2 dx$$

$$\Omega_P = -\frac{1}{2}P \int_0^{L} (y')^2 dx \tag{11.39}$$

$$y' = \sum_1^{\infty} \frac{n\pi}{L} a_n \cos \frac{n\pi x}{L}$$

$$\Omega_P = -\frac{\pi^2 P}{4L} \sum n^2 a_n^2$$

The total potential energy is

$$V = \frac{\pi^4 EI}{4L^3} \sum n^4 a_n^2 - \frac{\pi^2 P}{4L} \sum n^2 a_n^2 \tag{11.40}$$

For stationary potential energy, $\dfrac{\partial V}{\partial a_n} = 0$.

$$\frac{\pi^4 EI}{4L^3} \sum 2n^4 a_n - \frac{\pi^2 P}{4L} \sum 2n^2 a_n = 0 \tag{11.41}$$

For stability, $\dfrac{\partial^2 V}{\partial a_n^2} \succ 0$; therefore for $n = 1$, stability occurs only when

$$\frac{\pi^2 EI}{L^2} \succ P$$

So

$$P_{cr} = \frac{\pi^2 EI}{L^2} \tag{11.42}$$

which is Euler's buckling equation.

Principle of Complementary Energy

This principle is derived from the theory of stationary potential energy, which requires that

$$\frac{\partial V}{\partial x_i} = 0 \tag{11.43}$$

for equilibrium. This means equilibrium is established when

$$\frac{\partial U}{\partial x_i} = -\frac{\partial \Omega}{\partial x_i} \tag{11.44}$$

The work of external forces over a displacement of dx_i is

$$d\Omega = -[P_1\, dx_1 + P_2\, dx_2 + P_3\, dx_3 + P_4\, dx_4 + \cdots] \tag{11.45}$$

Following the chain rule of differentiation

$$d\Omega = \frac{\partial \Omega}{\partial x_1}\, dx_1 + \frac{\partial \Omega}{\partial x_2}\, dx_2 + \cdots \tag{11.46}$$

From Eqs. (11.45) and (11.46)

$$P_i = -\frac{\partial \Omega}{\partial x_i} \tag{11.47}$$

It follows from Eq. (11.44) that

$$\frac{\partial U}{\partial x_i} = P_i \tag{11.48}$$

This equation is not restricted to small deflections nor limited to systems with linear elastic properties.

A.M. Legendre (1752–1833) showed that Eq. (11.48) can be transformed into a conjugate form [1]

$$\frac{\partial \gamma}{\partial P_i} = x_i \tag{11.49}$$

where γ is complementary energy per volume as illustrated in Figure 11.11. Complementary energy is the shaded area.

Force displacement relationships typically are given in terms of stress and strain. In this case

$$d\gamma = d\sigma(\varepsilon)$$

$$d\gamma = d\left(\frac{P}{A}\right)\frac{x}{L}$$

Engineering Application

The application of Eq. (11.49) to trusses is summarized below. The total complementary energy for a truss is

$$\gamma = \sum \gamma_i, \quad \gamma_i - \text{complementary energy of } i\text{th member} \tag{11.50}$$

Let P be an external force acing on a joint and x be the component of the displacement of the joint in the direction of P. Then

$$x = \sum \frac{\partial \gamma_i}{\partial P} \tag{11.51}$$

where $\gamma_i = f(N_i)$ and N_i – tension in ith member.

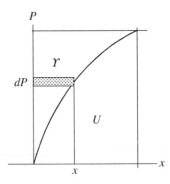

Figure 11.11 Complementary energy.

$$dy = dPx$$

$$\gamma = \int x\,dP$$

It is not necessary to determine γ_i for each member because we use only Eq. (11.51) as shown below.

Accounting for energy throughout the structure

$$x = \sum \frac{d\gamma_i}{dN}\frac{\partial N_i}{\partial P} \tag{11.52}$$

where

$$\frac{d\gamma_i}{dN} = e_i \quad \text{(according to Eq.(11.49))} \tag{11.53}$$

e_i – extension of ith member due to loads (N_i) in the ith member.

$$x = \sum e_i \frac{\partial N_i}{\partial P} = \sum e_i n_i \tag{11.54}$$

For statically determinant trusses, the total force in any member is the superposition of the effects of all forces applied throughout the structure.

$$N_i = a_1 P_1 + a_2 P_2 + a_3 P_3 + \cdots + aP \tag{11.55}$$

By differentiation

$$\frac{\partial N_i}{\partial P} = a = n_i \tag{11.56}$$

The interpretation of n_i is therefore the tension in ith member if $P = 1$ and all other loads are removed.

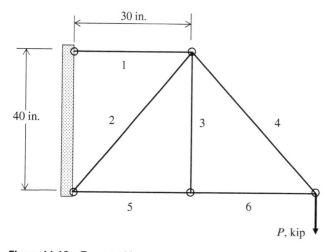

Figure 11.12 Truss problem.

where

m	– moment	"load in the *i*th member due to a unit load (or couple) acting alone on the
n	– tension (compression)	system at the point where the displacement δ is to be determined" (called
t	– torque	dummy variable)

are loads in the *i*th member due to a unit load (or couple) P acting alone on the system at point where δ is to be determined.

M	– moment	"load in the *i*th member due to the actual loads
N	– tension (compression)	(or couples) acting on the system"
T	– torque	

The displacement, δ, will be a linear (or angular) displacement depending on the type of loading. Several examples of the application of Eq. (11.69) are given below.

Example   Consider the split circle hoop fixed at one end and displaced by a force, F, normal to the plane of the hoop (Figure 11.13). The problem is to find the displacement, δ, in the direction of the applied force, P. This problem will be solved using Eq. (11.69).
Moment and torque are defined by

$$M = PR\sin\theta \tag{11.70}$$

$$T = PR(1 - \cos\theta) \tag{11.71}$$

$$\delta = \int_0^{2\pi} \left[\frac{Mm}{EI} + \frac{Tt}{GJ}\right] R\, d\theta \tag{11.72}$$

The dummy variables, m and t, are simply

$$m = R\sin\theta$$

$$t = R(1 - \cos\theta)$$

After substitution

$$\delta = R\int_0^{2\pi} \left[\frac{R^2\sin^2\theta}{EI}P + \frac{R^2(1 - \cos\theta)^2}{GJ}P\right] d\theta \tag{11.73}$$

$$\delta = R^3 \int_0^{2\pi} \left[\frac{\sin^2\theta}{EI}P + \frac{(1 - 2\cos\theta + \cos^2\theta)}{GJ}P\right] d\theta$$

Recall

$$\int_0^{2\pi} \sin^2\theta\, d\theta = \left(\frac{\theta}{2} - \frac{\sin 2\theta}{4}\right)_0^{2\pi} = \pi$$

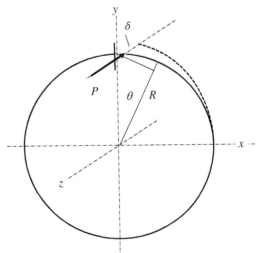

Figure 11.13   Transverse displacement of a split ring.

$$\int_{0}^{2\pi} \cos^2\theta \, d\theta = \left(\frac{\theta}{2} + \frac{\sin 2\theta}{4}\right)_{0}^{2\pi} = \pi$$

and

$$I = \frac{\pi r^4}{4} \quad \text{and} \quad J = \frac{\pi r^4}{2}$$

Collecting terms and integrating gives

$$\delta = \frac{2PR^3}{r^4}\left[\frac{2}{E} + \frac{3}{G}\right] \tag{11.74}$$

Example In this case, we wish to determine the twist angle at point b where a torque (T) is applied (Figure 11.14). Applying Eq. (11.69)

$$\delta = \sum_{1,2}\left(\int_{0}^{\ell}\frac{Mm}{EI} + \int_{0}^{\ell}\frac{Tt}{GJ}\right) dx$$

$$\delta = \int_{0}^{\ell}\frac{Tt}{GJ}dx + \int_{0}^{\ell}\frac{Mn}{EI}dx$$

$$T = M = T \quad \text{and} \quad t = m = 1$$

Bring everything together

$$\theta_b = \frac{Tt}{GJ}l + \frac{Mm}{EI}l \tag{11.75}$$

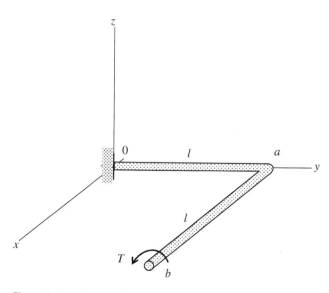

Figure 11.14 Bent rod.

$$\theta_b = \frac{T(1)}{GJ}l + \frac{T(1)}{EI}l$$

By substitution

$$\theta_b = \frac{2T\ell}{\pi r^4}\left(\frac{1}{G} + \frac{1}{E}\right) \tag{11.76}$$

Example   Here we want to find displacement at the end of a cantilevered beam due to a uniform load, w (Figure 11.15). This is a good example of the dummy load concept. While there is no applied force at the displacement point, dummy force, n, is still applied producing dummy moment, m.

Applying Eq. (11.69)

$$\delta = \int_0^L \frac{Mm}{EI}dx \tag{11.77}$$

where

$$m = -(L-x)(1)$$

$$M = -\left(\frac{L-x}{2}\right)(L-x)w \tag{11.78}$$

By substitution

$$\delta = \int_0^L \frac{(L-x)^3}{2EI}w\,dx$$

$$\delta = \frac{L^4w}{8EI} \tag{11.79}$$

which agrees with the solution in Chapter 2.

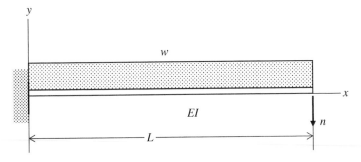

Figure 11.15   Beam deflection.

Example   Here is a good example of multiple sections in a structure (Figure 11.16). Total strain energy is sum of both sections. The applied force, F, is parallel to the x axis.

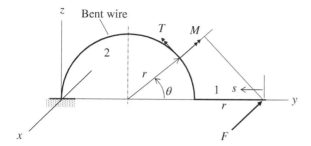

Figure 11.16 Multiple sections.

Equation (11.70) for this example is

$$\delta = \sum_{1,2,3} \int \left[\frac{Mm}{EI} + \frac{Tt}{GJ} \right] ds \qquad (11.80)$$

Member 1

$$m_1 = (1)s \quad M_1 = Fs$$

Member 2

$$m_2 = 2r \sin \theta \qquad M_2 = 2rF \sin \theta$$
$$t_2 = r(2 \cos \theta - 1) \quad T_2 = Fr(2 \cos \theta - 1)$$

Bring all terms together

$$\delta = \int_0^r \frac{s(Fs)}{EI} ds + \frac{4Fr^2}{EI} \int_0^\pi \sin^2 \theta r d\theta + \frac{r^2 F}{GJ} \int_0^\pi (2 \cos \theta - 1)^2 r d\theta \qquad (11.81)$$

The integrals are expanded as follows

$$\frac{F}{EI} \int_0^r s^2 ds = \frac{Fr^3}{3EI}$$

$$\frac{4Fr^3}{EI} \int_0^\pi \sin^2 \theta d\theta = \frac{4Fr^3}{EI} \left[\frac{\theta}{2} - \frac{\sin 2\theta}{4} \right]_0^\pi = \frac{2\pi Fr^3}{EI}$$

$$\frac{Fr^3}{GJ} \int_0^\pi (2 \cos \theta - 1)^2 d\theta = \frac{Fr^3}{GJ} \int_0^\pi (1 - 4 \cos \theta + 4 \cos^2 \theta) d\theta$$

$$= \frac{Fr^3}{GJ} \left(\theta - 4 \sin \theta + 4 \left(\frac{\theta}{2} + \frac{\sin 2\theta}{4} \right) \right)_0^\pi$$

$$= \frac{3\pi Fr^3}{GJ}$$

Bringing everything together

$$\delta = \frac{Fr^3}{EI}\left(\frac{1}{3} + 2\pi\right) + \frac{3\pi Fr^3}{GJ} \tag{11.82}$$

Chemically Induced Deflections

The theory and application of energy methods is an established engineering tool. Its main applications occur in predicting deflections and loads in engineering structures as previously described. Recent studies show how it is also used to predict deflections at the molecular and nano levels.

Microcantilever Sensors

Recently, microcantilevers have attracted much attention due to their potential as an extremely sensitive sensor platform for chemical and biological detection. Microcantilever sensors have many advantages over the competing technologies because of high sensitivity and easy production at low cost. Adsorption-induced cantilever bending is observed when the adsorption is confined to a single side of a cantilever. The adsorption-induced cantilever bending is ideally suited for measurements in air or under solution.

The exact molecular mechanism involved in adsorption-induced stress is not completely understood. Typically, deflections of microcantilevers are explained in terms of energy transfer between surface free energy and elastic energy associated with structural bending of the cantilevers [2–4]. Surface stress involved in adsorption-induced stress is often calculated by Stoney's equation [5].

$$Pt = \frac{Ed^2}{6r} \tag{11.83}$$

where

P – tension force per unit cross-sectional area of film
t – thickness of film
E – modulus of elasticity of beam material
d – thickness of beam
r – radius of curvature of deflected beam

Dareing and Thundat [6] take a different approach by explaining the mechanism of bending in terms of atomic and elastic energies.

Their simulation relates atomic (or molecular) interactive potential of the adsorbent to elastic energy in the cantilever. Total energy is the sum of this atomic potential plus elastic energy. Beam deflection is determined by minimizing the total potential energy expression in terms of beam curvature.

Simulation Model
The bending model is based on energy potential in the first layer of atoms attached to one surface of a cantilever and elastic potential in the microcantilever itself. Atoms are situated against the cantilever surface as shown in Figure 11.17.

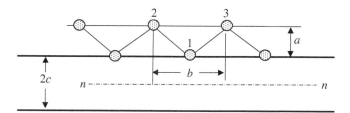

Figure 11.17   Arrangement of atoms (or small molecules) on cantilever surface.

This assumption that the first atomic layer on the beams surface plays a dominant role in micro-cantilever deflections is supported by the experimental works of Martinez et al. [7] and Schell-Sorokin and Tromp [8], who measured changes in curvature in cantilevered thin plates due to adsorption of submonolayer of different atoms in ultra-high vacuum conditions. Per the model, atoms in the attached film are attracted and repulsed per the Lennard-Jones potential formula [9].

$$w(r) = \frac{-A}{r^6} + \frac{B}{r^{12}} \tag{11.84}$$

$$F = \frac{dw}{dr} = \frac{6A}{r^7} - \frac{12B}{r^{13}} \tag{11.85}$$

where r is the spacing between atoms. Part of this potential is transferred into the cantilever as elastic strain energy causing the beam to deflect. The equilibrium configuration of the cantilever is determined by minimizing the total potential function, which is made up of the Lennard-Jones potential and the elastic energy in the cantilever. Both potential components will now be expressions in terms of the local curvature of the beam.

Assuming the following Lennard-Jones constants,

$$A = 10^{-77} \, J \, m^6 = 10^{-5} \, nN\text{-}nm \, nm^6$$

$$B = 10^{-134} \, J \, m^6 = 10^{-8} \, nN\text{-}nm \, nm^{12}$$

then atomic potential and force of attraction, vs. separation distance, vary as shown in Figures 11.18 and 11.19.

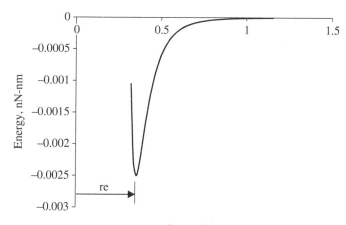

Separation, nm

Figure 11.18   Potential energy between two atoms.

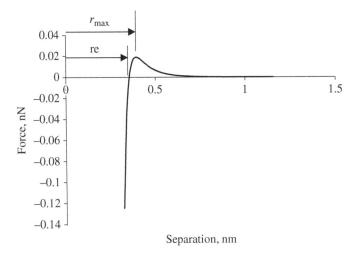

Figure 11.19 Resultant force between two atoms.

Figure 11.20 shows the curvature of the beam over a distance, b, which is taken as the atomic space between two atoms on the surface of the beam. Three atoms are involved in the atomic potential expression. Here, we assume that the adsorbate distribution is uniform over the surface since the surface is chemically homogeneous. Since the distribution of surface atoms is assumed to be uniform, the curvature will be uniform along the cantilever. The curvature can be established by considering the elastic energy in the beam over the length, d, and the atomic potential between the three atoms, one on the cantilever surface and two in the attached film or coating. It is assumed

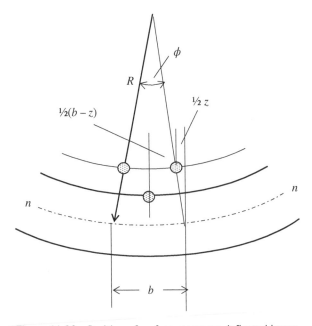

Figure 11.20 Position of surface atoms on deflected beam.

in this simulation that the second and higher layers of atoms play a minor role in deflecting the beam.

Other assumptions made in creating this model are typical of those used in beam theory, i.e. a cross-sectional plane before bending remains a plane after bending. In addition, the near-surface layer of atoms 2 and 3 also remain in the same plane after beam deformation. Based on these assumptions, the energy potential in the near surface layer of atoms can be expressed in terms of beam curvature. Note that the movement of 2 toward 3 is

$$z = \phi(c + a) \tag{11.86}$$
$$b = R\phi$$

so

$$z = \frac{b}{R}(c + a) \tag{11.87}$$

Molecular and Elastic Potential Energies

The atomic potential, U_s, in terms of beam curvature $(1/R)$ according to the Lennard-Jones expression is

$$U_s = \frac{-A}{(b-z)^6} + \frac{B}{(b-z)^{12}} + 2\left\{ \frac{-A}{\left[\frac{1}{4}(b-z)^2 + a^2\right]^3} + \frac{B}{\left[\frac{1}{4}(b-z)^2 + a^2\right]^6} \right\} \tag{11.88}$$

The elastic bending potential, U_b, over the atomic length, b, is

$$U_b = 1/2EI\left(\frac{1}{R}\right)^2 b \tag{11.89}$$

For stable equilibrium, the total potential energy

$$U = U_s + U_b \tag{11.90}$$

must be a relative minimum, which is determined from

$$\frac{dU}{d\left(\frac{c}{R}\right)} = 0 \tag{11.91}$$

The resulting equation is

$$\frac{EI}{c^2}\left(\frac{c}{R}\right) = \frac{\frac{A}{b^7}6\left(1 + \frac{a}{c}\right)}{\left[1 - \frac{c}{R}\left(1 + \frac{a}{c}\right)\right]^7} - \frac{\frac{B}{b^{13}}12\left(1 + \frac{a}{c}\right)}{\left[1 - \frac{c}{R}\left(1 + \frac{a}{c}\right)\right]^{13}}$$
$$+ 2\left\{ \frac{\frac{A}{b^7}\frac{3}{2}\left[1 - \frac{c}{R}\left(1 + \frac{a}{c}\right)\right]\left(1 + \frac{a}{c}\right)}{\left[\frac{1}{4}\left[1 - \frac{c}{R}\left(1 + \frac{a}{c}\right)\right]^2 + \left(\frac{a}{b}\right)^2\right]^4} - \frac{\frac{B}{b^{13}}3\left[1 - \frac{c}{R}\left(1 + \frac{a}{c}\right)\right]\left(1 + \frac{a}{c}\right)}{\left[\frac{1}{4}\left[1 - \frac{c}{R}\left(1 + \frac{a}{c}\right)\right]^2 + \left(\frac{a}{b}\right)^2\right]^7} \right\} \tag{11.92}$$

The values of c/R, which satisfy this expression, define the curvature of the microcantilever beam. This curvature is constant along the beam or the deflection of the beam is a circular arc. The transverse deflection of the end of the microcantilever can be determined from simple trigonometry using the known beam curvature.

$$\delta = R\left(1 - \cos\theta\right) \tag{11.93}$$

We have applied the above simulation to a typical microcantilever beam of 200 μm length with a beam cross section of, $30\,\mu m \times 1\,\mu m$. The material modulus is 1.79×10^{11} Pa. The thickness of atomic monolayer, $a = 0.5$ nm, and the spacing of the atoms on the cantilever surface, $b = 0.5$ nm. The Lennard-Jones constants, $A = 10^{-77}\,J\,m^6$ and $B = 10^{-134}\,J\,m^{12}$.

The computed curvature for this set of conditions is, $\dfrac{c}{R} = 3 \times 10^{-7}$. This curvature is positive meaning that the beam curves upward cupping the adsorbate monolayer. The corresponding angle, $\theta = 0.000\,119\,7$ rad or $0.006\,86°$. The corresponding transverse deflection of the end of the beam is predicted as 11.97 nm or approximately 12 nm.

References

1 Langhaar, H.L. (1962). *Energy Methods in Applied Mechanics*. New York: Wiley.

2 Ibach, H. (1997). *Surf. Sci. Rep.* **29**: 193–263.

3 Dahmen, K., Lehwald, S., and Ibach, H. (2000). Bending of crystalline plates under the influence of surface stress – a finite element analysis. *Surf. Sci.* **446**: 161–173.

4 Raiteri, R., Grattarola, M., Butt, H.-J., and Skladal, P. (2001). Micromechanical cantilever-based biosensors. *Sens. Actuat.* **B 79**: 115–126, Elsevier.

5 Stoney, G.G. (1909). *Proc. R. Soc. Lond.* **82**: 172.

6 Dareing, D.W. and Thundat, T. (2005). Simulation of adsorption-induced stress of a microcantilever sensor. *J. Appl. Phys.* **043526**: 97.

7 Martinez, R.E., Augustyniak, W.M., and Golovchenko, A. (1990). *Phys. Rev. Lett.* **64**: 1035.

8 Schell-Sorokin, A.J. and Tromp, R.M. (1990). *Phys. Rev. Lett.* **64**: 1039.

9 Israelachivili, J.N. (1991). *Intermolecular and Surface Forces*, 2e. Academic Press (Lennard-Jones constants, A and B).

Index

Engineering Practice with Oilfield and Drilling Applications, First Edition. Donald W. Dareing.
© 2022 John Wiley & Sons, Inc. Published 2022 by John Wiley & Sons, Inc.